Time-Dependent Behaviour of Concrete Structures

Serviceability failures of concrete structures involving excessive cracking or deflection are relatively common, even in structures that comply with code requirements. This is often as a result of a failure to adequately account for the time-dependent deformations of concrete in the design of the structure. Design for serviceability is complicated by the non-linear and inelastic behaviour of concrete at service loads. The serviceability provisions embodied in codes of practice are relatively crude and, in some situations, may be unreliable and often do not adequately model the in-service behaviour of structures. In particular, they fail to realistically account for the effects of creep and shrinkage of the concrete.

Providing detailed information, this book assists engineers to rationally predict the time-varying deformation of concrete structures under typical in-service conditions. It gives analytical methods to help anticipate time-dependent cracking, the gradual change in tension stiffening with time, creep-induced deformations and the load independent strains caused by shrinkage and temperature changes. Calculation procedures are illustrated with many worked examples.

A vital guide for practising engineers and advanced students of structural engineering on the design of concrete structures for serviceability, providing a penetrating insight into the time-dependent behaviour of reinforced and prestressed concrete structures.

Raymond Ian Gilbert is Professor of Civil Engineering at the University of New South Wales and currently holds an Australian Research Council Australian Professorial Fellowships. He has over 35 years experience in structural design and is a specialist in the analysis and design of reinforced and prestressed concrete structures.

Gianluca Ranzi is a Senior Lecturer of Structural Engineering at the University of Sydney, specialising in the analysis and design of concrete and composite steel-concrete structures.

Time-Dependent Behaviour of Concrete Structures

Raymond Ian Gilbert and Gianluca Ranzi

Spon Press
Taylor & Francis Group
LONDON AND NEW YORK

First published 2011
by Spon Press
2 Park Square, Milton Park, Abingdon, Oxon OX14 4RN

Simultaneously published in the USA and Canada by Spon Press
270 Madison Avenue, New York, NY 10016 USA

Spon Press is an imprint of the Taylor & Francis Group, an informa business

Typeset in Sabon by Glyph International Ltd
Printed and bound in Great Britain by CPI Antony Rowe, Chippenham, Wiltshire

British Library Cataloguing in Publication Data
A catalogue record for this book is available from the British Library

Library of Congress Cataloging-in-Publication Data
Gilbert, R. I., 1950–
Time-dependent behaviour of concrete structures / Raymond Ian Gilbert and Gianluca Ranzi.
p. cm.
Includes bibliographical references and index.
1. Concrete–Deterioration. 2. Service life (Engineering) 3. Concrete structures. I. Ranzi, Gianluca, 1972– II. Title.
TA440.G449 2010
624.1′834–dc22 2010007786

ISBN13: 978-0-415-49384-0 (hbk)
ISBN13: 978-0-203-87939-9 (ebk)

Contents

Preface

In the design of concrete structures, the two main design objectives are *strength* and *serviceability*. A structure must be strong enough and sufficiently ductile to resist, without collapsing, the overloads and environmental extremes that may be imposed on it. It must also perform satisfactorily under the day-to-day service loads without deforming, cracking or vibrating excessively. This book is concerned with the *serviceability* of concrete structures, in particular the prediction of instantaneous and time-dependent deformation under in-service conditions. The factors that most affect structural behaviour at the serviceability limit states are considered in considerable depth, including cracking of the tensile concrete, tension stiffening, and the time-dependent deformations caused by creep and shrinkage of the concrete.

The prediction of the final deformation of a concrete structure, and the final extent and width of cracks, are perhaps the most uncertain and least well-understood aspect of structural design. The long-term behaviour of the structure depends primarily on the deformational characteristics of the concrete, including its creep and shrinkage characteristics, and these are highly variable, depending not only on the mix proportions and types of cement and aggregates, but also on the local environment and the load history. In most undergraduate civil and structural engineering courses, the effects of concrete creep and shrinkage on structural behaviour are treated superficially, with only the simplified code-oriented procedures for deflection and crack control being presented. The same is true of most books on the design of concrete structures. Failure to recognise and quantify the non-linear effects of cracking, creep and shrinkage is a common cause of serviceability failure.

Creep and shrinkage of concrete and their effects on structural behaviour have been actively researched for over 100 years. Much has been written on the subject and many outstanding contributions have been made. However, much of the information and many of the analytical procedures that have been developed are not, in general, known or used by the profession. Structural designers often rely on the simplified procedures in codes of practice to estimate service load behaviour, and this often oversimplifies the problem and can be unreliable.

This book is an attempt to provide practising engineers and post-graduate students with a practical and useful treatment of the serviceability analysis of concrete structures. For this purpose, most sections included in each chapter are self-contained so that they can be read independently from the rest of the book. The analytical techniques are illustrated by numerous worked examples.

In Chapter 1, an introductory discussion of concrete creep and shrinkage and their effects on the deformation of concrete structures is presented, while typical material properties and a simple procedure for estimating the creep and shrinkage characteristics of concrete are provided in Chapter 2. No attempt has been made to provide a detailed description of the mechanisms of creep and shrinkage or the factors affecting them. The material properties and material constitutive relationships are only of interest insofar as they affect the methods of structural analysis.

Chapter 3 contains a qualitative discussion of structural behaviour at service loads and the simplified methods for the control of deflection and cracking contained in several of the major codes of practice are presented and critically reviewed, including ACI318-08, Eurocode 2 and the Australian Standard AS3600-2009. Alternative and more reliable procedures for deflection calculation and crack control are also provided.

Chapters 4–7 deal with the time-dependent analysis of cross-sections, exploring various procedures for predicting the gradual change of strain and curvature with time and the gradual redistribution of stress between the concrete and the bonded steel reinforcement. Chapter 4 examines cross-sections loaded in axial compression. Chapter 5 considers a wide range of uncracked reinforced, prestressed and composite cross-sections carrying axial force and uniaxial bending. Chapter 6 deals with uncracked cross-sections subject to combined axial force and biaxial bending. Cracked cross-sections subjected to axial force and bending are analysed in Chapter 7.

In Chapter 8, a range of different structural applications are considered, drawing on the cross-sectional analyses in Chapters 4–7 and the treatment of tension stiffening and cracking in Chapter 3. The techniques described and illustrated may be extended readily to include a wide range of additional structural applications. The procedures are not daunting and require little more than an elementary background in mechanics and structural analysis.

Chapter 9 illustrates how the time-dependent effects of creep and shrinkage can be readily included in the computer analysis of structures using the stiffness and the finite-element methods. These have been described and applied considering the common Euler–Bernoulli beam model as a case study, but the approach is general and can be extended to more complex models. The analytical formulations required to implement the computer analysis of Chapter 9 are presented in Appendix A.

It is hoped that this book will provide structural engineers not only with useful analytical tools to predict and control in-service performance, but also with a clearer picture of the interaction between concrete and reinforcement at service loads and a better understanding of why concrete structures behave as they do.

Raymond Ian Gilbert and Gianluca Ranzi
Sydney, 2010.

Acknowledgements

The authors acknowledge the support given by their respective institutions and by the Australian Research Council through its Discovery and Linkage schemes and the award of an ARC Australian Professorial Fellowship for the period 2005 to 2010 to the first author. Thanks are also extended to Associate Professor Graziano Leoni and Dr Andrew Kilpatrick for their comments on the drafts of each chapter and to Associate Professor Rob Wheen for providing the cover photograph of the Anzac Bridge during construction.

Notation and sign convention

All symbols are defined in the text where they first appear. The more frequently used symbols and those that appear throughout the book are listed below. It is assumed that tension is positive and compression is negative, and positive bending about a horizontal axis causes tension in the bottom fibres of a cross-section.

A, B, I	area, first moment of area, and second moment of area, respectively, calculated about the reference axis of the cross-section
A_c, B_c, I_c	properties A, B and I of the concrete part of the cross-section
$\overline{A}_e$, $\overline{B}_e$, $\overline{I}_e$	properties A, B and I of the age-adjusted transformed cross-section
A_{ss}, B_{ss}, I_{ss}	properties A, B and I of the structural steel part of a composite cross-section
A_{ct}	area of tensile concrete in the tension chord (Eq. 3.45)
$A_{s(i)}$, $A_{p(i)}$	areas of the ith layer of non-prestressed and prestressed steel, respectively
A_{st}, A_{sc}	areas of tensile and compressive steel on a reinforced section, respectively
b, b_w	flange width and web width of a T-section
b^*	width of cross-section at the level of the centroid of the tensile reinforcement
C	specific creep (also Celsius)
$C(t, \tau_0)$	specific creep at time t produced by a sustained unit stress first applied at τ_0
c	concrete cover to reinforcement
$\mathbf{D}_0$, $\mathbf{D}_j$, $\mathbf{D}_k$	matrices of cross-sectional rigidities at τ_0, τ_j and τ_k, respectively
D	overall depth of a cross-section
d	effective depth of the tensile reinforcement
d_b	bar diameter
$d_{s(i)}$, $d_{p(i)}$	depths to the i-th layer of the non-prestressed and prestressed steel, respectively
d_n $(=kd)$	depth to the neutral axis on a fully-cracked cross-section
d_o	depth from extreme compressive fibre to the centroid of the outermost layer of tensile reinforcement
d_{ref}	depth of reference axis below the top fibre of the cross-section

E_c, E_s, E_p, E_{ss}	elastic moduli of concrete, non-prestressed steel, prestressed steel, and structural steel, respectively
$E_{c,j}$	elastic modulus of concrete at time τ_j
$E_e, \overline{E}_e$	effective modulus of concrete (Eq. 4.3) and age-adjusted effective modulus of concrete (Eq. 4.35), respectively
$E_{e,j,i}$	effective modulus of concrete at time τ_j due to a stress first applied at τ_i
e	eccentricity; base of the natural logarithm
e_{AB}	axial deformation of member AB
e_L	axial shortening of a column of length, L
$\mathbf{f}_{cr,j}, \mathbf{f}_{cr,k}$	vectors of actions at times τ_j and τ_k, respectively, accounting for creep during previous time periods
$\mathbf{f}_{set,j}, \mathbf{f}_{set,k}$	vectors of corrected internal actions to be used at times τ_j and τ_k, respectively, when a new element is added to a cross-section
$\mathbf{f}_{p,init}$	vector of initial prestressing forces
$\mathbf{f}_{sh,j}, \mathbf{f}_{sh,k}$	vectors of actions at times τ_j and τ_k, respectively, accounting for shrinkage during previous time periods
$\mathbf{F}$	flexibility matrix
$\mathbf{F}_0, \mathbf{F}_j, \mathbf{F}_k$	matrices relating strain to internal actions on cross-section at τ_0, τ_j and τ_k, respectively
$F_{e,j,i}$	creep factor at time τ_j for a stress applied at τ_i (Eq. 4.26b)
$\overline{F}_{e,0}$	age-adjusted creep factor (Eq. 4.46)
f_c, f_t	strength of concrete in compression and in tension, respectively
f'_c	characteristic strength of concrete in compression
f_{cm}, f_{cmi}	mean concrete cylinder strength and mean concrete *in-situ* strength, respectively
f_{ct}	uniaxial tensile strength of concrete
$f'_{ct,f}$	characteristic flexural tensile strength of concrete (modulus of rupture)
f_{ji}	a flexibility coefficient (the displacement at release j due to a unit value of the ith redundant force)
f_p	breaking strength of the prestressing steel
f_{st}	steel stress on a cracked section corresponding to a crack width of w^*
f_y	yield stress of the non-prestressed steel
G, g	dead load (kN) and dead load per unit area (kPa), respectively
I_{cr}, I_{uncr}	second moments of area of the fully-cracked and uncracked transformed cross-section about the centroidal axis, respectively
I_{ef}	effective second moment of area after cracking
I_g	second moment of area of gross cross-section about centroidal axis
i, j, k	integers
h	slab thickness
J	creep function
$J(t, \tau_0)$	creep function at time t for concrete first loaded at τ_0 (Eq. 1.10)
$J_{j,i}$	creep function at time τ_j due to a stress first applied at τ_i
$\mathbf{K}$	structural stiffness matrix

kd	depth to neutral axis from extreme compressive fibre
k_{cs}	long-term to short-term deflection multiplication factor
$\mathbf{k}_e$	element stiffness matrix
k_1, k_2, k_3, k_4, k_5	factors describing the magnitude and rate of development of creep and shrinkage
L, ℓ	span of a beam or length of a column
L_x, L_y	orthogonal span lengths for two-way slabs
ln	the natural logarithm
ℓ_{ef}	the effective span (the lesser of the centre-to-centre distance between the supports and the clear span plus the member depth)
M	moment
M_{cr}	cracking moment
M_s^*	design in-service moment at the critical cross-section
M_u	ultimate flexural strength
M_{xe}, M_{ye}	applied moments about the x- and y-axes, respectively
m_c	number of layers of concrete in a discretised cross-section
m_p, m_s	number of layers of tendons and non-prestressed reinforcement on a cross-section
$N_{c,0}, N_{s,0}, N_{p,0}$	internal forces resisted by concrete, reinforcement and tendons at time τ_0, respectively
$N_{c,k}, N_{s,k}, N_{p,k}$	internal forces resisted by concrete, reinforcement and tendons at time τ_k, respectively
N_{cr}	restraining force after direct tension cracking
N_e, M_e	external axial force and moment applied to a cross-section
$N_{i,0}, M_{i,0}$	internal axial force and moment on a cross-section at time τ_0
$N_{i,k}, M_{i,k}$	internal axial force and moment on a cross-section at time τ_k
N_{sus}, M_{sus}	sustained axial force and moment applied to a cross-section
n	modular ratio (E_s/E_c)
$n_e, \overline{n}_e$	effective and age-adjusted effective modular ratios (E_s/E_e, $E_s/\overline{E}_e$), respectively
$n_{e,j,i}$	effective modular ratio ($E_s/E_{e,j,i}$)
$\mathbf{P}$	vector of nodal loads for whole structure
P	axial load; also prestressing force
P_{cr}	cracking load
$P_{p,init}$	initial prestressing force prior to transfer
$\mathbf{p}$	vector of nodal loads for element
Q, q	live load (kN) and live load per unit area (kPa), respectively
R	design relaxation force in the prestressing steel
$R_{A,0}, R_{B,0}, R_{I,0}$	cross-sectional rigidities at time τ_0
$R_{A,k}, R_{B,k}, R_{I,k}$	cross-sectional rigidities at time τ_k
R_b	basic relaxation
$\mathbf{R}_P$	vector of reactions caused by loads on the primary structure
$\mathbf{R}_R$	vector of redundant forces
$\mathbf{r}_{e,k}$	vector of external actions at time τ_k
$\mathbf{r}_{i,k}$	vector of internal actions at time τ_k
S	first moment of area of the steel reinforcement about the centroid of the section

s	spacing between cracks in the tensile zone
s_b	maximum bar spacing
s_{max}	maximum crack spacing immediately after loading
s_o	distance over which the stresses vary on each side of a direct tension crack
s^*	final maximum crack spacing
T	temperature; also tensile force
t	time
t_h	hypothetical thickness
$\mathbf{U}$	vector of nodal displacements for whole structure
$\mathbf{u}$	vector of nodal displacements for element
u_e	portion of section perimeter exposed to the atmosphere
v	deflection
v_{cx}	deflection of a column strip in the x-direction
v_{my}	deflection of a middle strip in the y-direction
v_{sus}	deflection due to sustained loading
W	wind load
w	crack width; also uniformly distributed load
w^*	maximum design crack width
w_s	uniformly distributed sustained service load
x	reference axis
y	transverse distance from the reference axis in the plane of the section
$y_{n,0}$	distance of neutral axis from x-axis on a cracked section at time τ_0
y_s, y_p	y-coordinates of steel reinforcement and tendons, respectively
Z	section modulus
z	direction of member axis
α	coefficient of thermal expansion; also a parameter to account for cracking and reinforcement on the change in deformation due to creep (Eqs 3.32)
$\alpha_1, \alpha_2, \alpha_3$	coefficients associated with creep and shrinkage models (Sections 2.1.4 and 2.1.5)
β	tension stiffening coefficient to account for duration of loading
Δ	an increment or a change
$\Delta\varepsilon,\ \Delta\kappa$	change in strain and curvature with time, respectively
δ	lateral displacement of a slender column
δ_B	settlement at support B
ε	strain
$\varepsilon_{top}, \varepsilon_{btm}$	strains at the top and bottom fibres of a cross-section, respectively
$\dot{\varepsilon}$	rate of change of strain with time ($d\varepsilon/dt$)
ε_{cr}	creep strain
$\varepsilon_{cr}(t, \tau_0)$	creep strain at time t due to a stress first applied at τ_0
ε^*_{cr}	final creep strain at time infinity
ε_e	instantaneous (elastic) strain
ε_r	strain at reference axis
$\varepsilon_{r0},\ \varepsilon_{rk}$	strain at reference axis at time τ_0 and at time τ_k, respectively
ε_p	strain in the prestressing steel

$\varepsilon_{p,init}$	initial strain in the prestressing steel before transfer
$\varepsilon_{p.rel,k}$	relaxation strain in the prestressing steel at time τ_k
ε_s	strain in the steel reinforcement
ε_{sh}	shrinkage strain
ε_{shd}	drying shrinkage strain
ε_{she}	endogenous (chemical plus thermal) shrinkage strain
ε^*_{sh}	final shrinkage strain at time infinity
ε_T	temperature component of strain
θ	slope
θ_M	angle between moment axis and the x-axis
θ_{NA}	angle between the neutral axis and the x-axis
κ	curvature
κ_{cr}, κ_{uncr}	curvatures on fully-cracked and uncracked cross-section, respectively
κ_0, κ_k	curvatures at time τ_0 and at time τ_k, respectively
κ_{sh}	curvature induced by shrinkage
κ_x, κ_y	curvature with respect to the x-axis and y-axis, respectively
$\lambda_r R$	reduced relaxation in the prestressing steel accounting for creep and shrinkage
λ_1, λ_2, λ_3	factors affecting bond stress
$\mu\varepsilon$	microstrain (10^{-6})
ρ, ρ'	tensile reinforcement ratio A_{st}/bd (or A_{st}/A_c) and compressive reinforcement ratio, A_{sc}/bd, respectively or; also, density
ρ_w, ρ_{cw}	web reinforcement ratio for the tensile steel $((A_{st}+A_{pt})/b_w d)$ and compressive steel $(A_{sc}/b_w d)$, respectively
Σ	the sum of
σ	stress
σ_c, σ_s	concrete and steel stresses, respectively
$\sigma_{c,0}$, $\sigma_{s,0}$	initial concrete and steel stresses, respectively, at first loading, τ_0
$\sigma_{c,k}$, $\sigma_{s,k}$	final concrete and steel stresses, respectively, at time τ_k
σ_{cs}	shrinkage-induced tensile stress in the concrete
$\sigma_{p,init}$	initial stress in the prestressing steel prior to transfer
$\dot{\sigma}$	rate of change of concrete stress with time, $d\sigma/dt$
$\sigma_{p(i)}$, $\sigma_{s(i)}$	stresses in the i-th layer of prestressed and non-prestressed steel, respectively
σ_{st1}	tensile steel stress at a crack
σ_{top}	concrete stress at top fibre of cross-section
τ	time instant; also shear stress
τ_0	age at first loading
τ_b	bond shear stress
τ_d	age at the commencement of drying
τ_i	i-th time instant
τ_k	final time instant at the end of the period of the time analysis
φ	creep coefficient
$\varphi(t, \tau_0)$	creep coefficient at time t for concrete loaded at τ_0
φ_{basic}	basic creep coefficient
$\varphi_{j,i}$	creep coefficient at time τ_j due to a stress first applied at τ_i

$\dot{\varphi}$	rate of change of the creep coefficient with time, $d\varphi/dt$
φ_p	creep coefficient for the prestressing steel
φ^*	final creep coefficient at time infinity
χ	ageing coefficient
$\chi(t, \tau_0)$	ageing coefficient at time t for concrete first loaded at τ_0
χ^*	final ageing coefficient at time infinity
ψ_1, ψ_2	short- and long-term serviceability load factors
ζ	distribution factor to account for tension stiffening (Eq. 3.24)
∞	infinity

1 Time-dependent deformation

1.1 Background

When a concrete specimen is subjected to load, its response is both immediate and time-dependent. Under sustained load, the deformation of a specimen gradually increases with time and eventually may be many times greater than its initial value. In order to satisfy the design objective of *serviceability* in a structural design, accurate and reliable predictions of the instantaneous and time-dependent deformation of the concrete structure are required.

If the temperature and stress remain constant, the gradual development of strain with time is caused by creep and shrinkage. Creep strain is produced by sustained stress, while shrinkage strain is independent of stress. These inelastic and time-dependent strains cause increases in deformation and curvature, losses of prestress and redistribution of stresses and internal actions. Creep and shrinkage are often responsible for excessive deflection at service loads. Creep frequently causes excessive camber and/or shortening in prestressed members. In addition, restraint to shrinkage may cause time-dependent cracking that could lead to serviceability or durability failures.

To accurately and efficiently predict these effects, the following two basic prerequisites are necessary:

(i) reliable data for the creep and shrinkage characteristics of the particular concrete mix; and
(ii) analytical and/or numerical procedures for the inclusion of these time effects in the analysis and design of the structure.

A number of sources are available from which to obtain information on the properties of concrete (see Sections 2.1.2–2.1.5). However, a comparison of the data obtained from these sources indicates significant differences. Laboratory tests may be undertaken to determine time-dependent material properties but this is often not a practical alternative. Designers seldom have the time or the inclination for long-term tests and then often cannot be sure that the concrete that has been tested in the laboratory is the same as that which will later be used in the structure. Even if testing is undertaken, the variability in the measured deformational properties of concrete is usually large. Coefficients of variation of 20 per cent or more can be expected. To best account for the variability of material properties in structural design, probabilistic methods are required. However, existing design methods are usually deterministic and the inexact nature of the results of such procedures must be kept in mind.

A number of analytical and numerical techniques are available for serviceability analysis and the design of reinforced and prestressed concrete structures. These range from simplified approaches for deflection and crack control contained in codes of practice, such as those presented and evaluated in Chapter 3, to the more refined methods contained in Chapters 4–9. Each level of refinement has its own set of simplifying assumptions and each has its advantages and disadvantages. Some are more suited to particular situations than others. Throughout the book, many worked examples are provided and a wide range of structural applications and situations are considered. The worked examples are also self-contained and may be read by practitioners who are more interested in using the final equations rather than following their detailed derivations. However, for most applications, the mathematics associated with each analysis is quite straightforward and there are obvious advantages in understanding the development of the procedures.

Methods of analysis that include the effects of creep and shrinkage in concrete structures are important in the design for serviceability. But perhaps even more importantly, such analyses can provide a clear picture of the interaction between concrete and steel and an otherwise unavailable insight into the mechanism of structural behaviour. A study of time effects can provide the structural designer with a far better *feel* for the task and a clearer understanding of why concrete structures behave as they do.

1.1.1 Concrete strain components

At any time t, the total concrete strain $\varepsilon(t)$ in an uncracked, uniaxially-loaded specimen consists of a number of components that include the instantaneous strain $\varepsilon_e(t)$, creep strain $\varepsilon_{cr}(t)$, shrinkage strain $\varepsilon_{sh}(t)$ and temperature strain $\varepsilon_T(t)$. Although not strictly correct, it is usual to assume that all four components are independent and may be calculated separately and combined to obtain the total strain. When calculating the in-service behaviour of a concrete structure at constant temperature, it is usual to express the concrete strain at a point as the sum of the instantaneous, creep and shrinkage components:

$$\varepsilon(t) = \varepsilon_e(t) + \varepsilon_{cr}(t) + \varepsilon_{sh}(t) \tag{1.1}$$

The strain components in a drying specimen held at constant temperature and subjected to a constant sustained compressive stress σ_{c0} first applied at time τ_0 are illustrated in Fig. 1.1. Immediately after the concrete sets or at the end of moist curing ($t = \tau_d$ in Fig. 1.1), shrinkage strains begin to develop and continue to increase at a decreasing rate. On the application of stress, a sudden jump in the strain diagram (instantaneous or elastic strain) is followed by an additional gradual increase in strain due to creep.

The prediction of the time-dependent behaviour of a concrete member requires the accurate prediction of each of these strain components at critical locations. This requires knowledge of the stress history, in addition to accurate data for the

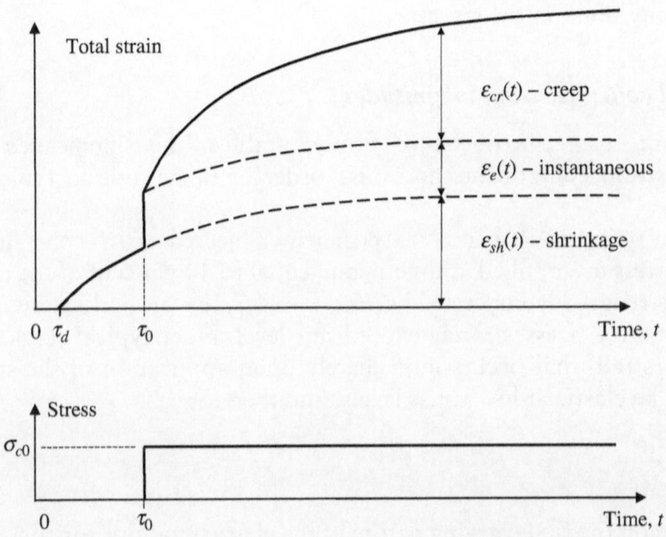

Figure 1.1 Concrete strain components under sustained load.

material properties. The stress history depends on both the applied load and the boundary conditions of the member.

It has been common practice in previous studies of the deformation of concrete to distinguish between *creep problems* and *relaxation problems*. Creep problems are those in which the gradual increase in strain under a sustained stress is calculated. Relaxation problems are those in which the total strain is held constant with time and the gradual change in stress is calculated.

Creep and relaxation problems differ only in their boundary conditions. Both involve the estimation of the individual strain components in Eq. 1.1. Relaxation problems have the additional constraint that the sum of the change of each strain component within any time interval is zero, that is, $\Delta\varepsilon_e(t) + \Delta\varepsilon_{cr}(t) + \Delta\varepsilon_{sh}(t) = 0$. When the total strain is held constant, the change of stress with time (relaxation) is simply the result of the gradual development of creep and shrinkage and the consequent equal and opposite change in the instantaneous strain. If creep and shrinkage increase with time, then the instantaneous strain $\varepsilon_e(t)$ must decrease.

There is little to be gained from separate studies of creep and relaxation type problems. In fact, in reinforced concrete structures, it is uncommon to find situations which fall neatly into the categories of either creep or relaxation problems. Concrete is almost never free to deform without some form of restraint. Reinforcement provides internal restraint to creep and shrinkage, while real support conditions often provide significant external restraint. On the other hand, a member is rarely restrained so completely that no deformation of the concrete is possible. Real problems often involve elements of both *creep* and *relaxation*, and fall between these two extremes. For the time analysis of concrete structures, attention should be focused squarely on the problem of establishing suitable constitutive relationships for concrete that can be used in the analysis of practical problems for the determination of the strain components of Eq. 1.1 at any point in the structure.

1.1.2 Typical concrete strain magnitudes

It is important from the outset to firmly establish the significance of each of the concrete strain components and their order of magnitude in typical in-service situations.

Consider a point in a concrete specimen subjected to a constant, sustained compressive stress σ_{c0} applied at time τ_0 and equal to 40 per cent of the characteristic compressive strength of concrete, i.e. $\sigma_{c0} = 0.4f'_c$. In properly designed concrete structures, concrete stresses rarely exceed this level under typical service loads. The instantaneous strain that occurs immediately upon application of the stress may be considered to be elastic at low stress levels, and therefore:

$$\varepsilon_e(t) = \frac{\sigma_{c0}}{E_c(\tau_0)} \tag{1.2}$$

Information on the time-varying nature of the elastic modulus for concrete, $E_c(t)$, is given in Section 2.1.3. However, for comparison between the order of magnitude of each of the strain components in Eq. 1.1, the elastic modulus may be taken as constant with time and approximately equal to 1000 f'_c. In this case, therefore, the elastic strain is approximated by:

$$\varepsilon_e \approx \frac{-0.4f'_c}{1000f'_c} = -0.0004 = -400\,\mu\varepsilon$$

where the negative sign indicates that the applied stress and elastic strain are compressive.

Under normal conditions, the final creep strain at time infinity $\varepsilon_{cr}(\infty)$ (denoted here as ε^*_{cr}) that is produced by a constant sustained stress of this magnitude is usually between 1.5 and 4.0 times the initial elastic strain. The higher end of this range is for low-strength concrete loaded at early ages and at a low relative humidity. A typical final creep strain for structural concrete subjected to constant stress of this level is 2.5 ε_e and, for this case,

$$\varepsilon^*_{cr} \approx 2.5\,\varepsilon_e = -1000\,\mu\varepsilon$$

The final shrinkage strain at time infinity $\varepsilon_{sh}(\infty)$ (denoted here as ε^*_{sh}) depends on the concrete composition, the environment and the size and shape of the specimen. For concrete elements in building structures, a typical value is:

$$\varepsilon^*_{sh} \approx -600\,\mu\varepsilon$$

The thermal strain produced by a change in temperature $\Delta T = T - T_0$ is calculated from:

$$\varepsilon_T = \int_{T_0}^{T} \alpha\, dT \tag{1.3}$$

where T_0 is the initial temperature and α is the coefficient of thermal expansion that is approximately $10 \times 10^{-6}/°C$ (but actually depends on temperature and moisture content). For a 20°C change in temperature, the change in thermal strain is approximately $\alpha\,\Delta T = 200\,\mu\varepsilon$.

If the specimen under consideration is held at constant temperature, the total concrete strain at time infinity (which may be considered as several years after loading) is:

$$\varepsilon^* = \varepsilon_e + \varepsilon_{cr}^* + \varepsilon_{sh}^* \approx -400 - 1000 - 600 = -2000\,\mu\varepsilon \tag{1.4}$$

Thus the magnitude of the final strain is about five times the magnitude of the instantaneous elastic strain. It must be emphasised that the magnitudes of the creep and shrinkage considered here are typical and not extreme values. Therefore, when calculating the deformation of concrete structures, time effects must be included in a rational and systematic way. Elastic analyses that ignore the effects of creep and shrinkage may grossly underestimate final deformations and, in the design of concrete structures for serviceability, are of little value.

1.2 Creep of concrete

1.2.1 Creep mechanisms and influencing factors

When concrete is subjected to a sustained stress, creep strain develops gradually with time as shown in Fig. 1.1. Creep increases with time at a decreasing rate. In the period immediately after initial loading, creep develops rapidly, but the rate of increase slows appreciably with time. Creep is generally thought to approach a limiting value as the time after first loading approaches infinity. About 50 per cent of the final creep develops in the first 2–3 months and about 90 per cent after 2–3 years. After several years under load, the rate of change of creep with time is very small.

Throughout this book, creep and shrinkage are treated as separate and independent phenomena. In reality, this is not the case. Creep is significantly greater when accompanied by shrinkage. In a loaded specimen that is in hygral equilibrium with the ambient medium (i.e. no drying), the time-dependent deformation caused by stress is known as *basic creep*. The additional creep that occurs in a drying specimen is sensibly known as *drying creep*. Creep is usually calculated as the difference between the total time-dependent deformation of a loaded specimen and the shrinkage of a similar unloaded specimen. Creep is treated here as the time-dependent deformation in excess of shrinkage.

Creep of concrete originates in the hardened cement paste that consists of a solid cement gel containing numerous capillary pores. The cement gel is made up of colloidal sheets of calcium silicate hydrates separated by spaces containing absorbed water. Creep is thought to be caused by several different and complex mechanisms not yet fully understood (Refs 1–7). Neville *et al.* (Ref. 4) identified the following mechanisms for creep:

(i) sliding of the colloidal sheets in the cement gel between the layers of absorbed water – *viscous flow*;

(ii) expulsion and decomposition of the interlayer water within the cement gel – *seepage*;
(iii) elastic deformation of the aggregate and the gel crystals as viscous flow and seepage occur within the cement gel – *delayed elasticity*;
(iv) local fracture within the cement gel involving the breakdown (and formation) of physical bonds – *microcracking*;
(v) *mechanical deformation theory*; and
(vi) *plastic flow*.

Recent research relates the creep response to the packaging density distributions of calcium-silicate-hydrates (Ref. 7). At high stress levels, additional deformation occurs due to the breakdown of the bond between the cement paste and aggregate particles. Further references on creep mechanisms can be found in Refs 1 and 8.

Many factors influence the magnitude and rate of development of creep. Some are properties of the concrete mix, while others depend on the environmental and loading conditions. In general, the capacity of the concrete to creep decreases as the concrete strength increases. For a particular stress level, creep in higher-strength concrete is less than that in lower-strength concrete. An increase in either the aggregate content or the maximum aggregate size reduces creep, as does the use of a stiffer aggregate type. Creep also decreases as the water-to-cement ratio is reduced.

Creep also depends on the environment. It increases as the relative humidity decreases. Creep is also greater in thin members with large surface-area-to-volume ratios, such as slabs. However, the dependence of creep on both the relative humidity and the size and shape of the specimen decreases as the concrete strength increases. Near the surface of a member, creep takes place in a drying environment and is therefore greater than in regions remote from a drying surface. In addition to the relative humidity, creep is dependent on the ambient temperature. A rise in temperature increases the deformability of the cement paste and accelerates drying, and thus increases creep. The dependence of creep on temperature is more pronounced at elevated temperatures and is far less significant for temperature variations between 0°C and 20°C. However, creep in concrete at a mean temperature of 40°C is perhaps 25 per cent higher than that at 20°C (Ref. 9).

In addition to the environment and the characteristics of the concrete mix, creep depends on the loading history, in particular the magnitude and duration of the stress and the age of the concrete when the stress was first applied. The age of the concrete, τ_0, when the stress was first applied has a marked influence on the final magnitude of creep. Concrete loaded at an early age creeps more than concrete loaded at a later age. Concrete is therefore a time-hardening material, although even in very old concrete the tendency to creep never entirely disappears (Ref. 10).

When the sustained concrete stress is less than about $0.5f'_c$, creep is approximately proportional to the stress and is known as *linear creep*. At higher stress levels creep increases at a faster rate and becomes non-linear with respect to stress. This non-linear behaviour of creep at high stress levels is thought to be related to an increase in micro-cracking. Compressive stresses rarely exceed $0.5f'_c$ in concrete structures at service loads, and creep may be taken as proportional to stress in most situations in the design for serviceability. In this book, the effects of non-linear creep at high stress levels are not examined.

1.2.2 Creep components

To best understand the physical nature of creep and to describe its characteristics, creep strain is often subdivided into several components. In Fig. 1.2a, the creep strain produced by the compressive stress history of Fig. 1.2b is shown. Under sustained stress σ_o, creep increases at a decreasing rate. When the stress is removed at time τ_1, there is no sudden change in creep strain, but a gradual reduction occurs with time as shown. There is of course a sudden change in the total strain at τ_1 due to the elimination of instantaneous strain. A portion of the creep strain is recoverable, while a usually larger portion is irrecoverable or permanent.

The recoverable part of creep is often referred to as the *delayed elastic strain*, $\varepsilon_{cr.d}(t)$. This delayed elasticity is thought to be caused by the elastic aggregate acting on the viscous cement paste after the applied stress is removed. If a concrete specimen is unloaded after a long period under load, the magnitude of the recoverable creep is in the order of 40–50 per cent of the elastic strain (between 10 and 20 per cent of the total creep strain). Although the delayed elastic strain is observed only as recovery when the load is removed, it is generally believed to be of the same magnitude under load and to develop rapidly in the period immediately after loading. Rüsch *et al.* (Ref. 8) suggest that the shape of the delayed elastic strain curve is independent of the age or dimensions of the specimen and is unaffected by the composition of the concrete.

The majority of creep strain is irrecoverable and is often referred to as *flow*, $\varepsilon_{cr.f}(t)$. The flow component of creep is sometimes further subdivided into *rapid initial flow*, $\varepsilon_{cr.fi}(t)$, that occurs in the first 24 hours after loading, and the remaining flow that develops gradually with time (Ref. 9). The rapid initial flow is irrecoverable and highly dependent on the age at first loading, that is, on the degree of hydration. The younger concrete is when first loaded, the higher $\varepsilon_{cr.fi}(t)$. The remaining flow that occurs after the first day under load is dependent on the relative humidity and, as discussed earlier, may be divided into a basic flow component $\varepsilon_{cr.fb}(t)$ and a drying flow component $\varepsilon_{cr.fd}(t)$. The drying flow is the additional irrecoverable creep that occurs when the specimen is loaded in a drying environment. The basic flow $\varepsilon_{cr.fb}(t)$ depends on the composition of the concrete mix (aggregate type, size and quantity, concrete strength, etc.) and the age of the concrete at the time of loading. The drying flow $\varepsilon_{cr.fd}(t)$ depends on the moisture content and gradient, and the size and shape of the specimen.

In structural analysis, it is unusual to subdivide the flow component of creep into all of these sub-components. However, consideration of the recoverable and irrecoverable

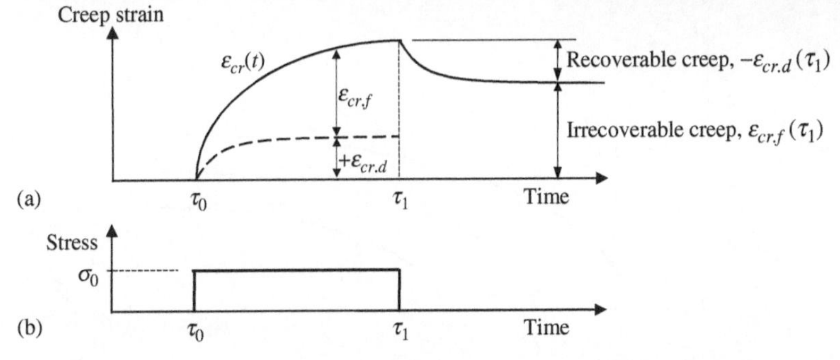

Figure 1.2 Recoverable and irrecoverable creep components.

components of creep (i.e. the delayed elastic and flow components, respectively) is appropriate and necessary if concrete is subjected to a time-varying stress history.

1.2.3 *Effects of ageing*

In the previous sections, we saw that all sub-components of creep are affected to some extent by the degree of hydration, i.e. the age of the concrete at the time of first loading, τ. Fig. 1.3 shows the effect of age at first loading on the creep–time curves of identical specimens first loaded at ages τ_0, τ_1 and τ_2.

Although not proven, it is usually assumed that the creep strain $\varepsilon_{cr}(t, \tau)$ approaches a limiting value as time approaches infinity, i.e. $\varepsilon_{cr}(\infty, \tau) = \varepsilon^*_{cr}(\tau)$. All else being equal:

$$\varepsilon^*_{cr}(\tau_i) > \varepsilon^*_{cr}(\tau_j) \quad \text{provided} \quad \tau_i < \tau_j \tag{1.5}$$

This time-hardening or ageing of concrete complicates the prediction of creep strain under time-varying stress histories.

1.2.4 *The creep coefficient, $\varphi(t,\tau)$, and the creep function, $J(t, \tau)$*

The capacity of concrete to creep is usually measured in terms of the creep coefficient, $\varphi(t,\tau)$. In a concrete specimen subjected to a constant sustained compressive stress, $\sigma_c(\tau)$, first applied at age τ, the creep coefficient at time t is the ratio of the creep strain to the instantaneous strain and is given by:

$$\varphi(t, \tau) = \frac{\varepsilon_{cr}(t, \tau)}{\varepsilon_e(\tau)} \tag{1.6}$$

Therefore, the creep strain at time t caused by a constant sustained stress $\sigma_c(\tau)$ first applied at age τ is:

$$\varepsilon_{cr}(t, \tau) = \varphi(t, \tau)\, \varepsilon_e(\tau) = \varphi(t, \tau) \frac{\sigma_c(\tau)}{E_c(\tau)} \tag{1.7}$$

where $E_c(\tau)$ is the elastic modulus at time τ. For concrete subjected to a constant sustained stress, knowledge of the creep coefficient allows the rapid determination of the creep strain at any time.

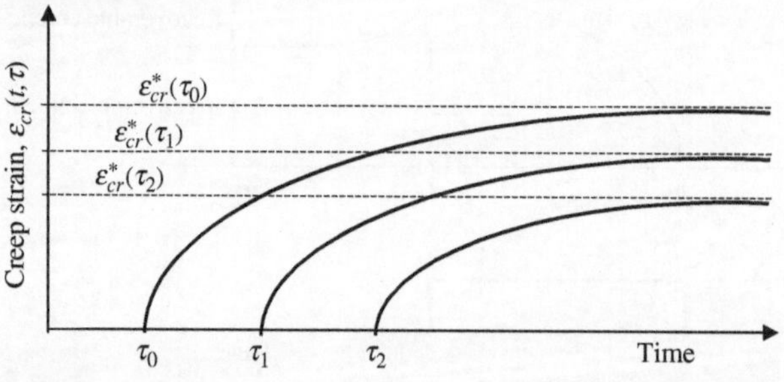

Figure 1.3 Effect of age at first loading on creep strains.

Since both the creep and instantaneous strain components are proportional to stress for compressive stress levels less than about $0.5f'_c$, the creep coefficient $\varphi(t,\tau)$ is a pure time function and is independent of the applied stress. The creep coefficient increases with time at an ever-decreasing rate. As time approaches infinity, the creep coefficient is assumed to approach a final value $\varphi^*(\tau) = \varepsilon^*_{cr}(\tau)/\varepsilon_e(\tau)$ that usually falls within the range 1.5–4.0. This final creep coefficient is a useful measure of the capacity of concrete to creep. Since creep strain depends on the age of the concrete at the time of first loading, so too does the creep coefficient. A family of creep coefficient versus time curves may be drawn that are similar to the creep–time curves shown in Fig. 1.3.

Another frequently used time function is known as *specific creep*, $C(t,\tau)$, and is the proportionality factor relating stress to linear creep, i.e.:

$$\varepsilon_{cr}(t,\tau) = C(t,\tau)\,\sigma_c(\tau) \quad \text{or} \quad C(t,\tau) = \frac{\varepsilon_{cr}(t,\tau)}{\sigma_c(\tau)} \tag{1.8}$$

$C(t,\tau)$ is the creep strain at time t produced by a sustained *unit* stress first applied at age τ.

The relationship between the creep coefficient and specific creep can be obtained from Eqs 1.6 and 1.8:

$$\varphi(t,\tau) = C(t,\tau)\,E_c(\tau) \tag{1.9}$$

The sum of the instantaneous and creep strains at time t produced by a sustained unit stress applied at τ is defined as the *creep function*, $J(t,\tau)$, and is given by:

$$J(t,\tau) = \frac{1}{E_c(\tau)} + C(t,\tau) = \frac{1}{E_c(\tau)}[1+\varphi(t,\tau)] \tag{1.10}$$

The stress-produced strains (i.e. the instantaneous plus creep strains) caused by a constant sustained stress $\sigma_c(\tau)$ first applied at age τ (also called the stress-dependent strains) may therefore be determined from:

$$\varepsilon_e(t) + \varepsilon_{cr}(t,\tau) = J(t,\tau)\sigma_c(\tau) = \frac{\sigma_c(\tau)}{E_c(\tau)}[1+\varphi(t,\tau)] = \frac{\sigma_c(\tau)}{E_e(t,\tau)} \tag{1.11}$$

where $E_e(t,\tau)$ is known as the *effective modulus* and is given by:

$$E_e(t,\tau) = \frac{E_c(\tau)}{[1+\varphi(t,\tau)]} \tag{1.12}$$

The variation with time of the creep strain caused by a constant sustained stress $\sigma_c(\tau_0)$ ($\equiv\sigma_{c0}$) is shown in Fig. 1.4.

In the remainder of this book, the creep coefficient and the creep function are used to mathematically define and quantify creep.

1.2.5 The principle of superposition

Because the load-dependent strains in concrete at service loads are proportional to stress, the principle of superposition is frequently used to estimate the deformation

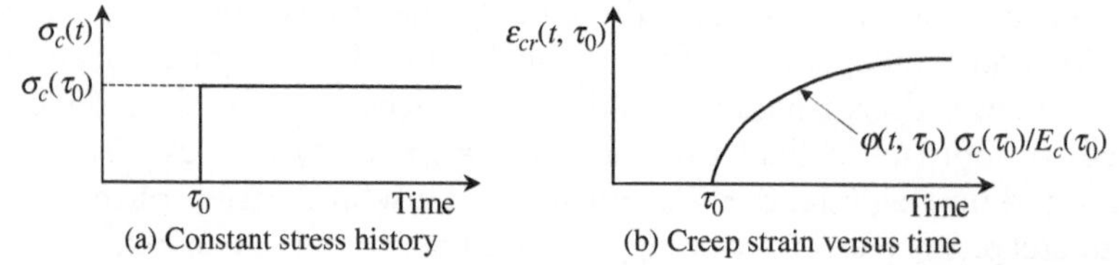

Figure 1.4 Creep strain produced by a constant sustained stress.

caused by a time-varying stress history. The principle of superposition was first applied to concrete by McHenry (Ref. 11) who stated that the strain produced by a stress increment applied at any time τ_i is not affected by any stress applied either earlier or later. This principle is illustrated in Fig. 1.5.

For increasing stress histories, such as that shown in Fig. 1.5c, the principle of superposition agrees well with experimental observations. The creep curve produced by the increasing stress history is assumed to be equal to the sum of the creep curves produced by each stress increment acting independently. However, for decreasing stress histories, the principle of superposition overestimates creep recovery, as shown in Fig. 1.5d. This can easily be seen when one divides creep into delayed elastic and flow components. The principle of superposition incorrectly assumes that at any time $t \gg \tau_1$, the delayed elastic strain that develops between τ_0 and τ_1, $\varepsilon_{cr.d}(\tau_1, \tau_0)$, is equal to the

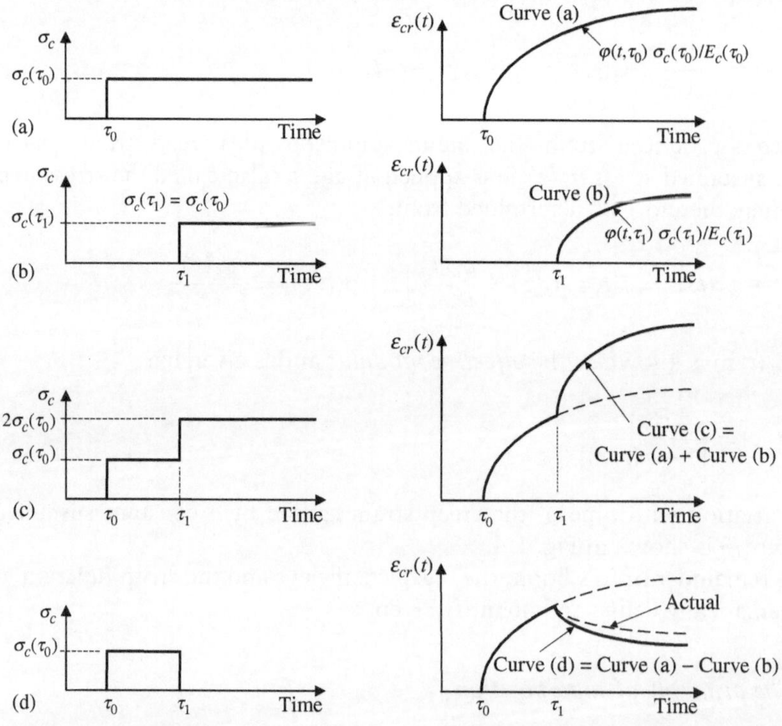

Figure 1.5 The principle of superposition of creep strains.

creep strain at time t produced by $\sigma_c(\tau_1) = \sigma_c(\tau_0)$ applied at τ_1 (i.e. $\varepsilon_{cr}(t, \tau_1)$). In fact, $\varepsilon_{cr.d}(\tau_1, \tau_0)$ is significantly less than $\varepsilon_{cr}(t, \tau_1)$. Nevertheless, for most practical purposes, the principle of superposition provides a good approximation of the time-dependent strains in concrete caused by a time-varying stress history.

To illustrate the principle of superposition, consider the stress history shown in Fig. 1.6a consisting of two stress increments $\Delta\sigma_c(\tau_0)$ and $\Delta\sigma_c(\tau_1)$ applied at times τ_0 and τ_1, respectively. In this case, two creep coefficients $\varphi(t, \tau_0)$ and $\varphi(t, \tau_1)$ are required to determine the creep strain at any time $t > \tau_1$. Graphs of the two creep coefficients versus time curves are shown in Fig. 1.6b and the corresponding creep function versus time curves are given in Fig. 1.6c.

According to the principle of superposition, the total stress-dependent strain in concrete at time t (elastic plus creep strains) can be written as:

$$\varepsilon_e(t) + \varepsilon_{cr}(t) = \frac{\Delta\sigma_c(\tau_0)}{E_c(\tau_0)}[1 + \varphi(t, \tau_0)] + \frac{\Delta\sigma_c(\tau_1)}{E_c(\tau_1)}[1 + \varphi(t, \tau_1)]$$

$$= \sum_{i=0}^{1} \frac{\Delta\sigma_c(\tau_i)}{E_c(\tau_i)}[1 + \varphi(t, \tau_i)] \quad (1.13)$$

where the argument τ has been eliminated from the elastic strain $\varepsilon_e(t)$ and creep strain $\varepsilon_{cr}(t)$ as the stress history varies with time. This notation is more appropriate for realistic time-varying stress histories and will be used throughout the remainder of the book.

Eq. 1.13 can also be written in terms of creep functions as:

$$\varepsilon_e(t) + \varepsilon_{cr}(t) = J(t, \tau_0)\Delta\sigma_c(\tau_0) + J(t, \tau_1)\Delta\sigma_c(\tau_1) = \sum_{i=0}^{1} J(t, \tau_i)\Delta\sigma_c(\tau_i) \quad (1.14)$$

The variations with time of the elastic and creep deformations are plotted in Fig. 1.7.

A useful graphic interpretation of the elastic and creep deformations is shown in Fig. 1.8, where Figs 1.6a and 1.6c have been combined. The stress increments of Fig. 1.6a are plotted on the horizontal axis of Fig. 1.8b and Fig. 1.6c has been reproduced for ease of reference in Fig. 1.8a. At any time $t > \tau_1$ the area under the

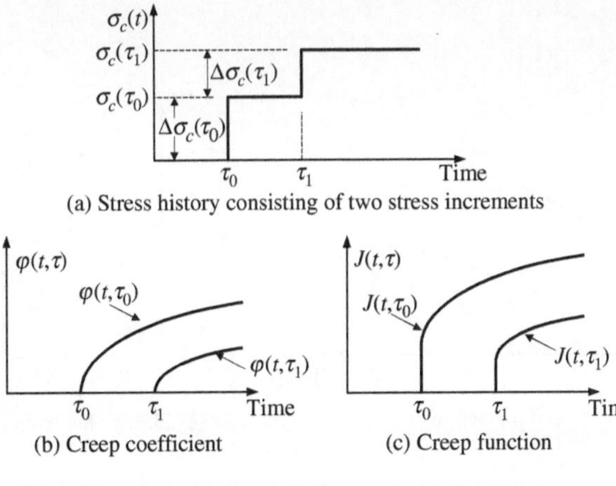

Figure 1.6 Creep coefficients and creep functions associated with two stress increments.

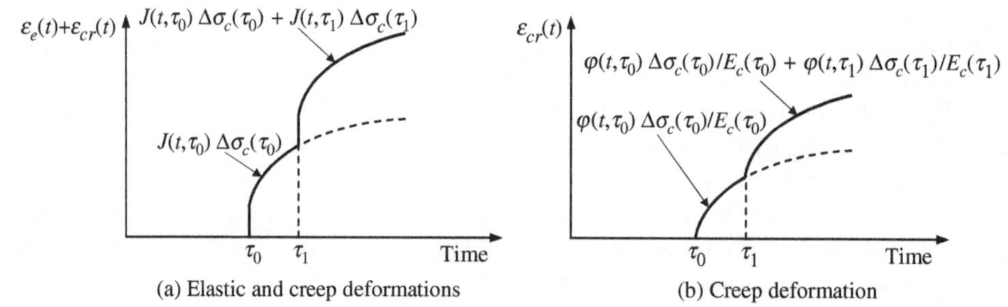

Figure 1.7 Creep and elastic deformations due to two stress increments (Fig. 1.6a).

curve of Fig. 1.8b represents the actual stress-dependent deformation (sum of elastic and creep strains) at time t as defined in Eq. 1.14.

Consider the stress history shown in Fig. 1.9a. By dividing the time under load into n time steps, the continuously varying stress may be approximated by a series of small stress increments applied at the end of each time step. The stress-produced strains at time t are given by:

$$\varepsilon_e(t)+\varepsilon_{cr}(t)=\sum_{i=0}^{n}\frac{\Delta\sigma_c(\tau_i)}{E_c(\tau_i)}\left[1+\varphi(t,\tau_i)\right]=\sum_{i=0}^{n}J(t,\tau_i)\Delta\sigma_c(\tau_i) \tag{1.15}$$

and a different creep coefficient is required for each small stress increment, $\Delta\sigma(\tau_i)$. According to the principle of superposition, the total strain in concrete at time t subjected to the stress history shown in Fig. 1.9 is obtained by summing the stress-produced strains and the shrinkage strain:

$$\varepsilon(t)=\varepsilon_e(t)+\varepsilon_{cr}(t)+\varepsilon_{sh}(t)$$

$$\varepsilon(t)=\sum_{i=0}^{n}J(t,\tau_i)\Delta\sigma_c(\tau_i)+\varepsilon_{sh}(t)=\sum_{i=0}^{n}\frac{\Delta\sigma_c(\tau_i)}{E_c(\tau_i)}+\sum_{i=0}^{n}\frac{\Delta\sigma_c(\tau_i)\varphi(t,\tau_i)}{E_c(\tau_i)}+\varepsilon_{sh}(t) \tag{1.16}$$

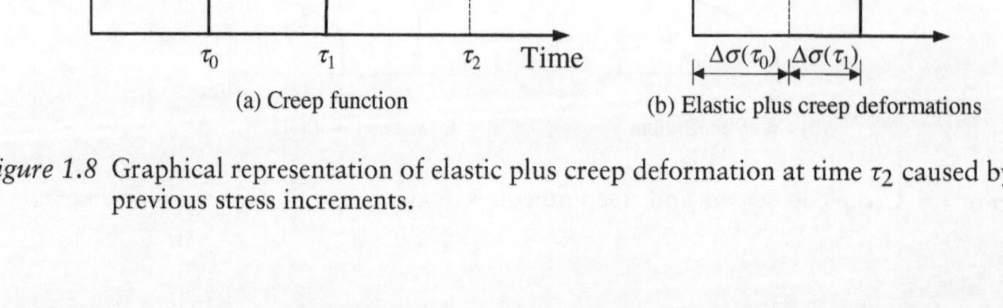

Figure 1.8 Graphical representation of elastic plus creep deformation at time τ_2 caused by two previous stress increments.

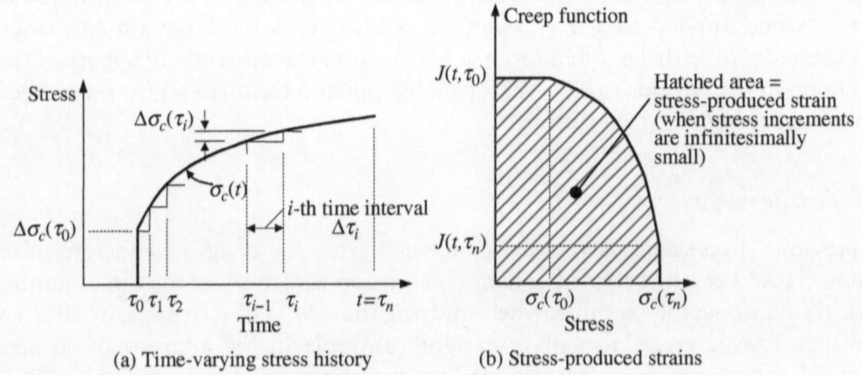

Figure 1.9 Time-varying stress history and stress-produced strains.

In the limit, considering infinitesimal stress increments $d\sigma_c(\tau)$, the summation included in Eq. 1.16 can be replaced by an integral:

$$\varepsilon(t) = \int_{\tau_0}^{t} J(t,\tau)\mathrm{d}\sigma_c(\tau) + \varepsilon_{sh}(t) = \int_{\tau_0}^{t} \frac{1+\varphi(t,\tau)}{E_c(\tau)}\mathrm{d}\sigma_c(\tau) + \varepsilon_{sh}(t) \tag{1.17}$$

Equation 1.17 is referred to as the integral-type creep law (Ref. 12). The integral in Eq. 1.17 for the stress-produced strains is the area under the graph of creep function versus concrete stress in Fig. 1.9b.

The stress at a point in a concrete structure under constant load often decreases with time as a result of creep. Consider the time-varying stress history shown in Fig. 1.10a where the initial stress $\sigma_c(\tau_0)$ is compressive and decreases with time (i.e. all subsequent stress increments $\Delta\sigma_c(\tau_i)$ are tensile). Notwithstanding the inaccuracies when the principle of superposition is applied to a decreasing stress history (mentioned previously), the stress produced strain may still be obtained using Eqs 1.16 or 1.17, with the integral in Eq. 1.17 represented by the hatched area under the curve in Fig. 1.10b.

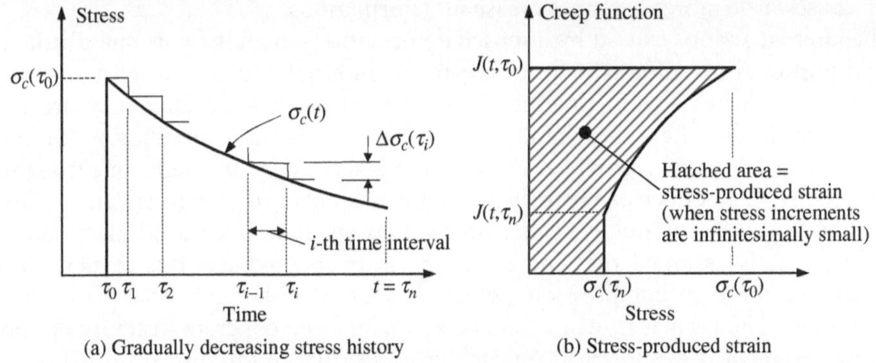

Figure 1.10 Gradually decreasing stress history and stress-produced strain.

The main disadvantage of the method of superposition, apart from its reduced accuracy when applied to a decreasing stress history, is the large amount of creep data required. Nevertheless, in a more general sense, the principle of superposition is relied upon in many fields of structural engineering and the analysis for time effects in concrete structures is no exception.

1.2.6 Tensile creep

The previous discussions have been concerned with the creep of concrete in compression. However, the creep of concrete in tension is also of interest in a number of practical situations; for example, when studying the effects of restrained or differential shrinkage. Tensile creep also plays a significant role in the analysis of suspended reinforced concrete slabs at service loads where stress levels are generally low and typically much of the slab is initially uncracked.

Comparatively little attention has been devoted to the study of tensile creep (Ref. 1) and only limited experimental results are available in the literature, e.g. Ref. 13. Some researchers have multiplied the creep coefficients measured for compressive stresses by factors in the range of 1 to 3 to produce equivalent coefficient describing tensile creep (e.g. Refs 14 and 15).

It appears that the mechanisms of creep in tension are different to those in compression. The magnitudes of both tensile and compressive creep increase when loaded at earlier ages. However, the rate of change of tensile creep with time does not decrease in the same manner as for compressive creep, with the development of tensile creep being more linear (Ref. 13). Drying tends to increase tensile creep in a similar manner to compressive creep and tensile creep is in part recoverable upon removal of the load. Further research is needed to provide clear design guidance. In this book, it is assumed that the magnitude and rate of development of tensile creep are similar to that of compressive creep at the same low stress levels.

1.2.7 The effects of creep on structural behaviour

Creep is the time-dependent strain that develops in concrete due to sustained stress. When a loaded concrete member is unrestrained by steel reinforcement or by the external support conditions (such as a statically determinate plain concrete member), creep causes little more than an increase in deformations.

The internal actions caused by imposed deformations in an indeterminate structure are proportional to stiffness. Due to creep, the internal actions caused by imposed deformations decrease with time in all concrete structures whether they are plain or reinforced. On the other hand, creep will not cause redistribution of load-induced internal actions provided the creep characteristics are uniform throughout the structure. The effect of creep in this case is similar to a gradual and uniform change in the elastic modulus. Such a situation exists in an uncracked plain concrete member, if it is assumed that the creep characteristics of concrete in tension are the same as those in compression (which is a usual although not strictly correct assumption). The same is true in an uncracked reinforced concrete structure provided the reinforcement is symmetrical on each cross-section and uniform throughout the structure.

When portions of a statically indeterminate structure are made of different materials or of concrete with a different composition or age, and hence different creep characteristics, creep causes a redistribution of the load-induced internal actions. The internal actions are redistributed from the regions with the higher creep rate to the regions with the lower creep rate. In practical reinforced concrete flexural members, the creep characteristics are rarely uniform throughout. Some regions are usually cracked and some are not. Reinforcement quantities usually vary along the length of the member and sections are seldom symmetrically reinforced. Nevertheless, the redistribution caused by creep of the load-induced bending moments and shear forces in statically indeterminate members is generally small and is quite often neglected in design.

Creep can dramatically change the stress distribution on a reinforced concrete section. For example, stresses caused by shrinkage or temperature changes are relieved by creep. Creep also causes a redistribution of stresses between concrete and the bonded reinforcement on a cross-section. For example, consider a reinforced concrete column section subjected to a constant sustained axial compressive load. The concrete and steel are bonded together so that at any time compatibility requires that the concrete and steel strains are identical. As the compressive concrete creeps (contracts), the steel is compressed and the compressive stress in the steel gradually increases. Shrinkage causes a similar effect. Equilibrium requires that the increase in the compressive force in the steel is balanced by an equal and opposite tensile force on the concrete. Therefore, the compressive stress in the concrete reduces with time, while the steel stress increases rapidly. Load is thus transferred from the concrete to the steel, with the proportion of the external load carried by the reinforcement increasing with time. Fig. 1.11 shows the typical variation in compressive stresses in the concrete and steel caused by creep on the cross-section of an axially loaded reinforced concrete column subjected to a constant sustained compressive force, P.

It is quite common, in fact, for the stress in the longitudinal reinforcement in typically proportioned reinforced concrete columns to be close to the yield stress at service loads,

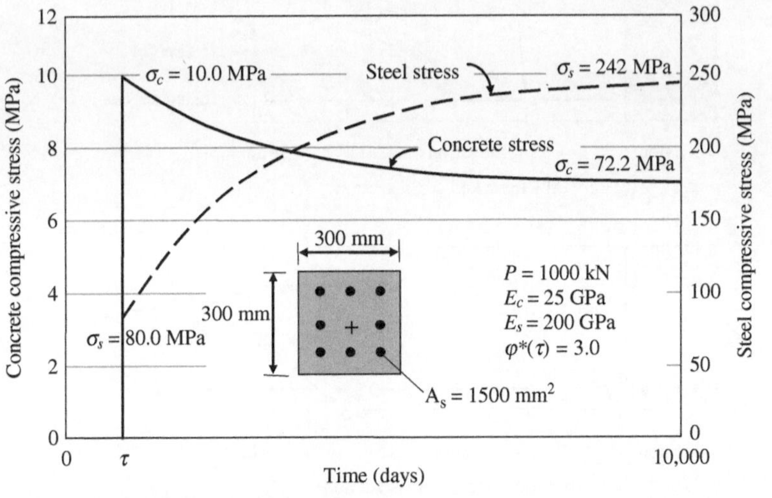

Figure 1.11 Variation of stresses with time due to creep on a symmetrically reinforced axially loaded column section.

due to the effects of creep and shrinkage (particularly for low-strength steel grades). The use of closely spaced lateral reinforcement in the form of closed ties or helices to prevent these highly stressed bars from buckling is therefore essential at the service load condition and not just at the ultimate load condition.

The gradual development of creep strain on a reinforced concrete beam section causes an increase of curvature, as shown in Fig. 1.12, and a consequent increase in the beam deflection. For a plain concrete member, the increase in curvature is proportional to the creep coefficient (i.e. the increase in curvature is proportional to the uniform increase in strain at every point on the section). For an *uncracked*, singly reinforced section (Fig. 1.12a), creep is restrained in the tensile zone by the reinforcement. Depending on the quantity of steel, the increase in curvature due to creep is proportional to a large fraction of the creep coefficient (usually between $0.6\varphi(t)$ and $0.9\varphi(t)$).

On a *cracked*, singly reinforced beam section (Fig. 1.12b), the initial curvature is comparatively large and the cracked tensile concrete below the neutral axis can be assumed to carry no stress and therefore does not creep. Creep in the compression zone causes a lowering of the neutral axis and a consequent reduction in the compressive stress level. Creep is slowed down as the compressive stress reduces, and the increase in curvature is proportional to a small fraction of the creep coefficient (usually less than one quarter). The relative increase in deflection caused by creep is therefore greater in an uncracked beam than in a cracked beam, although the total deflection in the cracked beam is significantly greater.

In prestressed concrete construction, in addition to causing increases in deflection (and camber), both creep and shrinkage cause shortening of the concrete member that in turn causes shortening of the prestressing tendons and a consequent reduction in the prestressing force. This loss of prestress may adversely affect the performance of the member at service loads and should be accounted for in its design.

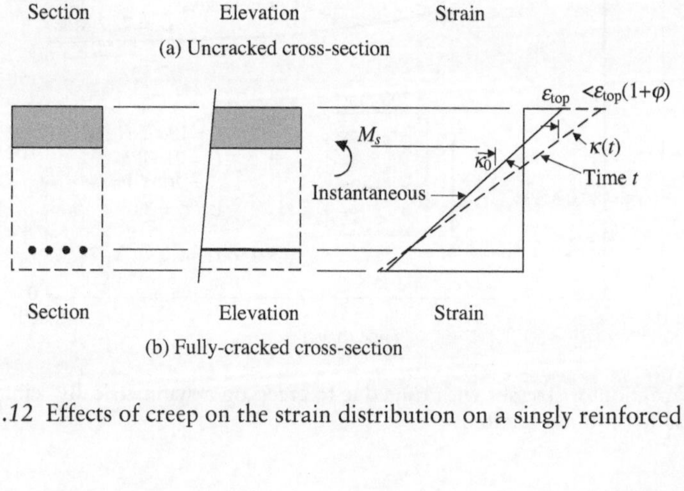

Figure 1.12 Effects of creep on the strain distribution on a singly reinforced section.

Creep does not normally affect the strength of a structural member. The magnitude of creep strain is usually small compared to the peak strains at the ultimate load condition. An exception is creep-induced buckling of slender columns, arches or domes. In a slender column, for example, creep increases the lateral deflection caused by the initial eccentricity resulting in additional secondary moments that in turn may eventually lead to instability. Another example where creep deformation may affect strength is a sagging floor or roof system subjected to (water) ponding loads.

In summary, creep significantly affects structural behaviour at service loads. In some instances, its effects are detrimental, while in others, creep is beneficial. Creep causes losses of prestress and increases in deformations and deflections that may impair the serviceability of the structure. Creep also adversely affects a favourable initial stress distribution that may be introduced intentionally by an imposed deformation (e.g. preflexed girders). On the other hand, creep reduces undesirable stresses in concrete caused by unintentionally imposed deformations such as support settlements, shrinkage and thermal gradients and so on. Creep relieves concrete stress concentrations and imparts deformability to concrete. In fact, the success of concrete as a structural material is due, in no small way, to its ability to creep.

1.3 Shrinkage of concrete

1.3.1 Types of shrinkage

Shrinkage of concrete is the time-dependent strain in an unloaded and unrestrained specimen at constant temperature. It is important from the outset to distinguish between *plastic* shrinkage, *chemical* shrinkage, *thermal* shrinkage and *drying* shrinkage. Plastic shrinkage occurs in the wet concrete before setting, whereas chemical, thermal and drying shrinkage all occur in the hardened concrete after setting. Some high-strength concretes are prone to plastic shrinkage that occurs in the wet concrete and may result in significant cracking during the setting process. This cracking occurs due to capillary tension in the pore water and is best prevented by taking measures during construction to avoid the rapid evaporation of bleed water. At this stage, the bond between the plastic concrete and the reinforcement has not yet developed, and the steel is ineffective in controlling such cracks.

Drying shrinkage is the reduction in volume caused principally by the loss of water during the drying process. It increases with time at a gradually decreasing rate and takes place in the months and years after casting. The magnitude and rate of development of drying shrinkage depend on all the factors that affect the drying of concrete, including the relative humidity, the mix characteristics (in particular, the type and quantity of the binder, the water content and water-to-cement ratio, the ratio of fine-to-coarse aggregate, and the type of aggregate), and the size and shape of the member.

Chemical shrinkage results from various chemical reactions within the cement paste and includes hydration shrinkage that is related to the degree of hydration of the binder in a sealed specimen with no moisture exchange. Chemical shrinkage (often called *autogenous shrinkage*) occurs rapidly in the days and weeks after casting and is less dependent on the environment and the size of the specimen than drying shrinkage. Thermal shrinkage is the contraction that results in the first few hours (or days) after setting as the heat of hydration gradually dissipates. The term *endogenous shrinkage* is sometimes used to refer to that part of the shrinkage of the hardened concrete that is not associated with drying (i.e. the sum of autogenous and thermal shrinkage).

1.3.2 Factors affecting shrinkage

The shrinkage of concrete is defined here as the time-dependent strain measured at constant temperature in an unloaded and unrestrained specimen. The shrinkage strain, ε_{sh}, is usually considered to be the sum of the drying shrinkage component, ε_{shd}, (which is the reduction in volume caused principally by the loss of water during the drying process) and the endogenous shrinkage component, ε_{she} (which is mainly due to chemical reactions within the cement paste, such as carbonation). Drying shrinkage in high-strength concrete is smaller than in normal-strength concrete due to the smaller quantities of free water after hydration. However, thermal and chemical shrinkage may be significantly higher. Although drying and endogenous shrinkage are quite different in nature, there is no need to distinguish between them from a structural engineering point of view.

Shrinkage continues to increase with time at a decreasing rate, as illustrated in Fig. 1.1. Shrinkage is assumed to approach a final value, ε^*_{sh}, as time approaches infinity. Drying shrinkage is affected by all of the factors that affect the drying of concrete, in particular the water content and the water–cement ratio of the mix, the size and shape of the member and the ambient relative humidity. All else being equal, drying shrinkage increases when the water–cement ratio increases, the relative humidity decreases and the ratio of the exposed surface area to volume increases. Temperature rises accelerate drying and therefore increase shrinkage. By contrast, endogenous shrinkage increases as the cement content increases and the water–cement ratio decreases. In addition, endogenous shrinkage is not affected by the ambient relative humidity.

The effect of a member's size on drying shrinkage should be emphasised. For a thin member, such as a slab, the drying process may be complete after several years, but for the interior of a larger member, the drying process may continue throughout its lifetime. For uncracked mass concrete structures, there is no significant drying (shrinkage) except for about 300 mm from each exposed surface. By contrast, the chemical shrinkage is less affected by the size and shape of the specimen.

Shrinkage is also affected by the volume and type of aggregate. Aggregate provides restraint to shrinkage of the cement paste, so that an increase in the aggregate content reduces shrinkage. Shrinkage is also smaller when stiffer aggregates are used, i.e. aggregates with higher elastic moduli. Thus shrinkage is considerably higher in lightweight concrete than in normal weight concrete (by up to 50 per cent).

1.3.3 The effects of shrinkage on structural behaviour

Shrinkage is greatest at the member surfaces exposed to drying and decreases towards the interior of the member. In Fig. 1.13a, the shrinkage strains through the thickness of a plain concrete slab that is drying on both the top and bottom surfaces are shown. The slab is unloaded and unrestrained. The mean shrinkage strain, ε_{sh} in Fig. 1.13a, is the average contraction. The strain marked $\Delta\varepsilon_{sh}$ is the portion of the shrinkage strain that causes the internal stresses required to restore strain compatibility (i.e. to ensure that plane sections remain plane). These self-equilibrating internal stresses (called *eigenstresses*) occur in all concrete structures and are tensile near the drying surface and compressive in the interior of the member. Because the shrinkage-induced internal stresses develop gradually with time, they are relieved by creep. Nevertheless, soon after the commencement of drying, the tensile stresses near the drying surfaces may overcome the tensile strength of the concrete (particularly in poorly cured concrete) and result in surface cracking.

The elastic plus creep strains caused by the internal stresses are equal and opposite to $\Delta\varepsilon_{sh}$, as shown in Fig. 1.13b. The total strain distribution, obtained by summing the elastic, creep and shrinkage components, is linear, as shown in Fig. 1.13c, and therefore satisfies compatibility.

If the drying conditions are the same at both the top and bottom surfaces, the total strain is uniform over the depth of the slab and equal to the mean shrinkage strain, ε_{sh}. It is this quantity that is usually of significance in the analysis of concrete structures. The currently available procedures for estimating shrinkage strains (such as the procedure outlined in Section 2.1.5) are empirical formulae for the mean shrinkage on a section (ε_{sh} in Fig. 1.13). Procedures for making reliable estimates of the variation of shrinkage strain through the depth of a section are at present unavailable. If drying occurs at a different rate from the top and bottom surfaces, the total strain distribution is no longer uniform over the depth of the section; a curvature develops on the section and deflection of the member results.

In concrete structures, unrestrained contraction and unrestrained rotation are unusual. Reinforcement embedded in the concrete provides restraint to shrinkage. If the reinforcement is not symmetrically placed on a section, a shrinkage-induced curvature develops with time. Consider the singly reinforced member shown in Fig. 1.14a, and the small segment of length, Δz. The shrinkage-induced stresses and strains on an uncracked and on a cracked cross-section are shown in Figs 1.14b and c, respectively. As the concrete shrinks, it compresses the steel reinforcement that, in turn, imposes an equal and opposite tensile force, ΔT, on the concrete at the level of the steel. This gradually increasing tensile force, acting at some eccentricity to the centroid of the concrete cross-section produces elastic plus creep strains and a resulting curvature on the section. The shrinkage-induced curvature often leads to significant load independent deflection of the member. The magnitude of

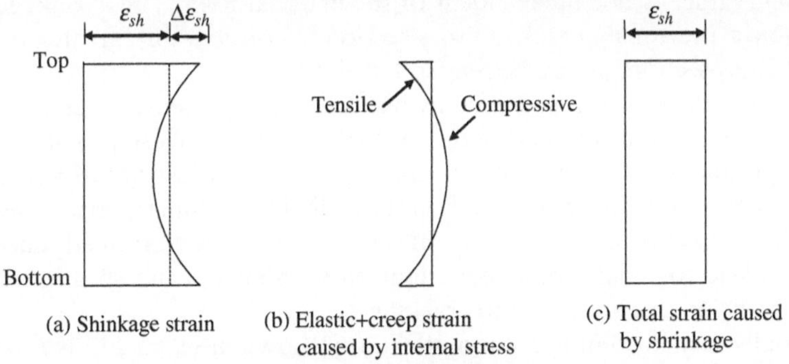

Figure 1.13 Strain components caused by shrinkage in plain concrete.

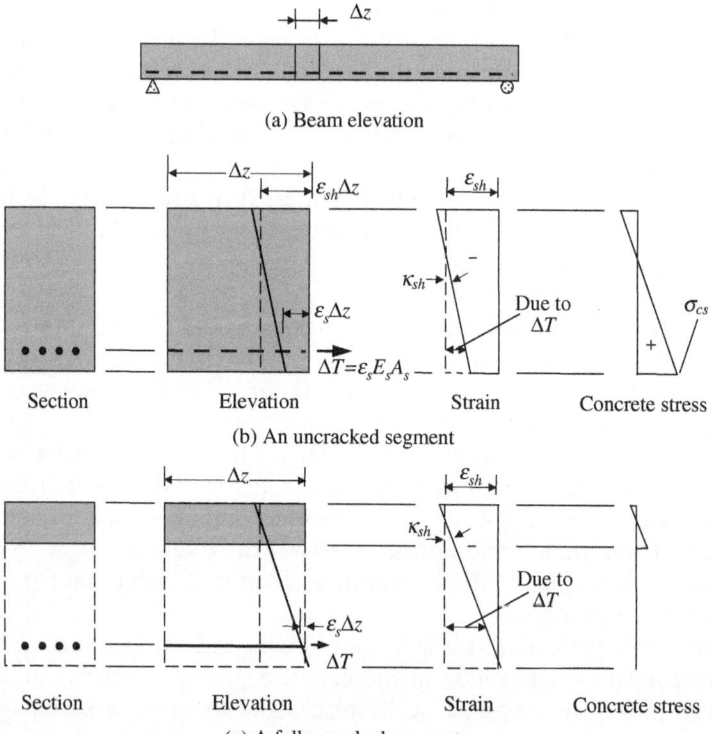

Figure 1.14 Shrinkage-induced deformation and stresses in a singly reinforced beam.

ΔT (and hence the shrinkage induced curvature) depends on the quantity and position of the reinforcement and on whether or not the cross-section has cracked.

The curvature caused by ΔT obviously depends on the size of the (uncracked) concrete part of the cross-section, and hence on the extent of cracking, and this in turn depends on the magnitude of the applied moment and the quantity of reinforcement. Although shrinkage strain is independent of stress, it appears that shrinkage curvature is not independent of the external load. The shrinkage induced curvature on a previously cracked cross-section is considerably greater than on an uncracked cross-section, as can be seen in Fig. 1.14.

In addition to the restraint provided by bonded reinforcement, the connections of a member to other portions of the structure or to the foundation also provide restraint to the shortening and rotation caused by shrinkage. Both of these forms of restraint create internal stresses and deformations and, in statically indeterminate structures, internal actions. The shrinkage-induced change in the reactions of a restrained indeterminate member can lead to a significant redistribution of moments and shears, as well as a build-up of tension that may lead to cracking.

If the axially loaded column section previously shown in Fig. 1.11 also commenced shrinking at time τ (i.e. $\tau_d = \tau$), the redistribution of internal stresses between the concrete and the steel is exacerbated by the shrinkage as illustrated in Fig. 1.15.

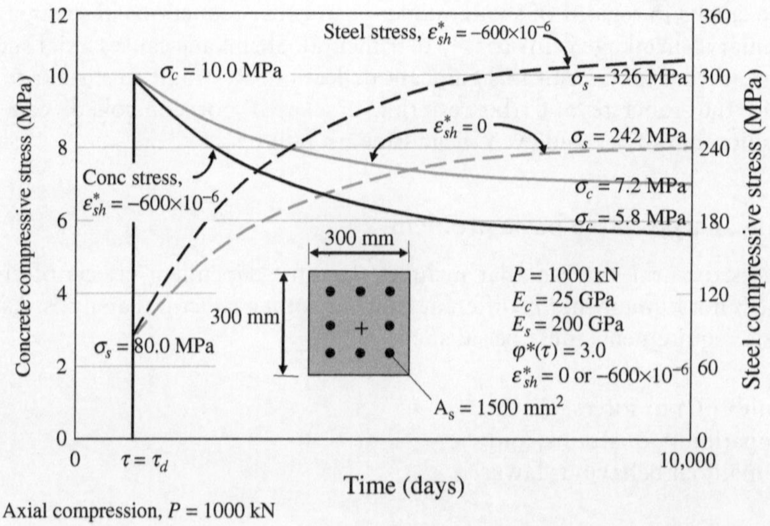

Figure 1.15 Variation of stresses with time due to shrinkage in a symmetrically reinforced section subjected to axial compression.

If the member was unloaded (i.e. $P = 0$), shrinkage causes a gradual build up of tension in the concrete as illustrated in Fig. 1.16.

Shrinkage is probably the most common cause of cracking in concrete structures. Direct tension cracks caused by restrained shrinkage tend to be more parallel sided than flexural cracks and often penetrate completely through the member. Such cracks are difficult to control and are often difficult to anticipate. In addition to the obvious

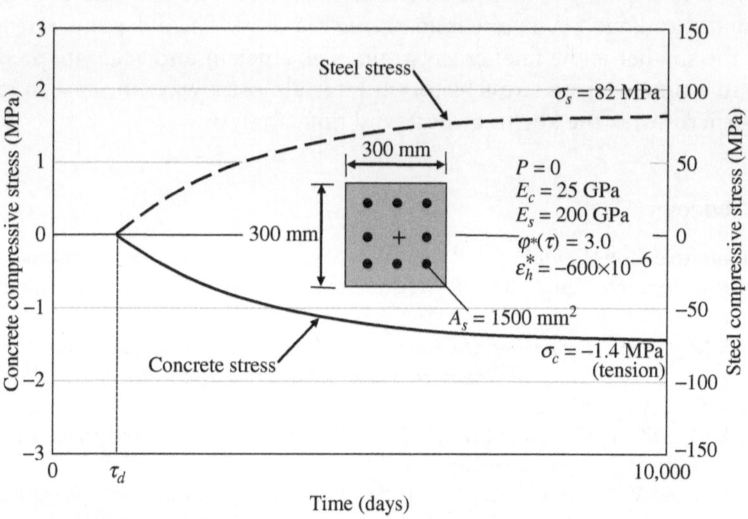

Figure 1.16 Variation of stresses with time due to shrinkage on an unloaded symmetrically reinforced cross-section.

serviceability and durability problems that shrinkage cracking can create, full depth shrinkage cracks in regions of small moment can cause reductions in shear strength.

In summary, shrinkage is invariably detrimental. Shrinkage causes axial shortening and rotations that may result in significant deflection. Restraint to shrinkage induces tension in the concrete and the resulting cracks, if not controlled, can lead to serviceability, durability and even shear strength failures.

1.4 Time analysis – the basic problem

An elastic structural analysis that includes the time-dependent effects of creep and shrinkage is not fundamentally different from any other structural analysis. As always, three basic requirements must be satisfied:

(i) equilibrium of forces;
(ii) compatibility of strains; and
(iii) the material behaviour laws.

In a short-term analysis of a concrete structure, suitable stress versus strain relationships for concrete are used to model material behaviour in both tension and compression. If linear-elastic behaviour is assumed, Hooke's Law may be adopted. To include time effects in the analysis, a suitable stress versus strain versus time relationship (i.e. a suitable constitutive relationship) must be used.

Two main complicating factors must be overcome to reliably predict time-dependent behaviour. The first is the change in section properties brought about by time-dependent cracking resulting from the combined effects of external load and restraint to shrinkage and temperature changes. The second complicating factor is the interdependence between creep strain and stress history. The magnitude of the creep strain at a point in a reinforced concrete member depends on the previous stress history; but the stress history depends to a large extent on the magnitude and rate of development of creep (and shrinkage). Concrete is an *ageing* material. The older concrete is when it is loaded, the smaller is the final creep strain. The efficient and accurate prediction of creep due to a time-varying stress history (that is almost always the case in reinforced concrete structures) is the key to a successful time analysis.

1.5 References

1. ACI Committee 209 (2008). *Guide for modeling and calculating shrinkage and creep in hardened concrete (ACI 209.2R-8).* American Concrete Institute, Farmington Hills, Michigan.
2. Neville, A.M. (1995). *Properties of Concrete.* 4th Edition, Pearson Education Limited, UK.
3. Neville, A.M. (1970). *Creep of Concrete: Plain, Reinforced and Prestressed.* North Holland Publishing Co., Amsterdam.
4. Neville, A.M., Dilger, W.H. and Brooks, J.J. (1983). *Creep of Plain and Structural Concrete.* Construction Press (Longman Group Ltd).
5. Bazant, Z.P. and Wittmann, F.H. (eds) (1983). *Creep and Shrinkage in Concrete Structures.* John Wiley and Sons Ltd.
6. Gilbert, R.I. (1988). *Time Effects in Concrete Structures.* Elsevier Science Publishers, Amsterdam.

7. Vandamme, M. and Ulm, F.-J. (2009). Nanogranular origin of concrete creep. *Proceedings of the National Academy of Sciences*, 106(26), 10552–10557.
8. ACI Committee 209 (2009). Analysis of creep and shrinkage effects in concrete structures (ACI 209.3R-XX). *Draft. American Concrete Institute.* Michigan.
9. Rüsch, H., Jungwirth, D. and Hilsdorf, H.K. (1983). *Creep and Shrinkage – Their Effect on the Behaviour of Concrete Structures.* Springer-Verlag, New York.
10. Trost, H. (1978). Creep and Creep Recovery of Very Old Concrete. *RILEM Colloqulum on Creep of Concrete*, Leeds, England, April.
11. McHenry, D. (1943). A new aspect of creep in concrete and its application to design. *Proceedings of the ASTM, 43*, pp 1069–1084.
12. Comité Euro-International du Béton (1984). *CEB Design manual on structural effects of time-dependent behaviour of concrete.* M.A. Chiorino (ed.). Georgi Publishing, Saint-Saphorin, Switzerland.
13. Ostergaard, L., Lange, D.A., Altouabat, S.A. and Stang, H. (2001). Tensile basic creep of early-age concrete under constant load. *Cement and Concrete Research*, 31, 1895–1899.
14. Chu, K.-H. and Carreira, D.J. (1986). Time-dependent cyclic deflections in R/C beams. *Journal of Structural Engineering*, 112(5), 943–959.
15. Bazant, Z.P. and Oh, B.H. (1984). Deformation of progressively cracking reinforced concrete beams. *ACI Journal*, 81(3), 268–278.

2 Material properties

2.1 Concrete

2.1.1 Introductory remarks

To adequately predict deflections and crack widths in the design for serviceability, methods of analysis that realistically account for cracking and the time-dependent deformations caused by creep and shrinkage of the concrete are required, as are appropriate material modelling rules. The properties and deformation characteristics of concrete that are most often required in serviceability calculations are the tensile strength, elastic modulus, creep coefficient and shrinkage strain.

The elastic modulus is needed in the analysis of structures to estimate the stiffness of each member and to determine the internal actions. It is also required to estimate the instantaneous deformations caused by internal actions and the stresses induced by imposed deformations. The tensile strength of concrete is required to determine the extent of cracking due to both applied load and applied deformation. The creep coefficient associated with a particular time period and a particular loading regime is needed to estimate the time-dependent deformation of the structure, and the magnitude and rate of shrinkage strain is required to predict the development of load-independent deformations with time and the onset of time-dependent cracking.

In this chapter, simple models are presented for predicting the tensile strength, elastic modulus, creep coefficient and shrinkage strain for concretes with a compressive strength in the range 20 MPa $\leq f'_c \leq$ 100 MPa. These models were originally developed by Gilbert (Ref. 1) and have been incorporated into the most recent edition of the Australian Standard for Concrete Structures, AS3600-2009 (Ref. 2). They have been included here because the equations will be used in subsequent chapters to obtain typical material properties. Other alternative material modelling rules are available in the literature, for example, in Refs 3–11.

2.1.2 Compressive and tensile strength

The strength of concrete is usually specified in terms of the lower characteristic compressive cylinder (or cube) strength at 28 days, f'_c. This is the value of compressive strength exceeded by 95 per cent of all standard cylinders or cubes tested at age 28 days after curing under standard laboratory conditions. The mean compressive strength of the sample cylinders or cubes at 28 days (f_{cm}) is about 25 per cent higher than the characteristic strength when $f'_c = 20$ MPa, reducing to about 10 per cent higher than the characteristic strength when $f'_c = 100$ MPa. The *in-situ* strength of concrete (i.e. the strength of the concrete in the structure on site) is often taken to be about 90 per cent of the cylinder strength (Ref. 2).

The tensile strength, f_{ct}, is defined here as the maximum stress that the concrete can withstand when subjected to uniaxial tension. Direct uniaxial tensile tests are difficult to perform and tensile strength is usually measured from either flexural tests on prisms or indirect splitting tests on cylinders. In flexure, the apparent tensile stress at the extreme tensile fibre of the critical cross-section under the peak load is calculated assuming linear elastic behaviour and taken to be the flexural tensile strength (or modulus of rupture), $f_{ct.f}$. The flexural tensile strength $f_{ct.f}$ is significantly higher than f_{ct} due to the strain gradient and the post-peak unloading portion of the stress–strain curve for concrete in tension. Typically, f_{ct} is about 50–60 per cent of $f_{ct.f}$. The indirect tensile strength measured from a split cylinder test is also higher than f_{ct} (usually by about 10 per cent) due to the confining effect of the bearing plate in the standard test.

For design purposes, the lower characteristic flexural tensile strength, $f'_{ct.f}$, and the lower characteristic uniaxial tensile stress, f'_{ct}, may be taken as:

$$f'_{ct.f} = 0.6\sqrt{f'_c} \tag{2.1a}$$

and

$$f'_{ct} = 0.36\sqrt{f'_c} \tag{2.1b}$$

The mean and upper characteristic values may be estimated by multiplying the lower characteristic values by 1.4 and 1.8, respectively (Ref. 2). In serviceability calculations, mean values of tensile strength should be used rather than characteristic values in most situations.

2.1.3 Elastic modulus

The value of the elastic modulus, E_c, increases with time as the concrete gains strength and stiffness. It is common practice to assume that E_c is constant with time and equal to its value calculated at the time of first loading. For stress levels less than about $0.4f_{cm}$ for normal strength concrete ($f'_c \leq 50$ MPa) and about $0.6f_{cm}$ for high strength concrete ($50 < f'_c \leq 100$ MPa), and for stresses applied over a relatively short period (say up to 5 minutes), a numerical estimate of the *in-situ* elastic modulus may be made from:

For $f_{cmi} \leq 40$ MPa:

$$E_c = \rho^{1.5} 0.043\sqrt{f_{cmi}} \text{ (in MPa)} \tag{2.2a}$$

For $40 < f_{cmi} \leq 100$ MPa:

$$E_c = \rho^{1.5}\,[0.024\sqrt{f_{cmi}} + 0.12] \text{ (in MPa)} \tag{2.2b}$$

where ρ is the density of the concrete in kg/m^3 (not less than 2400 kg/m^3 for normal weight concrete) and f_{cmi} is the mean *in-situ* compressive strength in MPa at the time of first loading. Eq. 2.2a was originally proposed by Pauw (Ref. 12). Values for E_c obtained using Eqs 2.2 for *in-situ* normal weight concrete ($\rho = 2400$ kg/m^3) at age 28 days for different values of f'_c are given in Table 2.1. The mean *in-situ* strength compressive strength, f_{cmi}, in Table 2.1 is taken to be 90 per cent of the standard mean cylinder strength and for 100 MPa concrete is actually smaller than the characteristic cylinder strength, f'_c.

Table 2.1 The elastic modulus for *in-situ* concrete, E_c

f'_c (MPa)	20	25	32	40	50	65	80	100
f_{cmi} (MPa)	22.5	27.9	35.4	43.7	53.7	68.2	81.9	99.0
E_c (MPa)	24,000	26,700	30,100	32,750	34,800	37,400	39,650	42,200

The magnitude of E_c given by Eqs 2.2 has an accuracy of $\pm$ 20 per cent depending, among other things, on the type and quantity of aggregate and the rate of application of the load. In general, the faster the load is applied, the larger the value of E_c. For stresses applied over a longer time period (say up to one day), significant increases in deformation occur due to the rapid early development of creep. Yet in a broad sense, loads of one day duration are usually considered to be short-term and the effects of creep are often ignored. This may lead to significant error. If the *short-term* deformation after 1 day of loading is required, it is suggested that E_c be reduced by about 20 per cent to account for early creep (Ref. 13).

Eq. 2.2c provides an estimate of the variations of the elastic modulus with time (Ref. 7):

$$E_c(t) = \left(e^{s(1-\sqrt{28/t})}\right)^{0.5} E_c(28) \qquad (2.2c)$$

where the coefficient s is taken as 0.38 for Ordinary Portland Cement and 0.25 for High Early Strength Cement) and $E_c(28)$ is the 28-day value of the elastic modulus. Ref. 9 adopts an exponent of 0.3 in Eq. 2.2c (instead of 0.5). Typical variations in E_c with time t are shown in Table 2.2.

Table 2.2 Increase in elastic modulus with age of concrete $t - (E_c(t)/E_c(28))$

Cement type	*Age of concrete in days (t)*					
	3	*7*	*28*	*90*	*360*	*30,000*
Ordinary Portland Cement	0.68	0.83	1.0	1.09	1.15	1.20
High Early Strength Cement	0.77	0.88	1.0	1.06	1.09	1.13

2.1.4 *Creep coefficient*

In Section 1.2.4, the creep coefficient at time, t, associated with a constant stress first applied at age, τ, was defined as the ratio of the creep strain at time, t, to the (initial) elastic strain and given the symbol, $\varphi(t, \tau)$.

The most accurate way of determining the final creep coefficient is by testing or using results obtained from measurements on similar local concretes. However, testing is often not a practical option for the structural designer. In the absence of long-term test results, the final creep coefficient may be determined by extrapolation from short-term test results, where creep is measured over a relatively short period (say 28 days) in specimens subjected to constant stress. Various mathematical expressions for the shape of the creep coefficient versus time curve are available from which long-term values may be predicted from the short-term measurements (Ref. 13). The longer the period of measurement, the more accurate are the long-term predictions.

If testing is not an option, numerous analytical methods are available for predicting the creep coefficient. These predictive methods vary in complexity. Some are simple and easy to use and provide a quick and approximate estimate of $\varphi(t,\tau)$. Such a method is included in the Australian Standard AS3600-2009 (Ref. 2) and an improved version of the method is described below. Some other methods are more complicated and attempt to account for the many parameters that affect the magnitude and rate of development of creep. Unfortunately, an increase in complexity does not necessarily result in an increase in accuracy, and predictions made by some of the more well-known methods differ widely (Refs 3, 14–16).

The simple approach described here does not account for such factors as aggregate type, cement type, cement replacement materials and more, but it does provide a 'ball-park' estimate of the creep coefficient for concrete suitable for routine use in structural design.

The creep coefficient at any time may be calculated from:

$$\varphi(t,\tau) = k_2\, k_3\, k_4\, k_5\, \varphi_{basic} \tag{2.3}$$

The basic creep coefficient φ_{basic} is given in Table 2.3, where the values have been updated from those given in Ref. 1 and are based on more recent test data.

The factor k_2 in Eq. 2.3 describes the development of creep with time. It depends on the hypothetical thickness, t_h, the environment and the time after loading and is given in Fig. 2.1. The hypothetical thickness is defined as $t_h = 2A/u_e$, where A is the cross-sectional area of the member and u_e is that portion of the section perimeter exposed to the atmosphere plus half the total perimeter of any voids contained within the section.

The factor k_3 depends on the age at first loading τ (in days) and is given by:

$$k_3 = \frac{2.7}{1+\log(\tau)} \tag{2.4}$$

The factor k_4 accounts for the environment, with $k_4 = 0.7$ for an arid environment, $k_4 = 0.65$ for an interior environment, $k_4 = 0.60$ for a temperate environment and $k_4 = 0.5$ for a tropical/coastal environment.

The factor k_5 is given in Eq. 2.5 and accounts for the reduced influence of both the relative humidity and the specimen size on the creep of concrete as the concrete strength increases (or more precisely, as the water-binder ratio decreases).

Table 2.3 The basic creep coefficient φ_{basic}

f'_c (MPa)	20	25	32	40	50	65	80	100
φ_{basic}	4.5	3.8	3.0	2.4	2.0	1.7	1.5	1.3

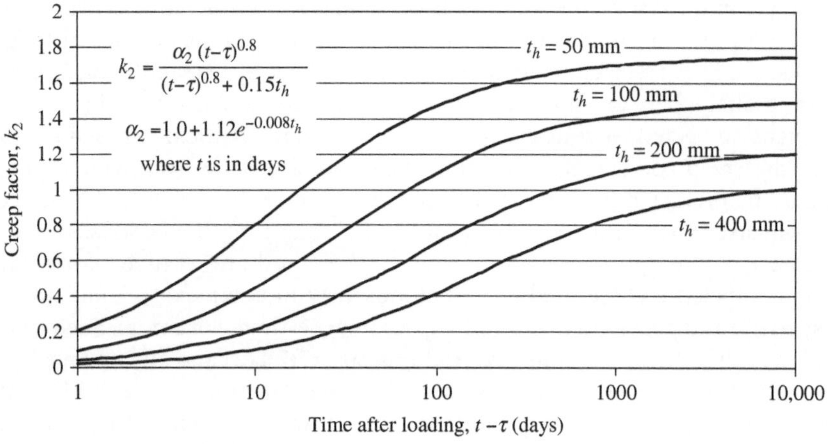

Figure 2.1 The factor k_2 versus time (Ref. 1).

When $f'_c \leq 50$ MPa:

$$k_5 = 1.0 \tag{2.5a}$$

When 50 MPa $< f'_c \leq 100$ MPa:

$$k_5 = (2.0 - \alpha_3) - 0.02\,(1.0 - \alpha_3)\,f'_c \tag{2.5b}$$

where $\alpha_3 = 0.7/(k_4 \alpha_2)$.

A family of creep coefficient versus duration of loading curves obtained using Eq. 2.3 is shown in Fig. 2.2 for a concrete specimen located in a temperate environment, with a hypothetical thickness $t_h = 150$ mm, concrete strength $f'_c = 40$ MPa and loaded at different ages, τ. The final creep coefficients φ^* (after 30 years) predicted by the above

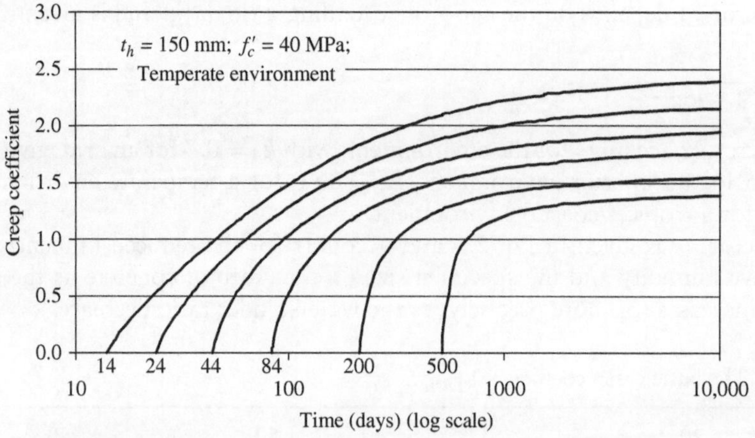

Figure 2.2 Typical creep coefficient versus time curves (from Eq. 2.3).

method are given in Table 2.4 for concrete first loaded at 28 days, for characteristic strengths of 25–100 MPa, for three hypothetical thicknesses (t_h = 100, 200 and 400 mm) and for concrete located in different environments.

The above discussion is concerned with compressive creep. In many practical situations, creep of concrete in tension is also of interest. Tensile creep plays an important part in delaying the onset of cracking caused by restrained shrinkage. In design, it is usual to assume that the creep coefficients in tension and in compression are identical. Although not strictly correct, this assumption simplifies calculations and does not usually introduce serious inaccuracies.

It must be emphasised that creep of concrete is highly variable with significant differences in the measured creep strains in seemingly identical specimens tested under identical conditions (both in terms of load and environment). The creep coefficient predicted by Eq. 2.3 should be taken as an average value with a range of ± 30 per cent.

Table 2.4 Final creep coefficients (after 30 years) φ^* for concrete first loaded at 28 days

	Final creep coefficient, φ^*											
f'_c *(MPa)*	*Arid environment*			*Interior environment*			*Temperate inland environment*			*Tropical and near-coastal environment*		
	t_h*(mm)*			t_h*(mm)*			t_h*(mm)*			t_h*(mm)*		
	100	*200*	*400*	*100*	*200*	*400*	*100*	*200*	*400*	*100*	*200*	*400*
25	4.37	3.53	2.96	4.06	3.28	2.75	3.75	3.03	2.53	3.12	2.52	2.11
32	3.45	2.79	2.33	3.20	2.59	2.17	2.96	2.39	2.00	2.46	1.99	1.67
40	2.76	2.23	1.87	2.56	2.07	1.73	2.37	1.91	1.60	1.97	1.59	1.33
50	2.30	1.86	1.56	2.14	1.73	1.45	1.97	1.59	1.33	1.64	1.33	1.11
65	1.76	1.49	1.31	1.66	1.41	1.24	1.56	1.33	1.17	1.37	1.18	1.04
80	1.38	1.24	1.14	1.33	1.20	1.10	1.28	1.16	1.07	1.18	1.08	1.00
100	0.99	0.99	0.97	0.99	0.99	0.97	0.99	0.99	0.97	0.99	0.99	0.97

2.1.5 Shrinkage strain

The model presented below for estimating the magnitude of shrinkage strain in normal and high strength concrete was proposed by Gilbert (Ref. 1) and is included in the Australian Standard AS3600-2009 (Ref. 2). Many other approaches are available in the literature (e.g. Refs 3–11).

The model divides the total shrinkage strain, ε_{sh}, into two components: endogenous shrinkage, ε_{she}, and drying shrinkage, ε_{shd}, as given in Eq. 2.6. Endogenous shrinkage is taken to be the sum of chemical (or autogenous) shrinkage and thermal shrinkage and is assumed to develop relatively rapidly and to increase with concrete strength. Drying shrinkage develops more slowly and decreases with concrete strength.

$$\varepsilon_{sh} = \varepsilon_{she} + \varepsilon_{shd} \tag{2.6}$$

At any time t (in days) after casting, the endogenous shrinkage is given by:

$$\varepsilon_{she} = \varepsilon^*_{she}(1.0 - e^{-0.1t}) \tag{2.7}$$

where ε^*_{she} is the final endogenous shrinkage and may be taken as:

$$\varepsilon^*_{she} = (0.06f'_c - 1.0) \times 50 \times 10^{-6} \ (f'_c \text{ in MPa}) \qquad (2.8)$$

The basic drying shrinkage $\varepsilon_{shd.b}$ is given by:

$$\varepsilon_{shd.b} = (1.0 - 0.008\, f'_c) \times \varepsilon^*_{shd.b} \qquad (2.9)$$

where $\varepsilon^*_{shd.b}$ depends on the quality of the local aggregates and may be taken as 800×10^{-6} when the aggregate quality is known to be good and 1000×10^{-6} when aggregate quality is uncertain.

At any time after the commencement of drying $(t - \tau_d)$, the drying shrinkage may be taken as:

$$\varepsilon_{shd} = k_1 k_4 \varepsilon_{shd.b} \qquad (2.10)$$

where k_1 is given in Fig. 2.3.

The factor k_4 depends on the environment and is equal to 0.7 for an arid environment, 0.65 for an interior environment, 0.6 for a temperate inland environment and 0.5 for a tropical or near-coastal environment.

As expressed in Eq. 2.6, the design shrinkage at any time is therefore the sum of the endogenous shrinkage (Eq. 2.7) and the drying shrinkage (Eq. 2.10). The proposed model provides good agreement with available shrinkage measurements on Australian concretes. For specimens located in arid, temperate and tropical environments with average quality aggregate (i.e. with $\varepsilon^*_{shd.b} = 1000 \times 10^{-6}$) and with a hypothetical thickness $t_h = 200$ mm, the shrinkage strain components predicted by the above model at 28 days after the commencement of drying and after 30 years (i.e. at $t - \tau_d = 28$ days and $t - \tau_d = 10{,}950$ days) are given in Table 2.5.

For each environment identified by the model, typical final design shrinkage strains for specimens with average quality aggregate ($\varepsilon^*_{shd.b} = 1000 \times 10^{-6}$) and with hypothetical thicknesses $t_h = 50$, 100, 200 and 400 mm are given in Table 2.6.

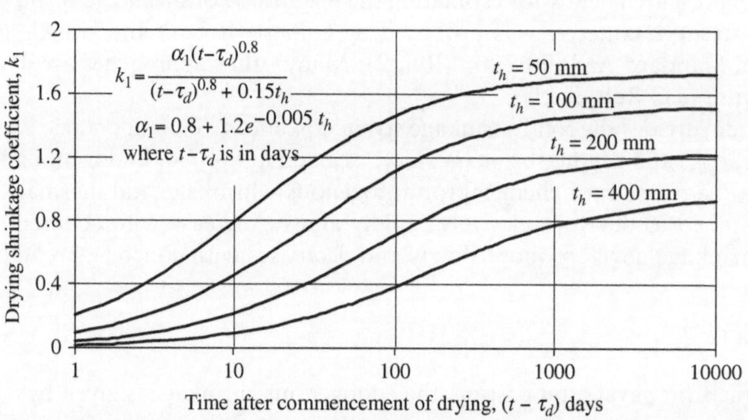

Figure 2.3 Drying shrinkage strain coefficient k_1 for various values of t_h (Ref. 1).

Table 2.5 Design shrinkage strain components ($t_h = 200$ mm and $\varepsilon^*_{shd.b} = 1000 \times 10^{-6}$)

Environment	f'_c *(MPa)*	*Shrinkage strain* $\varepsilon_{sh}(\times 10^{-6})$ *and shrinkage strain components* $\varepsilon_{she}(\times 10^{-6})$ *and* $\varepsilon_{shd}(\times 10^{-6})$					
		t = 28 days			*t = 10,950 days (30 years)*		
		ε_{she}	ε_{shd}	ε_{sh}	ε_{she}	ε_{shd}	ε_{sh}
Arid	25	25	225	250	25	685	710
	32	45	210	255	45	635	680
	40	65	190	255	70	580	650
	50	95	170	265	100	510	610
	65	135	135	270	145	415	560
	80	180	100	280	190	310	500
	100	235	55	290	250	170	420
Temperate inland	25	25	195	220	25	585	610
	32	45	180	225	45	545	590
	40	65	165	230	70	500	570
	50	95	145	240	100	440	540
	65	135	115	250	145	355	500
	80	180	85	265	190	260	450
	100	235	50	285	250	150	400
Tropical	25	25	160	185	25	485	510
	32	45	150	195	45	455	500
	40	65	140	205	70	420	490
	50	95	120	215	100	370	470
	65	135	100	235	145	295	440
	80	180	70	250	190	220	410
	100	235	40	275	255	115	370

2.2 Steel reinforcement

2.2.1 General

The strength of a reinforced or prestressed concrete element in bending, shear, torsion, or direct tension is primarily dependent on the properties of the steel reinforcement. However, at service loads, the steel stresses are usually in the elastic range and the non-linear properties of the concrete most affect structural behaviour. In this book, it is the day-to-day, in-service behaviour of concrete structures which is of most interest and, therefore, it is the modelling of the properties of concrete that creates the most difficulties. Nevertheless, it is also necessary to adequately model the various types of steel reinforcement and their material properties.

Steel reinforcement is used in concrete structures to provide strength, ductility and serviceability. With regard to serviceability, non-prestressed reinforcement can be strategically placed to reduce both immediate and time-dependent deformations. Adequate quantities of bonded, non-prestressed steel also provide crack control, wherever cracks occur in the concrete, and for whatever reason (with the exception of plastic shrinkage cracking in the wet concrete prior to setting).

Prestressing steel is the means whereby an initial compressive force is exerted on the concrete in order to reduce or eliminate cracking. By changing the drape of the

Table 2.6 Typical design shrinkage strains after 30 years ($\varepsilon^*_{shd.b} = 1000 \times 10^{-6}$)

Environment	f'_c *(MPa)*	*Final shrinkage strain* $\varepsilon^*_{sh}(\times 10^{-6})$			
		$t_h = 50$ *mm*	$t_h = 100$ *mm*	$t_h = 200$ *mm*	$t_h = 400$ *mm*
Arid	25	990	870	710	550
	32	950	840	680	530
	40	890	790	650	510
	50	830	740	610	490
	65	730	650	560	460
	80	630	570	500	420
	100	490	460	420	380
Interior	25	920	810	660	510
	32	880	780	640	500
	40	830	740	610	480
	50	770	690	580	460
	65	680	620	530	440
	80	590	540	480	410
	100	480	450	410	370
Temperate inland	25	850	750	610	470
	32	820	720	590	460
	40	780	690	570	450
	50	720	650	540	440
	65	640	580	500	410
	80	560	520	450	390
	100	460	430	400	360
Tropical or near-coastal	25	720	630	510	400
	32	690	610	500	390
	40	660	590	490	390
	50	620	550	470	380
	65	560	510	440	370
	80	500	460	410	360
	100	420	400	370	340

prestressing steel along the length of a member, a transverse force will be exerted on the member that may counteract the external transverse loads and thus reduce and help control both short-term and long-term deflection. The design for serviceability is therefore very much associated with the determination of suitable types and quantities of reinforcement, wherever cracking or deformation is to be controlled.

2.2.2 *Conventional, non-prestressed reinforcement*

The non-prestressed reinforcement commonly used in both reinforced and prestressed concrete structures takes the form of bars, cold-drawn wires or welded wire mesh. Types and sizes vary from country to country. In Australia, for example, reinforcing bars are available in two grades, Grade R250N and Grade D500N (that correspond to characteristic yield stresses f_y of 250 MPa and 500 MPa, respectively). Grade R250N bars are hot-rolled plain round bars of 6 or 10 mm diameter (designated R6 and R10 bars) and are commonly used for fitments such as ties and stirrups. Grade D500N bars are hot-rolled deformed bars with sizes ranging from 12 to 40 mm diameter (in 4 mm increments).

Regularly spaced rib-shaped deformations on the surface of a deformed bar improve the bond between the concrete and the steel and greatly improve the anchorage potential of the bar. It is for this reason that deformed bars are used as longitudinal reinforcement in most reinforced and partially-prestressed concrete members.

In design calculations, non-prestressed steel is usually assumed to be elastic-plastic. Before yielding, the reinforcement is elastic, with steel stress, σ_s, proportional to the steel strain, ε_s, that is, $\sigma_s = E_s\varepsilon_s$, where E_s is the elastic modulus of the steel. After yielding, the stress–strain curve is usually assumed to be horizontal (perfectly plastic) and the steel stress $\sigma_s = f_y$ at all values of strain exceeding the yield strain $\varepsilon_{sy} = f_y / E_s$. The yield stress, f_y, is taken to be the *strength* of the material and strain hardening is most often ignored. The stress-strain curve in compression is also assumed to be linear-elastic similar to that in tension.

At service loads, the stress in the non-prestressed steel is usually less than the yield stress and behaviour is linear-elastic. Throughout this book, the elastic modulus for non-prestressed reinforcing steel is taken to be $E_s = 200$ GPa.

2.2.3 *Prestressing steel*

2.2.3.1 *Types of prestressing steel*

The time-dependent shortening of the concrete caused by creep and shrinkage in a prestressed concrete member causes a corresponding shortening of the prestressing steel that is physically attached to the concrete either by bond or by anchorages at the ends of the steel tendon. This shortening, together with relaxation of the steel, usually results in a time-dependent loss of stress in the steel of between 100 and 350 MPa (i.e. between about 7 and 25 per cent of the initial prestress). Significant additional losses of prestress can result from other sources, such as elastic shortening of the member when the prestress is first applied, or friction along a post-tensioned tendon, or draw-in at an anchorage.

For an efficient and practical design, the total loss of prestress should be a relatively small portion of the initial prestressing force. The steel used to prestress concrete must therefore be capable of carrying a relatively high initial stress. A tensile strength f_p of between 1000 and 1900 MPa is typical for modern prestressing steels. Early attempts to prestress concrete with low-strength steels failed because a large proportion of the prestressing force was rapidly lost due to the time-dependent deformations of the relatively poor quality concrete in use at that time.

There are three basic types of high-strength steel commonly used as tendons in modern prestressed concrete construction, *viz.* cold-drawn stress-relieved round wire (usually indented or crimped), stress-relieved strand and high-strength alloy steel bars.

The stress–strain curves for the various types of prestressing steel exhibit similar characteristics. There is no well-defined yield point (as exists for some lower strength steels). Each curve is initially linear-elastic (with elastic modulus, E_p) and with a relatively high proportional limit. When the curves become non-linear as deformation increases, the stress gradually increases monotonically until the steel eventually fractures. The strain at fracture is usually about 5 per cent. High-strength steel is therefore considerably less ductile than conventional, hot-rolled non-prestressed reinforcing steel. For design purposes, the *yield stress* is the stress corresponding to the 0.2 per cent offset strain and is often taken to be 85 per cent of the minimum tensile strength (i.e. $0.85f_p$). For cold-drawn wires, E_p may be taken as 205 ± 10 GPa (Ref. 2).

Stress-relieved strand is perhaps the most commonly used prestressing steel. Strand is fabricated from a number of prestressing wires, usually seven (although 19-wire strand is also available in some countries). Seven-wire strand consists of six wires tightly wound around a seventh, slightly larger diameter central wire. The pitch of the six spirally-wound wires is between 12 and 16 times the nominal diameter of the strand. After stranding, the cable is further stress-relieved. *Low relaxation* (or stabilised) strand is most often used by the prestressing industry today. The mechanical properties of the strand are slightly different from those of the wire from which it is made. This is because the stranded wires tend to straighten slightly when subjected to tension. For design purposes, the yield stress of stress-relieved strand may be taken to be $0.85f_p$ and the elastic modulus to be $E_p = 200 \pm 5$ GPa (Ref. 2).

The high strength of alloy steel bars is obtained by introducing alloying elements in the manufacture of the steel. The bars can be hot rolled or cold worked (stretched). The elastic modulus for cold-worked bars is generally lower than for strand or wire. For design purposes, E_p may be taken to be 170 ± 10 GPa for a cold-worked bar or 205 ± 10 GPa for a hot-rolled bar.

2.2.3.2 *Steel relaxation*

The initial stress level in the prestressing steel after the prestress is transferred to the concrete is usually high, often in the range of 60 to 75 per cent of the tensile strength of the material. At such high stress levels, high-strength steel creeps. At lower stress levels, such as is typical for non-prestressed steel, the creep of steel is negligible. If a tendon is stretched and held at a constant length (constant strain), the development of creep strain in the steel is exhibited as a loss of elastic strain, and hence a loss of stress. This loss of stress in a specimen subjected to constant strain is known as *relaxation*. Relaxation in steel is highly dependent on the stress level and increases at an increasing rate as the stress level increases. Relaxation in steel also increases rapidly as temperature increases.

In recent years, low relaxation steel has normally been used in order to minimise the losses of prestress resulting from relaxation. AS3600-2009 (Ref. 2) specifies that the *design relaxation* R of a low relaxation tendon (as a percentage of the initial prestress) be determined from:

$$R = k_4 k_5 k_6 R_b \qquad (2.11)$$

where k_4 is a coefficient depending on the duration of the prestressing force and given by $k_4 = \log[5.4(t)^{1/6}]$; t is the time in days after prestressing; k_5 is a coefficient that is dependent on the stress in the tendon as a proportion of the characteristic minimum breaking strength of the tendon, f_p, and is given in Fig. 2.4; k_6 depends on the average temperature T(°C) over the time period t and may be taken as $T/20$, but not less than 1.0; and R_b is the basic relaxation of the tendon after 1000 hours at 20°C and at $0.7f_p$, which may be determined by testing or taken as $R_b = 2\%$ for low relaxation wire; $R_b = 2.5\%$ for a low relaxation strand and $R_b = 4\%$ for alloy steel bars.

Typical final (30 year) values of the relaxation loss of low-relaxation wire, strand and bars at an average temperature of 20°C are given in Table 2.7.

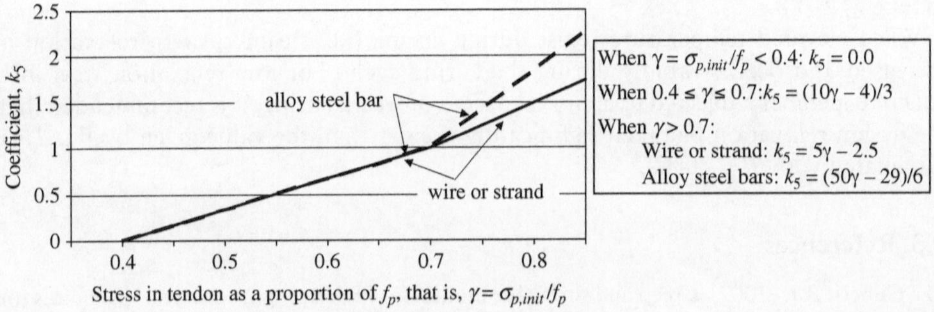

Figure 2.4 Coefficient, k_5 (Ref. 2).

Creep in the prestressing steel may also be defined in terms of a creep coefficient $\varphi_p(t, \sigma_{p,init})$ rather than as a relaxation loss. If the creep coefficient for the prestressing steel $\varphi_p(t, \sigma_{p,init})$ is the ratio of creep strain in the steel to the initial elastic strain, then the final creep coefficients for low relaxation wire, strand and bar are also given in Table 2.7 and may be approximated using Eq. 2.12.

$$\varphi_p(t, \sigma_{p,init}) = \frac{R}{1 - R} \tag{2.12}$$

Note that the creep of high-strength steels is non-linear with respect to stress.

As has already been emphasised, creep (relaxation) of the prestressing steel depends on the stress level. In a prestressed concrete member, the stress in a tendon is gradually reduced with time as a result of creep and shrinkage in the concrete. This gradual reduction of stress results in a reduction of creep in the steel and hence smaller relaxation losses. To determine relaxation losses in a concrete structure therefore, the final relaxation loss obtained from Eq. 2.11 (or Table 2.7) should be multiplied by a reduction factor, λ_r, that accounts for the time-dependent shortening of the concrete due to creep and shrinkage. The factor λ_r depends on the creep and shrinkage characteristics of the concrete, the initial prestressing force, and the stress in the concrete at the level of the steel, and can be determined by iteration (Ref. 17). However, because relaxation losses in modern prestressed concrete

Table 2.7 Long-term relaxation losses and corresponding final creep coefficients for low relaxation wire, strand and bar ($T = 20°C$)

	Type of tendon	*Tendon stress as a proportion of f_p*		
		0.6	*0.7*	*0.8*
Wire	Relaxation loss, $R(\%)$	1.9	2.8	4.2
	Creep coefficient, $\varphi_p(t, \sigma_{p,init})$	0.019	0.029	0.044
Strand	Relaxation loss, $R(\%)$	2.4	3.5	5.3
	Creep coefficient, $\varphi_p(t, \sigma_{p,init})$	0.025	0.036	0.056
Bar	Relaxation loss, $R(\%)$	3.7	5.6	10.3
	Creep coefficient, $\varphi_p(t, \sigma_{p,init})$	0.039	0.059	0.115

structures (employing low-relaxation steels) are relatively small, it is usually sufficient to take $\lambda_r \approx 0.8$.

When elevated temperatures exist during curing (i.e. steam curing), relaxation is increased and occurs rapidly during the curing cycle. For low relaxation steel in a concrete member subjected to an initial period of steam curing, it is recommended that the design relaxation be taken significantly greater than the value given by Eq. 2.11 (calculated with $T = 20°C$).

2.3 References

1. Gilbert, R.I. (2002). Creep and shrinkage models for high strength concrete – proposals for inclusion in AS3600. *Australian Journal of Structural Engineering*, IEAust, 4(2), 95–106.
2. Standards Australia (2009). Australian Standard for Concrete Structures, *AS3600–2009*. Sydney.
3. ACI Committee 209 (2008). Guide for modeling and calculating shrinkage and creep in hardened concrete *(ACI 209.2R-8)*. American Concrete Institute, Farmington Hills, Michigan.
4. ACI Committee 209 (1992). Prediction of creep, shrinkage, and temperature effects in concrete structures *(ACI 209R-92)*. American Concrete Institute, Farmington Hills, Michigan.
5. Bazant, Z.P. and Baweja, S. (2000). Creep and shrinkage prediction model for analysis and design of concrete structures: model B3. The Adam Neville Symposium: Creep and Shrinkage – Structural Design Effects, *ACI SP-194*. American Concrete Institute, Farmington Hills, Michigan, 236–260.
6. Bazant, Z.P. and Baweja, S. (1995). Creep and shrinkage prediction model for analysis and design of concrete structures – model B3. *Materials and Structures*, 28, 357–365.
7. Comité Euro-International du Béton (1993). *CEB-FIB Model Code 1990: Design Code*. Thomas Telford, London.
8. Comité Euro-International du Béton (1999). Structural concrete – Textbook on behaviour, design and performance – Volume 2: Basis of design. *Fib Bulletin No. 2*, Federation Internationale du Béton, Lausanne, Switzerland, 37–52.
9. British Standard (2004). *BS EN 1992, Eurocode 2: Design of concrete structures - Part 1-1: general rules and rules for buildings*. British Standards Institute, Milton Keynes.
10. Gardner, J. and Lockman, M.J. (2001). Design provisions for drying shrinkage and creep of normal-strength concrete. *ACI Materials Journal*, 98(2), 159–167.
11. Comité Euro-International du Béton (1984). *CEB Design manual on structural effects of time-dependent behaviour of concrete*. M.A. Chiorino (ed.), Georgi Publishing, Saint-Saphorin, Switzerland.
12. Pauw, A. (1960). Static Modulus of Elasticity of Concrete as Affected by Density. *ACI Journal*, 57(6), 679–687.
13. Gilbert, R.I. (1988). *Time Effects in Concrete Structures*. Elsevier Science Publishers, Amsterdam.
14. Bazant, Z.P. and Li, G.-H. (2008). Unbiased statistical comparison of creep and shrinkage prediction models. *ACI Materials Journal*, 105(6), 610–619.
15. Chiorino, M.A. (2005). A rational approach to the analysis of creep structural effects. *Shrinkage and Creep of Concrete*, *SP-227*, American Concrete Institute, Farmington Hills, Michigan, 107–141.
16. Brooks, J.J. (2005). 30-year creep and shrinkage of concrete. *Magazine of Concrete Research*, 57(9), 545–556.
17. Ghali, A., Favre, R., and Eldbadry, M. (2002). *Concrete Structures: Stresses and Deformations*. Third edition, Spon Press, London.

3 Design for serviceability

Deflection and crack control

3.1 Introduction

The tensile capacity of concrete is usually neglected when calculating the strength of a reinforced concrete member, even though the concrete continues to carry tensile stress between the cracks due to the transfer of forces from the tensile reinforcement to the concrete through bond. This contribution of the tensile concrete is known as *tension stiffening* and it affects the stiffness of the member after cracking and hence its deflection and the width of the cracks.

Reinforced concrete members often contain relatively small quantities of tensile reinforcement, in some situations close to the minimum amount permitted by the relevant building code. This is particularly so in the case of floor slabs. For such members, the flexural rigidity of a fully cracked cross-section ($E_c I_{cr}$) is many times smaller than that of an uncracked cross-section ($E_c I_{uncr}$) and tension stiffening contributes greatly to the member's stiffness after cracking. In structural design, deflection and crack control at service load levels are usually the governing considerations and accurate modelling of the stiffness after cracking is required.

In-service deflections depend primarily on the properties of the concrete and these are often not known reliably at the design stage. The non-linear behaviour that complicates serviceability calculations is due to cracking, tension stiffening, creep and shrinkage of the concrete. Of these, shrinkage is perhaps the most problematic. Shrinkage may cause time-dependent cracking that reduces member stiffness and gradually reduces the beneficial effects of tension stiffening. It results in a gradual widening of existing cracks and, in flexural members, a significant increase in deflections with time. The problem is particularly difficult in the case of slabs that are typically shallow, with relatively large span to depth ratios, and are therefore deflection sensitive.

The final deflection of a slab depends very much on the extent of initial cracking which, in turn, depends on the construction procedure (shoring and re-shoring), the amount of early shrinkage, the temperature gradients in the first few weeks after casting, the degree of curing and so on. Many of these parameters are, to a large extent, outside of the control of the designer. In field measurements of the deflection of many identical slab panels (Refs 1 and 2), large variability was reported. Deflections of identical panels after one year differed by over 100 per cent in some cases. These differences can be attributed to the different conditions (both in terms of load and environment) that existed in the first few weeks after the casting of each slab.

Serviceability failures of concrete structures involving excessive cracking and/or excessive deflection are relatively common. Numerous cases have been reported of structures that complied with the local code requirements but still deflected or cracked excessively (Refs 1–3). In many of these cases, shrinkage of the concrete was primarily responsible or, probably more precisely, failure to adequately account for shrinkage (and creep) in the design was primarily responsible.

The need for a more reliable deflection calculation procedure has been exacerbated by the introduction, in recent years, of higher strength reinforcing steels. The use of higher strength steel usually means that less steel is required for strength and, consequently, less stiffness is available after cracking, leading to greater deflection and wider cracks under service loads. Design for serviceability has increasingly assumed a more prominent role in the design of both beams and slabs.

This chapter describes the behaviour of reinforced concrete elements under service loads and outlines the simplified approaches for deflection and crack control that are specified in modern design codes, including ACI 318-08 (Ref. 4), Eurocode 2 (Refs 5 and 6) and AS 3600-2009 (Ref. 7). The limitations of these simplified approaches are discussed and the need for the more refined methods of analysis at the serviceability limit states that are described in subsequent chapters is demonstrated.

3.2 Design objectives and criteria

The broad design objective for a concrete structure is that it should satisfy the needs for which it was contrived. In doing so, the structural designer must ensure that it is safe and serviceable, so that the chance of it failing during its design lifetime is sufficiently small. Structural failure can take a variety of forms. The structure must be strong enough and sufficiently ductile to resist, without collapse, the overloads and environmental extremes that may be imposed upon it. It must also perform satisfactorily under day-to-day service loads without deforming, cracking or vibrating excessively. The two primary structural design objectives are therefore *strength* and *serviceability*. Other structural design objectives are *stability* and *durability* – a structure must be stable and resist overturning or sliding; reinforcement must not corrode; concrete must resist abrasion and spalling, and the structure must not suffer a significant reduction of strength or serviceability with time.

Another non-structural, but important, objective is *aesthetics* and, of course, an overarching design objective is *economy*. Ideally, the structure should be in harmony with, and enhance, the environment and this often requires collaboration between the structural engineer, environmental engineer and architect. The aim is to achieve, at minimum cost, an aesthetically pleasing and functional structure that satisfies the required structural objectives.

In order for design calculations to proceed for a particular structure, the design objectives must be translated into quantitative terms called *design criteria*. For example, the designer needs to determine the maximum acceptable deflection for a particular beam or slab, or the maximum crack width that can be tolerated in a concrete floor or wall. Also required are minimum numerical values for the strength of individual elements or connections. It is also necessary to identify and quantify all of the loads or actions that will be applied to the structure and the probability of their occurrence.

The design criteria are specified in codes of practice and provide a suitable margin of safety (called the *safety index*) against a structure becoming unfit for service in any way.

The specific form of the design criteria depends on the philosophy and method of design adopted by the relevant code and the manner in which the inherent variability in both the load and the structural performance is considered. Modern design codes for structures have generally adopted the *limit states method* of design, whereby a structure must be designed to simultaneously satisfy a number of different *limit states* or design requirements, including adequate strength and serviceability. Minimum performance criteria are specified for each of these limit states and any one may become critical and govern the design of a particular member.

If a structure ceases to fulfil its intended function in any way, it is said to enter a limit state. Each possible mode of failure is a limit state. For each limit state, codes of practice specify both load combinations and methods of predicting the actual structural performance that together ensure an acceptably low probability of failure, depending on the consequences and cost of the failure. For the strength limit states, the consequences of failure are high and so the probability of failure must be very low. For the serviceability limit states, such as excessive deflection or excessive cracking, the consequences of failure are not as great and a higher probability of failure is justifiable.

Design for the serviceability limit states involves making reliable predictions of the time-dependent deformation of the concrete structure. This is complicated by the non-linear material behaviour of concrete, caused mainly by cracking, tension stiffening, creep and shrinkage. This book deals primarily with the inclusion of these material non-linearities in the analysis of concrete structures, and so it is primarily concerned with designing for serviceability.

In order to satisfy the serviceability limit states, a concrete structure must be serviceable and perform its intended function throughout its working life. Excessive deflection should not impair the function of the structure or be aesthetically unacceptable. Excessive deflection should also not cause unintended load paths, such as occurs when a deflecting slab begins to bear on a non-load-bearing partition. Cracks should not be unsightly or wide enough to lead to durability problems and vibration should not cause distress to the structure or discomfort to its occupants.

3.3 Design actions

In the design of concrete structures for the serviceability limit states, the internal actions and deformations that arise from appropriate combinations of the day-to-day service loads should be considered, including, where applicable: dead load (G); live load (Q); wind load (W); prestress (P); earthquake load (F_{eq}); earth pressure (F_{ep}); liquid pressure (F_{lp}); and snow load (F_s). In addition, any accidental loading and loads arising during construction should be considered where they may adversely affect the various limit states' requirements. Other actions that may affect the serviceability of the structure include, creep of concrete, shrinkage of concrete and other imposed deformations, such as may result from temperature changes and gradients, support settlements and foundation movements.

Dead loads are generally defined as those loads imposed by both the structural and non-structural components of a structure. Dead loads include the self-weight of the structure and the forces imposed by all walls, floors, roofs, ceilings, permanent partitions, service machinery and other permanent construction. Dead loads are usually permanent, fixed in position and can be estimated reasonably accurately from the mass of the relevant material or type of construction. For example, normal weight concrete weighs about 24 kN/m^3. Lightweight reinforced concrete weighs between 15 and 20 kN/m^3.

Live loads are the loads that are attributed to the intended use or purpose of the structure and are generally specified by regional or national codes and specifications, such as Refs 8–13. The specified live load depends on the expected use or occupancy of the structure and usually includes allowances for impact and inertia loads (where applicable) and for possible overload. Both uniformly distributed and concentrated live loads are normally specified. The magnitude and distribution of the actual live load is never known accurately at the design stage, and it is by no means certain that the specified live load will not be exceeded at some stage during the life of the structure. Live loads may or may not be present at any particular time; they are not constant and their position can vary. Although part of the live load is transient, some portion may be permanently applied to the structure and have effects similar to dead loads. Live loads also arise during construction due to the stacking of building materials, the use of particular equipment or the construction procedure (such as the loads induced by floor-to-floor propping in multi-storey construction).

The specified wind, earthquake, snow and temperature loads depend on the geographical location and relative importance of the structure (the mean return period). Wind loads also depend on the surrounding terrain and the height of the structure above the ground. These *environmental* loads are also usually specified by regional or national codes and specifications, such as Refs 13–16.

The design loads to be used in serviceability calculations are the day-to-day *service loads* and these may be considerably less than the *specified loads*. For example, the specified live load, Q, has a built-in allowance for overload and impact. There is a low probability that it will be exceeded. It is usually not necessary, therefore, to ensure acceptable deflections and crack widths under the full specified loads. Use of the actual load combinations under normal conditions of service – the expected loads – is more appropriate.

Often, codes of practice differentiate between the specified (or characteristic) loads and the expected loads. Depending on the type of structure, the expected loads may be significantly less than the specified loads. If the aim of the serviceability calculation is to produce a best estimate of the likely behaviour, then expected loads should be considered. Often the magnitudes of the expected loads are not defined and are deemed to be a matter of engineering judgment. In some codes, serviceability load factors ($\psi \leq 1.0$) are nominated to determine the expected load from the specified load for both short-term ($\psi_1 Q$) and long-term loading ($\psi_2 Q$). If the aim is to satisfy a particular serviceability limit state and the consequences of failure to do so are high, then the specified loads may be more appropriate. Once again, the decision should be based on engineering judgment and experience.

With regard to dead loads, the expected and specified values are the same and therefore the specified dead load should be used in all serviceability calculations. For live loads, the expected values should be used when a best estimate is required. For building structures, the expected live load for short-term serviceability load combinations is usually about 70 per cent of the specified live load for dwellings (i.e. $\psi_1 = 0.7$), 60 per cent for retail areas, 50 per cent for offices and parking areas and 100 per cent for storage areas. For bridges, the live load for short-term serviceability load combinations is usually taken to be about 70 per cent of the specified live load for spans of less than about 10 m, reducing to 50 per cent for spans greater than 100 m.

For long-term serviceability calculations, the following percentages of the specified live load are usually considered as permanent or sustained (with the remainder being transitory in nature): 30 per cent for dwellings and retail stores (i.e. $\psi_2 = 0.3$), 25 per cent for parking areas, 20 per cent for offices, and 50–80 per cent for storage areas.

With the appropriate design actions for serviceability determined, the response of the structure must satisfy the relevant design criteria for deflection and crack control. These are usually also embodied in the codes of practice and are discussed in the following sections.

3.4 Design criteria for serviceability

3.4.1 Deflection limits

The design for serviceability, particularly the control of deflections, is frequently the primary consideration when determining the cross-sectional dimensions of beams and floor slabs in concrete buildings. This is particularly so in the case of slabs, as they are typically thin in relation to their spans and are therefore deflection sensitive. It is stiffness rather than strength that usually controls the design of slabs, particularly in the cases of flat slabs and flat plates.

Most concrete design codes, including ACI 318-08 (Ref. 4), Eurocode 2 (Ref. 6) and the Australian Standard AS 3600-2009 (Ref. 7), specify two basic approaches for deflection control. The first and simplest approach is deflection control by the satisfaction of a minimum depth requirement or a maximum span-to-depth ratio (see Section 3.5). The second approach is deflection control by the calculation of deflection, where the deflection (or camber) is calculated using realistic models of material and structural behaviour. This calculated deflection should not exceed the *deflection limits* that are appropriate to the structure and its intended use. The deflection limits should be selected by the designer and are often a matter of engineering judgment.

Codes of practice give general guidance for both the selection of the maximum deflection limits and the calculation of deflection. However, the simplified procedures for calculating the deflections of beams and slabs in most codes are necessarily design-oriented and simple to use, involving crude approximations of the complex effects of cracking, tension stiffening, concrete creep, concrete shrinkage and the load history (see Section 3.6). Generally, they have been developed and calibrated for simply-supported reinforced concrete beams (Ref. 17) and often produce inaccurate and non-conservative predictions when applied to lightly reinforced concrete slabs (Ref. 18). In addition, most existing code procedures do not provide real guidance on how to adequately model the time-dependent effects of creep and shrinkage in deflection calculations.

There are three main types of deflection problem that may affect the serviceability of a concrete structure:

1. where excessive deflection causes either aesthetic or functional problems;
2. where excessive deflection results in unintended load paths or damage to either structural or non-structural elements attached to the member;
3. where dynamic effects due to insufficient stiffness cause discomfort to occupants.

Examples of deflection problems of type 1 include visually unacceptable sagging (or hogging) of slabs and beams and ponding of water on roofs. Type 1 problems are generally overcome by limiting the magnitude of the final long-term deflection (here called *total deflection*) to some appropriately low value. The total deflection of a beam or slab in a building is usually the sum of the short-term and time-dependent deflections caused by the dead load (including self-weight), the prestress (if any), the expected in-service live load and the load-independent effects of shrinkage and temperature change. When the total deflection exceeds about span/200 below the horizontal, it may become visually unacceptable. Total deflection limits that are appropriate for the particular member and its intended function must be selected by the designer. For example, a total deflection limit of span/200 may be appropriate for the floor of a car park, but would be totally inadequate for a gymnasium floor that is required to remain essentially plane under service conditions and where functional problems arise at very small total deflections. AS 3600-2009 (Ref. 7) requires that a limit on the total deflection be selected that is appropriate to the structure and its intended use, but that limit should not be greater than span/250 for a span supported at both ends and span/125 for a cantilever (see Table 3.1).

Examples of type 2 problems include deflection-induced damage to ceiling or floor finishes, cracking of masonry walls and other brittle partitions, improper functioning of sliding windows and doors, tilting of storage racking and so on. To avoid these problems, a limit must be placed on that part of the total deflection that occurs after the attachment of the non-structural elements in question, that is, the *incremental deflection*. This incremental deflection is the sum of the long-term deflection due to all of the sustained loads and shrinkage, the short-term deflection due to the transitory live load and the short-term deflection due to any dead load applied to the structure after the attachment of the non-structural elements under consideration, together with any temperature-induced deflection.

For roof or floor construction supporting or attached to non-structural elements that are likely to be damaged by large deflection, ACI 318-08 (Ref. 4) limits the incremental deflection to span/480 (and span/240 when the non-structural elements are unlikely to be damaged by deflection). Incremental deflections of span/480 can, in fact, cause cracking in supported masonry walls, particularly when doorways or corners prevent arching and when no provisions are made to minimise the effect of movement. AS 3600-2009 limits the incremental deflection for members supporting masonry partitions or other brittle finishes to between span/500 and span/1000 depending on the provisions made to minimise the effect of movement (see Table 3.1). Eurocode 2 (Ref. 6) recommends limits on total (incremental) deflection of span/250 (span/500).

Type 3 deflection problems include the perceptible springy vertical motion of floor systems and other vibration-related problems. Very little quantitative information for controlling this type of deflection problem is available in codes of practice. For a floor that is not supporting or attached to non-structural elements likely to be damaged by large deflection, ACI 318-08 (Ref. 4) places a limit of span/360 on the short-term deflection due to live load (and span/180 for a flat roof). This limit provides a minimum requirement on the stiffness of members that may, in some cases, be sufficient to avoid type 3 problems. Such problems are potentially the most common for prestressed concrete floors, where load balancing is often employed to produce a nearly horizontal floor under the sustained load and the bulk of the final deflection is due to the transient

Table 3.1 Limits for calculated deflections of beams and slabs with effective span ℓ_{ef}

Type of member	*Deflection to be considered*	*Deflection limitation for spans (Notes 2 and 3)*	*Deflection limitation for cantilevers (Note 5)*
All members	The total deflection	$\ell_{ef}/250$	$\ell_{ef}/125$
Members supporting masonry partitions	The deflection that occurs after the addition or attachment of the partitions	$\ell_{ef}/500$ where provision is made to minimise the effect of movement, otherwise $\ell_{ef}/1000$	$\ell_{ef}/250$ where provision is made to minimise the effect of movement, otherwise $\ell_{ef}/500$
Members supporting other brittle finishes	The deflection that occurs after the addition or attachment of the finish	Manufacturer's specification but not more than $\ell_{ef}/500$	Manufacturer's specification but not more than $\ell_{ef}/250$
Members subjected to vehicular or pedestrian traffic	The imposed action (live load and dynamic impact) deflection	$\ell_{ef}/800$	$\ell_{ef}/400$
Transfer members	Total deflection	$\ell_{ef}/500$ where provision is made to minimize the effect of deflection of the transfer member on the supported structure, otherwise $\ell_{ef}/1000$	$\ell_{ef}/250$

1 The effective span ℓ_{ef} is the lesser of the centre-to-centre distance between the supports and the clear span plus the member depth. For a cantilever ℓ_{ef} is the clear projection plus half the member depth.
2 In general, deflection limits should be applied to all spanning directions. This includes, but is not limited to, each individual member and the diagonal spans across each design panel. For flat slabs with uniform loadings, only the column strip deflections in each direction need be checked.
3 If the location of masonry partitions or other brittle finishes is known and fixed, these deflection limits need only be applied to the length of the member supporting them. Otherwise, the more general requirements of Note 2 should be followed.
4 Deflection limits given may not safeguard against ponding.
5 For cantilevers, the deflection limitations given in this table apply only if the rotation at the support is included in the calculation of deflection.
6 Consideration should be given by the designer to the cumulative effect of deflections, and this should be taken into account when selecting a deflection limit.
7 When checking the deflections of transfer members and structures, allowance should be made in the design of the supported members and structure for the deflection of the supporting members. This will normally involve allowance for settling supports and may require continuous bottom reinforcement at settling columns.

live load. Such structures are generally uncracked at service loads, the total deflection is small and type 1 and 2 deflection problems are easily avoided.

Where a structure supports vibrating machinery (or any other significant dynamic load) or where a structure may be subjected to ground motion caused by earthquake, blast or adjacent road or rail traffic, vibration control becomes an important design requirement. This is particularly so for slender structures, such as tall buildings or long-span beams or slabs.

Vibration is best controlled by isolating the structure from the source of vibration. Where this is not possible, vibration may be controlled by limiting the frequency of the fundamental mode of vibration of the structure to a value that is significantly different from the frequency of the source of vibration. When a structure is subjected only to pedestrian traffic, 5 Hz is often taken as the minimum frequency of the fundamental mode of vibration of a beam or slab (Refs 19 and 20) and a method for vibration analysis is described in Ref. 20.

In modern concrete structures, serviceability failures are relatively common. The tendency towards higher strength materials and the use of ultimate strength design procedures for the proportioning of structures has led to shallower, more slender elements, and consequently, an increase in deformations at service loads. As far back as 1967 (Ref. 21), the most common cause of damage in concrete structures was due to excessive slab deflections. If the incidence of serviceability failure is to decrease, design for serviceability must play a more significant part in routine structural design and the structural designer must resort more often to analytical tools that are more accurate than those found in most building codes. The analytical models outlined in the remainder of this book provide designers with reliable and rational means for predicting both the short-term and time-dependent deformations in concrete structures.

3.4.2 Crack width limits

In routine structural design, the calculation of crack widths is often not required. Crack control is deemed to be provided by appropriate detailing of the reinforcement and by limiting the stress in the reinforcement crossing the crack to some appropriately low value (see Section 3.7.1). The limiting steel stress depends on the maximum acceptable crack width for the structure that in turn depends on the structural requirements and the local environment. Recommended maximum crack widths are given in Table 3.2.

3.5 Maximum span-to-depth ratio – minimum thickness

In the design of a reinforced concrete beam or slab, the designer is first confronted with the problem of selecting a suitable depth or thickness of the member. In the case of floor slabs, a reasonable first estimate is desirable since, in many cases, the self-weight of the slab is a large proportion of the total service load. Strength considerations alone may result in a slab thickness that leads to excessive deflection at service loads. The relatively low tensile reinforcement quantities commonly used in slabs (A_{st}/bd is typically in the range 0.0025–0.006) are evidence of the importance of serviceability in the selection of slab thickness.

Many codes of practice specify a minimum thickness, h, or a maximum span-to-effective depth ratio, ℓ/d. The implication is that deflection will be acceptable if the beam or slab thickness is greater than the minimum value, and thus the deflection need not be calculated.

For example, for normal weight concrete members containing steel reinforcement with a yield stress of $f_y = 500$ MPa and for members that are not supporting or attached to partitions or other construction likely to be damaged by large deflection, ACI-318-08 (Ref. 4) specifies that deflections need not be calculated if the thickness of a beam or one-way slab is greater than the minimum thickness given in Table 3.3

Table 3.2 Recommended maximum final design crack width, w^*

Environment	*Design requirement*	*Maximum final crack width, w^* (mm)*
Sheltered environment (where crack widths will not adversely affect durability)	Aesthetic requirement • where cracking could adversely affect the appearance of the structure	
	• close in buildings	0.3
	• distant in buildings	0.5
	• where cracking will not be visible and aesthetics is not important.	0.7
Environment	Durability requirement • where wide cracks could lead to corrosion of reinforcement	0.3
Aggressive environment	Durability requirement • where wide cracks could lead to corrosion of reinforcement	0.30 (when c* ≥ 50 mm) 0.25 (otherwise)

*c is the concrete cover to the nearest steel reinforcement.

Table 3.3 Minimum thickness for non-prestressed beams or one-way slabs – $f_y = 500$ MPa (Ref. 4)

	Minimum thickness, h			
	Simply-supported	*One end continuous*	*Both ends continuous*	*Cantilever*
Solid one-way slab	$\ell/18$	$\ell/21.5$	$\ell/25$	$\ell/9$
Beam or ribbed slab	$\ell/14.5$	$\ell/16.5$	$\ell/19$	$\ell/7$

ℓ is the span measured centre to centre of supports or the clear projection of a cantilever.

or if the thickness of a two-way flat slab or flat plate is greater than the value given in Table 3.4. This deemed-to-comply approach is attractive because of its simplicity and, if it always led to appropriately proportioned and serviceable slabs, it would be ideal for use in routine design. However, in some situations, the use of the minimum thicknesses in Tables 3.3 and 3.4 leads to slabs that are far thicker than they need to be. In other situations, some heavily loaded slabs with the minimum thicknesses specified in Tables 3.3 and 3.4 deflect excessively. Indeed, a single value for minimum thickness that does not account for the load level, the steel quantities and location, the occupancy of the structure (and therefore the desired deflection limit), the load duration, the quality of the concrete and the environment cannot possibly ensure deflection control in every situation, unless, of course, it is grossly conservative in many situations and therefore entirely unsuitable for use in structural design.

Based on the work of Rangan (Ref. 22) and Gilbert (Ref. 23), the Australian Standard AS3600-2009 (Ref. 7) specifies a maximum span to effective depth ratio

Table 3.4 Minimum thickness of two-way slabs without interior beams (flat slabs or flat plates) – $f_y = 500$ MPa (Ref. 4)

Minimum thickness, h					
Without drop panels			*With drop panels*		
Exterior panels		*Interior panels*	*Exterior panels*		*Interior panels*
Without edge beams	With edge beams		Without edge beams	With edge beams	
$\ell_n/28$	$\ell_n/31$	$\ell_n/31$	$\ell_n/31$	$\ell_n/34$	$\ell_n/34$

ℓ_n is the clear span measured face to face of supports in long span direction.

for slabs given by:

$$\ell_{ef}/d = k_3 k_4 \left[\frac{(\Delta/\ell_{ef})1000E_c}{F_{d.ef}} \right]^{1/3} \tag{3.1}$$

where ℓ_{ef} is the lesser of the clear span plus slab thickness and the centre-to-centre distance between supports; d is the effective depth from the compressive surface of the slab to the centroid of the tensile reinforcement; Δ is the deflection limit selected in design (either total or incremental deflection); E_c is the elastic modulus of the concrete (in MPa); $F_{d.ef}$ is the effective design service load (in kPa) and is equal to $3g + q_1 + 2q_2$ for total deflection and $2g + q_1 + 2q_2$ for incremental deflection; g is the dead load (in kPa); q_1 is the expected short-term live load (in kPa); q_2 is the sustained part of the live load (in kPa); and k_3 is a slab system factor given by:

- $k_3 = 1.0$ for one-way slabs and two-way edge-supported slabs;
- $k_3 = 0.95$ for a two-way flat slab without drop panels; and
- $k_3 = 1.05$ for a two-way flat slab with drop panels.

The term k_4 is a factor that depends on the continuity at the supports of the slab and, for a one-way slab or two-way flat slab, k_4 equals 1.4 for a simply-supported span, 1.75 for the end span of a continuous slab and 2.1 for an interior span of a continuous slab. For a two-way edge-supported slab, k_4 is given in Table 3.5.

Alternative expressions for the limiting span-to-depth ratios are specified in other codes. For example, according to Eurocode 2 (Ref. 6), if the span-to-depth ratio of a member does not exceed the following limiting values, deflections need not be calculated for members where the maximum total deflection is not to exceed span/250 and where the incremental deflection is not to exceed span/500:

$$\frac{\ell}{d} = K\left[11 + 1.5\sqrt{f'_c}\frac{\rho_o}{\rho} + 3.2\sqrt{f'_c}\left(\frac{\rho_o}{\rho} - 1\right)^{3/2}\right] \quad \text{if } \rho \le \rho_o \tag{3.2a}$$

$$\frac{\ell}{d} = K\left[11 + 1.5\sqrt{f'_c}\frac{\rho_o}{\rho - \rho'} + \frac{1}{12}\sqrt{f'_c}\sqrt{\frac{\rho'}{\rho_o}}\right] \quad \text{if } \rho > \rho_o \tag{3.2b}$$

Table 3.5 Continuity factor k_4 for a rectangular slab panel supported on four sides – (Ref. 7)

Edge condition	*Continuity factor, k_4*			
	Ratio of long to short span (ℓ_y/ℓ_x)			
	1.0	1.25	1.5	2.0
Four edges continuous	3.60	3.10	2.80	2.50
One short edge discontinuous	3.40	2.90	2.70	2.40
One long edge discontinuous	3.40	2.65	2.40	2.10
Two short edges discontinuous	3.20	2.80	2.60	2.40
Two long edges discontinuous	3.20	2.50	2.00	1.60
Two adjacent edges discontinuous	2.95	2.50	2.25	2.00
Three edges discontinuous (one long edge continuous)	2.70	2.30	2.20	1.95
Three edges discontinuous (one short edge continuous)	2.70	2.10	1.90	1.60
Four edges discontinuous	2.25	1.90	1.70	1.50

where K depends on the structural system, with $K = 1.0$ for simply supported beams and one-way or two-way spanning slabs; $K = 1.3$ for end spans of continuous beams and one-way or two-way slabs continuous over one long side; $K = 1.5$ for interior spans of beams and one-way or two-way spanning slabs; $K = 1.2$ for flat slabs supported on columns without beams (based on longer span); and $K = 0.4$ for a cantilever; ρ_o is a reference reinforcement ratio given by $\rho_o = 0.001\sqrt{f'_c}$; ρ is the tension reinforcement ratio at mid-span required to resist the moment due to the design loads (at the support for cantilevers); ρ' is the compression reinforcement ratio at mid-span (at the support for cantilevers); f'_c is in MPa.

For flanged sections where the ratio of the flange width to the web width exceeds 3, the values obtained from Eqs 3.2 should be multiplied by 0.8. For beams and slabs, other than flat slabs, supporting partitions likely to be damaged by excessive deflection and with spans exceeding 7 m, the values of ℓ/d given by Eqs 3.2 should be multiplied by $7/\ell_{ef}$, where ℓ_{ef} is the effective span in metres. For flat slabs supporting partitions likely to be damaged by excessive deflection and where the longer span exceeds 8.5 m, the values of ℓ/d given by Eqs 3.2 should be multiplied by $8.5/\ell_{ef}$.

An iterative procedure is required to use Eqs 3.2. An initial estimate of the effective depth d must be made in order to calculate the reinforcement ratios ρ and ρ' required to resist the design moment at the critical sections. These reinforcement ratios are in turn required in the calculation of the limiting span-to-depth ratio using Eqs 3.2. Depending on the initial estimate of d, the required reinforcement ratios may need to be recalculated, together with a revised slab depth, to ensure deflection control. Similarly, the use of Eq. 3.1 requires an initial estimate of slab thickness to determine the self-weight and an iteration may be required if the initial estimate is poor. On the other hand, the ACI 318-08 (Ref. 4) minimum thicknesses do not require any iteration and can be used to select an initial member thickness at the beginning of the design.

Designers should be aware that the use of either the ACI 318-08 minimum thicknesses (Tables 3.3 and 3.4) or the AS3600-2009 (Ref. 7) maximum span-to-depth ratio (Eq. 3.1) is inevitably conservative for most practical situations and leads to beam and slab thicknesses that may be considerably larger than they need to be. In contrast, the Eurocode 2 (Ref. 6) limiting span-to-depth ratios may result in beam or slab thicknesses that are unacceptably small, particularly in situations where the sustained load is a large proportion of the total load, and should not be regarded as a guarantee that deflections will not be excessive.

Example 3.1

The thickness of the end span of a one-way slab is to be estimated for the cases where:

1) the maximum long-term mid-span deflection is not to exceed span/250; and
2) the incremental deflection is not to exceed span/480.

The span is $\ell = \ell_{ef} = 5.0$ m; the dead load $g = 2.0$ kPa + self-weight; and the live load is $q = 3.0$ kPa. The short-term expected live load is $q_1 = 2.1$ kPa and the sustained part of the live load $q_2 = 1.2$ kPa. The specified concrete and reinforcement strengths are $f'_c = 25$ MPa and $f_y = 500$ MPa; the clear concrete cover to the longitudinal reinforcement is 20 mm; and the reinforcement bar diameter is $d_b = 12$ mm.

ACI318-08

From Table 3.3, if deflections are not to be calculated, the minimum thickness according to ACI318-08 is $h = 5000/21.5 = 230$ mm, irrespective of the deflection requirements for the slab and irrespective of the loads to be applied to the slab.

AS3600-2009

From Table 2.1, $E_c = 26700$ MPa. To calculate the maximum effective span-to-depth ratio using Eq. 3.1, an estimate of self-weight of the slab is required. Take self-weight = $0.23 \times 24 = 5.5$ kPa and the dead load is therefore $g = 2.0 + 5.5 = 7.5$ kPa.

(i) If the maximum total deflection is $\Delta = 5000/250 = 20$ mm, then the effective design service load is $F_{d.ef} = 3g + q_1 + 2q_2 = 3 \times 7.5 + 2.1 + 2 \times 1.2 = 27.0$ kPa and, from Eq. 3.1, the maximum span to effective depth ratio is:

$$\ell_{ef}/d = 1.0 \times 1.75 \left[\frac{(20/5000) \times 1000 \times 26700}{27.0} \right]^{1/3} = 27.7$$

The minimum effective depth according to AS 3600-2009 is therefore $d = 5000/27.7 = 181$ mm and the minimum slab thickness is $h = d + d_b/2 +$ cover $= 181 + 6 + 20 = 207$ mm.

(ii) If the incremental deflection is $\Delta = 5000/480 = 10.4$ mm, then the effective design service load is $F_{d.ef} = 2g + q_1 + 2q_2 = 2 \times 7.5 + 2.1 + 2 \times 1.2 = 19.5$ kPa and, from Eq. 3.1, the maximum span to effective depth ratio is:

$$\ell_{ef}/d = 1.0 \times 1.75 \left[\frac{(10.4/5000) \times 1000 \times 26700}{19.5} \right]^{1/3} = 24.8$$

The minimum effective depth according to AS 3600-2009 is therefore $d = 5000/24.8 = 201$ mm and the minimum slab thickness is $h = d + d_b/2 +$ cover $= 201 + 6 + 20 = 227$ mm.

Eurocode 2

If the design moment at mid-span is $w^*\ell^2/12$, where w^* is the factored design load for the strength limit state, and if after several iterations the initial estimate of d is 150 mm, the reinforcement ratio required to resist the design moment is $\rho = 0.0035$ and $\rho' = 0$. The reference reinforcement ratio is $\rho_o = 0.001\sqrt{f'_c} = 0.005$ and the factor $K = 1.3$. From Eq. 3.2a:

$$\frac{\ell}{d} = 1.3 \left[11 + 1.5\sqrt{25}\frac{0.005}{0.0035} + 3.2\sqrt{25}\left(\frac{0.005}{0.0035} - 1\right)^{3/2} \right] = 34.1$$

The minimum effective depth according to Eurocode 2 is therefore $d = 5000/34.1 = 147$ mm and the minimum slab thickness is $h = d + d_b/2 +$ cover $= 147 + 6 + 20 = 173$ mm.

Clearly, in this example, the minimum slab thickness obtained using Eq. 3.2a from Eurocode 2 is significantly smaller than the value obtained using either the ACI 318-08 minimum thickness provisions or the maximum span to depth ratio specified in AS 3600-2009.

3.6 Deflection control by simplified calculation

3.6.1 Calculation of deformation

If the axial strain and curvature are known at regular intervals along a member, it is a relatively simple task to determine the deformation of that member. Consider the statically determinate member subjected to the axial and transverse loads shown in Fig. 3.1. The axial deformation of the member e_{AB} (either elongation or shortening) is obtained by integrating the axial strain at the centroid of the member, $\varepsilon_a(z)$, over the length of the member. That is:

$$e_{AB} = \int_0^\ell \varepsilon_a(z)\,dz \qquad (3.3)$$

where z is measured along the member, as shown.

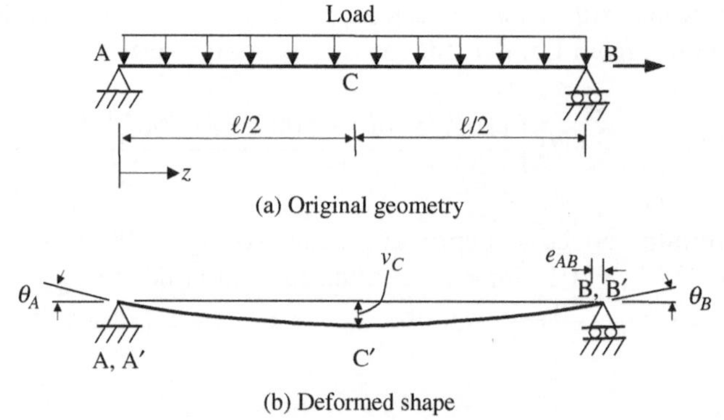

Figure 3.1 Deformation of a statically determinate member.

Provided that deflections are small and that simple beam theory is applicable, the slope θ and deflection v at any point z along the member are obtained by integrating the curvature $\kappa(z)$ along the member as follows:

$$\theta = \int \kappa(z)\, \mathrm{d}z \tag{3.4}$$

$$v = \int\int \kappa(z)\, \mathrm{d}z\, \mathrm{d}z \tag{3.5}$$

Eqs 3.4 and 3.5 are quite general and apply to both elastic and inelastic material behaviour.

If the axial strain and curvature are calculated at any time after loading at a pre-selected number of points along the member shown in Fig 3.1 and if a reasonable variation of strain and curvature is assumed between adjacent points, it is a simple matter of geometry to determine the deformation of the member. For convenience, some simple equations are given below for the determination of the deformation of a single span and of a cantilever. If the axial strain ε_a and the curvature κ are known at the mid-span and at each end of the member shown in Fig 3.1 (i.e. at supports A and B and at point C), the axial deformation e_{AB}, the slope at each support θ_A and θ_B and the deflection at mid-span v_C are given by Eqs 3.6 and 3.7.

For a linear variation of strain and curvature:

$$e_{AB} = \frac{\ell}{4}(\varepsilon_{aA} + 2\,\varepsilon_{aC} + \varepsilon_{aB}) \tag{3.6a}$$

$$v_C = \frac{\ell^2}{48}(\kappa_A + 4\,\kappa_C + \kappa_B) \tag{3.6b}$$

$$\theta_A = \frac{\ell}{24}(5\,\kappa_A + 6\,\kappa_C + \kappa_B) \tag{3.6c}$$

$$\theta_B = -\frac{\ell}{24}(\kappa_A + 6\,\kappa_C + 5\,\kappa_B) \tag{3.6d}$$

For a parabolic variation of strain and curvature:

$$e_{AB} = \frac{\ell}{6}\,(\varepsilon_{aA} + 4\,\varepsilon_{aC} + \varepsilon_{aB}) \tag{3.7a}$$

$$v_C = \frac{\ell^2}{96}\,(\kappa_A + 10\,\kappa_C + \kappa_B) \tag{3.7b}$$

$$\theta_A = \frac{\ell}{6}\,(\kappa_A + 2\,\kappa_C) \tag{3.7c}$$

$$\theta_B = -\frac{\ell}{6}\,(2\kappa_C + \kappa_B) \tag{3.7d}$$

In addition to the simple span shown in Fig. 3.1, Eqs 3.6 and 3.7 also apply to any member in a statically indeterminate frame, provided the strain and curvature at each support and at mid-span are known.

Consider the fixed-end cantilever shown in Fig. 3.2. If the curvatures at A, B and C are known, then the slope and deflection at the free end of the member are given by Eqs 3.8 and 3.9.

For a linear variation of curvature:

$$\theta_C = -\frac{\ell}{4}(\kappa_A + 2\kappa_B + \kappa_C) \tag{3.8a}$$

$$v_C = -\frac{\ell^2}{24}(5\kappa_A + 6\,\kappa_B + \kappa_C) \tag{3.8b}$$

For a parabolic variation of curvature:

$$\theta_C = -\frac{\ell}{6}(\kappa_A + 4\kappa_B + \kappa_C) \tag{3.9a}$$

$$v_C = -\frac{\ell^2}{6}\,(\kappa_A + 2\kappa_B) \tag{3.9b}$$

If the curvatures at the fixed and free ends of the cantilever only are known (i.e. at A and C), then the slope and deflection at the free end C are given by Eqs 3.10 and 3.11.

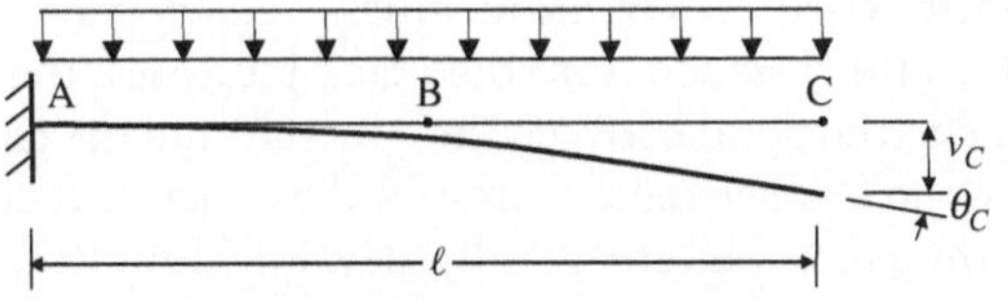

Figure 3.2 Deformation of a fixed-end cantilever.

For a linear variation of curvature:

$$\theta_C = -\frac{\ell}{2}\,(\kappa_A + \kappa_C) \tag{3.10a}$$

$$v_C = -\frac{\ell^2}{6}\,(2\kappa_A + \kappa_C) \tag{3.10b}$$

For a parabolic variation of curvature (typical of what occurs in a uniformly loaded cantilever):

$$\theta_C = -\frac{\ell}{3}\,(\kappa_A + 2\kappa_C) \tag{3.11a}$$

$$v_C = -\frac{\ell^2}{4}\,(\kappa_A + \kappa_C) \tag{3.11b}$$

3.6.2 Load versus deflection response of a reinforced concrete member

The short-term or instantaneous deformation of a cracked reinforced concrete cross-section subjected to combined bending and axial force can be readily determined using simple *modular ratio theory* (Section 3.6.3). After cracking, the properties of both the fully-cracked section and the uncracked section are often combined empirically to model tension stiffening and to approximate the average properties of the cracked region.

Consider the load-deflection response of a simply-supported, singly reinforced concrete beam or one-way slab shown in Fig. 3.3. At loads less than the cracking load, P_{cr}, the member is uncracked and behaves homogeneously and elastically. The slope of the load-deflection plot (OA in Fig. 3.3) is proportional to the second moment of area of the uncracked transformed section, I_{uncr}. The member first cracks at P_{cr} when the extreme fibre tensile stress in the concrete at the section of maximum moment reaches the flexural tensile strength of the concrete, $f_{ct.f}$. There is a sudden change in the local stiffness at, and immediately adjacent to, this first crack. At the section containing the crack, the flexural stiffness drops significantly, but the rest of the member remains uncracked. As load increases, more cracks form and the average flexural stiffness of the entire member decreases. If the tensile concrete in the cracked regions of the beam carried no stress, the load-deflection relationship would follow the dashed line ACD. If the average extreme fibre tensile stress in the concrete remained at $f_{ct.f}$ after cracking, the load-deflection relationship would follow the dashed line AE. In reality, the actual response lies between these extremes and is shown in Fig. 3.3 as the solid line AB. The difference between the actual response and the zero tension response is the tension stiffening effect (that reduces the instantaneous deflection by $\delta\Delta$ as shown).

Tension stiffening is the contribution of the intact tensile concrete between the cracks to the post-cracking stiffness of the member. At each crack, the tensile concrete carries no stress, but as the distance from the crack increases, the tensile stress in the concrete increases due to the bond between the concrete and the tensile reinforcement. As the load increases, the average tensile stress in the concrete reduces as more cracks develop and, when the crack pattern is fully developed and the number of cracks has stabilised, the *actual response* becomes approximately parallel to the *no tension* response (OD in Fig. 3.3). For slabs containing small quantities of tensile reinforcement

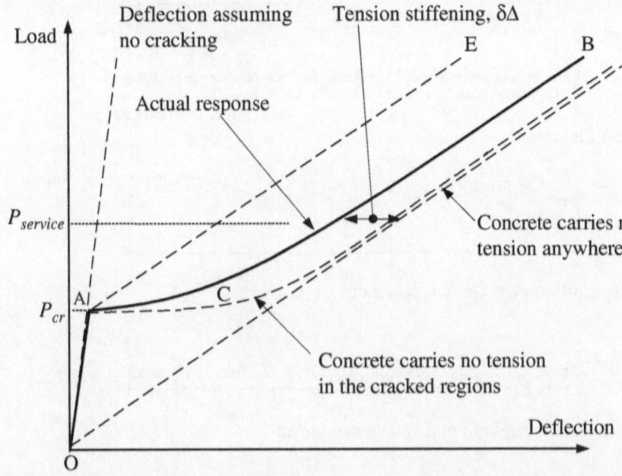

Figure 3.3 Typical load versus deflection relationship.

(typically in floor slabs $A_{st}/bd < 0.005$), tension stiffening may be responsible for more than 50 per cent of the stiffness of the cracked member at service loads and $\delta\Delta$ remains significant up to and beyond the point where the steel yields and the ultimate load is approached.

Figure 3.4a shows an elevation of a singly reinforced concrete flexural member subjected to a uniform bending moment M of sufficient magnitude to establish the primary flexural cracks. The variation of stress in the tensile reinforcement along the member is shown in Fig. 3.4b and the variation of tensile stress in the concrete at the steel level is shown in Fig. 3.4c. Over a gauge length containing several cracks, the average concrete tensile stress $\sigma_{c.avg}$ at typical in-service levels of applied moment is a significant percentage of the tensile strength of the concrete.

The keys to predicting the instantaneous deflection are first to evaluate the load required to cause first cracking or, more precisely, the moment to cause first cracking at the critical cross-section, and secondly to model tension stiffening accurately. Both of these tasks are not straightforward. Restraint to shrinkage provided by the bonded reinforcement and restraint to shrinkage at the member's ends can cause significant tension in the concrete in the first few days after casting. Cracking may therefore occur at loads far less than that required to produce an extreme fibre tensile stress equal to the modulus of rupture $f_{ct.f}$ in a member without shrinkage.

One commonly used approach for modelling tension stiffening in deflection calculations involves determining an average effective second moment of area (I_{ef}) for a cracked member. For a prismatic member, the effective second moment of area after cracking is less than the second moment of area of the uncracked transformed section (I_{uncr}) and greater than the second moment of area of the fully-cracked cross-section (I_{cr}). Several different empirical equations are available for I_{ef}, including the well-known equation developed by Branson (Ref. 17) that is included in ACI 318-08. A modified and significantly more realistic version of Branson's equation is specified in AS3600-2009. Another model for I_{ef} was recently proposed by Bischoff (Ref. 24) and may be derived from the method specified in Eurocode 2 for

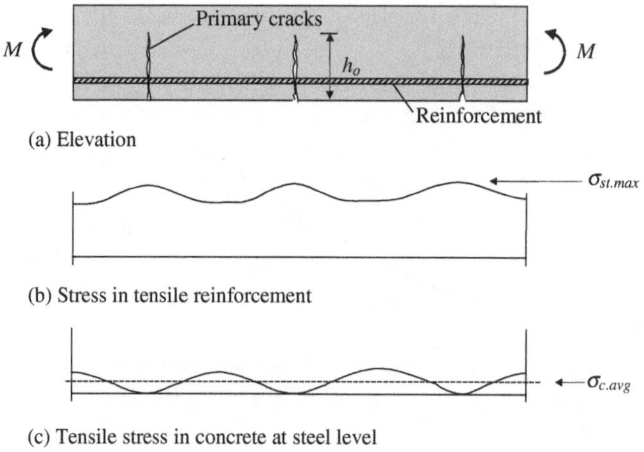

Figure 3.4 Stress distributions at the steel level in a cracked reinforced concrete member.

the calculation of deflection. These approaches are presented and reviewed in the following sub-sections.

3.6.3 Modular ratio theory

The second moment of area of a reinforced concrete cross-section is conventionally calculated using linear elastic analysis with the following assumptions:

(i) plane sections remain plane (i.e. the strain distribution is linear);
(ii) perfect bond exists between the reinforcement and the concrete; and
(iii) the stress-strain relationships for concrete and steel are linear and elastic.

The steel reinforcement is transformed into an equivalent area of concrete by multiplying the reinforcement area by the modular ratio n (where $n = E_s/E_c$) and the properties of the cross-section are determined using the principles of mechanics of solids. After cracking, the concrete in tension is assumed to carry no stress. This simple procedure is often called *modular ratio theory* and it is used to calculate the second moment of area of the section before and after cracking. In itself, it does not account for tension stiffening and does not include the time-dependent deformations caused by creep and shrinkage.

Figure 3.5 shows the strains, stresses and resultant forces on a cracked, singly reinforced, rectangular cross-section subjected to an applied moment M. With regard to the stresses and deformations, there are two unknowns – the top fibre compressive strain ε_{top} and the depth to the neutral axis $d_n = kd$. The depth to the neutral axis may be determined by equating the resultant compressive and tensile forces (i.e. by enforcing the requirement of equilibrium of longitudinal forces):

$$C = T$$

$$0.5\,\sigma_{top}\,b\,kd = \sigma_{st}A_{st}$$

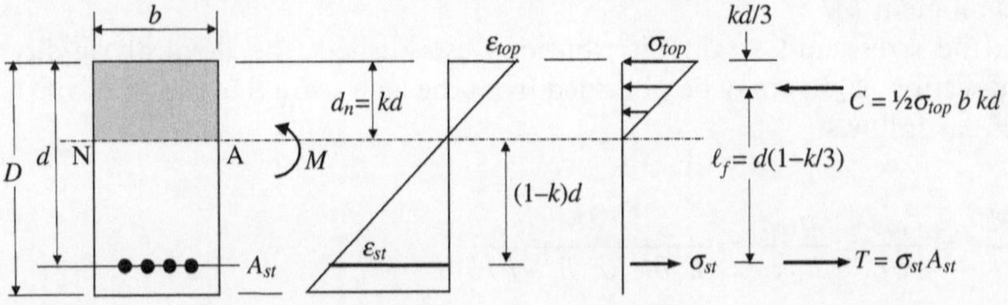

(a) Cross-section (b) Strains (c) Stresses (d) Forces

Figure 3.5 Strains, stresses and forces on a cracked section in bending.

and, assuming linear-elastic stress-strain laws for the reinforcement and the compressive concrete, this becomes:

$$0.5k = \frac{\sigma_{st}}{\sigma_{top}} \frac{A_{st}}{bd} = \frac{\varepsilon_{st} E_s}{\varepsilon_{top} E_c} \frac{A_{st}}{bd} = \frac{\varepsilon_{st}}{\varepsilon_{top}} n\rho \tag{3.12}$$

where ρ is the tensile reinforcement ratio, A_{st}/bd. Compatibility requires that the strain diagram is linear and therefore:

$$\frac{\varepsilon_{st}}{\varepsilon_{top}} = \frac{(1-k)d}{kd} = \frac{1-k}{k} \tag{3.13}$$

Substituting Eq. 3.13 into Eq. 3.12 gives:

$$0.5k = \frac{1-k}{k} n\rho$$

and solving this quadratic equation for k gives:

$$k = \sqrt{(n\rho)^2 + 2n\rho} - n\rho \tag{3.14}$$

It should be noted that k (and hence the depth to the neutral axis kd) depends only on the modular ratio n and the reinforcement ratio ρ and is independent of the applied moment M. The depth to the neutral axis remains constant after cracking as the moment increases, until either the concrete compressive stress distribution becomes curvilinear or the reinforcing steel yields.

With k determined from Eq. 3.14, the top fibre concrete stress σ_{top} and the steel stress σ_{st} may be found from the moment equilibrium equation:

$$M = C\,\ell_f = T\,\ell_f = ½\sigma_{top}\, b\, kd^2(1-k/3) = \sigma_{st} A_{st}\, d(1-k/3)$$

and hence

$$\sigma_{top} = \frac{2M}{b\,kd^2(1-k/3)} \text{ and } \sigma_{st} = \frac{M}{A_{st} d(1-k/3)} \tag{3.15}$$

Clearly, the stresses in both the concrete and the steel are linear functions of the applied moment M.

With the stress and strain distributions established, the flexural rigidity of the cracked section ($E_c I_{cr}$) may be obtained from the curvature (i.e. the slope of the strain diagram) as follows:

$$\frac{M}{E_c I_{cr}} = \frac{\varepsilon_{top}}{kd} = \frac{\sigma_{top}}{E_c kd} = \frac{2M}{E_c b k^2 d^3 (1 - k/3)} \quad (3.16)$$

From Eq. 3.16, the second moment of area of the cracked section is:

$$I_{cr} = \tfrac{1}{2}\, b\, d^3 k^2 (1 - k/3) \quad (3.17)$$

For convenience, values of the neutral axis parameter k and the cracked second moment of area I_{cr} of singly reinforced rectangular sections are given in Table 3.6.

Table 3.6 Neutral axis depth (kd) and second moment of area ($I_{cr} = \lambda b d^3$) for singly reinforced rectangular sections ($n = E_s/E_c$)

$\rho = A_{st}/bd$	$n=4$		$n=5$		$n=6$		$n=7$		$n=8$		$n=9$		$n=10$	
	k	λ	k	λ	k	λ	k	λ	k	λ	k	λ	k	λ
0.002	0.119	0.0068	0.132	0.0083	0.143	0.0098	0.154	0.0112	0.164	0.0127	0.173	0.0140	0.181	0.0154
0.0025	0.132	0.0083	0.146	0.0102	0.159	0.0119	0.170	0.0137	0.181	0.0154	0.191	0.0170	0.200	0.0187
0.003	0.143	0.0098	0.159	0.0119	0.173	0.0140	0.185	0.0161	0.196	0.0180	0.207	0.0199	0.217	0.0218
0.0035	0.154	0.0112	0.170	0.0137	0.185	0.0161	0.198	0.0183	0.210	0.0206	0.221	0.0227	0.232	0.0248
0.004	0.164	0.0127	0.181	0.0154	0.196	0.0180	0.210	0.0206	0.223	0.0230	0.235	0.0254	0.246	0.0277
0.0045	0.173	0.0140	0.191	0.0170	0.207	0.0199	0.221	0.0227	0.235	0.0254	0.247	0.0280	0.258	0.0305
0.005	0.181	0.0154	0.200	0.0187	0.217	0.0218	0.232	0.0248	0.246	0.0277	0.258	0.0305	0.270	0.0332
0.0055	0.189	0.0167	0.209	0.0202	0.226	0.0236	0.242	0.0268	0.256	0.0299	0.269	0.0329	0.281	0.0358
0.006	0.196	0.0180	0.217	0.0218	0.235	0.0254	0.251	0.0288	0.266	0.0321	0.279	0.0353	0.292	0.0384
0.0065	0.204	0.0193	0.225	0.0233	0.243	0.0271	0.260	0.0308	0.275	0.0343	0.289	0.0376	0.301	0.0408
0.007	0.210	0.0206	0.232	0.0248	0.251	0.0288	0.268	0.0327	0.283	0.0363	0.298	0.0399	0.311	0.0433
0.0075	0.217	0.0218	0.239	0.0263	0.258	0.0305	0.276	0.0345	0.292	0.0384	0.306	0.0421	0.319	0.0456
0.008	0.223	0.0230	0.246	0.0277	0.266	0.0321	0.283	0.0363	0.299	0.0404	0.314	0.0442	0.328	0.0479
0.0085	0.229	0.0242	0.252	0.0291	0.272	0.0337	0.291	0.0381	0.307	0.0423	0.322	0.0463	0.336	0.0501
0.009	0.235	0.0254	0.258	0.0305	0.279	0.0353	0.298	0.0399	0.314	0.0442	0.330	0.0483	0.344	0.0523
0.0095	0.240	0.0266	0.264	0.0319	0.285	0.0369	0.304	0.0416	0.321	0.0461	0.337	0.0503	0.351	0.0544
0.010	0.246	0.0277	0.270	0.0332	0.292	0.0384	0.311	0.0433	0.328	0.0479	0.344	0.0523	0.358	0.0565
0.011	0.256	0.0299	0.281	0.0358	0.303	0.0413	0.323	0.0465	0.341	0.0514	0.357	0.0561	0.372	0.0605
0.012	0.266	0.0321	0.292	0.0384	0.314	0.0442	0.334	0.0497	0.353	0.0548	0.369	0.0597	0.384	0.0644
0.013	0.275	0.0343	0.301	0.0408	0.325	0.0470	0.345	0.0527	0.364	0.0581	0.381	0.0633	0.396	0.0681
0.014	0.283	0.0363	0.311	0.0433	0.334	0.0497	0.355	0.0557	0.374	0.0613	0.392	0.0667	0.407	0.0717
0.015	0.292	0.0384	0.319	0.0456	0.344	0.0523	0.365	0.0585	0.384	0.0644	0.402	0.0699	0.418	0.0752
0.016	0.299	0.0404	0.328	0.0479	0.353	0.0548	0.374	0.0613	0.394	0.0674	0.412	0.0731	0.428	0.0785
0.017	0.307	0.0423	0.336	0.0501	0.361	0.0573	0.383	0.0640	0.403	0.0703	0.421	0.0762	0.437	0.0817
0.018	0.314	0.0442	0.344	0.0523	0.369	0.0597	0.392	0.0667	0.412	0.0731	0.430	0.0791	0.446	0.0848
0.019	0.321	0.0461	0.351	0.0544	0.377	0.0621	0.400	0.0692	0.420	0.0758	0.438	0.0820	0.455	0.0878
0.020	0.328	0.0479	0.358	0.0565	0.384	0.0644	0.407	0.0717	0.428	0.0785	0.446	0.0848	0.463	0.0908

3.6.4 AS 3600-2009 and ACI 318-08

3.6.4.1 Instantaneous deflection

According to AS 3600-2009, the instantaneous or short-term deflection of a beam may be calculated using the mean value of the elastic modulus of concrete at the time of first loading, E_c, together with *the effective second moment of area* of the span I_{ef}. The effective second moment of area involves an empirical adjustment of the second moment of area of a cracked member to account for tension stiffening. For a given cross-section, I_{ef} is calculated using Branson's formula (Ref. 17):

$$I_{ef} = I_{cr} + (I_{uncr} - I_{cr})(M_{cr}/M_s^*)^3 \leq I_{ef.max} \tag{3.18}$$

where I_{cr} is the second moment of area of the fully-cracked section (calculated using modular ratio theory); I_{uncr} is the second moment of area of the uncracked cross-section about its centroidal axis; M_s^* is the maximum bending moment at the section, based on the short-term serviceability design load or the construction load; M_{cr} is the cracking moment given by:

$$M_{cr} = Z(f'_{ct.f} - \sigma_{cs} + P/A) + Pe \geq 0.0 \tag{3.19}$$

where Z is the section modulus of the uncracked section, referred to the extreme fibre at which cracking occurs; $f'_{ct.f}$ is the characteristic flexural tensile strength of concrete specified as $f'_{ct.f} = 0.6\sqrt{f'_c}$; P is the effective prestressing force (if any); e is the eccentricity of prestress measured to the centroidal axis of the section; A is the area of the uncracked cross-section; and σ_{cs} is the maximum shrinkage-induced tensile stress on the uncracked section at the extreme fibre at which cracking occurs. In the absence of more refined calculation, σ_{cs} may be taken as:

$$\sigma_{cs} = \left(\frac{2.5\rho_w - 0.8\rho_{cw}}{1 + 50\rho_w} E_s \varepsilon^*_{sh}\right) \tag{3.20}$$

where ρ_w is the web reinforcement ratio for the tensile steel $(A_{st} + A_{pt})/b_w d$; ρ_{cw} is the web reinforcement ratio for the compressive steel, if any $(A_{sc}/b_w d)$; A_{st} is the area on non-prestressed tensile reinforcement; A_{pt} is the area of prestressing steel in the tensile zone; A_{sc} is the area of non-prestressed compressive reinforcement; E_s is the elastic modulus of the steel in MPa; and ε^*_{sh} is the final design shrinkage strain (after 30 years). For non-prestressed members, the maximum value of I_{ef} at any cross-section in Eq. 3.18 is $I_{ef.max} = I_{uncr}$ when $\rho = A_{st}/bd \geq 0.005$ and $I_{ef.max} = 0.6 I_{uncr}$ when $\rho < 0.005$.

For a simple-supported beam or slab, the value of I_{ef} for the member is determined from the value of I_{ef} at mid-span. For interior spans of continuous beams or slabs, I_{ef} is half of the value of I_{ef} at mid-span plus one quarter of the value of I_{ef} at each support, while for end spans of continuous beams or slabs, I_{ef} is half the mid-span value plus half the value at the continuous support. For a cantilever, I_{ef} is the value of I_{ef} at the support.

The term σ_{cs} is introduced into Eq. 3.19 to allow for the reduction in the cracking moment that inevitably occurs because of the restraint to shrinkage provided by the bonded tensile reinforcement. Shrinkage-induced tension often causes extensive time-dependent cracking, particularly in lightly loaded members. This time-dependent cracking may occur weeks and months after the member is first loaded and must be accounted for if a meaningful estimate of deflection is required. Eq. 3.20 is based on the expression originally proposed by Gilbert (Ref. 26) for singly reinforced rectangular sections, where conservative values were assumed for the elastic modulus and the creep coefficient of concrete and about 70 per cent of the final shrinkage was considered in the calculation of M_{cr}. The expression was modified to accommodate the inclusion of compressive reinforcement (Ref. 27). The calculation of σ_{cs} for more general cross-sections is discussed in detail in Ref. 28. Where appropriate, the additional tension that may arise due to restraint to shrinkage provided by the supports of a beam or slab should also be taken into account.

The allowance for shrinkage-induced tension is particularly important in the case of lightly reinforced members (including slabs) where the tension induced by the full service moment alone might not be enough to cause cracking. In such cases, failure to account for shrinkage may lead to deflection calculations in which cracking is not adequately taken into account and this may lead to gross underestimates of the actual deflection. For heavily reinforced sections, the problem is not as significant, because the service loads are usually well in excess of the cracking load and tension stiffening is not as significant.

ACI 318-08 also specifies Eq. 3.18, except that the gross second moment of area, I_g, is used rather than I_{uncr} and, with M_{cr} given by $Z(0.62\sqrt{f_c'})$, it does not include any allowance for shrinkage-induced tension or the loss of stiffness caused by cracking due to early shrinkage.

For many slabs, cracking will occur within weeks of casting due to early drying shrinkage and temperature changes, often well before the slab is exposed to its full service loads. In a recent comparison of the instantaneous deflection predicted by ACI 318-08 and the measured deflections of lightly reinforced concrete slabs (Ref. 18), ACI 318-08 significantly underestimated the instantaneous deflection after cracking of every slab and, for the very lightly reinforced slabs ($\rho < 0.003$), the deflection was grossly underestimated. In addition, ACI 318 does not model the abrupt change in direction of the moment-deflection response at first cracking and does not predict the correct shape of the moment-deflection plot after cracking (Ref. 18).

3.6.4.2 *Time-dependent deflection*

For the calculation of long-term deflection, one of two approaches is specified in AS 3600-2009. For reinforced or prestressed beams, the creep and shrinkage deflections can be calculated separately (using the material data specified in the standard and the principles of mechanics). Alternatively, for reinforced concrete beams and slabs, the additional long-term deflection caused by creep and shrinkage may be crudely approximated by multiplying the short-term or immediate deflection caused by the sustained load by a multiplier k_{cs} given by:

$$k_{cs} = [2 - 1.2(A_{sc}/A_{st})] \geq 0.8 \tag{3.21}$$

where A_{sc} is the area of steel in the compressive zone of the cracked section between the neutral axis and the extreme compressive fibre and the ratio A_{sc}/A_{st} is taken at mid-span for a simple or continuous span and at the support for a cantilever.

ACI 318-08 also specifies that unless a more comprehensive analysis is undertaken the additional long-term deflection resulting from the creep and shrinkage of flexural members may be determined by multiplying the immediate deflection caused by the sustained load by a multiplier λ_Δ given by:

$$\lambda_\Delta = \frac{\xi}{(1 + 50A_{sc}/bd)} \tag{3.22}$$

where A_{sc}/bd is taken at mid-span for a simple or continuous span and at the support for a cantilever; and the time-dependent factor ξ depends on the duration of loads, with $\xi = 2.0$ for 5 years or more; $\xi = 1.4$ for 1 year; $\xi = 1.2$ for 6 months; and $\xi = 1.0$ for 3 months.

The use of a deflection multiplier (k_{cs} or λ_Δ) to calculate time-dependent deflections is simple and convenient and, provided the section is initially cracked under short-term loads, it may provide a 'ball-park' estimate of the final deflection. However, to calculate the shrinkage-induced deflection by multiplying the load-induced short-term deflection by a long-term deflection multiplier is fundamentally incorrect. Shrinkage can cause significant deflection even in unloaded members. The approach ignores many of the factors that influence the final deflection, including the creep and shrinkage characteristics of the concrete, the environment and the age at first loading. At best, it must be seen as providing a very approximate estimate. At worst, it is misleading and not worth the time involved in making the calculation. In addition, when using the ACI 318-08 multiplier, no account is taken of the loss of stiffness caused by shrinkage-induced cracking at any stage of the calculation procedure and, in the case of slabs, this may be very significant.

It is, however, not too much more complicated to calculate long-term creep and shrinkage deflections separately and this is the subject of much of the rest of this book. A recently proposed simplified method for deflection calculation that does calculate the creep and shrinkage deflections separately (and more accurately) is outlined in Section 3.6.6.

Example 3.2

The final long-term deflection $(v_C)_{max}$ at the mid-point C in the end span of a continuous one-way slab is to be calculated using the simplified procedures in AS 3600-2009 and ACI 318-08. The slab is 180 mm thick. The span is $\ell = \ell_{ef} = 5.0$ m, the dead load $g = 2.0$ kPa + self-weight = 6.4 kPa, and the live load is $q = 3.0$ kPa. The short-term expected live load is $q_1 = 2.1$ kPa and the sustained part of the live load is $q_2 = 1.2$ kPa. The deflection requirement is that $(v_C)_{max} \leq \ell_{ef}/250 = 20$ mm.

The concrete and steel strengths are $f'_c = 25$ MPa, $f'_{ct.f} = 0.6\sqrt{f'_c} = 3.0$ MPa and $f_y = 500$ MPa. The elastic moduli for concrete and steel are $E_c = 26{,}700$ MPa and $E_s = 200{,}000$ MPa, and the modular ratio is therefore $n = E_s/E_c = 7.5$. The final long-term shrinkage strain is taken to be $\varepsilon^*_{sh} = 0.0007$ and the clear concrete cover to the reinforcement is 20 mm.

The bottom longitudinal reinforcement throughout the span consists of 12 mm diameter bars at 250 mm centres ($A_{st} = 440$ mm²/m at $d = 154$ mm) and the top reinforcement at the interior support consists of 12 mm bars at 200 mm centres ($A_{st} = 550$ mm²/m at $d = 154$ mm), as shown in Fig. 3.6. The reinforcement has been designed to provide adequate strength. Also shown in Fig. 3.6 is the bending moment diagram under the maximum uniformly distributed *service* load, $w_{max} = g + q_1 = 8.5$ kPa.

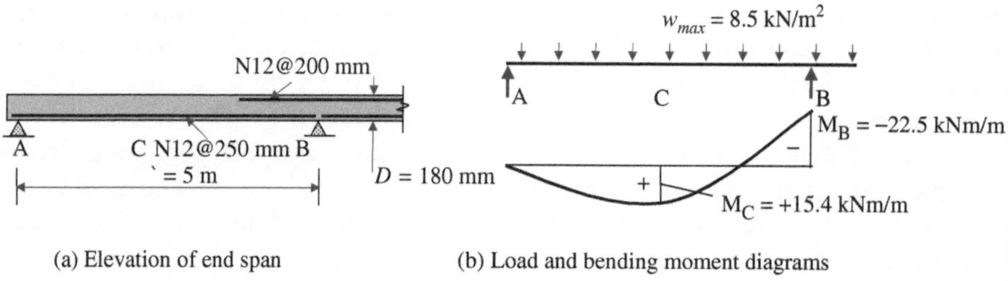

Figure 3.6 Slab elevation, load and bending moment diagrams (Example 3.2).

AS3600-2009

Section at mid-span C

The maximum in-service moment at mid-span is $M_s^* = 15.4$ kNm/m. With $A_{st} = 440$ mm²/m and $d = 154$ mm, the reinforcement ratio $\rho = A_{st}/bd = 0.00286$ and $n\rho = 0.0214$, the centroidal axis of the uncracked transformed cross-section is 91.0 mm below the top surface and the second moment of area of the uncracked transformed section about its centroidal axis is $I_{uncr} = 498 \times 10^6$ mm⁴/m. The depth to the neutral axis for the cracked section is calculated using Eq. 3.14:

$$k = \sqrt{(0.0214)^2 + 2 \times 0.0214} - 0.0214 = 0.187$$

and therefore $d_n = kd = 28.7$ mm. The second moment of area of the fully-cracked transformed section, I_{cr}, is next obtained from Eq. 3.17 as:

$$I_{cr} = \tfrac{1}{2} b d^3 k^2 (1 - k/3) = 59.6 \times 10^6 \text{ mm}^4/\text{m}.$$

The bottom fibre section modulus of the uncracked section is $Z = I_{uncr}/y_b = 5.59 \times 10^6$ mm³/m. From Eq. 3.20, the shrinkage-induced tension in the bottom fibre of the uncracked section is:

$$\sigma_{cs} = \left(\frac{2.5 \times 0.00286}{1 + 50 \times 0.00286} \times 2 \times 10^5 \times 0.0007 \right) = 0.875 \text{ MPa}$$

and the time-dependent cracking moment is obtained from Eq. 3.19 is:

$$M_{cr} = 5.59 \times 10^6 (3.00 - 0.875) \times 10^{-6} = 11.9 \text{ kNm/m}$$

From Eq. 3.18, the effective second moment of area at mid-span is:

$$(I_{ef})_C = 59.6 \times 10^6 + (498 \times 10^6 - 59.6 \times 10^6)(11.9/15.4)^3$$
$$= 261 \times 10^6 \text{ mm}^4/\text{m}$$

Section at continuous B

The maximum in-service moment at support B is $M_s^* = -22.5$ kNm/m. At this critical section, $A_{st} = 550$ mm^2/m, $d = 154$ mm, $\rho = 0.00357$ and $n\rho = 0.0268$. The section properties calculated as above are $I_{uncr} = 500 \times 10^6$ mm^4/m, $k = 0.206$, and $I_{cr} = 72.2 \times 10^6$ mm^4/m. From Eq. 3.20, $\sigma_{cs} = 1.06$ MPa and the cracking moment is $M_{cr} = 10.9$ kNm/m. From Eq. 3.18, $(I_{ef})_B = 121 \times 10^6$ mm^4/m.

For entire span

The average effective second moment of area for the entire end span is taken as the average of the values at mid-span and at the continuous support. That is:

$$I_{ef} = 0.5 \times ((I_{ef})_C + (I_{ef})_B) = 0.5 \times (261 \times 10^6 + 121 \times 10^6)$$
$$= 191 \times 10^6 \text{ mm}^4/\text{m}$$

Instantaneous deflection

Assuming a uniform average rigidity for the span of $E_c I_{ef}$, the instantaneous curvatures at the sections at mid-span and at the continuous support under the full service load after the effects of all cracking have been included are:

$$(\kappa_i)_C = \frac{M_s^*}{E_c I_{ef}} = \frac{15.4 \times 10^6}{26{,}700 \times 191 \times 10^6} = 3.02 \times 10^{-6} \text{ mm}^{-1}$$

and $(\kappa_i)_B = -4.41 \times 10^{-6}$ mm^{-1}

The instantaneous deflection at mid-span due to the full service load after accounting for all load and shrinkage-induced cracking is obtained from Eq. 3.7b as:

$$(v_C)_{i.\max} = \frac{\ell^2}{96} (\kappa_A + 10\,\kappa_C + \kappa_B)$$
$$= \frac{5000^2}{96} (0 + 10 \times 3.02 \times 10^{-6} - 4.41 \times 10^{-6}) = 6.7 \text{ mm}$$

Time-dependent deflection

In this example, the sustained load is $w_{sus} = g + q_2 = 7.6$ kPa. The instantaneous deflection due to the sustained loads is therefore:

$$(v_C)_{i.sus} = (v_C)_{i.max}(w_{sus}/w_{max}) = 6.0 \text{ mm}$$

With no compressive reinforcement, the time-dependent deflection multiplication factor given by Eq. 3.21 is $k_{cs} = 2.0$ and the time-dependent deflection at mid-span is:

$$(v_C)_{time} = k_{cs}(v_C)_{i.sus} = 2.0 \times 6.0 = 12.0 \text{ mm}$$

The final long-term deflection

The final long-term deflection at mid-span $(v_C)_{max}$ is the sum of the instantaneous deflection due to the maximum service load and the time-dependent deflection due to creep and shrinkage and is given by:

$$(v_C)_{max} = (v_C)_{i.max} + (v_C)_{time} = 6.7 + 12.0 = 18.7 \text{ mm} = \text{span}/267$$

Discussion

According to AS3600-2009, this slab will deflect a little less than the maximum deflection limit of span/250 and so the slab is just serviceable. A slightly thinner slab may be possible with the inclusion of some compressive reinforcement in the positive moment region.
It is of interest to note that the deemed to comply maximum span-to-depth ratio in AS 3600-2009 is more conservative requiring minimum slab thickness of 207 mm, if deflections were not to be checked by calculation (see Example 3.1).

ACI318-08

Section at mid-span C

The second moment of area of the gross cross-section is $I_g = 486 \times 10^6$ mm^4/m. As calculated previously, with A_{st} = 440 mm^2/m and d = 154 mm, the second moment of area of the fully-cracked transformed section is I_{cr} = 59.6 × 10^6 mm^4/m. The bottom fibre section modulus of the gross section is $Z = I_g/y_b = 5.4 \times 10^6$ mm^3/m and the cracking moment is $M_{cr} = Z(0.62\sqrt{f'_c}) = 5.4 \times 10^6(0.62\sqrt{25}) \times 10^{-6} = 16.7$ kNm/m. According to ACI 318-08, with $M_s^* = 15.4$ kNm/m, the slab will not crack in the positive moment region at any stage throughout its life and $(I_{ef})_C = I_g = 486 \times 10^6$ mm^4/m – a conclusion that experience would suggest is unsupportable.

Section at continuous support B

With $M_s^* = -22.5$ kNm/m, A_{st} = 550 mm^2/m, d= 154 mm, I_{cr} = 72.2×10^6 mm^4/m and M_{cr} = 16.7 kNm/m, Eq. 3.18 gives $(I_{ef})_B$ = 243 × 10^6 mm^4/m.

For entire span

According to ACI 318-08, the average effective second moment of area for the entire span may be taken as either the value at mid-span or the average of the values obtained at the positive and negative moment regions. For this typical example, taking the value at mid-span would be unconservative (and unwise), as the effects of any cracking would be completely ignored. Assuming that I_{ef} is the average of the values at mid-span and at the continuous support gives $I_{ef} = 364 \times 10^6$ mm^4/m.

Instantaneous deflection

Assuming a uniform average rigidity for the span of $E_c I_{ef}$, the instantaneous curvatures at the section at mid-span and at the continuous support under the full service load are $(\kappa_i)_C = 1.58 \times 10^{-6}$ mm^{-1} and $(\kappa_i)_B = -2.31 \times 10^{-6}$ mm^{-1}. From Eq. 3.7b, the instantaneous deflection at mid-span due to the full service load is:

$$(v_C)_{i.\max} = \frac{\ell^2}{96}(\kappa_A + 10\,\kappa_C + \kappa_B)$$

$$= \frac{5000^2}{96}(0 + 10 \times 1.58 \times 10^{-6} - 2.31 \times 10^{-6}) = 3.5 \text{ mm}$$

Time-dependent deflection

In this example, $(v_C)_{i.sus} = (v_C)_{i.max}(w_{sus}/w_{max}) = 3.1$ mm and with no compressive reinforcement, the time-dependent deflection multiplication factor given by Eq. 3.22 for a load duration of more than 5 years is $\lambda_\Delta = 2.0$ and the time-dependent deflection at mid-span is:

$$(v_C)_{time} = \lambda_\Delta (v_C)_{i.sus} = 2.0 \times 3.1 = 6.2 \text{ mm}$$

The final long-term deflection

The final long-term deflection at mid-span $(v_C)_{max}$ is therefore:

$$(v_C)_{max} = (v_C)_{i.max} + (v_C)_{time} = 3.5 + 6.2 = 9.7 \text{ mm} = \text{span}/515$$

Discussion

According to ACI 318-08, the maximum final long-term deflection of this 180 mm thick slab is less than half of that calculated by AS 3600-2009 and a significantly thinner slab would be possible if the maximum final deflection were to be limited to span/250. Clearly, the loss of stiffness due to shrinkage-induced cracking is ignored by ACI 318-08. Comparison of the ACI 318-08 predictions with the observed deflections in real slabs confirms that often this method grossly underestimates deflection.

By contrast, the minimum thickness specified for this slab in ACI 318-08 is 230 mm (see Example 3.1) and the selection of a slab thickness that met this requirement would be a very conservative design decision.

3.6.5 Eurocode 2

3.6.5.1 Instantaneous curvature

An alternative approach for the calculation of deflection in a cracked reinforced concrete member is specified in Eurocode 2 (Ref. 6). This approach involves the calculation of the curvature at particular cross-sections and then twice integrating the curvatures along the member to obtain its deflection. The instantaneous curvature at a section after cracking, κ, is calculated as a weighted average of the values calculated on a fully-cracked section (κ_{cr}) and on an uncracked section (κ_{uncr}) as follows:

$$\kappa = \zeta\kappa_{cr} + (1-\zeta)\kappa_{uncr} \tag{3.23}$$

where ζ is a distribution coefficient that accounts for the moment level and the degree of cracking and, for members containing deformed bars, is given by:

$$\zeta = 1 - \beta\left(\frac{\sigma_{sr}}{\sigma_s}\right)^2 \tag{3.24a}$$

where β is a coefficient to account for the effects of duration of loading or repeated loading on the average deformation and equals 1.0 for a single, short-term load and 0.5 for sustained loading or for many cycles of repeated loading; σ_{sr} is the stress in the tensile reinforcement at the load causing first cracking that is calculated by ignoring concrete in tension and ignoring any shrinkage-induced tension; σ_s is stress in the reinforcement at the load under consideration that is calculated by ignoring concrete in tension; κ_{cr} is the curvature at the section calculated by ignoring concrete in tension; and κ_{uncr} is the curvature on the uncracked transformed section. If the compressive concrete and the reinforcement are both linear and elastic, the ratio σ_{sr}/σ_s in Eq. 3.24a can be replaced by M_{cr}/M_s^* and therefore:

$$\zeta = 1 - \beta\left(\frac{M_{cr}}{M_s^*}\right)^2 \tag{3.24b}$$

The cracking moment M_{cr} is the moment required to produce an extreme fibre tensile stress equal to the mean uniaxial tensile strength of concrete, f_{ctm}. For a reinforced concrete section in pure bending $M_{cr} = Zf_{ctm}$, where Z is the section modulus of the uncracked section, referred to the extreme fibre at which cracking occurs, and $f_{ctm} = 0.3(f'_c)^{2/3}$ when $f'_c \leq 50$ MPa.

The introduction of $\beta = 0.5$ in Eqs 3.24 for long-term deflection calculations is tantamount to reducing the cracking moment by about 30 per cent and is a crude way of accounting for shrinkage-induced tension and time-dependent cracking.

3.6.5.2 *Time-dependent curvature*

For sustained loads, Eurocode 2 specifies that the total load-induced deformation including creep may be calculated by using an *effective modulus* for concrete, E_e, given by:

$$E_e = \frac{E_c}{1+\varphi(\infty,\tau)} \qquad (3.25)$$

where $\varphi(\infty,\tau)$ is the creep coefficient at time infinity due to a load first applied at time τ.

The shrinkage-induced curvature κ_{sh} is obtained from the following expression:

$$\kappa_{sh} = \varepsilon_{sh} n_e \frac{S}{I} \qquad (3.26)$$

where ε_{sh} is the free shrinkage strain; n_e is the effective modular ratio (E_s/E_e); S is the first moment of area of the reinforcement about the centroid of the section; and I is the second moment of area of the section. S and I and the corresponding curvatures should be calculated for both the uncracked condition and for the fully-cracked condition, and the final average shrinkage-induced curvature in the cracked region is then calculated using Eq. 3.23.

Example 3.3

Using the Eurocode 2 approach, the final long-term deflection $(v_C)_{max}$ is to be calculated at the mid-point C in the end span of the one-way slab of Example 3.2 that is illustrated in Fig. 3.6. For convenience, details of the slab are included here: slab thickness, $D = 180$ mm; span $\ell = \ell_{ef} = 5.0$ m; clear cover = 20 mm; dead load (including self-weight), $g = 6.4$ kPa; live load, $q = 3.0$ kPa; short-term expected live load, $q_1 = 2.1$ kPa; sustained part of the live load, $q_2 = 1.2$ kPa; the deflection requirement, $(v_C)_{max} \leq \ell_{ef}/500 = 20$ mm; and the reinforcement details are outlined in Example 3.2. The material properties are: $f'_c = 25$ MPa; $f_y = 500$ MPa; $E_c = 26{,}700$ MPa; $E_s = 200{,}000$ MPa; $n = E_s/E_c = 7.5$; $\varepsilon^*_{sh} = 0.0007$ and the final creep coefficient is $\varphi^*(\tau) = 2.5$. The mean uniaxial tensile strength of concrete is $f_{ctm} = 0.3(f'_c)^{2/3} = 2.57$ MPa.

Instantaneous deflection – Eurocode 2

Section at mid-span C

As given in Example 3.2, the maximum in-service moment at mid-span is $M^*_s = 15.4$ kNm/m. With $A_{st} = 440$ mm^2/m and $d = 154$ mm, the second moments of area of the uncracked transformed section and the fully-cracked transformed section are $I_{uncr} = 498 \times 10^6$ mm^4/m and $I_{cr} = 59.6 \times 10^6$ mm^4/m, respectively. The uncracked and fully cracked curvatures are therefore $\kappa_{uncr} = M^*_s/E_cI_{uncr} = 1.16 \times 10^{-6}$ mm^{-1} and $\kappa_{cr} = M^*_s/E_cI_{cr} = 9.68 \times 10^{-6}$ mm^{-1}, respectively.

The cracking moment is $M_{cr} = Zf_{ctm} = 5.59 \times 10^6 \times 2.57 \times 10^{-6} = 14.34$ kNm/m and, with $\beta = 1.0$ for short-term loading, the distribution coefficient ζ is obtained from Eq. 3.24b:

$$\zeta = 1 - \beta\left(\frac{M_{cr}}{M_s^*}\right)^2 = 1 - 1.0 \times (14.34/15.4)^2 = 0.133$$

The instantaneous curvature at mid-span is obtained from Eq. 3.23 as:

$$(\kappa_i)_C = 0.133 \times 9.68 \times 10^{-6} + (1 - 0.133) \times 1.16 \times 10^{-6}$$
$$= 2.29 \times 10^{-6}\ \text{mm}^{-1}$$

Section at continuous support B

At this critical section, $M_s^* = -22.5$ kNm/m, $A_{st} = 550$ mm^2/m, $d = 154$ mm, $I_{uncr} = 500 \times 10^6$ mm^4/m, $I_{cr} = 72.2 \times 10^6$ mm^4/m and $M_{cr} = 14.46$ kNm/m. The uncracked and fully-cracked curvatures at B are therefore $\kappa_{uncr} = -1.68 \times 10^{-6}$ mm^{-1} and $\kappa_{cr} = -11.67 \times 10^{-6}$ mm^{-1}, respectively. From Eq. 3.24b, $\zeta = 1 - 1.0 \times (14.46/22.5)^2 = 0.587$ and the instantaneous curvature at B is:

$$(\kappa_i)_B = 0.587 \times (-11.67 \times 10^{-6}) + (1 - 0.587) \times (-1.68 \times 10^{-6})$$
$$= -7.54 \times 10^{-6}\ \text{mm}^{-1}$$

Deflection at mid-span

The instantaneous deflection at mid-span due to the full service load is again obtained from Eq. 3.7b as:

$$(v_C)_{i.\max} = \frac{\ell^2}{96}(\kappa_A + 10\,\kappa_C + \kappa_B)$$
$$= \frac{5000^2}{96}(0 + 10 \times 2.29 \times 10^{-6} - 7.54 \times 10^{-6}) = 4.0\ \text{mm}$$

of which, the instantaneous deflection due to the sustained load is 3.6 mm and the remaining instantaneous deflection due to the transient live load is $(v_C)_{i.(q1-q2)} = 0.4$ mm.

Time-dependent deflection due to sustained loads – Eurocode 2

The effective modulus of the concrete is obtained from Eq. 3.25 as:

$$E_e = \frac{26700}{1+2.5} = 7630\ \text{MPa}$$

and the effective modular ratio $n_e = E_s/E_e = 26.2$. With the effective modular ratio, the revised section properties of the uncracked and cracked transformed cross-sections at the mid-span and at the continuous support are required.

Section at mid-span C

With the reinforcement area transformed into an equivalent area of the softened concrete (due to creep) using the effective modular ratio, the second moments of area of the uncracked and fully-cracked cross-sections are $I_{uncr} = 529 \times 10^6$ mm^4/m and $I_{cr} = 166 \times 10^6$ mm^4/m, respectively. Due to the sustained loads ($w_{sus} = g + q_2 = 7.6$ kPa), the moment at the section at mid-span is $(M_C)_{sus} = 13.77$ kNm/m and the time-dependent curvatures allowing for creep are $\kappa_{uncr} = (M_C)_{sus}/E_e I_{uncr} = 3.41 \times 10^{-6}$ mm^{-1} and $\kappa_{cr} = (M_C)_{sus}/E_e I_{cr} = 10.8 \times 10^{-6}$ mm^{-1}. With $M_{cr} = 14.34$ kNm/m and with $\beta = 0.5$ for long-term loading, the distribution coefficient ζ is obtained from Eq. 3.24b:

$$\zeta = 1 - \beta \left(\frac{M_{cr}}{(M_C)_{sus}} \right)^2 = 1 - 0.5 \times (14.34/13.77)^2 = 0.458$$

The long-term load-induced curvature at mid-span (including creep) is obtained from Eq. 3.23 as:

$$(\kappa_{sus})_C = 0.458 \times 10.8 \times 10^{-6} + (1 - 0.458) \times 3.41 \times 10^{-6}$$

$$= 6.82 \times 10^{-6} \text{ mm}^{-1}$$

Section at continuous support B

At this critical section, the revised second moments of area are $I_{uncr} = 539 \times 10^6$ mm^4/m and $I_{cr} = 197 \times 10^6$ mm^4/m. The sustained moment at B is $(M_B)_{sus} = -20.12$ kNm/m and the time-dependent curvatures allowing for creep are $\kappa_{uncr} = (M_B)_{sus}/E_e I_{uncr} = -4.89 \times 10^{-6}$ mm^{-1} and $\kappa_{cr} = -13.4 \times 10^{-6}$ mm^{-1}. From Eq. 3.24b, $\zeta = 1 - 0.5 \times (14.46/20.12)^2 = 0.742$ and, the long-term load-induced curvature at B is:

$$(\kappa_{sus})_B = 0.742 \times (-13.4 \times 10^{-6}) + (1 - 0.742) \times (-4.89 \times 10^{-6})$$

$$= -11.2 \times 10^{-6} \text{ mm}^{-1}$$

Deflection at mid-span

The time-dependent deflection at mid-span due to the sustained loads plus creep is obtained from Eq. 3.7b as:

$$(v_C)_{sus} = \frac{\ell^2}{96} \left((\kappa_{sus})_A + 10\,(\kappa_{sus})_C + (\kappa_{sus})_B \right)$$

$$= \frac{5000^2}{96} (0 + 10 \times 6.82 \times 10^{-6} - 11.2 \times 10^{-6}) = 14.8 \text{ mm}$$

Time-dependent deflection due to shrinkage – Eurocode 2

Section at mid-span C

As for the creep analysis, the second moments of area of the uncracked and fully-cracked cross-sections are $I_{uncr} = 529 \times 10^6$ mm^4/m and $I_{cr} = 166 \times 10^6$ mm^4/m, respectively, and the first moment of area of the reinforcement about the centroidal axis of the uncracked and fully-cracked cross-sections are $S_{uncr} = 26.5 \times 10^3$ mm^3/m and $S_{cr} = 46.1 \times 10^3$ mm^3/m, respectively. The shrinkage-induced curvatures are obtained from Eq. 3.26:

$$(\kappa_{sh})_{uncr} = \varepsilon_{sh} n_e \frac{S_{uncr}}{I_{uncr}} = 0.92 \times 10^{-6}\ \text{mm}^{-1}$$

and

$$(\kappa_{sh})_{cr} = \varepsilon_{sh} n_e \frac{S_{cr}}{I_{cr}} = 5.09 \times 10^{-6} \text{mm}^{-1}$$

With $\zeta = 0.458$, as for in the creep analysis, the shrinkage-induced curvature at mid-span is obtained from Eq. 3.23 as:

$$(\kappa_{sh})_C = 0.458 \times 5.09 \times 10^{-6} + (1 - 0.458) \times 0.92 \times 10^{-6}$$

$$= 2.82 \times 10^{-6}\ \text{mm}^{-1}$$

At the supports

At the discontinuous support A, the member is uncracked and so $(\kappa_{sh})_A = 0.92 \times 10^{-6}$ mm^{-1}. In the negative moment region, adjacent to support B, the slab contains both top and bottom reinforcement and $S_{uncr} = -6.2 \times 10^3$ mm^3/m, $I_{uncr} = 588 \times 10^6$ mm^4/m, $S_{cr} = -46.9 \times 10^3$ mm^3/m and $I_{cr} = 204 \times 10^6$ mm^4/m. Therefore, from Eq. 3.26: $(\kappa_{sh})_{uncr} = -0.19 \times 10^{-16}$ mm^{-1} and $(\kappa_{sh})_{cr} = -4.22 \times 10^{-6}$ mm^{-1}. With $\zeta = 0.742$, Eq. 3.23 gives $(\kappa_{sh})_B = -3.18 \times 10^{-6}$ mm^{-1}.

Shrinkage-induced deflection at mid-span

An estimate of the shrinkage-induced deflection is obtained by assuming a parabolic distribution of shrinkage curvature between the end sections and mid-span. Eq. 3.7b gives:

$$(v_C)_{sh} = \frac{\ell^2}{96}\ ((\kappa_{sh})_A + 10\ (\kappa_{sh})_C + (\kappa_{sh})_B)$$

$$= \frac{5000^2}{96}(0.92 + 10 \times 2.82 - 3.18) \times 10^{-6} = 6.8\ \text{mm}$$

The final long-term deflection – Eurocode 2

The final long-term deflection at mid-span $(v_C)_{max}$ is therefore:

$$(v_C)_{max} = (v_C)_{i.q1-q2} + (v_C)_{sus} + (v_C)_{sh}$$

$$= 0.4 + 14.8 + 6.8 = 22.0\ \text{mm} = \text{span}/227$$

Discussion

This result is in reasonable agreement with the final deflection calculated in Example 3.2 using AS 3600-2009, although a little more conservative. Unlike AS 3600-2009 (and ACI 318-08), the approach provides a rational consideration of the long-term deflection due to creep and shrinkage of the concrete. Of course, the final deflection here is heavily dependent on input values of the creep coefficient and the shrinkage strain of concrete – as is the case in real slabs.

The use of the mean concrete tensile strength f_{ctm} instead of the flexural strength indirectly compensates for the tension induced by restrained shrinkage (Ref. 24).

It is of interest to note that deemed-to-comply maximum span-to-depth ratios in Eurocode 2 (Eqs 3.2) suggest that, if a slab thickness greater than 173 mm was selected (see Example 3.1), then the maximum final long-term deflection will not exceed span/250 and deflections need not be calculated. This example suggests that this is not the case and that Eqs 3.2 cannot be relied on to guarantee adequate serviceability.

3.6.6 Recommended simplified approach

It is well known that Branson's Equation (Eq. 3.18) overestimates stiffness after cracking for members containing relatively small quantities of tensile reinforcement (Refs 18, 25 and 26). A much better model for I_{ef} can be developed from the deflection calculation approach in Eurocode 2 (as outlined in Section 3.6.5). Substituting Eq. 3.24b into Eq. 3.23, the instantaneous curvature at a particular section in the cracked region of a member may be calculated from:

$$\kappa = \left[1 - \beta\left(\frac{M_{cr}}{M_s^*}\right)^2\right]\kappa_{cr} + \left[\beta\left(\frac{M_{cr}}{M_s^*}\right)^2\right]\kappa_{uncr} \qquad (3.27)$$

Using the notation of Eq. 3.18, the curvatures in Eq. 3.27 are $\kappa = M_s^*/E_cI_{ef}$, $\kappa_{cr} = M_s^*/E_cI_{cr}$ and $\kappa_{uncr} = M_s^*/E_cI_{uncr}$ and Eq. 3.27 can be re-expressed as:

$$\left(\frac{M_s^*}{E_cI_{ef}}\right) = \left[1 - \beta\left(\frac{M_{cr}}{M_s^*}\right)^2\right]\left(\frac{M_s^*}{E_cI_{cr}}\right) + \left[\beta\left(\frac{M_{cr}}{M_s^*}\right)^2\right]\left(\frac{M_s^*}{E_cI_{uncr}}\right) \qquad (3.28)$$

Eq. 3.28 can be rearranged to give the following expression for I_{ef}:

$$I_{ef} = \frac{I_{cr}}{1 - \beta\left(1 - \dfrac{I_{cr}}{I_{uncr}}\right)\left(\dfrac{M_{cr}}{M_s^*}\right)^2} \le I_{ef.max} \qquad (3.29)$$

This alternative expression for I_{ef} was first proposed by Bischoff (Ref. 25).

3.6.6.1 Instantaneous deflection

The instantaneous curvature κ_i due to the service moment M_s^* at a particular cross-section is $\kappa_i = M_s^*/E_cI_{ef}$, with I_{ef} calculated from Eq. 3.29, where I_{cr} is the second moment of area of the fully-cracked section about its centroidal axis; I_{uncr} is the second moment of area of the uncracked transformed cross-section about its centroidal axis; M_s^* is the maximum bending moment at the section, based on the short-term serviceability design load or the construction load; M_{cr} is the cracking moment given by

$$M_{cr} = Z(f'_{ct.f} + P/A_g) + Pe \geq 0.0 \qquad (3.30)$$

where Z is the section modulus of the uncracked section, referring to the extreme fibre at which cracking occurs; and $f'_{ct.f}$ $(= 0.6\sqrt{f'_c})$ is the characteristic flexural tensile strength of concrete.

An upper limit $I_{ef.max}$ is recommended for the value of I_{ef} calculated from Eq. 3.29, where $I_{ef.max} = I_{uncr}$ when $p = A_{st}/bd \geq 0.005$ and $I_{ef.max} = 0.6I_{uncr}$ when $p < 0.005$. For lightly reinforced members (where $\rho < 0.005$), the value of I_{ef} is very sensitive to the calculated value of M_{cr}. The ratio I_{uncr}/I_{cr} is large and the maximum in-service moment M_s^* may be similar in magnitude to the cracking moment. An upper limit for I_{ef} of $0.6I_{uncr}$ is imposed on such members, because failure to account for cracking due to unanticipated shrinkage restraint, temperature gradients or construction loads can result in significant underestimates of deflection.

The term β in Eq. 3.29 is used to account for both shrinkage-induced cracking and the reduction in tension stiffening with time. Early shrinkage in the days and weeks after casting will cause tension in the concrete and a reduction in the cracking moment. As time progresses and the concrete continues to shrink, the level of shrinkage-induced tension increases in an uncracked member, further reducing the cracking moment. If shrinkage has not occurred before first loading, the deflection immediately after loading may be calculated with $\beta = 1.0$. However, in practice, significant shrinkage usually occurs before first loading and β is less than 1.0. When calculating the short-term or elastic part of the deflection, the following is recommended:

$\beta = 0.7$ at early ages (less than 28 days); and
$\beta = 0.5$ at ages greater than 6 months.

For long-term calculations, when the final deflection is to be estimated, $\beta = 0.5$ should be used.

The instantaneous deflection is calculated by assuming a uniform average rigidity for the span of $E_c(I_{ef})_{av}$. For a simple-supported beam or slab, the average value $(I_{ef})_{av}$ for the span is determined from the value of I_{ef} at mid-span. For interior spans of continuous beams or slabs, $(I_{ef})_{av}$ is taken to be half the mid-span value plus one quarter of the value at each support. For end spans of continuous beams or slabs, $(I_{ef})_{av}$ is taken to be half the mid-span value plus half the value at the continuous support. For a cantilever, $(I_{ef})_{av}$ is the value at the support.

3.6.6.2 *Time-dependent creep-induced curvature*

The creep-induced curvature $\kappa_{cr}(t)$ on a particular cross-section at any time t due to a sustained service load first applied at age τ_0 may be obtained from:

$$\kappa_{cr}(t) = \kappa_{sus,0}\frac{\varphi(t,\tau_0)}{\alpha} \qquad (3.31)$$

where $\kappa_{sus,0}$ is the instantaneous curvature due to the sustained service loads; $\varphi(t, \tau_0)$ is the creep coefficient at time t due to load first applied at age τ_0; and α is a creep modification factor that accounts for the effects of cracking and the braking action of the reinforcement on creep and may be estimated from either Eqs 3.32a, b or c (Refs 26 and 30).

For a cracked reinforced concrete section in pure bending ($I_{ef} < I_{uncr}$), $\alpha = \alpha_1$, where:

$$\alpha_1 = \left[0.48\rho^{-0.5}\right]\left[\frac{I_{cr}}{I_{ef}}\right]^{0.33}\left[1 + (125\rho + 0.1)\left(\frac{A_{sc}}{A_{st}}\right)^{1.2}\right] \tag{3.32a}$$

For an uncracked reinforced or prestressed concrete section ($I_{ef} = I_{uncr}$), $\alpha = \alpha_2$, where:

$$\alpha_2 = 1.0 + \left[45\rho - 900\rho^2\right]\left[1 + \frac{A_{sc}}{A_{st}}\right] \tag{3.32b}$$

and $\rho = A_{st}/bd_o$; A_{st} is the equivalent area of bonded reinforcement in the tensile zone at depth d_o (the depth from the extreme compressive fibre to the centroid of the outermost layer of tensile reinforcement); and A_{sc} is the area of the bonded reinforcement in the compressive zone between the neutral axis and the extreme compressive fibre. The area of any bonded reinforcement in the tensile zone (including bonded tendons) not contained in the outermost layer of tensile reinforcement (i.e. located at a depth d_1 less than d_o) should be included in the calculation of A_{st} by multiplying that area by d_1/d_o. For the purpose of the calculation of A_{st}, the tensile zone is that zone that would be in tension due to the applied moment acting in isolation.

For a cracked, partially prestressed section or for a cracked reinforced concrete section subjected to bending and axial compression, α may be taken as:

$$\alpha = \alpha_2 + (\alpha_1 - \alpha_2)\left(\frac{d_{n1}}{d_n}\right)^{2.4} \tag{3.32c}$$

where d_n is the depth of the intact compressive concrete on the cracked section and d_{n1} is the depth of the intact compressive concrete on the cracked section ignoring the axial compression and/or the prestressing force (i.e. the value of d_n for an equivalent cracked reinforced concrete section in pure bending containing the same quantity of bonded reinforcement).

3.6.6.3 *Time-dependent shrinkage-induced curvature*

The shrinkage-induced curvature on a reinforced concrete section is approximated by:

$$\kappa_{sh} = \left[\frac{k_r \varepsilon_{sh}}{D}\right] \tag{3.33}$$

where D is the overall depth of the section; ε_{sh} is the shrinkage strain; and k_r depends on the quantity and location of bonded reinforcement A_{st} and A_{sc}, and may be estimated from Eqs 3.34a–d, as appropriate (Ref. 30).

For a cracked reinforced concrete section in pure bending ($I_{ef} < I_{uncr}$), $k_r = k_{r1}$, where:

$$k_{r1} = 1.2\left(\frac{I_{cr}}{I_{ef}}\right)^{0.67}\left(1 - 0.5\frac{A_{sc}}{A_{st}}\right)\left(\frac{D}{d_o}\right) \tag{3.34a}$$

For an uncracked cross-section ($I_{ef} = I_{uncr}$), $k_r = k_{r2}$, where:

$$k_{r2} = (100\rho - 2500\rho^2)\left(\frac{d_o}{0.5D} - 1\right)\left(1 - \frac{A_{sc}}{A_{st}}\right)^{1.3} \quad \text{when } \rho = A_{st}/bd_o \leq 0.01 \tag{3.34b}$$

$$k_{r2} = (40\rho + 0.35)\left(\frac{d_o}{0.5D} - 1\right)\left(1 - \frac{A_{sc}}{A_{st}}\right)^{1.3} \quad \text{when } \rho = A_{st}/bd_o > 0.01 \tag{3.34c}$$

In Eqs 3.34a, b and c, A_{sc} is defined as the area of the bonded reinforcement on the compressive side of the cross-section. This is a different definition to that provided under Eq. 3.32b. While bonded steel near the compressive face of a cracked cross-section that is at or below the neutral axis will not restrain compressive creep, it will provide restraint to shrinkage and will affect the shrinkage-induced curvature on a cracked section.

For a cracked, partially prestressed section or for a cracked reinforced concrete section subjected to bending and axial compression, k_r may be taken as:

$$k_r = k_{r2} + (k_{r1} - k_{r2})\left(\frac{d_{n1}}{d_n}\right) \tag{3.34d}$$

where d_n and d_{n1} are as defined after Eq. 3.32c.

Eqs 3.32, 3.33 and 3.34 have been developed (see Ref. 30) as empirical fits to results obtained from a parametric study of the creep- and shrinkage-induced changes in curvature on reinforced concrete cross-sections under constant sustained internal actions using the age-adjusted effective modulus method (AEMM) of analysis (as outlined subsequently in Chapter 5).

Example 3.4

Using the approach recommended above (Eqs 3.29–3.34), the final long-term deflection $(v_C)_{max}$ is to be calculated at the mid-point C in the end span of the one-way slab of Example 3.2 that was illustrated in Fig. 3.6. For convenience, details of the slab are included here: $h = 180$ mm; $\ell = \ell_{ef} = 5.0$ m; clear cover = 20 mm; dead load (including self-weight), $g = 6.4$ kPa; live load, $q = 3.0$ kPa; short-term expected live load, $q_1 = 2.1$ kPa; sustained live load, $q_2 = 1.2$ kPa; and $(v_C)_{max} \leq \ell_{ef}/500 = 20$ mm. The reinforcement details are outlined in Example 3.2 and shown in Fig. 3.6a; and the bending moment diagram under the maximum uniformly distributed *service* load, $w_{max} = g + q_1 = 8.5$ kPa is shown in Fig. 3.6b.

The material properties are $f'_c = 25$ MPa; $f'_{ct.f}$ 3.0 MPa; $f_y = 500$ MPa; $E_c = 26{,}700$ MPa; $E_s = 200{,}000$ MPa; $n = E_s/E_c = 7.5$; $\varepsilon^*_{sh} = 0.0007$ and $\varphi^*(\tau) = 2.5$.

Instantaneous deflection

Section at mid-span C

The maximum in-service moment at mid-span is $M^*_s = 15.4$ kNm/m and with $A_{st} = 440$ mm^2/m, $d = 154$ mm and $p = 0.00286$, the second moments of area of the uncracked transformed section and the fully-cracked transformed section are $I_{uncr} = 498 \times 10^6$ mm^4/m and $I_{cr} = 59.6 \times 10^6$ mm^4/m, respectively. The cracking moment is $M_{cr} = Zf'_{ct.f} = 5.59 \times 10^6 \times 3.0 \times 10^{-6} = 16.8$ kNm/m and, for the calculation of immediate deflection due to loads applied at any time after shrinkage-induced cracking has occurred, take $\beta = 0.5$. The value of I_{ef} for the section at mid-span is obtained from Eq. 3.29:

$$I_{ef} = \frac{59.6 \times 10^6}{1 - 0.5\left(1 - \dfrac{59.6 \times 10^6}{498 \times 10^6}\right)\left(\dfrac{16.8}{15.4}\right)^2} = 125 \times 10^6 \text{ mm}^4/\text{m}$$

Section at continuous support B

At this critical section, $M^*_s = -22.5$ kNm/m, $A_{st} = 550$ mm^2/m, $d = 154$ mm, $I_{uncr} = 500 \times 10^6$ mm^4/m, $I_{cr} = 72.2 \times 10^6$ mm^4/m, $M_{cr} = 16.9$ kNm/m and therefore:

$$I_{ef} = \frac{72.2 \times 10^6}{1 - 0.5\left(1 - \dfrac{72.2 \times 10^6}{500 \times 10^6}\right)\left(\dfrac{16.9}{22.5}\right)^2} = 95.3 \times 10^6 \text{mm}^4/\text{m}$$

Short-term deflection at mid-span

The average value $(I_{ef})_{av}$ for the span is taken as one half the value of I_{ef} at mid-span plus half the value at the continuous support. That is $(I_{ef})_{av} = 0.5\,(125 + 95.3) \times 10^6 = 110 \times 10^6$ mm^4/m. The instantaneous curvature at mid-span and at support B are $(\kappa_i)_C = M^*_s/E_c(I_{ef})_{av} = (15.4 \times 10^6)/(26{,}700 \times 110 \times 10^6) = 5.24 \times 10^{-6}$ mm^{-1} and $(\kappa_i)_B = M^*_s/E_c(I_{ef})_{av} = (-22.5 \times 10^6)/(26{,}700 \times 110 \times 10^6) = -7.66 \times 10^{-6}$ mm^{-1}, respectively. The instantaneous deflection at mid-span due to the full service load is obtained from Eq. 3.7b as:

$$(v_C)_{i.\max} = \frac{\ell^2}{96}\left((\kappa_i)_A + 10\,(\kappa_i)_C + (\kappa_i)_B\right)$$

$$= \frac{5000^2}{96}(0 + 10 \times 5.24 \times 10^{-6} - 7.66 \times 10^{-6}) = 11.7 \text{ mm}$$

Time-dependent deflection

Creep

The sustained service load is $w_{sus} = g + q_2 = 7.6$ kPa $= 0.894w_{max}$, and therefore the sustained moments at mid-span and at the continuous support are $M_{sus} = 13.8$ kNm/m and -20.1 kNm/m, respectively. The instantaneous curvatures at mid-span and at the continuous support due to the sustained loads ($\kappa_{sus,0} = M_{sus}/E_c I_{ef.av}$) are therefore 4.68×10^{-6} mm^{-1} and -6.85×10^{-6} mm^{-1}, respectively. The creep modification factor α for the sections at mid-span and at the continuous support are obtained from Eq. 3.32a as $\alpha = \alpha_1 = [0.48 \times 0.00286^{-0.5}] \times [59.6 \times 10^6/125 \times 10^6]^{0.33} = 7.0$ and $\alpha = \alpha_1 = [0.48 \times 0.00357^{-0.5}] \times [72.2 \times 10^6/95.3 \times 10^6]^{0.33} = 7.3$, respectively. For the cross-section at the continuous support, the bottom (compressive) steel is not located in the compressive zone of the cracked section and so A_{sc} in Eq. 3.32a is zero. From Eq. 3.31 the final creep-induced curvature at the mid-span C and at the continuous support B are $(\kappa_{cr})_C = 4.68 \times 10^{-6} \times 2.5/7.0 = 1.67 \times 10^{-6}$ mm^{-1} and $(\kappa_{cr})_B = -6.85 \times 10^{-6} \times 2.5/7.3 = -2.35 \times 10^{-6}$ mm^{-1}, respectively.

From Eq. 3.7b, the creep-induced deflection is:

$$(v_C)_{cr} = \frac{\ell^2}{96}\left((\kappa_{cr})_A + 10\,(\kappa_{cr})_C + (\kappa_{cr})_B\right)$$

$$= \frac{5000^2}{96}(0 + 10 \times 1.67 \times 10^{-6} - 2.35 \times 10^{-6}) = 3.7 \text{ mm}$$

Shrinkage

For the cracked cross-section at mid-span, Eq. 3.34a gives $k_r = k_{r1} = 1.2 \times (59.6 \times 10^6/125 \times 10^6)^{0.67} \times 180/154 = 0.85$ and the shrinkage-induced curvature is given by Eq. 3.33, $(\kappa_{sh})_C = 0.85 \times 0.0007/180 = 3.32 \times 10^{-6}$ mm^{-1}. For the cracked section at the continuous support, $A_{sc} = 440$ mm^2/m in Eq. 3.34a and $k_r = k_{r1} = 1.2 \times (72.2 \times 10^6/95.3 \times 10^6)^{0.67} \times (1 - 0.5 \times 440/550) \times (180/154) = 0.70$ and the shrinkage-induced curvature is given by Eq. 3.33: $(\kappa_{sh})_B = -0.70 \times 0.0007/180 = -2.72 \times 10^{-6}$ mm^{-1}. For the uncracked, singly-reinforced section at support A, Eq. 3.34b gives $k_r = k_{r2} = (100 \times 0.00286 - 2500 \times 0.00286^2) \times (154/90 - 1) = 0.19$ and, from Eq. 3.33, $(\kappa_{sh})_A = 0.19 \times 0.0007/180 = 0.74 \times 10^{-6}$ mm^{-1}.

If shrinkage-induced deflection is obtained by assuming a parabolic distribution of shrinkage curvature between the end sections and mid-span, Eq. 3.7b gives:

$$(v_C)_{sh} = \frac{\ell^2}{96}\left((\kappa_{sh})_A + 10\,(\kappa_{sh})_C + (\kappa_{sh})_B\right)$$

$$= \frac{5000^2}{96}(0.74 + 10 \times 3.32 - 2.72) \times 10^{-6} = 8.1 \text{ mm}$$

The final long-term deflection

The final long-term deflection at mid-span $(v_C)_{max}$ is therefore:

$$(v_C)_{max} = (v_C)_{i.max} + (v_C)_{cr} + (v_C)_{sh}$$
$$= 11.7 + 3.7 + 8.1 = 23.5 \text{ mm} = \text{span}/213$$

Discussion

The final deflection calculated here is similar to that calculated in Example 3.3 using the procedure in Eurocode 2. However, this is not necessarily the case in all situations. The procedure described in Section 3.6.6 is considered to be the most reliable of the simplified procedures presented in this chapter and will generally provide a conservative estimate of deflections.

3.7 Crack control

3.7.1 General

Reinforced concrete elements crack wherever the tensile stress in the concrete reaches the tensile strength of the concrete. Tensile stress at any location in a concrete structure may be caused by a number of factors, including applied loads, restrained shrinkage, temperature changes, settlement of supports and so on. Cracks formed by axial tensile forces and restrained shrinkage (*direct tension cracks*) often penetrate completely through a member. Cracks caused by bending (*flexural cracks*) occur at the tensile face when the extreme fibre tensile stress reaches the tensile strength of the concrete. Flexural cracks propagate from the extreme tensile fibre through the tensile zone and are arrested at or near the neutral axis. Flexural cracks increase in width as the distance from the tensile reinforcement increases and taper to zero width near the neutral axis. A linear relationship is generally assumed to exist between the crack width at the side or soffit of a member and the distance from the bar. In general, the spacing between flexural cracks is in the range 0.5 to 1.5 times the depth of the member.

Many variables influence the width and spacing of cracks, including the magnitude and duration of loading, the quantity, orientation and distribution of the reinforcement, the cover to the reinforcement, the slip between the reinforcement and the concrete in the vicinity of the crack (that depends on the bond characteristics of the reinforcement), the deformational properties of the concrete (including its creep and shrinkage characteristics) and the size of the member. Considerable variations exist in the crack width from crack to crack and the spacing between adjacent cracks because of random variations in the properties of concrete.

The control of cracking in concrete structures is usually achieved by limiting the stress in the bonded reinforcement crossing a crack to some appropriately low value and ensuring that the bonded reinforcement is suitably distributed within the tensile zone. The limit on the tensile steel stress imposed in design depends on the maximum acceptable crack width. Typical values for maximum acceptable crack widths were given in Table 3.2 in Section 3.4.2. If the maximum acceptable crack width is increased, the maximum permissible tensile steel stress also increases. Building codes usually specify maximum bar spacings for the bonded reinforcement and maximum concrete cover requirements. Some codes specify deterministic procedures for calculating crack widths, with the intention to control cracking by limiting the calculated crack width to some appropriately low value. However, current design procedures to control cracking using conventional steel reinforcement are often overly simplistic and fail to adequately account for the gradual increase in crack widths with time due to shrinkage.

In Chapter 1, Section 1.3.3, the restraint provided to shrinkage by the bonded reinforcement in a reinforced concrete member was discussed, with the concrete compressing the reinforcement as it shrinks and the reinforcement imposing an equal and opposite tensile force on the concrete at the level of the steel (as shown in Fig. 1.14b). This internal tensile restraining force is often significant enough to cause time-dependent cracking. In addition, the connections of a concrete member to other parts of the structure or to the foundations also provide restraint to shrinkage. The tensile restraining force that develops rapidly with time at the restrained ends of the member usually leads to cracking, often within days of the commencement of drying. In a restrained flexural member, shrinkage also causes a gradual widening of flexural cracks and a gradual build-up of tension in the uncracked regions that may lead to additional cracking. The influence of shrinkage on crack widths is not properly considered in the major building codes and is therefore not adequately considered in structural design. As a consequence, excessively wide cracks are a relatively common problem for many reinforced concrete structures.

3.7.2 *Simplified code-oriented approaches for flexural crack control*

Many simplified approaches for crack control in reinforced concrete are available in the literature. Some of the more commonly used methods are presented and evaluated here.

3.7.2.1 *Gergely and Lutz (Ref. 31)*

Based on a statistical analysis of experimental data, Gergely and Lutz proposed the following expression for the maximum crack width w (mm) at the tensile face of a beam or slab containing deformed bars:

$$w = 0.011\frac{D-kd}{d-kd}\sigma_{st1}(c_1 A)^{0.33} \times 10^{-3} \tag{3.35}$$

where c_1 is the concrete cover measured from the tensile face to the centre of the longitudinal bar closest to that face (mm); A is the concrete area surrounding each longitudinal bar and may be taken as the total effective area of the concrete in the tensile zone of the cross-section having the same centroid as the tensile reinforcement divided by the number of longitudinal bars (mm^2); σ_{st1} is the tensile steel stress at the crack (MPa); D is the overall depth of the cross-section; d is the effective depth defined as the distance from the compressive edge of the cross-section to the centroid of the tensile reinforcement; and kd is the depth from the compressive edge to the neutral axis of the cracked section.

3.7.2.2 Frosch (Ref. 32)

The following alternative expression for the maximum crack width w (mm) at the tensile face of a beam or slab containing deformed bars was proposed by Frosch:

$$w = 2\frac{\sigma_{st1}}{E_s}\frac{D - kd}{d - kd}\sqrt{c_1^2 + (s_b/2)^2} \tag{3.36}$$

where s_b is the maximum bar spacing (mm); and E_s is the elastic modulus of the steel (MPa).

3.7.2.3 Eurocode 2 – 1992 (Ref. 5)

According to the 1992 version of Eurocode 2, the design crack width may be calculated from:

$$w = \beta\ s_{rm}\frac{\sigma_{st1}}{E_s}\left(1 - \beta_1\beta_2\left(\frac{\sigma_{sr}}{\sigma_{st1}}\right)^2\right) \tag{3.37}$$

where $\beta_1 = 1.0$ for deformed bars and 0.5 for plain bars; $\beta_2 = 1.0$ for a single, short-term load and 0.5 for repeated or sustained loading; σ_{sr} is the stress in the tensile reinforcement at the loading causing first cracking calculated ignoring the concrete in tension; σ_{st1} is the stress in the reinforcement at the loading under consideration calculated ignoring the concrete in tension; β is the ratio of the design crack width to the average crack width taken as $\beta = 1.7$ for load-induced cracking and, for cracking due to restrained shrinkage, $\beta = 1.3$ when the minimum dimension of the member is less than 300 mm and increases to $\beta = 1.7$ when the minimum dimension is 800 mm or greater; s_{rm} is the average crack spacing given by:

$$s_{rm} = 50 + 0.25\ k_1 k_2 d_b/\rho_{eff} \tag{3.38}$$

in which k_1 is a coefficient equal to 0.8 for deformed bars and 1.6 for plain round bars; k_2 is a coefficient that takes into account the form of the strain distribution on the cross-section and equals 0.5 for bending and 1.0 for direct tension; d_b is the average bar diameter for the tensile reinforcement; and ρ_{eff} is the effective reinforcement ratio, $A_{st}/A_{c.eff}$; A_{st} is the area of tensile reinforcement; and $A_{c.eff}$ is the effective area of the tensile concrete surrounding the tensile reinforcement of depth equal to 2.5 times the distance from the tension face of the section to the centroid of the tensile reinforcement (i.e. $2.5(D - d)$), but not greater than $(D - kd)/3$ or $D/2$.

3.7.2.4 Eurocode 2 – 2004 (Ref. 6)

In the 2004 version of Eurocode 2, the crack width in a reinforced concrete member is calculated from:

$$w = s_{r,\max}(\varepsilon_{sm} - \varepsilon_{cm}) \tag{3.39}$$

where $s_{r,max}$ is the maximum crack spacing; ε_{sm} is the mean strain in the reinforcement at the design loads, including the effects of tension stiffening and any imposed deformations; ε_{cm} is the mean strain in the concrete between the cracks. The difference between the mean strain in the reinforcement and the mean strain in the concrete may be taken as:

$$\varepsilon_{sm} - \varepsilon_{cm} = \frac{\sigma_{st1}}{E_s} - k_t \frac{f_{ct,eff}}{E_s \rho_{eff}}(1 + n\rho_{eff}) \geq 0.6\frac{\sigma_{st1}}{E_s} \tag{3.40}$$

where k_t is a factor that depends on the duration of load and equals 0.6 for short-term loading and 0.4 for long-term loading; n is the modular ratio E_s/E_c; $f_{ct,eff}$ is the mean value of the axial tensile strength of concrete at the time cracking is expected; ρ_{eff} $(= A_s/A_{c,eff})$ and $A_{c,eff}$ are defined under Eq. 3.38. For reinforced concrete sections with bonded reinforcement fixed at reasonably close centres, the maximum final crack width may be calculated from:

$$s_{r,max} = 3.4c + 0.425\ k_1 k_2 d_b/\rho_{eff} \tag{3.41}$$

in which c is the clear cover to the longitudinal reinforcement; and k_1 and k_2 are defined under Eq. 3.38.

Alternatively, cracking is deemed to be controlled by Eurocode 2 (2004) if the quantity of tensile reinforcement in a beam or slab is greater than the minimum value given in Eq. 3.42 and if either the bar diameter and/or the bar spacing is limited to the maximum values given in Tables 3.7 and 3.8, respectively. The minimum area of steel for crack control is

$$A_{s,min} = k_c f_{ct,eff} A_{ct}/f_s \tag{3.42}$$

where k_c depends on the stress distribution prior to cracking and equals 1.0 for pure tension and 0.4 for pure bending; A_{ct} is the cross-sectional area of concrete in the tensile zone, i.e. the area in tension just before the formation of the first crack; and f_s is the maximum stress permitted in the reinforcement immediately after crack formation and is the lesser of the yield stress f_y and the value given in Table 3.7.

If the area of steel in the tension zone exceeds the minimum value given by Eq. 3.42, cracking is deemed to be controlled if either Tables 3.7 or 3.8 are satisfied. The steel stress used in these tables is the steel stress on the cracked section due to the quasi-permanent loads.

Table 3.7 Maximum bar diameters d_b for crack control (Ref. 6)

Steel stress (MPa)	*Maximum bar diameter (mm)*		
	Crack width, $w = 0.4$ mm	*Crack width, $w = 0.3$ mm*	*Crack width, $w = 0.2$ mm*
160	40	32	25
200	32	25	16
240	20	16	12
280	16	12	8
320	12	10	6
360	10	8	5
400	8	6	4
450	6	5	–

Table 3.8 Maximum bar spacing for crack control (Ref. 6)

Steel stress (MPa)	*Maximum bar spacing (mm)*		
	Crack width, $w = 0.4$ mm	*Crack width, $w = 0.3$ mm*	*Crack width, $w = 0.2$ mm*
160	300	300	200
200	300	250	150
240	250	200	100
280	200	150	50
320	150	100	–
360	100	50	–

3.7.2.5 ACI 318-08 (Ref. 4)

For the control of flexural cracking in beams and one-way slabs in normal exposure conditions, ACI 318-08 requires the following.

(i) The area of tensile steel provided at every cross-section should be greater than $A_{st.min}$ where:

$$A_{st.\min} = \frac{0.25\sqrt{f'_c}}{f_y} b_w d \geq 1.4 b_w d / f_y \tag{3.43}$$

(ii) The flexural tensile reinforcement is well distributed within the tensile zone of the cross-section and the spacing of the reinforcing bars closest to the tensile face is less than that given by:

$$s = 380\left(\frac{280}{\sigma_{st1}}\right) - 2.5c \leq 300\left(\frac{280}{\sigma_{st1}}\right) \tag{3.44}$$

where c is the smallest distance from the surface of the reinforcement to the tensile face (mm); σ_{st1} is the calculated stress at the crack in the reinforcement closest to the tensile face at the maximum service load (MPa), but it is permissible to take $\sigma_{st1} = 0.67 f_y$.

For two-way slabs, ACI 318-08 requires that the spacing of the reinforcement in each direction in a solid slab not exceed two times the slab thickness.

Example 3.5

The maximum crack width determined using Eqs 3.35, 3.36, 3.37 and 3.39 are here compared with the measured maximum final crack widths for 12 prismatic, one-way, singly reinforced concrete specimens (six beams and six slabs) that were tested by Gilbert and Nejadi (Ref. 35) under sustained service loads for periods in excess of 400 days. The specimens were simply-supported over a span of 3.5 m with cross-sections shown in Fig. 3.7. All specimens were cast from the same batch of concrete and moist cured prior to first loading at age 14 days. Details of each test specimen are given in Table 3.9.

The time-dependent development of cracking, including the crack spacing and crack width, was measured in each specimen throughout the test. The measured elastic modulus and compressive strength of the concrete at the age of first loading were $E_c = 22{,}820$ MPa and $f_c = 18.3$ MPa, whilst the measured creep coefficient and shrinkage strain associated with the 400-day period of sustained loading were $\varphi(t,\tau) = 1.71$ and $\varepsilon_{sh} = -0.000825$.

Two identical specimens 'a' and 'b' were constructed for each combination of parameters as indicated in Table 3.9, with the 'a' specimens loaded more heavily than the 'b' specimens. The 'a' specimens were subjected to a constant sustained load sufficient to cause a maximum moment at mid-span of between 40 and 50 per cent of the calculated ultimate moment and the 'b' specimens were subjected to a constant sustained mid-span moment of between 25 and 40 per cent of the calculated ultimate moment.

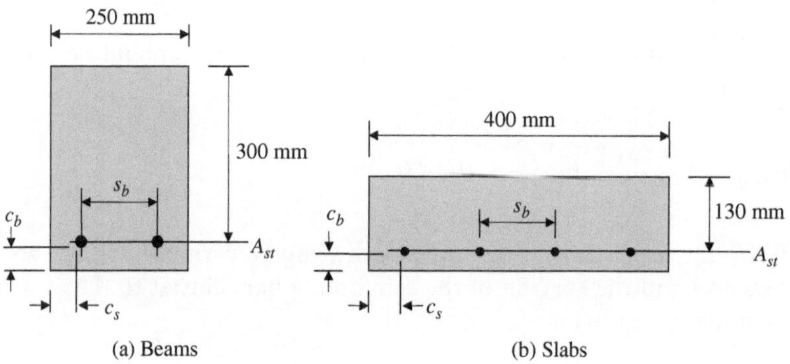

Figure 3.7 Cross-sections of test specimens (Ref. 35).

Table 3.9 Details of the test specimens (Ref. 35)

Beam	*No. of bars*	d_b *(mm)*	A_{st} *(mm²)*	c_b *(mm)*	c_s *(mm)*	s_b *(mm)*	*Slab*	*No. of bars*	d_b *(mm)*	A_{st} *(mm²)*	c_b *(mm)*	c_s *(mm)*	s_b *(mm)*
B1-a	2	16	400	40	40	154	S1-a	2	12	226	25	40	308
B1-b	2	16	400	40	40	154	S1-b	2	12	226	25	40	308
B2-a	2	16	400	25	25	184	S2-a	3	12	339	25	40	154
B2-b	2	16	400	25	25	184	S2-b	3	12	339	25	40	154
B3-a	3	16	600	25	25	92	S3-a	4	12	452	25	40	103
B3-b	3	16	600	25	25	92	S3-b	4	12	452	25	40	103

The loads on all specimens were sufficient to cause primary cracks to develop in the region of maximum moment at first loading. In Table 3.10, the sustained in-service moment at mid-span, M_{sus}, is presented, together with the stress in the tensile steel at mid-span, σ_{st1}, due to M_{sus} (calculated on the basis of a fully cracked section); the calculated ultimate flexural strength, M_u (assuming a characteristic yield stress of the reinforcing steel of 500 MPa); the ratio M_{sus}/M_u; and the cracking moment, M_{cr} (calculated assuming a tensile strength of concrete of $0.6\sqrt{f_c(t)}$, where $f_c(t)$ is the measured compressive strength at the time of loading in MPa).

Table 3.10 Moments and steel stresses in the test specimens (Ref. 35)

Beam	M_{cr} (kNm)	M_{sus} (kNm)	σ_{st1} (MPa)	M_u (kNm)	M_{sus}/M_u (%)	*Slab*	M_{cr} (kNm)	M_{sus} (kNm)	σ_{st1} (MPa)	M_u (kNm)	M_{sus}/M_u (%)
B1-a	14.0	24.9	227	56.2	44.3	S1-a	4.65	6.81	252	13.9	49.0
B1-b	14.0	17.0	155	56.2	30.2	S1-b	4.65	5.28	195	13.9	38.0
B2-a	13.1	24.8	226	56.2	44.1	S2-a	4.75	9.87	247	20.3	48.6
B2-b	13.1	16.8	153	56.2	29.8	S2-b	4.75	6.81	171	20.3	33.6
B3-a	13.7	34.6	214	81.5	42.4	S3-a	4.86	11.4	216	26.4	43.0
B3-b	13.7	20.8	129	81.5	25.5	S3-b	4.86	8.34	159	26.4	31.6

At first loading, a regular pattern of primary cracks developed in each test specimen. With time, the cracks gradually increased in width and additional cracks developed between some of the primary cracks. Thus, the average crack spacing reduced with time. The ratio of final to initial crack spacing ranged from 0.57 to 0.85, with an average value of 0.70. Crack widths increased rapidly in the first few weeks after loading, but the rate of increase slowed significantly after about 2 months. For all specimens, there was little change in the maximum crack width after about 200 days under load. Typical calculations for the maximum crack width are provided here for Beam B2-a.

Beam B2-a:

Typical maximum crack width calculations are provided here for Beam B2-a, with $b = 250$ mm, $D = 333$ mm, $d = 300$ mm, $A_{st} = 400$ mm^2, $E_c = 22820$ MPa, $n = E_s/E_c = 8.76$. A cracked section analysis gives $kd = 78.8$ mm and $I_{cr} = 212 \times 10^6$ mm^4. The applied moment at mid-span is $M_{sus} = 24.8$ kN and the stress in the tensile steel on the cracked section is:

$$\sigma_{st1} = \frac{nM_{sus}(d - kd)}{I_{cr}} = \frac{8.76 \times 24.8 \times 10^6 \times (300 - 78.8)}{212 \times 10^6} = 226 \text{ MPa}$$

Gergely and Lutz:

For B2-a, $c_1 = 33$ mm, $A = (250 \times 66)/2 = 8250$ mm^2 and from Eq. 3.35:

$$w = 0.011 \times \frac{333 - 78.8}{300 - 78.8} \times 226 \times (33 \times 8250)^{0.33} = 0.178 \text{ mm}$$

Frosch:

For B2-a, the bar spacing is $s_b = 184$ mm and from Eq. 3.36:

$$w = 2 \times \frac{226}{200{,}000} \times \frac{333 - 78.8}{300 - 78.8} \times \sqrt{33^2 + (184/2)^2} = 0.254 \text{ mm}$$

Eurocode 2 (1992):

For B2-a, the additional parameters required for the Eurocode 2 (1992) approach are: $\beta = 1.7$; $\beta_1 = 1.0$; $\beta_2 = 0.5$ for sustained loading; $f_{ctm} = 2.08$ MPa; $\sigma_{sr} = (M_{cr}/M_s)\ \sigma_{st1} = 96.8$ MPa; $k_1 = 0.8$; $k_2 = 0.5$; $A_{c.eff} = 2.5 \times 33 \times 250 = 20{,}625$ mm^2; $\rho_{eff} = 400/20{,}625 = 0.0194$; and $s_{rm} = 50 + 0.25 \times 0.8 \times 0.5 \times 16/0.0194 = 132$ mm. From Eq. 3.37:

$$w = 1.7 \times 132 \times \frac{226}{200{,}000}\left(1 - 1.0 \times 0.5 \times \left(\frac{96.8}{226}\right)^2\right) = 0.232 \text{ mm}$$

Eurocode 2 (2004):

For B2-a, the additional parameters required for the Eurocode 2 (2004) approach are: $c = 25$ mm; $f_{ct,eff} = 2.08$ MPa; and from Eq. 3.41, $s_{r,max} = 3.4 \times 25 + 0.425 \times 0.8 \times 0.5 \times 16/0.0194 = 225$ mm. Now $\sigma_{st1}/E_s = 0.00113$ and from Eq. 3.40: $\varepsilon_{sm} - \varepsilon_{cm} = 0.00113 - 0.4 \times 2.08 \times (1 + 8.76 \times 0.0194)/(200{,}000 \times 0.0194) = 0.000881$. From Eq. 3.39:

$$w = 225 \times 0.000881 = 0.198 \text{ mm}$$

Calculated versus measured maximum crack width:

For this specimen, all four calculation methods significantly underestimate the measured maximum final crack width of 0.36 mm. The measured and calculated maximum final crack widths for all 12 beam and slab specimens are compared in Table 3.11. The mean of the ratios of predicted to measured crack widths and the coefficient of variation for each calculation method are also provided for both the beam and the slab specimens.

For the beam specimens, all four calculation methods significantly underestimate the measured maximum crack width. The Eurocode 2 (1992) approach has the lowest coefficient of variation. None of the calculation methods adequately accounts for the very significant time-dependent increase in crack width due to drying shrinkage. In fact, the crack width predicted by each of the methods is independent of the level of drying shrinkage or the creep coefficient.

For the slab specimens, the Gergely and Lutz method (Eq. 3.35) provides a reasonable estimate of maximum crack width for this set of test data. Both Eurocode 2 approaches again have the lowest coefficients of variation, but underestimate the measured test data. The expression proposed by Frosch (Eq. 3.36) tends to overestimate the measured maximum crack width, particularly when the bar spacing is wide (such as in specimens S1-a and S1-b).

Table 3.11 Comparison of measured and predicted maximum final crack widths (mm)

Specimen	*Measured (mm)*	*Gergely and Lutz (Eq. 3.35)*		*Frosch (Eq. 3.36)*		*Eurocode 2 1992 (Eq. 3.37)*		*Eurocode 2 2004 (Eq. 3.39)*	
		Predicted (mm)	*Predicted/ Measured*	*Predicted (mm)*	*Predicted/ Measured*	*Predicted (mm)*	*Predicted/ Measured*	*Predicted (mm)*	*Predicted/ Measured*
B1-a	0.38	0.242	0.638	0.251	0.661	0.242	0.637	0.250	0.658
B1-b	0.18	0.165	0.919	0.171	0.952	0.143	0.796	0.146	0.811
B2-a	0.36	0.178	0.494	0.254	0.707	0.232	0.644	0.198	0.551
B2-b	0.18	0.121	0.670	0.172	0.957	0.138	0.769	0.116	0.645
B3-a	0.28	0.149	0.531	0.141	0.503	0.179	0.638	0.157	0.562
B3-b	0.13	0.089	0.688	0.085	0.651	0.097	0.747	0.082	0.634
	Mean		0.657		0.739		0.705		0.644
Coefficient of variation (%)			22.8		24.5		10.4		14.5
S1-a	0.25	0.254	1.015	0.520	2.081	0.257	1.029	0.218	0.872
S1-b	0.20	0.197	0.984	0.403	2.016	0.176	0.878	0.150	0.749
S2-a	0.23	0.221	0.962	0.274	1.192	0.211	0.917	0.184	0.800
S2-b	0.18	0.153	0.848	0.189	1.051	0.132	0.735	0.113	0.629
S3-a	0.20	0.179	0.895	0.176	0.881	0.160	0.801	0.142	0.708
S3-b	0.15	0.131	0.873	0.129	0.859	0.111	0.738	0.096	0.639
	Mean		0.930		1.346		0.849		0.733
Coefficient of variation (%)			7.2		41.4		13.5		12.8

3.7.3 Tension chord model for flexural cracking in reinforced concrete

A model for predicting the maximum final crack width (w^*) in reinforced concrete flexural members based on the tension chord model of Marti *et al.* (Ref. 32) was recently proposed (Ref. 34). A modified version of that model is presented here and has been shown to provide good agreement with the measured final spacing and width of cracks in reinforced concrete beams and slabs under sustained loads. The notation associated with the model is shown in Fig. 3.8.

Consider a segment of a singly reinforced beam of rectangular section subjected to an in-service bending moment, M_s, greater than the cracking moment, M_{cr}. The spacing between the primary cracks is s, as shown in Fig. 3.8a. A typical cross-section between the cracks is shown in Fig. 3.8b and a cross-section at a primary crack is shown in Fig. 3.8c. The cracked beam is idealised as a compression chord of depth kd and width b and a cracked tension chord consisting of the tensile reinforcement of area A_{st} surrounded by an area of tensile concrete (A_{ct}) as shown in Fig. 3.8d. The centroids of A_{st} and A_{ct} are assumed to coincide at a depth d below the top fibre of the section.

For the sections containing a primary crack (Fig. 3.8c), $A_{ct} = 0$ and the depth of the compressive zone, kd, and the second moment of area about the centroidal axis (I_{cr}) may be determined from a cracked section analysis using modular ratio theory (Eqs 3.14 and 3.17).

Away from the crack, the area of the concrete in the tension chord of Fig. 3.8d (A_{ct}) is assumed to carry a uniform tensile stress (σ_{ct}) that develops due to the bond stress (τ_b) that exists between the tensile steel and the surrounding concrete.

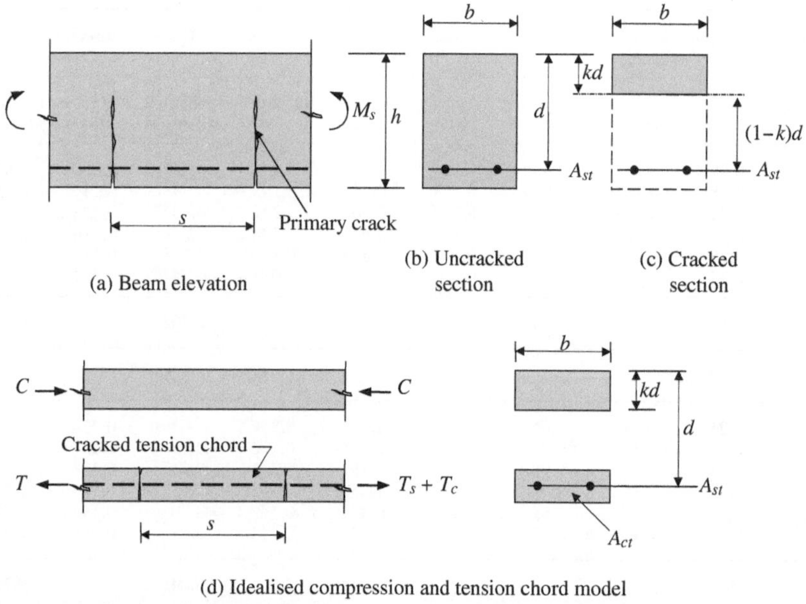

Figure 3.8 Cracked reinforced concrete beam and idealised tension chord model (Ref. 34).

For the tension chord, the area of concrete between the cracks, A_{ct}, may be taken as:

$$A_{ct} = 0.5(D - kd)b^* \tag{3.45}$$

where b^* is the width of the section at the level of the centroid of the tensile steel (i.e. at the depth d) but not greater than the number of bars in the tension zone multiplied by $12d_b$. At each crack, the concrete carries no tension and the tensile steel stress is $\sigma_{st1} = T/A_{st}$, where:

$$T = \frac{nM_s(d - kd)}{I_{cr}} A_{st} \tag{3.46}$$

As the distance z from the crack in the direction of the tension chord increases, the stress in the steel reduces due to the bond shear stress τ_b between the steel and the surrounding tensile concrete. For reinforced concrete under service loads, where σ_{st1} is less than the yield stress f_y, Marti *et al.* (Ref. 33) assumed a rigid-plastic bond shear stress-slip relationship, with $\tau_b = 2.0f_{ct}$ at all values of slip, where f_{ct} is the direct tensile strength of the concrete. To account for the reduction in bond stress with time due to tensile creep and shrinkage, Gilbert (Ref. 34) took the bond stress to be $\tau_b = 2.0f_{ct}$ for short-term calculations and $\tau_b = 1.0f_{ct}$ when the final long-term crack width was to be determined. Experimental observations (Refs 35 and 36) indicate that τ_b reduces as the stress in the reinforcement increases and, consequently, the tensile stresses in the concrete between the cracks reduces (i.e. tension stiffening reduces with increasing steel stress). In reality, the magnitude of τ_b is affected by many factors, including steel stress, concrete cover, bar spacing, transverse reinforcement (stirrups), lateral pressure, compaction of the concrete, size of bar deformations, tensile creep and shrinkage. It is recommended here that in situations where the concrete cover and the clear spacing between the bars are greater than the bar diameter, the bond stress τ_b may be taken as:

$$\tau_b = \lambda_1 \lambda_2 \lambda_3 f_{ct} \tag{3.47}$$

where λ_1 accounts for the load duration with $\lambda_1 = 1.0$ for short-term calculations and $\lambda_1 = 0.7$ for long-term calculations; λ_2 is a factor that accounts for the reduction in bond stress as the steel stress σ_{st1} (in MPa) increases and is given by (Ref. 36):

$$\lambda_2 = 1.6\dot{6} - 0.00\dot{3}\sigma_{st1} \geq 0.0 \tag{3.48}$$

and λ_3 is a factor that accounts for the very significant increase in bond stress that has been observed in laboratory tests for small diameter bars (Ref. 35) and may be taken as:

$$\lambda_3 = 7.0 - 0.3 d_b \geq 2.0 \ \ (d_b \text{ in mm}) \tag{3.49}$$

An elevation of the tension chord is shown in Fig. 3.9a and the stress variations in the concrete and steel in the tension chord are illustrated in Figs 3.9b and 3.9c, respectively. Following the approach of Marti *et al.* (Ref. 33), the concrete and steel tensile stresses in Figs 3.9b and 3.9c, where $0 < z \leq s/2$, may be expressed as:

$$\sigma_{stz} = \frac{T}{A_{st}} - \frac{4\tau_b z}{d_b} \tag{3.50a}$$

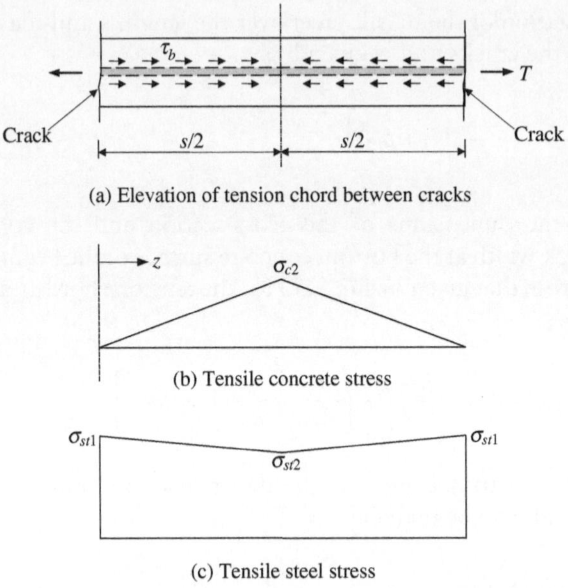

Figure 3.9 Tension chord – actions and stresses (Ref. 33).

and

$$\sigma_{cz} = \frac{4\tau_b \rho_{tc} z}{d_b} \tag{3.50b}$$

where ρ_{tc} is the reinforcement ratio of the tension chord ($= A_{st}/A_{ct}$) and d_b is the reinforcing bar diameter. Mid-way between the cracks, at $z = s/2$, the stresses are:

$$\sigma_{st2} = \frac{T}{A_s} - \frac{2\tau_b s}{d_b} \tag{3.51a}$$

and

$$\sigma_{c2} = \frac{2\tau_b \rho_{tc} s}{d_b} \tag{3.51b}$$

The maximum crack spacing immediately after loading, $s = s_{max}$, occurs when $\sigma_{c2} = f_{ct}$, and from Eq. 3.51b:

$$s_{max} = \frac{f_{ct} d_b}{2\tau_b \rho_{tc}} \tag{3.52}$$

with $\lambda_1 = 1.0$ in Eq. 3.47. If the spacing between two adjacent cracks just exceeds s_{max}, the concrete stress mid-way between the cracks will exceed f_{ct} and another crack will form between the two existing cracks. It follows that the minimum crack spacing is half the maximum value, that is, $s_{min} = s_{max}/2$.

The instantaneous crack width $(w_i)_{tc}$ in the fictitious tension chord is the difference between the elongation of the tensile steel over the length s and the elongation of the concrete between the cracks and is given by:

$$(w_i)_{tc} = \frac{s}{E_s}\left[\frac{T}{A_{st}} - \frac{\tau_b s}{d_b}(1 + n\rho_{tc})\right] \tag{3.53}$$

Depending on the dimensions of the cross-section and the concrete cover, the instantaneous crack width at the bottom concrete surface of the beam or slab, $(w_i)_{soffit}$, may be different from that given by Eq. 3.53 for the tension chord and may be obtained from:

$$(w_i)_{soffit} = k_{cover}(w_i)_{av} = \frac{k_{cover} s}{E_s}\left[\frac{T}{A_{st}} - \frac{\tau_b s}{d_b}(1 + n\rho_{tc})\right] \tag{3.54}$$

where k_{cover} is a term to account for the dependence of crack width on the clear concrete cover c and may be taken as:

$$k_{cover} = \left(\frac{D - kd}{d - kd}\right)\left(\frac{5c}{(D - kd) - 2d_b}\right)^{0.3} \tag{3.55}$$

Under sustained load, additional cracks occur between widely spaced cracks (usually when $0.67s_{max} < s \leq s_{max}$). The additional cracks are due to the combined effect of tensile creep rupture and shrinkage. As a consequence, the number of cracks increases and the maximum crack spacing reduces with time. The final maximum crack spacing s^* is only about two-thirds of that given by Eq. 3.52, but the final minimum crack spacing remains about half of the value given by Eq. 3.52.

As previously mentioned, experimental observations indicate that τ_b decreases with time, probably as a result of shrinkage-induced slip and tensile creep. Hence, the stress in the tensile concrete between the cracks gradually reduces. Furthermore, although creep and shrinkage will cause a small increase in the resultant tensile force T in the real beam and a slight reduction in the internal lever arm, this effect is relatively small and is ignored in the tension chord model presented here. The final crack width is the elongation of the steel over the distance between the cracks minus the extension of the concrete caused by σ_{cz} plus the shortening of the concrete between the cracks due to shrinkage. For a final maximum crack spacing of s^*, the final maximum crack width at the member soffit is:

$$(w^*)_{soffit} = \frac{k_{cover}s^*}{E_s}\left[\frac{T}{A_{st}} - \frac{\tau_b s^*}{d_b}(1 + n_e\rho_{tc}) - \varepsilon_{sh}E_s\right] \tag{3.56}$$

where ε_{sh} is the shrinkage strain in the tensile concrete (and is a negative value); $n_e = E_s/E_e =$ *the effective modular ratio*; E_e is the effective modulus given by $E_e = E_c/(1+\varphi(t,\tau))$; E_c and E_s are the elastic moduli of the concrete and steel respectively; and $\varphi(t,\tau)$ is the creep coefficient of the concrete.

A good estimate of the final maximum crack width is given by Eq. 3.56, where s^* is the maximum crack spacing after all time-dependent cracking has taken place, that is, $s^* = 0.67s_{max}$, and s_{max} is given by Eq. 3.52. By rearranging Eq. 3.56, the steel stress on a cracked section corresponding to a particular maximum final crack width (w^*) is given by:

$$f_{st} = \frac{w^* E_s}{s^* k_{cover}} + \frac{\tau_b s^*}{d_b}(1 + n_e\rho_{tc}) + \varepsilon_{sh}E_s \tag{3.57}$$

By substituting τ_b (from Eq. 3.47) and $s^* = 0.67s_{max}$ into Eq. 3.57 and by selecting a maximum desired crack width in a particular structure, w^*, the maximum permissible tensile steel stress can be determined.

Example 3.6

The maximum final crack widths determined using Eq. 3.56 are compared with the measured maximum final crack widths for the 12 prismatic, one-way singly reinforced concrete specimens (six beams and six slabs) tested by Gilbert and Nejadi (Ref. 35). The specimens were described in Example 3.5 and their cross-sections are shown in Fig. 3.7 and details provided in Tables 3.9 and 3.10. Typical maximum crack width calculations are provided here for Beam B2-a.

Beam B2-a

Relevant dimensions and properties are: $b = 250$ mm, $D = 333$ mm, $d = 300$ mm, $A_{st} = 400$ mm^2, $E_c = 22{,}820$ MPa, $n = E_s/E_c = 8.76$, $\varphi(t, \tau) = 1.71$, $\varepsilon_{sh} = -0.000825$, $E_e = 8420$ MPa and $n_e = 23.8$. A cracked section analysis gives $kd = 78.8$ mm and $I_{cr} = 212 \times 10^6$ mm^4.
From Eq. 3.45, the area of concrete in the tension chord is $A_{ct} = 0.5 \times (333 - 78.8) \times 250 = 31{,}780$ mm^2 and the reinforcement ratio of the tension chord $\rho_{tc} = A_{st}/A_{ct} = 0.0126$. When $M_s = 24.8$ kNm, the tensile force in the steel on the cracked section is Eq. 3.46:

$$T = \frac{8.76 \times 24.8 \times 10^6(300 - 78.8)}{212 \times 10^6} \times 400 = 90.6 \text{ kN}$$

$$\text{and } \sigma_{st1} = \frac{T}{A_{st}} = 226 \text{ MPa}$$

For short-term calculations, $\lambda_1 = 1.0$ and, from Eqs 3.48 and 3.49, $\lambda_2 = 1.6\dot{6} - 0.00\dot{3} \times 226 = 0.91$ and $\lambda_3 = 7.0 - 0.3 \times 16 = 2.2$. With $f_{ct} = 0.6\sqrt{f_c}(t) = 2.57$ MPa, the instantaneous bond stress is obtained from Eq. 3.47 as $\tau_b = 1.0 \times 0.91 \times 2.2 \times 2.57 = 5.15$ MPa and the maximum crack spacing immediately after loading is given by Eq. 3.52:

$$s_{max} = \frac{2.57 \times 16}{2 \times 5.15 \times 0.0126} = 317 \text{ mm}$$

For the calculation of the maximum final crack width, the maximum crack spacing s^* is taken as 2/3 of the instantaneous value and therefore $s^* = 2/3 \times 317 = 211$ mm. From Eq. 3.47, for long-term calculations, $\tau_b = 0.7 \times 0.91 \times 2.2 \times 2.57 = 3.60$ MPa and, from Eq. 3.55, $k_{cover} = 0.967$.
The maximum final (long-term) crack width at the soffit of the beam specimen B2-a is obtained from Eq. 3.56:

$$(w^*)_{soffit} = \frac{0.967 \times 211}{200{,}000}\left[\frac{90{,}600}{400} - \frac{3.60 \times 211}{16}(1 + 23.8 \times 0.0126)\right.$$

$$\left. - (-0.000825) \times 200{,}000\right] = 0.337 \text{ mm}$$

The measured maximum final crack width on this specimen after 400 days under load was 0.36 mm.
The measured and calculated maximum final crack widths for all 12 test specimens are compared in Table 3.12. The mean of the ratios of predicted to measured crack widths for the six beam specimens is 1.059, with a coefficient of variation of 22.8 per cent while for the six slab specimens the mean is 0.945, with a coefficient of variation of 17.7 per cent. The agreement between the calculated and measured maximum final crack width for this set of test data is good.

Table 3.12 Comparison of measured and predicted maximum final crack widths (mm)

Specimen	*B1-a*	*B1-b*	*B2-a*	*B2-b*	*B3-a*	*B3-b*	*S1-a*	*S1-b*	*S2-a*	*S2-b*	*S3-a*	*S3-b*
Measured, w_{max}	0.38	0.18	0.36	0.18	0.28	0.13	0.25	0.20	0.23	0.18	0.20	0.15
Predicted, $(w^*)_{soffit}$	0.425	0.262	0.337	0.207	0.212	0.122	0.287	0.196	0.258	0.155	0.162	0.111
$(w^*)_{soffit}/w_{max}$	1.119	1.457	0.935	1.149	0.758	0.936	1.148	0.981	1.123	0.863	0.812	0.742

Example 3.7

A 150 mm thick simply-supported one-way slab located inside a building is to be considered. With appropriate regard for durability, the concrete strength is selected to be $f'_c = 32$ MPa and the cover to the tensile reinforcement is 20 mm. The final shrinkage strain is taken to be $\varepsilon_{sh} = -0.0006$. Other relevant material properties are $E_c = 28{,}600$ MPa; $n = E_s/E_c = 7.0$; $\varphi(t,\tau) = 2.5$; $f_{ct} = 2.04$ MPa and $E_s = 200$ GPa. The effective modulus is therefore $E_e = E_c/(1+\varphi(t,\tau)) =$ 8170 MPa and the effective modular ratio $n_e = E_s/E_e = 24.5$. The tensile face of the slab is to be exposed and the maximum final crack width is to be limited to $w^* = 0.3$ mm.

After completing the design for strength and deflection control, the required minimum area of tensile steel is 650 mm^2/m. Under the full service loads, the maximum in-service sustained moment at mid-span is 20.0 kNm/m. The bar diameter and bar spacing must be determined so that the requirements for crack control are also satisfied.

Case 1 – Use 10 mm bars at 120 mm centres

$A_{st} = 655$ mm^2/m at $d = 125$ mm and, referring to Figs 3.5 and 3.8, an elastic analysis of the cracked section gives $kd = 29.6$ mm and $I_{cr} = 50.3 \times 10^6$ mm^4. The maximum in-service tensile steel stress on the fully-cracked section at mid-span is calculated using Eq. 3.46 as $\sigma_{st1} = T/A_s = 265$ MPa.

The area of concrete in the tension chord is $A_{ct} = 60{,}200$ mm^2 (Eq. 3.45) and the reinforcement ratio of the tension chord is $\rho_{tc} = A_s/A_{ct} = 0.0109$. From Eqs 3.48 and 3.49, $\lambda_2 = 0.78$ and $\lambda_3 = 4.0$ and from Eq. 3.47, $\tau_b = 6.39$ MPa for short-term calculations ($\lambda_1 = 1.0$) and $\tau_b = 4.47$ MPa for long-term calculations ($\lambda_1 = 0.7$). The maximum final crack spacing s^* is obtained using Eq. 3.52 as $s^* = 0.67 s_{max} = 97.6$ mm and Eq. 3.55 gives $k_{cover} = 1.26$. The maximum permissible steel stress required for crack control is obtained from Eq. 3.57:

$$f_{st} = \frac{0.3 \times 200{,}000}{97.6 \times 1.26} + \frac{4.47 \times 97.6}{10}(1 + 24.5 \times 0.0109)$$

$$+ (-0.0006 \times 200{,}000) = 422 \text{ MPa}$$

The actual stress at the crack $\sigma_{st} = 265$ MPa is much less than $f_{st} = 422$ MPa and, therefore, cracking is easily controlled.

Case 2 – Use 12 mm bars at 170 mm centres

$A_s = 665$ mm^2/m at $d = 124$ mm and, for this section, $kd = 29.6$ mm and $I_{cr} = 50.1 \times 10^6$ mm^4. The maximum in-service tensile steel stress on the fully-cracked section at mid-span is $\sigma_{st1} = T/A_s = 264$ MPa.
The area of concrete in the tension chord is $A_{ct} = 50{,}980$ mm^2 and $\rho_{tc} = A_s/A_{ct} = 0.0130$. Now $\lambda_2 = 0.79$ and $\lambda_3 = 3.4$ and $\tau_b = 5.47$ MPa for short-term calculations ($\lambda_1 = 1.0$) and $\tau_b = 3.83$ MPa for long-term calculations ($\lambda_1 = 0.7$). The maximum final crack spacing is $s^* = 114$ mm and from Eq. 3.55, $k_{cover} = 1.29$. The maximum permissible steel stress required for crack control is obtained from Eq. 3.57 as $f_{st} = 335$ MPa – which is also significantly greater than the actual maximum stress at the crack $\sigma_{st} = 264$ MPa.
Therefore, the final maximum crack width will be less than the maximum permissible value of 0.3 mm.

Case 3 – Use 16 mm bars at 300 mm centres

$A_s = 670$ mm^2/m at $d = 122$ mm and, for this section, $kd = 29.4$ mm and $I_{cr} = 48.6 \times 10^6$ mm^4. The maximum in-service tensile steel stress on the fully-cracked section at mid-span is $\sigma_{st1} = T/A_s = 266$ MPa. The area of concrete in the tension chord is $A_{ct} = 38{,}580$ mm^2 and $\rho_{tc} = A_s/A_{ct} = 0.0174$. Now $\lambda_2 = 0.78$ and $\lambda_3 = 2.2$ and $\tau_b = 3.50$ MPa for short-term calculations ($\lambda_1 = 1.0$) and $\tau_b = 2.45$ MPa for long-term calculations ($\lambda_1 = 0.7$). The maximum final crack spacing is $s^* = 179$ mm and from Eq. 3.55, $k_{cover} = 1.35$. The maximum permissible steel stress required for crack control is $f_{st} = 167$ MPa (from Eq. 3.57), which is much less than the actual steel stress due to the sustained moment of $\sigma_{st} = 266$ MPa.
Therefore, crack control is *not* adequate and the maximum final crack width will exceed 0.3 mm.

Simplified approaches – case 3

The deemed-to-comply crack control provisions of Eurocode 2 (2004) are in reasonable agreement with the above calculations, while the ACI 318-08 (Ref. 4) provisions are not. Considering first the Eurocode 2 provisions, for case 3 above, the area of tensile steel provided exceeds the minimum steel area given by Eq. 3.42 and Table 3.7 suggests that, for a steel stress of $\sigma_{st1} = 266$ MPa, if the maximum crack width is to be limited to 0.3 mm, the maximum bar diameter for crack control is 12 mm.
Considering the ACI 318-08 requirement for Case 3, the area of tensile reinforcement exceeds the minimum value $A_{st.min}$ given by Eq. 3.43, and with a steel stress of $\sigma_{st1} = 266$ MPa, the maximum bar spacing permitted by Eq. 3.44 is 315 mm. With the bar spacing of 300 mm, Case 3 satisfies the requirements for crack control in ACI 318-08.
The maximum crack widths predicted by the Gergely and Lutz expression (Eq. 3.35) and the two Eurocode 2 equations (Eqs 3.37 and 3.39) for case 3 are 0.284 mm, 0.304 mm and 0.244 mm, respectively. These indicate that the

maximum final crack width will be less than (or in the case of Eurocode 2 (1992) only just greater than) the design crack width $w^* = 0.3$ mm. In a structural design situation, all three approaches suggest that case 3 is acceptable.
By contrast, the maximum crack width given by the equation proposed by Frosch (Eq. 3.36) is 0.528 mm, indicating that case 3 will lead to excessively wide cracks.

3.7.4 Model for direct tension cracking in restrained reinforced concrete

Consider the fully restrained member shown in Fig. 3.10a (Refs 37 and 38). As the concrete shrinks, the restraining force $N(t)$ gradually increases until the first crack occurs when $N(t) = A_c f_{ct}(t)$ (usually within a week of the commencement of shrinkage).

At first cracking, the restraining force reduces to N_{cr}, and the concrete stress away from the crack is less than the tensile strength of the concrete f_{ct}. The concrete on either side of the crack shortens elastically and the crack opens to a width $w(t)$ that depends on the area of reinforcement (see Fig. 3.10b). At the crack, the steel carries the entire force N_{cr} and the stress in the concrete is clearly zero. In the region adjacent to the crack, the concrete and steel stresses vary considerably and the bond stress at the steel-concrete interface is high. At some distance s_o on each side of the crack, the concrete and steel stresses are no longer influenced directly by the presence of the crack, as shown in Figs 3.10c and 3.10d.

Referring to Fig. 3.10, numerical values of N_{cr}, σ_{c1}, σ_{s1}, and σ_{s2} can be obtained from (Ref. 37):

$$N_{cr} = \frac{n\rho f_{ct} A_c}{C_1 + n\rho(1 + C_1)} \tag{3.58a}$$

$$\sigma_{c1} = \frac{N_{cr}(1 + C_1)}{A_c} \tag{3.58b}$$

$$\sigma_{s1} = -C_1 \sigma_{s2} \tag{3.58c}$$

$$\sigma_{s2} = \frac{N_{cr}}{A_s} \tag{3.58d}$$

where n is the modular ratio (E_s/E_c); ρ is the reinforcement ratio (A_s/A_c); f_{ct} is the direct tensile strength of the concrete at first cracking; and

$$C_1 = \frac{2s_o}{3L - 2s_o} \tag{3.59}$$

The distance s_o in which stresses vary on either side of a crack depends on those factors that affect the bond stress at the steel concrete interface and include the reinforcement quantity, the bar diameter d_b and the surface characteristics of the bar. In Ref. 37 s_o was taken to be:

$$s_o = \frac{d_b}{10\rho} \tag{3.60}$$

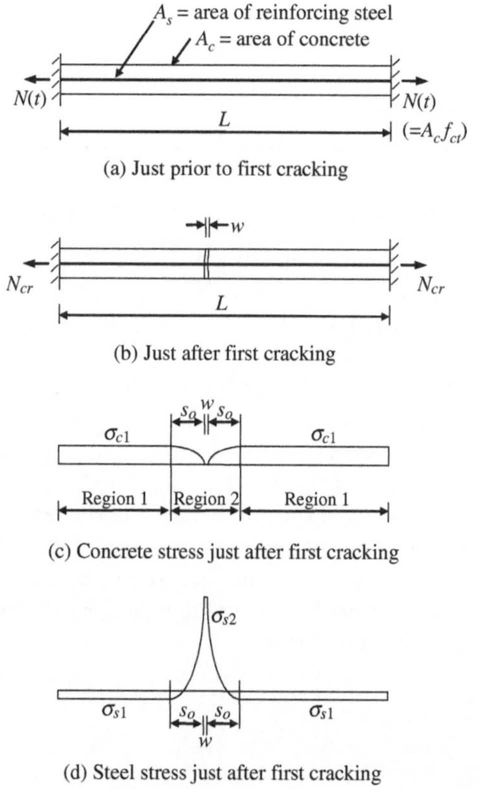

Figure 3.10 First cracking in a restrained direct tension member (Ref. 37).

This expression was used earlier by Favre *et al.* (Ref. 39) and others for a member containing deformed bars or welded wire mesh. Recent experimental results (Ref. 40) suggest that shrinkage causes a deterioration in bond at the steel-concrete interface and a gradual increase in s_o with time. For calculations at first cracking (Eqs 3.58 and 3.59), s_o may be taken from Eq. 3.60, and for final or long-term calculations, the value of s_o from Eq. 3.60 should be multiplied by 1.33 (Ref. 38) (i.e. ($s_o^* = 1.33\ s_o$).

The final number of cracks and the final average crack width depend on the length of the member, the quantity and distribution of reinforcement, the quality of bond between the concrete and steel, the amount of shrinkage and the concrete properties. In Fig. 3.11a, a portion of a restrained direct tension member is shown after all shrinkage has taken place and the final crack pattern is established. The average concrete and steel stresses caused by shrinkage are illustrated in Figs 3.11bbreak and 3.11c.

In many practical situations, the supports of a reinforced concrete member that provide the restraint to shrinkage are not immovable, but instead are adjacent parts of the structure that are themselves prone to shrinkage and other movements. If the supports of the restrained member in Fig. 3.10a suffer a relative movement Δu with time (in the direction of the length L), such that the final length of the member is $(L + \Delta u)$, the final restraining force $N(\infty)$ changes and this affects both the crack

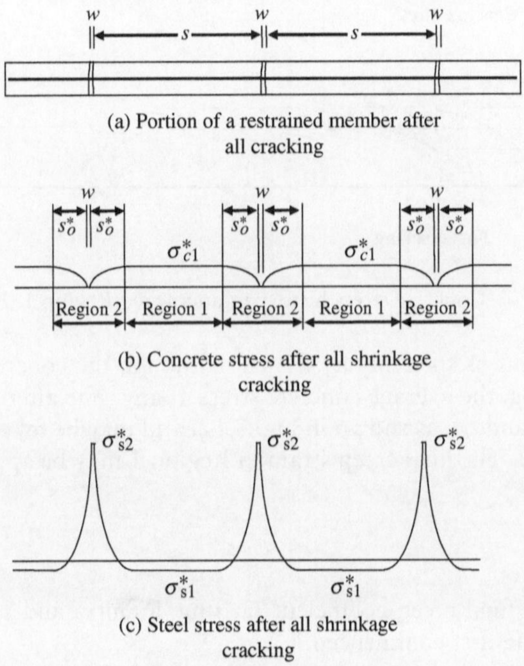

Figure 3.11 Final concrete and steel stresses after direct tension cracking (Ref. 37).

spacing and the crack width. If Δu increases, $N(\infty)$ increases and so too does the number and width of the cracks.

By enforcing the requirements of compatibility and equilibrium, expressions for the final average crack spacing (s) and final crack width (w) in a restrained member can be derived (Ref. 37). For a member containing m cracks and providing the reinforcing steel has not yielded, equating the overall elongation of the steel reinforcement to Δu gives:

$$\frac{\sigma_{s1}^*}{E_s}L + m\frac{\sigma_{s2}^* - \sigma_{s1}^*}{E_s}\left(\frac{2}{3}s_o^* + w\right) = \Delta u \tag{3.61}$$

and, as w is much less than s_o, rearranging gives:

$$\sigma_{s1}^* = \frac{-2\,s_o^* m}{3L - 2s_o^* m}\sigma_{s2}^* + \frac{3\,\Delta u\,E_s}{3L - 2s_o^* m} \tag{3.62}$$

At each crack:

$$\sigma_{s2}^* = N(\infty)/A_s \tag{3.63}$$

In Region 1 in Fig. 3.11, where the distance away from each crack exceeds s_o^* (with s_o^* taken as 1.33 times the value given by Eq. 3.60), the concrete stress history is shown diagrammatically in Fig. 3.12. The concrete tensile stress increases gradually with time as shrinkage progresses and approaches the direct tensile strength of the concrete f_{ct}. When cracking occurs elsewhere in the member, the tensile stress in the

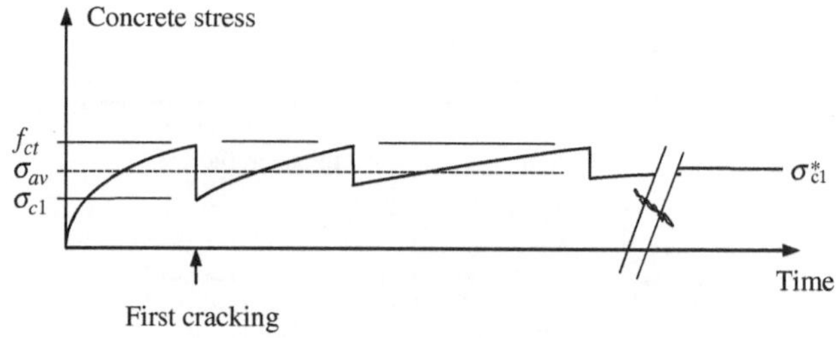

Figure 3.12 Concrete stress history in uncracked Region 1 (Ref. 37).

uncracked regions drops suddenly as shown. Although the concrete stress history is continually changing, the average concrete stress at any time after the start of drying, σ_{av}, is between σ_{c1} and f_{ct}, as shown in Fig. 3.12, and may be taken as the average of σ_{c1} and f_{ct} (Ref. 37). The final creep strain in Region 1 may be approximated by:

$$\varepsilon_c^* = \frac{\sigma_{av}}{E_c}\varphi^*(\tau_d) \tag{3.64}$$

where $\varphi^*(\tau_d)$ is the final creep coefficient (at time infinity) and τ_d is the age of the concrete when drying first commenced.

The final concrete strain in Region 1 is the sum of the elastic, creep, and shrinkage components and may be approximated as:

$$\varepsilon_1^* = \varepsilon_e + \varepsilon_c^* + \varepsilon_{sh}^* = \frac{\sigma_{av}}{E_c} + \frac{\sigma_{av}}{E_c}\varphi^*(\tau_d) + \varepsilon_{sh}^* \tag{3.65}$$

and the magnitude of the final creep coefficient $\varphi^*(\tau_d)$ is usually between 2 and 4, depending on the age at the commencement of drying and the quality of the concrete. Eq. 3.65 may be expressed as:

$$\varepsilon_1^* = \frac{\sigma_{av}}{E_e} + \varepsilon_{sh}^* \tag{3.66}$$

where E_e is the final effective modulus for concrete ($E_e = \dfrac{E_c}{1+\phi^*(\tau_d)}$).

In Region 1, at any distance from a crack greater than s_o, equilibrium requires that the sum of the force in the concrete and the force in the steel is equal to $N(\infty)$. That is:

$$\sigma_{c1}^* A_c + \sigma_{s1}^* A_s = N(\infty) \text{ or } \sigma_{c1}^* = \frac{N(\infty) - \sigma_{s1}^* A_s}{A_c} \tag{3.67}$$

The compatibility requirement is that the final concrete and steel strains are identical ($\varepsilon_{s1}^* = \varepsilon_1^*$). Using Eq. 3.66, this becomes:

$$\frac{\sigma_{s1}^*}{E_s} = \frac{\sigma_{av}}{E_e} + \varepsilon_{sh}^* \tag{3.68}$$

Substituting Eqs 3.62 and 3.63 into Eq. 3.68 and rearranging gives:

$$N(\infty) = \frac{3A_s E_s \Delta u}{2s_o^* m} - \frac{(3L - 2s_o^* m) n_e A_s}{2s_o^* m} (\sigma_{av} + \varepsilon_{sh}^* E_e) \tag{3.69}$$

With the restraining force $N(\infty)$ and the steel stress in Region 1 obtained from Eqs 3.69 and 3.62, respectively, the final concrete stress in Region 1 is

$$\sigma_{c1}^* = \frac{N(\infty) - \sigma_{s1}^* A_s}{A_c} \tag{3.70}$$

The number of cracks m is the lowest integer value of m for which $\sigma_{c1}^* \leq f_{ct}$, where the direct tensile strength f_{ct} should be taken as the mean 28-day value. The final average crack spacing is $s = L/m$.

The overall shortening of the concrete is an estimate of the sum of the crack widths. The final concrete strain at any point in Region 1 of Fig. 3.11 is given by Eq. 3.66, and in Region 2, the final concrete strain is:

$$\varepsilon_2^* = \frac{fn\sigma_{c1}^*}{E_e} + \varepsilon_{sh}^* \tag{3.71}$$

where fn varies between zero at a crack and unity at s_o from a crack. If a parabolic variation of stress is assumed in Region 2, the following expression for the average crack width w is obtained by integrating the concrete strain over the length of the member:

$$w = -\left[\frac{\sigma_{c1}^*}{E_e}\left(s - \frac{2}{3}s_o^*\right) + \varepsilon_{sh}^* s\right] \tag{3.72}$$

The preceding analysis is valid provided the assumption of linear-elastic behaviour in the steel is valid, that is, provided the steel has not yielded.

Example 3.8

The 140 mm thick slab of length 4 m shown in Fig. 3.13 is to be considered. The slab is rigidly held in position at each end support and, except for shrinkage, is unloaded. The slab is symmetrically reinforced with 12 mm diameter bars at 250 mm centres near both the top and bottom surfaces. Hence, $A_s = 900\ \text{mm}^2/\text{m}$ and $\rho = A_s/A_c = 0.00643$. The average final spacing between the restrained shrinkage cracks and the average final crack width are to be determined.

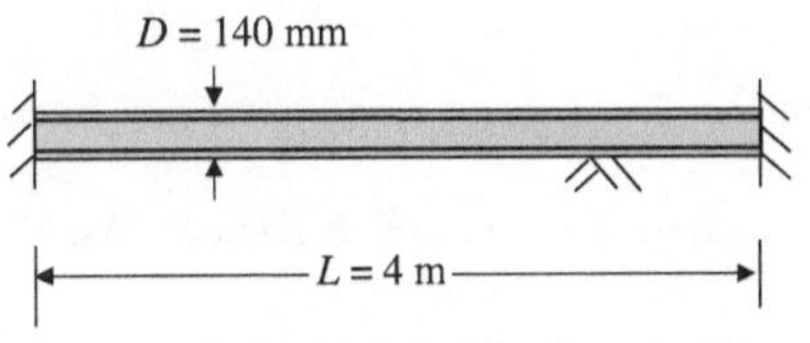

Figure 3.13 Slab of Example 3.8.

The material properties are:

$f_c' = 25$ MPa; $f_{ct} = 2.5$ MPa;
$E_c = 25000$ MPa; $n = E_s/E_c = 8.0$;
$\varphi(t,\tau) = 3.0$;
$E_e^* = E_c/(1+\varphi^*(\tau)) = 6250$ MPa;
$n_e^* = E_s/E_e^* = 32$; and
$\varepsilon_{sh}^* = -0.0007$.

From Eqs 3.60 and 3.59:

$$s_o = \frac{12}{10 \times 0.00643} = 187 \text{ mm} \quad \text{and}$$

$$C_1 = \frac{2 \times 187}{3 \times 4000 - 2 \times 187} = 0.0321.$$

From Eqs 3.58a and b:

$$N_{cr} = \frac{8 \times 0.00643 \times 2.5 \times 140{,}000}{0.0321 + 8 \times 0.00643 \times (1 + 0.0321)} = 211 \text{ kN};$$

$$\sigma_{c1} = \frac{211{,}000 \times (1 + 0.0321)}{140{,}000} = 1.55 \text{ MPa};$$

$$\text{and therefore } \sigma_{av} = \frac{\sigma_{c1} + f_{ct}}{2} = 2.02 \text{ MPa}.$$

For final shrinkage calculations: $s_o^* = 1.33 \times 187 = 248$ mm.
The tabulation below shows the final restraining force and concrete and steel stresses for different numbers of cracks:

Number of cracks, m	*N(∞) from Eq. 3.69 (kN)*	*σ_{s1}^* from Eq. 3.62 (MPa)*	*σ_{s2}^* from Eq. 3.63 (MPa)*	*σ_{c1}^* from Eq. 3.70 (MPa)*
3	477	−75.1	530	3.89
4	341	−75.1	379	2.92
5	259	−75.1	288	2.33
6	205	−75.1	227	1.94

After the fifth crack forms, the final concrete stress is less than the tensile strength of concrete and the final crack spacing is therefore about $s = L/m = 800$ mm. The final crack width is estimated from Eq. 3.72:

$$w = -\left[\frac{2.33}{6250}(800 - \frac{2}{3} \times 248) + (-0.0007 \times 800)\right] = 0.32 \text{ mm}$$

3.8 References

1. Sbarounis, J.A. (1984). Multi-storey flat plate buildings – measured and computer one-year deflections. *Concrete International*, 6(8), 31–35.
2. Jokinen, E.P. and Scanlon, A. (1985). Field Measured Two-way Slab Deflections. *CSCE Annual Conference*, Saskatoon, Canada.
3. Gilbert, R.I. (1999). The risk and responsibility of structural engineers – A case study of a structural failure. *International Congress – Creating with Concrete*, University of Dundee, Scotland, Sept. *Concrete Durability and Repair Technology*, Thomas Telford, pp 773–783.

4. American Concrete Institute (2008), *Building Code Requirements for Structural Concrete, ACI 318-08*, Detroit.
5. British Standards Institution (1992). *Eurocode 2: Design of Concrete Structures – Part 1: General rules and rules for buildings – DD ENV 1992-1-1:1992*. European Committee for Standardization, Brussels.
6. British Standards Institution (2004). *Eurocode 2: Design of Concrete Structures – part 1-1: General rules and rules for buildings – BS EN 1992-1-1:2004*. European Committee for Standardization, Brussels.
7. Standards Australia (2009). *Australian Standard for Concrete Structures, AS3600-2009*. Sydney.
8. Standards Australia (2002). *Structural design actions – Permanent, imposed and other actions AS/NZS 1170.1*. Amendment 1: 2005, Amendment 2: 2009 (Australian/New Zealand Standards).
9. Standards Australia (2004). Bridge Design Part 2: Design Loads *AS5100.2-2004*. Sydney.
10. National Research Council of Canada (2005). *National Building Code of Canada*. Ottawa, Ontario.
11. British Standards Institution (1996). *BS 6399: Part 1 – Loading for Buildings – Code of Practice for Dead and Imposed Loads*. London.
12. American Association of State Highway and Transportation Officials (2002). *Standard Specifications for Highway Bridges AASHTO HB-17*, Washington, DC.
13. American Society of Civil Engineers (2005). *ASCE Standard ASCE 7-05 - Minimum Design Loads for Buildings and Other Structures*. Reston, Virginia.
14. Standards Australia (2002). *Structural design actions – Wind actions AS/NZS 1170.2*. Amendment 1: 2005 (Australian/New Zealand Standards).
15. British Standards Institution (1997). *BS 6399: Part 2 – Loading for Buildings – Code of Practice for Wind Loads*. London.
16. Structural Engineers Association of California (2009). *SEAOC Blue Book – Seismic Design Recommendations*. Seismology Committee, Sacramento, CA.
17. Branson, D.E. (1965). *Instantaneous and time-dependent deflections of simple and continuous reinforced concrete beams. HPR Report No. 7. Part 1*. Alabama Highway Dept. Bureau of Public Roads, Alabama.
18. Gilbert, R.I. (2007). Tension stiffening in lightly reinforced concrete slabs. *Journal of Structural Engineering, ASCE*, 133(6), 899–903.
19. Irwin, A.W. (1978). Human response to dynamic motion of structures. *The Structural Engineer*, 56A(9).
20. Wilford, M. and Yound, P. (2006). *A Design Guide for Footfall induced Vibrations of Structures*. The Concrete Centre, Camberley, UK.
21. Mayer, H. and Rüsch, H. (1967). *Building Damage Caused by Deflection of Reinforced Concrete Building Components. Technical Translation 1412*, National Research Council Ottawa, Canada (Deutscher Auschuss für Stahlbeton, Heft 193, Berlin, West Germany, 1967).
22. Rangan, B.V. (1982). Control of beam deflections by allowable span-depth ratios. *ACI Journal*, 79(5), 372–377.
23. Gilbert, R.I. (1985). Deflection control of slabs using allowable span to depth ratios. *ACI Journal*, 82(1), 67–72.
24. The Concrete Society (2005). *Deflections in concrete slabs and beams*. Technical Report No. 58.
25. Bischoff, P.H. (2005). Reevaluation of deflection prediction for concrete beams reinforced with steel and FRP bars. *Journal of Structural Engineering, ASCE*, 131(5), 752–767.
26. Gilbert, R.I. (1999). Deflection calculations for reinforced concrete structures – why we sometimes get it wrong. *ACI Structural Journal*, 96(6), 1027–1032.

27. Patrick, M. (2009). Beam deflection by simplified calculation: including effect of compressive reinforcement in formula for σ_{cs} in draft AS3600. *BD-002 Committee Document.*
28. Gilbert R.I. (2003). Deflection by simplified calculation in AS3600-2001 – on the determination of f_{cs}. *Australian Journal of Structural Engineering, Engineers Australia*, 5(1), 61–71.
29. Gilbert, R.I. (2008). Instantaneous and time-dependent deflection of reinforced concrete flexural members. *Concrete Forum – Journal of the Concrete Institute of Australia*, 1(1), 7–17.
30. Gilbert, R.I. (2001). Deflection calculation and control – Australian code amendments and improvements. In *ACI International SP 203*, *Code Provisions for Deflection Control in Concrete Structures*, Chapter 4, American Concrete Institute. Editors E.G. Nawy and A. Scanlon, Detroit, 45–78.
31. Gergely, P. and Lutz, L.A. (1968). Maximum crack width in reinforced concrete flexural members. In *Causes, Mechanism, and Control of Cracking in Concrete*, SP-20, American Concrete Institute, Farmington Hills, MI, 1–17.
32. Frosch, R.J. (1999). Another look at cracking and crack control in reinforced concrete. *ACI Structural Journal*, 96(3), 437–442.
33. Marti, P., Alvarez, M., Kaufmann, W. and Sigrist, V. (1998). Tension chord model for structural concrete. *Structural Engineering International*, 4/98, 287–298.
34. Gilbert, R.I. (2008). Control of flexural cracking in reinforced concrete. *ACI Structural Journal*, 105(3), 301–307.
35. Gilbert, R.I. and Nejadi, S. (2004). An Experimental Study of Flexural Cracking in Reinforced Concrete Members under Sustained Loads. *UNICIV Report No. R-435,* School of Civil and Environmental Engineering, University of New South Wales, Sydney, Australia.
36. Wu, H.Q. and Gilbert, R.I. (2009). Modelling short-term tension stiffening in reinforced concrete prisms using a continuum-based finite element model. *Engineering Structures*, 31(10), 2380–2391.
37. Gilbert, R.I. (1992). Shrinkage cracking in fully-restrained concrete members. *ACI Structural Journal*, 89(2), 141–149.
38. Nejadi, S. and Gilbert, R.I. (2004). Shrinkage cracking and crack control in restrained reinforced concrete members. *ACI Structural Journal*, 101(6), 840–845.
39. Favre, R. *et al.* (1983). *Fissuration et Deformations.* Manual du Comite Euro-International du Beton (CEB), Ecole Polytechnique Federale de Lausanne, Switzerland.
40. Nejadi, S. and Gilbert, R.I. (2004). Shrinkage Cracking in Restrained Reinforced Concrete Members. *UNICIV Report No. R-433,* School of Civil and Environmental Engineering, The University of New South Wales, Sydney, Australia.

4 Uncracked sections

Axial loading

4.1 Preamble

In the previous chapter, simplified methods were presented for the determination of deflection and crack widths. While often suitable for design purposes, the simplified methods approximate structural deformation at service loads without necessarily providing an insight into the time-dependent behaviour of the structure. In situations where accurate and reliable estimates of the effects of creep and shrinkage are required, the simplified methods of Chapter 3 may not be adequate. More refined methods are available that realistically model the time-dependent deformation of concrete structures and also enable the calculation of the redistribution of stresses between the concrete and the reinforcement and the redistribution of internal actions in statically indeterminate members. Refined methods for the time analysis of cross-sections are introduced in this chapter and illustrated using the simple example of a symmetrically reinforced cross-section subjected to a constant sustained axial compressive load.

The time analysis of a concrete structure involves the determination of strains, stresses, curvatures and deflections at critical points and at critical times during the life of the structure. Often a structural designer is most interested in the final deformation and the final internal actions at time infinity after the effects of creep and shrinkage have taken place, i.e. the long-term behaviour.

The creep and shrinkage characteristics of concrete are highly variable and are never known exactly. In addition, the methods for the time analysis of concrete structures are plagued by simplifying assumptions and approximations. Accurate numerical predictions of time-dependent behaviour are therefore not possible. However, it is possible, and indeed necessary, to establish upper and lower limits to the final stresses and deformations in order to determine whether or not time effects are critical in any particular situation and, if required, to adjust the design in order to reduce undesirable long-term deformations.

If the concrete stress $\sigma_c(\tau_0)$ at a point in a structure, first applied at age τ_0, remains constant with time, each of the concrete strain components may be calculated readily as follows:

$$\varepsilon(t) = \varepsilon_e(t) + \varepsilon_{cr}(t) + \varepsilon_{sh}(t) = \frac{\sigma_c(\tau_0)}{E_c(\tau_0)} + \frac{\sigma_c(\tau_0)}{E_c(\tau_0)}\varphi(t, \tau_0) + \varepsilon_{sh}(t) \tag{4.1}$$

Numerical values of $\varphi(t, \tau_0)$ and $\varepsilon_{sh}(t)$ may be obtained from the models discussed in Sections 2.1.4 and 2.1.5, and the calculation of time-dependent structural behaviour is relatively straightforward.

If the concrete stress at a point varies with time, the determination of the creep strain becomes more involved. In reinforced concrete structures, even under constant sustained loads, stresses are rarely constant and Eq. 4.1 can no longer be used to predict deformation accurately. The stress history and the effects of ageing must be included.

In this chapter, several methods for the time analysis of cross-sections are discussed and applied to a simple practical problem. The mathematical description of each procedure is followed by a discussion of its strengths and weaknesses, and its advantages and disadvantages. To illustrate each procedure, the analysis of a symmetrically reinforced concrete column section under a constant axial compressive load is presented and numerical results obtained using each procedure are compared.

4.2 The effective modulus method

4.2.1 Formulation

The simplest and oldest technique for including creep in structural analysis is Faber's *effective modulus method* (EMM) (Ref. 1). In Section 1.2.4, the instantaneous and creep components of strain were combined and a reduced or effective modulus for concrete, $E_e(t, \tau_0)$, was defined in Eqs 1.11 and 1.12. In Section 1.2.5, the integral-type creep law was presented in Eq. 1.17 to describe the time-dependent deformation of a continuously varying stress history. In the EMM, Eq. 1.17 is approximated by assuming that the stress-dependent deformations are produced only by a sustained stress equal to the final value of the stress history, that is:

$$\varepsilon(t) = \int_{\tau_0}^{t} \frac{1+\varphi(t,\tau)}{E_c(\tau)} d\sigma_c(\tau) + \varepsilon_{sh}(t) \approx \frac{1+\varphi(t,\tau_0)}{E_c(\tau_0)} \sigma_c(t) + \varepsilon_{sh}(t)$$

$$= \frac{\sigma_c(t)}{E_e(t,\tau_0)} + \varepsilon_{sh}(t) \tag{4.2}$$

where $E_e(t, \tau_0)$ is the effective modulus for concrete defined in Eq. 1.12 (which for convenience is repeated here) as:

$$E_e(t,\tau_0) = \frac{E_c(\tau_0)}{1+\varphi(t,\tau_0)} \tag{4.3}$$

Creep is treated as a delayed elastic strain and is taken into account simply by reducing the elastic modulus of concrete with time. A time analysis using the effective modulus method is nothing more than an elastic analysis in which $E_e(t, \tau_0)$ is used instead of $E_c(\tau_0)$. Shrinkage may be included in this elastic time analysis in a similar way as a sudden temperature change in the concrete would be included in a short-term elastic analysis.

According to the EMM, the creep strain at time t (in Eq. 4.2) depends only on the current stress in the concrete $\sigma_c(t)$ and is therefore independent of the previous stress history. This, of course, is not the case. The ageing of the concrete has been ignored. For an increasing stress history, the EMM overestimates creep, while for a decreasing stress history, creep is underestimated. If the stress is entirely removed, the creep strains disappear. The EMM therefore predicts complete creep recovery, which is not correct.

Eq. 4.2 is valid only when the concrete stress is constant in time. In such cases, the EMM gives excellent results. Good results are also obtained if the concrete is old when first loaded and the effect of ageing is not great. Despite its shortcomings, the EMM is the simplest of all of the methods for the time analysis of concrete structures and its simplicity recommends it, particularly under the conditions mentioned above. In many practical problems, the method is sufficiently accurate for design purposes.

However, many practical situations involve rapidly changing stress histories in young concrete. In such cases, the EMM may be unsuitable and potentially misleading, and a more sophisticated method of analysis is required.

4.2.2 *Example application (EMM)*

Consider the short, axially-loaded, symmetrically reinforced column shown in Fig. 4.1. The column is subjected to a constant, sustained axial force P as shown. Creep and shrinkage cause a gradually decreasing stress history in the concrete similar to that shown in Fig. 1.15.

The redistribution of internal forces due to the gradual development of creep and shrinkage strains is to be examined and the time-dependent stresses and strains in both the concrete and the steel are to be calculated using the EMM.

The problem is solved by enforcing the three basic requirements of any time analysis: namely, equilibrium of forces, compatibility of strains and satisfaction of the constitutive relationships of the concrete and the steel.

The external compressive load P is applied at time τ_0 and is resisted by the internal forces in the concrete and steel $N_c(t)$ and $N_s(t)$, where at any time t after loading $N_c(t) = \sigma_c(t)A_c$ and $N_s(t) = \sigma_s(t)A_s$.

Equilibrium requires that the sum of the internal forces at time t equals the external load, that is:

$$P = N_c(t) + N_s(t) = \sigma_c(t)A_c + \sigma_s(t)A_s \tag{4.4}$$

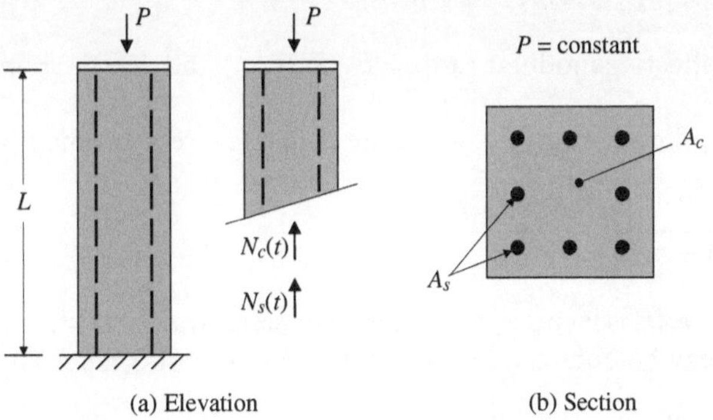

Figure 4.1 Axially-loaded short column.

Compatibility requires that the total concrete strain $\varepsilon(t)$ and the steel strain $\varepsilon_s(t)$ are identical at all times:

$$\varepsilon(t) = \varepsilon_s(t) \tag{4.5}$$

The concrete constitutive relationship used in the EMM is given by Eq. 4.2 and the steel is assumed to be linear and elastic, with modulus E_s. Therefore, at any time t:

$$\varepsilon(t) = \frac{\sigma_c(t)}{E_e(t, \tau_0)} + \varepsilon_{sh}(t) \tag{4.6a}$$

and

$$\varepsilon_s(t) = \frac{\sigma_s(t)}{E_s} \tag{4.6b}$$

Re-arranging Eqs 4.6 in terms of $\sigma_c(t)$ and $\sigma_s(t)$ gives:

$$\sigma_c(t) = \varepsilon(t)E_e(t, \tau_0) - \varepsilon_{sh}(t)E_e(t, \tau_0) \tag{4.7a}$$

and

$$\sigma_s(t) = \varepsilon_s(t)E_s \tag{4.7b}$$

Inserting these equations into the equilibrium equation (Eq. 4.4) and enforcing compatibility (Eq. 4.5) yields:

$$P = E_e(t, \tau_0)\varepsilon(t)A_c - E_e(t, \tau_0)\varepsilon_{sh}(t)A_c + \varepsilon(t)E_sA_s \tag{4.8}$$

which represents the governing equation of the problem. Solving Eq. 4.8 for the unknown total concrete strain $\varepsilon(t)$ gives:

$$\begin{aligned}\varepsilon(t) &= \frac{P}{A_cE_e(t, \tau_0) + A_sE_s} + \frac{A_cE_e(t, \tau_0)\varepsilon_{sh}(t)}{A_cE_e(t, \tau_0) + A_sE_s} \\ &= \frac{P}{A_cE_e(t, \tau_0)(1 + n_e\rho)} + \frac{\varepsilon_{sh}(t)}{1 + n_e\rho}\end{aligned} \tag{4.9}$$

where n_e = the effective modular ratio $= E_s/E_e(t, \tau_0)$ and ρ = the reinforcement ratio $= A_s/A_c$.

By substituting Eq. 4.9 into Eq. 4.7a, the concrete stress at time t is obtained:

$$\sigma_c(t) = \frac{P}{A_c(1 + n_e\rho)} - \frac{E_s\rho\,\varepsilon_{sh}(t)}{1 + n_e\rho} \tag{4.10}$$

The steel strain $\varepsilon_s(t)$ is identical to the total concrete strain $\varepsilon(t)$ (Eq. 4.5) and therefore the steel stress may be obtained by substituting Eq. 4.9 into Eq. 4.7b:

$$\sigma_s(t) = \frac{n_eP}{A_c(1 + n_e\rho)} + \frac{E_s\,\varepsilon_{sh}(t)}{1 + n_e\rho} \tag{4.11a}$$

The steel stress may also be obtained from the equilibrium equation (Eq. 4.4):

$$\sigma_s(t) = \frac{P - \sigma_c(t)A_c}{A_s} \tag{4.11b}$$

The elastic component of concrete strain at time t is usually taken to be $\varepsilon_e(t) = \sigma_c(t)/E_c(\tau_0)$ and the creep strain is therefore $\varepsilon_{cr}(t) = \varepsilon(t) - \varepsilon_e(t) - \varepsilon_{sh}(t)$.

Example 4.1

Consider the short, symmetrically reinforced column shown in Fig. 4.1. The column is subjected to a constant sustained axial force $P = -1000$ kN first applied at age $\tau_0 = 14$ days. The cross-sectional areas of the concrete and the steel reinforcement are $A_c = 90{,}000$ mm^2 and $A_s = 1800$ mm^2, respectively, and the reinforcement ratio is therefore $\rho = A_s/A_c = 0.02$. The column is located in a temperate environment and shrinkage is assumed to commence at the age of first loading. The mean *in-situ* compressive strength of concrete at the time of first loading is taken to be $f_{cmi}(\tau_0) = 28$ MPa and the characteristic 28-day compressive strength is $f'_c = 40$ MPa. The elastic modulus of concrete at first loading is $E_c(\tau_0) = 26.7$ GPa and the elastic modulus of steel is $E_s = 200$ GPa. The creep coefficient is obtained from Eq. 2.3 and shrinkage strain is obtained from Eq. 2.6, with final (long-term) values: $\varphi^*(\tau_0) = 2.39$ and $\varepsilon^*_{sh} = -510 \times 10^{-6}$. The variations of the creep coefficient and shrinkage with time are given in the following table and shown in Fig. 4.2. For illustrative purposes it has been assumed that shrinkage begins at the time of loading.

$(t - \tau_0)$ *in days*	0	10	30	70	200	500	10,000
$\varphi(t, \tau_0)$	0	0.53	0.98	1.38	1.83	2.10	2.39
$\varepsilon_{sh}(t - \tau_0) \times 10^{-6}$	0	−142	−246	−325	−407	−456	−510

The redistribution of internal forces due to the gradual development of creep and shrinkage strains is to be examined and the time-dependent changes in stresses and strains in both the concrete and the steel are to be calculated. Sample calculations are provided below.

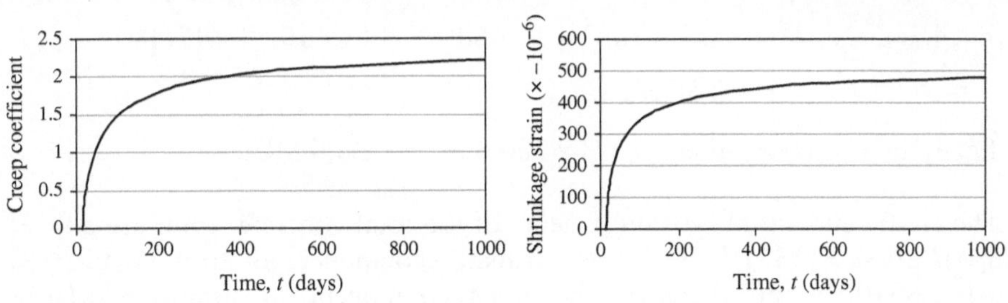

Figure 4.2 Creep coefficient and shrinkage strain versus time curves for Example 4.1.

At first loading τ_0, $(t-\tau_0)=0$:

$\varphi(t,\tau_0)=0$ and $\varepsilon_{sh}(t-\tau_0)=0$.
With $E_e(t,\tau_0)=E_c(\tau_0)=26.7$ GPa; and $n_e=n=E_s/E_c(\tau_0)=7.48$.

$$\text{Eq. 4.9:}\quad \varepsilon(\tau_0)=\frac{-1000\times10^3}{90{,}000\times26{,}700\times(1+7.48\times0.02)}+0=-361\times10^{-6};$$

$$\text{Eq. 4.5:}\quad \varepsilon_s(\tau_0)=-361\times10^{-6};$$

$$\text{Eq. 4.7a:}\quad \sigma_c(\tau_0)=-361\times10^{-6}\times26{,}700-0=-9.67\text{ MPa};$$

$$\text{Eq. 4.7b:}\quad \sigma_s(\tau_0)=-361\times10^{-6}\times200{,}000=-72.3\text{ MPa}.$$

The stresses in the concrete and steel could have also been calculated directly from Eqs 4.10 and 4.11:

$$\text{Eq. 4.10:}\quad \sigma_c(\tau_0)=\frac{-1000\times10^3}{90{,}000(1+7.48\times0.02)}-0=-9.67\text{ MPa};$$

$$\text{Eq. 4.11b:}\quad \sigma_s(\tau_0)=\frac{-1000\times10^3+9.67\times90{,}000}{1800}=-72.3\text{ MPa}.$$

With the creep and shrinkage components of the concrete strain both equal to zero at first loading, the concrete strain $\varepsilon(\tau_0)$ is entirely made up of the elastic or instantaneous component of strain $\varepsilon_e(\tau_0)$.

At $(t-\tau_0)=10$ days:

$\varphi(t,\tau_0)=0.53$ and $\varepsilon_{sh}(t)=-142\times10^{-6}$.
$E_e(t,\tau_0)=E_c(\tau_0)/(1+\varphi(t,\tau_0))=17.48$ GPa; and $n_e=E_s/E_e(t,\tau_0)=11.44$; and

$$\text{Eq. 4.9:}\quad \varepsilon(t)=\frac{-1000\times10^3}{90{,}000\times17{,}480\times(1+11.44\times0.02)}+\frac{-142\times10^{-6}}{1+11.44\times0.02}$$
$$=-633\times10^{-6};$$

$$\text{Eq. 4.5:}\quad \varepsilon_s(t)=-633\times10^{-6};$$

$$\text{Eq. 4.7a:}\quad \sigma_c(t)=-633\times10^{-6}\times17{,}480-(-142\times10^{-6})\times17{,}480$$
$$=-8.58\text{ MPa};$$

$$\text{Eq. 4.7b:}\quad \sigma_s(t)=-633\times10^{-6}\times200{,}000=-126.6\text{ MPa}.$$

The elastic and shrinkage components of the total concrete strain are $\varepsilon_e(t)=\sigma_c(t)/E_c(\tau_0)=-321\times10^{-6}$; the shrinkage component of strain is $\varepsilon_{sh}(t)=-142\times10^{-6}$; and therefore the creep component of strain is $\varepsilon_{cr}(t)=\varepsilon(t)-\varepsilon_e(t)-\varepsilon_{sh}(t)=-170\times10^{-6}$.

The results of the analyses at first loading and throughout the period of sustained loading are provided in the following table.

Time t (days)	*Duration of load* $t-\tau_0$ *(days)*	*Concrete stress* $\sigma_c(t)$ *(MPa)*	*Steel stress* $\sigma_s(t)$ *(MPa)*	*Total strain* $\varepsilon(t)$ $(\times 10^{-6})$	*Elastic concrete strain* $\varepsilon_e(t)$ $(\times 10^{-6})$	*Creep strain* $\varepsilon_{cr}(t)$ $(\times 10^{-6})$	*Shrinkage strain* $\varepsilon_{sh}(t)$ $(\times 10^{-6})$
14	0	−9.67	−72.3	−361	−361	0	0
24	10	−8.58	−127	−633	−321	−170	−142
44	30	−7.82	−165	−824	−292	−285	−246
84	70	−7.23	−194	−969	−270	−374	−325
214	200	−6.66	−222	−1112	−249	−456	−407
514	500	−6.35	−238	−1191	−237	−497	−456
10,014	10,000	−6.02	−255	−1273	−225	−538	−510

4.3 The principle of superposition – step-by-step method (SSM)

4.3.1 Formulation

The principle of superposition and its applicability for the determination of strains in concrete was discussed in Section 1.2.5. The creep strain produced by a stress increment applied at any time τ_i is assumed to be unaffected by any other stress increment applied either earlier or later. The superposition integral for the total concrete strain at any time t due to a time-varying stress history (Eq. 1.17) cannot be solved in closed form unless simplifying assumptions are made, such as the parallel creep curve assumption illustrated subsequently in Fig. 4.10.

In the step-by-step method (SSM), a solution technique is employed using the incremental form of the superposition equation (Eq. 1.16). The period of sustained stress is divided into k time intervals as shown in Fig. 4.3, with the age at first loading designated τ_0 and the end of the period of sustained stress designated τ_k.

Consider the time-varying stress history shown in Fig. 4.4. The actual stress history is approximated by a step-wise variation of stress, with an increment of stress $\Delta\sigma_c(\tau_j)$ applied at the end of each time interval, as shown. For this decreasing stress history, if the initial stress $\sigma_c(\tau_0)$ is compressive, the stress increments $\Delta\sigma_c(\tau_j)$ applied at the end of each time interval are tensile. The stress is assumed to remain constant during

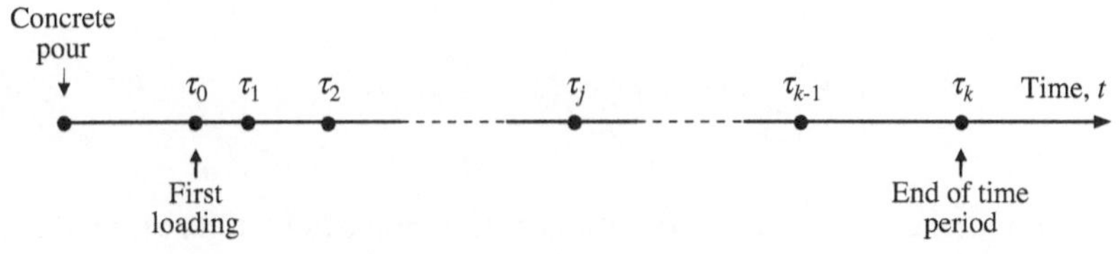

Figure 4.3 Time discretisation.

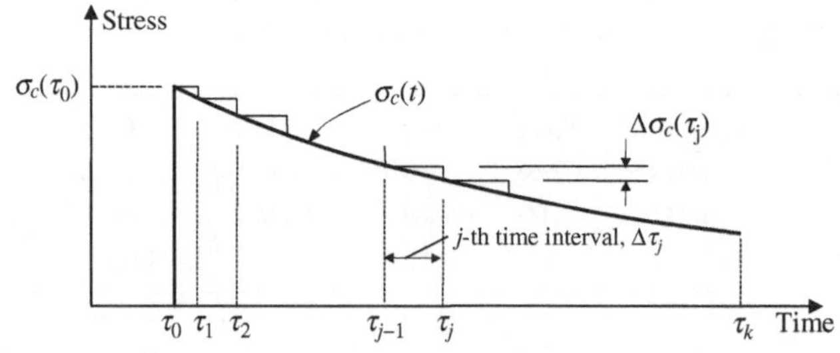

Figure 4.4 Time-varying stress history – SSM.

each time interval. The greater the number of time intervals, the more accurate is the final prediction. Various formulations of the SSM are available in the literature (for example, Refs 2–5), all of which give similar answers provided the time intervals are suitably small.

For a particular time interval $\tau_k - \tau_0$, the time discretisation should be such that an approximately equal portion of the creep coefficient $\varphi(\tau_k, \tau_0)$ develops during each time step. A convenient and effective time discretisation based on a geometrical progression is given by:

$$\tau_1 = \tau_0 + \frac{(\tau_k - \tau_0)}{\tau_k\, k} \tag{4.12a}$$

$$\tau_j = \tau_0 + \left(\tau_k\, k\right)^{1/(k-1)} \left(\tau_{j-1} - \tau_0\right) \tag{4.12b}$$

where τ_0 and τ_k define the beginning and end of the time period to be discretised; k represents the number of time intervals; and $j = 2, 3, \ldots, k$.

The SSM is perfectly general and can be used to predict behaviour due to any stress or strain history using any desired creep and shrinkage curves. Using the integral in the integral-type creep law (Eq. 1.17):

$$\varepsilon(t) = J(t, \tau_0)\sigma_c(\tau_0) + \int_{\tau_0^+}^{t} J(t, \tau)\mathrm{d}\sigma_c(\tau) + \varepsilon_{sh}(t)$$

and the total concrete strain at the end of the j-th time period ($t = \tau_j$) may be approximated by:

$$\varepsilon(\tau_j) = J(\tau_j, \tau_0)\sigma_c(\tau_0) + \sum_{i=1}^{j} J(\tau_j, \tau_i)\, \Delta\sigma_c(\tau_i) + \varepsilon_{sh}(\tau_j) \tag{4.13a}$$

where $J(\tau_j, \tau_i)$ represents the creep (compliance) function calculated at time τ_j related to a unit stress applied at time τ_i, while $\Delta\sigma_c(\tau_i)$ is calculated as $\sigma_c(\tau_i) - \sigma_c(\tau_{i-1})$ and depicts the stress variation that occurs between times τ_{i-1} and τ_i. A better approximation is obtained if the average creep function associated with the time step τ_i and τ_{i-1} is used within the summation in Eq. 4.13a, that is:

$$\varepsilon(\tau_j) = J(\tau_j, \tau_0)\sigma_c(\tau_0) + \sum_{i=1}^{j} \frac{J(\tau_j, \tau_i) + J(\tau_j, \tau_{i-1})}{2} \Delta\sigma_c(\tau_i) + \varepsilon_{sh}(\tau_j) \tag{4.13b}$$

Eq. 4.13a approximates the integral-type creep law by means of the so-called rectangular rule, while Eq. 4.13b represents the trapezoidal rule. Both approximations tend towards the same result as the number of time steps increases.

Simplifying the notation, Eqs 4.13a and 4.13b become, respectively:

$$\varepsilon_j - \varepsilon_{sh,j} = J_{j,0}\,\sigma_{c,0} + \sum_{i=1}^{j} J_{j,i}\,\Delta\sigma_{c,i} \tag{4.13c}$$

and

$$\varepsilon_j - \varepsilon_{sh,j} = J_{j,0}\,\sigma_{c,0} + \sum_{i=1}^{j} \frac{J_{j,i} + J_{j,i-1}}{2} \Delta\sigma_{c,i} \tag{4.13d}$$

where $\varepsilon_j = \varepsilon(\tau_j)$; $\varepsilon_{sh,j} = \varepsilon_{sh}(\tau_j)$; $J_{j,0} = J(\tau_j, \tau_0)$; $\sigma_{c,0} = \sigma_c(\tau_0)$; $J_{j,i} = J(\tau_j, \tau_i)$; $\sigma_{c,i} = \sigma_c(\tau_i)$ and $\Delta\sigma_c(\tau_i) = \sigma_{c,i} - \sigma_{c,i-1}$.

Eqs 4.13a and b can be expressed in terms of creep coefficients as follows:

$$\varepsilon(\tau_j) = \frac{1 + \varphi(\tau_j, \tau_0)}{E_c(\tau_0)} \sigma_c(\tau_0) + \sum_{i=1}^{j} \frac{1 + \varphi(\tau_j, \tau_i)}{E_c(\tau_i)} \Delta\sigma_c(\tau_i) + \varepsilon_{sh}(\tau_j) \tag{4.14a}$$

and

$$\varepsilon(\tau_j) = \frac{1+\varphi(\tau_j, \tau_0)}{E_c(\tau_0)} \sigma_c(\tau_0) + \sum_{i=1}^{j} \frac{1}{2} \left[\frac{1+\varphi(\tau_j, \tau_i)}{E_c(\tau_i)} + \frac{1+\varphi(\tau_j, \tau_{i-1})}{E_c(\tau_{i-1})} \right] \Delta\sigma_c(\tau_i) + \varepsilon_{sh}(\tau_j) \tag{4.14b}$$

Simplifying the notation gives:

$$\varepsilon_j - \varepsilon_{sh,j} = \frac{1 + \varphi_{j,0}}{E_{c,0}} \sigma_{c,0} + \sum_{i=1}^{j} \frac{1 + \varphi_{j,i}}{E_{c,i}} \Delta\sigma_{c,i} \tag{4.14c}$$

and

$$\varepsilon_j - \varepsilon_{sh,j} = \frac{1 + \varphi_{j,0}}{E_{c,0}} \sigma_{c,0} + \sum_{i=1}^{j} \frac{1}{2} \left(\frac{1 + \varphi_{j,i}}{E_{c,i}} + \frac{1 + \varphi_{j,i-1}}{E_{c,i-1}} \right) \Delta\sigma_{c,i} \tag{4.14d}$$

where $\varphi_{j,0} = \varphi(\tau_j, \tau_0)$, $\varphi_{j,i} = \varphi(\tau_j, \tau_i)$, $E_{c,0} = E_c(\tau_0)$ and $E_{c,i} = E_c(\tau_i)$.

As Eqs 4.14 indicate, a large amount of input data is required for the SSM. For each time step, $E_c(\tau_i)$ and $\varphi(\tau_j, \tau_i)$ must be specified and the previous stress history must be stored throughout the analysis. However, the SSM is not subject to many of the simplifying assumptions contained in other methods of analysis and generally leads to reliable results. For most practical problems, satisfactory results are obtained using as few as 6–10 time intervals. For illustrative purposes, Eqs 4.13a and 4.14a are used in the following.

The axially loaded, symmetrically reinforced column shown in Fig. 4.1, and analysed in Section 4.2.2, is re-analysed here using the SSM. Time is discretised into a pre-selected number of time intervals. Two approaches will be considered:

(i) Approach 1 obtains the solution by calculating the change in concrete stress that occurs at the end of each time step; and
(ii) Approach 2 expresses the governing system of equations in terms of the concrete stress at the time considered.

Both approaches are in fact identical and differ only in the manipulation of the governing equations.

4.3.2 *Example application (SSM) – Approach 1*

In Approach 1, Eq. 4.14a is used to calculate the change in stress at the end of each time step. For the continuously varying stress history shown in Fig. 4.4, the stress increment, $\Delta\sigma_c(\tau_i)$, is assumed to be applied at the end of the *i*-th time interval, as shown. The increments of instantaneous plus creep strain caused by $\Delta\sigma_c(\tau_i)$ are calculated at the end of each time interval using the appropriate elastic modulus and creep coefficient ($E_c(\tau_i)$ and $\varphi(\tau_j, \tau_i)$, respectively). The total strain at the end of each time interval is obtained by superposition of the strain increments caused by stress changes in all previous time intervals and by shrinkage.

4.3.2.1 *At first loading* ($t = \tau_0$)

Following the approach outlined in Section 4.2.2 for the EMM, the instantaneous stresses and strains at τ_0 immediately after the application of the axial force P, and before any creep and shrinkage has occurred, are obtained using Eqs 4.5–4.11 and are given by:

$$\varepsilon(\tau_0) = \frac{P}{A_c E_c(\tau_0)(1+n_0\rho)};\ \varepsilon_s(\tau_0) = \varepsilon(\tau_0);\sigma_c(\tau_0) = \varepsilon(\tau_0)E_c(\tau_0);\ \text{and}\ \sigma_s(\tau_0) = \varepsilon_s(\tau_0)E_s$$

where $n_0 = E_s/E_c(\tau_0)$ = the modular ratio at time τ_0 and $\rho = A_s/A_c$ = the reinforcement ratio.

The analysis at each subsequent time step is outlined in the following.

4.3.2.2 *At the end of the j-th time step (i.e.* $t = \tau_j$*)*

The stress in the concrete at the end of the *j*-th time step is:

$$\sigma_c(\tau_j) = \sigma_c(\tau_{j-1}) + \Delta\sigma_c(\tau_j) \tag{4.15}$$

where $\sigma_c(\tau_{j-1})$ is known as it has been calculated at the previous time step and $\Delta\sigma_c(\tau_j)$ is the unknown stress increment at τ_j. The equilibrium and compatibility equations are:

$$\left[\sigma_c(\tau_{j-1})+\Delta\sigma_c(\tau_j)\right]A_c+\sigma_s(\tau_j)A_s=P \tag{4.16}$$

and

$$\varepsilon(\tau_j)=\varepsilon_s(\tau_j) \tag{4.17}$$

The stress–strain relationships for the concrete and the steel at time τ_j are:

$$\varepsilon(\tau_j)=\frac{1+\varphi(\tau_j,\tau_0)}{E_c(\tau_0)}\sigma_c(\tau_0)+\sum_{i=1}^{j}\frac{1+\varphi(\tau_j,\tau_i)}{E_c(\tau_i)}\Delta\sigma_c(\tau_i)+\varepsilon_{sh}(\tau_j) \tag{4.18a}$$

and

$$\varepsilon_s(\tau_j)=\frac{\sigma_s(\tau_j)}{E_s} \tag{4.18b}$$

Rearranging Eqs 4.18 in terms of stresses at time τ_j gives:

$$\Delta\sigma_c(\tau_j)=E_c(\tau_j)\varepsilon(\tau_j)-\frac{E_c(\tau_j)}{E_e(\tau_j,\tau_0)}\sigma_c(\tau_0)-E_c(\tau_j)\sum_{i=1}^{j-1}\frac{\Delta\sigma_c(\tau_i)}{E_e(\tau_j,\tau_i)}-E_c(\tau_j)\varepsilon_{sh}(\tau_j) \tag{4.19a}$$

and

$$\sigma_s(\tau_j)=E_s\varepsilon_s(\tau_j) \tag{4.19b}$$

where $E_e(\tau_j,\tau_i)$ is the effective modulus of concrete at time τ_j associated with a stress increment first applied at time τ_i and is given by:

$$E_e(\tau_j,\tau_i)=\frac{E_c(\tau_i)}{1+\varphi(\tau_j,\tau_i)} \tag{4.20}$$

By substituting Eqs 4.19 and 4.17 into Eq. 4.16 and simplifying, the unknown strain, $\varepsilon(\tau_k)$, is readily determined as:

$$\varepsilon(\tau_j)=\frac{1}{1+n_j\rho}\left[\frac{P}{A_cE_c(\tau_j)}-\frac{\sigma_c(\tau_{j-1})}{E_c(\tau_j)}+\frac{n_{e,j,0}\sigma_c(\tau_0)}{E_s}+\sum_{i=1}^{j-1}\frac{n_{e,j,i}\,\Delta\sigma_c(\tau_i)}{E_s}+\varepsilon_{sh}(\tau_j)\right] \tag{4.21}$$

where $n_j = E_s/E_c(\tau_j)$ = the modular ratio at time τ_j; $n_{e,j,0} = E_s/E_e(\tau_j, \tau_0)$ and $n_{e,j,i} = E_s/E_e(\tau_j, \tau_i)$ are the effective modular ratios; and ρ = the reinforcement ratio $= A_s/A_c$. The steel strain $\varepsilon_s(\tau_j)$ is found using Eq. 4.17 and is identical to the total concrete strain $\varepsilon(\tau_j)$.

The concrete stress increment is obtained from Eq. 4.19a:

$$\Delta\sigma_c(\tau_j) = \frac{1}{1+n_j\rho}\left[\frac{P}{A_c} - \sigma_c(\tau_{j-1}) - n_{e,j,0}\rho\sigma_c(\tau_0) - \sum_{i=1}^{j-1} n_{e,j,i}\rho\Delta\sigma_c(\tau_i) - \varepsilon_{sh}(\tau_j)E_s\rho\right] \tag{4.22}$$

With $\Delta\sigma(\tau_j)$ calculated from Eq. 4.22, the concrete stress at τ_j is obtained from Eq. 4.15. The steel stress $\sigma_s(\tau_j)$ is determined from Eq. 4.19b. Alternatively, $\sigma_s(\tau_j)$ could also be calculated from the equilibrium equation (Eq. 4.16).

From the strain calculated at time instant τ_j ($\varepsilon(\tau_j) = \varepsilon_s(\tau_j)$), the axial deformation or axial shortening of the column can be readily calculated. For the column of length, L, shown in Fig. 4.1, the axial shortening of the column at time, τ_j, due to the applied load, P, and shrinkage is designated $e_L(\tau_j)$ and given by:

$$e_L(\tau_j) = L\varepsilon(\tau_j) \tag{4.23}$$

The approach is useful in the determination of time-dependent axial shortening in the columns and walls of multi-storey buildings where differential shortening may be an important design consideration.

Example 4.2

The symmetrically reinforced column analysed in Example 4.1 (refer Fig. 4.1) is to be re-analysed here using the SSM. As in the previous example, the column is subjected to an axial load $P = -1000$ kN first applied at age $\tau_0 = 14$ days and then held constant. The properties of the column cross-section are: $A_c = 90{,}000$ mm^2; $A_s = 1800$ mm^2; $\rho = A_s/A_c = 0.02$; and $E_s = 200$ GPa.

In this example, the six time intervals specified in the following table are considered for illustrative purposes. A significant amount of input data is required; specifically, a different creep coefficient and elastic modulus for each concrete stress increment applied at the end of each time interval. The following table contains the input data necessary to obtain the concrete and steel stresses and strains at the end of each time interval. The concrete properties are obtained from the material modelling procedures outlined in Chapter 2, that is, $E_c(\tau_j)$ from Eqs 2.2a and 2.2c, $\varphi(\tau_j, \tau_i)$ from Eq. 2.3 and $\varepsilon_{sh}(\tau_j)$ from Eq. 2.6. The concrete is mixed with high early strength cement. As in Example 4.1, shrinkage is taken to begin at the time of first loading.

j	0	1	2	3	4	5	6
τ_j (days)	14	24	44	84	214	514	10,014
$t-\tau_j$ (days)	0	10	30	70	200	500	10,000
$\varphi(\tau_j,\tau_0)$	0	0.53	0.98	1.38	1.83	2.10	2.39
$\varphi(\tau_j,\tau_1)$		0	0.72	1.18	1.63	1.89	2.15
$\varphi(\tau_j,\tau_2)$			0	0.90	1.44	1.69	1.94
$\varphi(\tau_j,\tau_3)$				0	1.22	1.51	1.75
$\varphi(\tau_j,\tau_4)$					0	1.26	1.54
$\varphi(\tau_j,\tau_5)$						0	1.38
$E_c(\tau_j)$ (MPa)	26,700	27,900	28,900	29,700	30,500	31,000	31,700
$\varepsilon_{sh}(\tau_j)\times 10^{-6}$	0	−142	−246	−325	−407	−456	−510

Note that the shrinkage strains and the creep coefficients associated with the age at first loading, τ_0, are the same as those adopted previously in Example 4.1 using the EMM. The family of creep coefficient versus the logarithm of time curves is shown in Fig. 4.5.

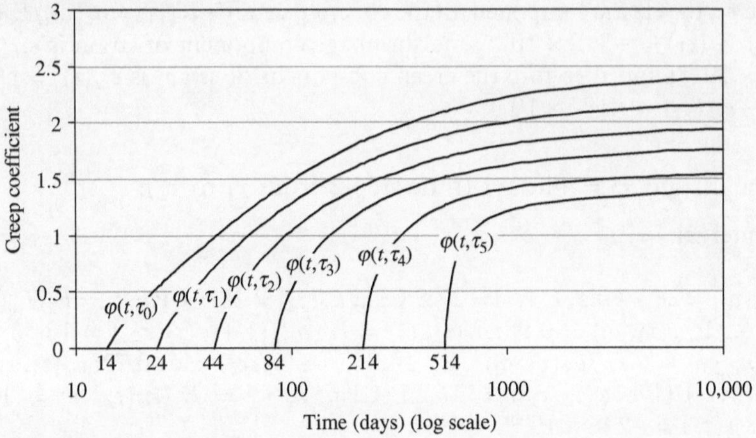

Figure 4.5 Creep coefficient versus log time curves for Example 4.2.

At first loading (τ_0):

As in Example 4.1, with $E_c(\tau_0) = E_e(\tau_0,\tau_0) = 26.7$ GPa and $n_0 = E_s/E_c(\tau_0) = 7.48$, the instantaneous stresses and strains immediately after first loading are:

$\varepsilon(\tau_0) = -361\times 10^{-6}$; $\varepsilon_s(\tau_0) = -361\times 10^{-6}$; $\sigma_c(\tau_0) = -9.67$ MPa and

$\sigma_s(\tau_0) = -72.3$ MPa.

At time instant $\tau_1 = 24$ days (time step 1 from τ_0 to τ_1):

For time interval, $j = 1$:

$E_c(\tau_0) = 26.7$ GPa; $E_c(\tau_1) = 27.9$ GPa; $n_1 = E_s/E_c(\tau_1) = 7.17$; $\varphi(\tau_1,\tau_0) = 0.53$; $E_e(\tau_1,\tau_0) = E_c(\tau_0)/(1+\varphi(\tau_1,\tau_0)) = 17.48$ GPa; $n_{e,1,0} = E_s/E_e(\tau_1,\tau_0) = 11.44$; and $\varepsilon_{sh}(\tau_1) = -142\times 10^{-6}$.

From Eq. 4.22:

$$\Delta\sigma_c(\tau_1) = \frac{1}{1+7.17\times 0.02}\left(\frac{-1000\times 10^3}{90{,}000} + 9.67 + 11.44\times 0.02\times 9.67 + 142\times 10^{-6}\times 200{,}000\times 0.02\right) = 1.17 \text{ MPa}$$

The concrete stress at τ_1 is therefore

$$\sigma_c(\tau_1) = \sigma_c(\tau_0) + \Delta\sigma_c(\tau_1) = -9.67 + 1.17 = -8.50 \text{ MPa}$$

and from Eq. 4.16: $\sigma_s(\tau_1) = \dfrac{-1000\times 10^3 + 8.50\times 90{,}000}{1800} = -131$ MPa

Eq. 4.18b and Eq. 4.17 give:

$$\varepsilon_s(\tau_1) = \frac{-131}{200{,}000} = -653\times 10^{-6}$$

$$\varepsilon(\tau_1) = -653\times 10^{-6}$$

At time τ_1, the elastic component of the concrete strain is $\varepsilon_e(\tau_1) = \sigma_c(\tau_0)/E_c(\tau_0) + \Delta\sigma(\tau_1)/E_c(\tau_1) = -319\times 10^{-6}$; the shrinkage component of strain is $\varepsilon_{sh}(\tau_1) = -142\times 10^{-6}$; and therefore the creep component of strain is $\varepsilon_{cr}(\tau_1) = \varepsilon(\tau_1) - \varepsilon_e(\tau_1) - \varepsilon_{sh}(\tau_1) = -192\times 10^{-6}$.

At time instant τ_2 = 44 days (time step 2 from τ_1 to τ_2):

Time interval, j= 2:

$E_c(\tau_0) = 26.7$ GPa; $E_c(\tau_1) = 27.9$ GPa; $E_c(\tau_2) = 28.9$ GPa; $n_2 = E_s/E_c(\tau_2) = 6.92$; $\varphi(\tau_2,\tau_0) = 0.98$; $E_e(\tau_2,\tau_0) = E_c(\tau_0)/(1+\varphi(\tau_2,\tau_0)) = 13.53$ GPa; $n_{e,2,0} = E_s/E_e(\tau_2,\tau_0) = 14.78$; $\varphi(\tau_2,\tau_1) = 0.72$; $E_e(\tau_2,\tau_1) = E_c(\tau_1)/(1+\varphi(\tau_2,\tau_1)) = 16.25$ GPa; $n_{e,2,1} = E_s/E_e(\tau_2,\tau_1) = 12.31$; $\varepsilon_{sh}(\tau_2) = -246\times 10^{-6}$.

Eq. 4.22:

$$\Delta\sigma(\tau_2) = \frac{1}{1+6.92\times 0.02}\left(\frac{-1000\times 10^3}{90{,}000} + 8.50 + 14.78\times 0.02\times 9.67 - 12.31\times 0.02\times 1.17 + 246\times 10^{-6}\times 200{,}000\times 0.02\right) = 0.83 \text{ MPa}$$

The concrete stress at τ_2 is therefore:

$$\sigma_c(\tau_2) = \sigma_c(\tau_1) + \Delta\sigma_c(\tau_2) = -8.50 + 0.83 = -7.67 \text{ MPa}$$

and from Eq. 4.16, the steel stress is:

$$\sigma_s(\tau_2) = \frac{-1000 \times 10^3 + 7.67 \times 90{,}000}{1800} = -172 \text{ MPa}$$

The steel strain is obtained from Eq. 4.18b and the total concrete strain from Eq. 4.17:

$$\varepsilon_s(\tau_2) = \frac{-172}{200{,}000} = -860 \times 10^{-6}$$

and

$$\varepsilon\,(\tau_2) = -860 \times 10^{-6}$$

At time τ_2, the elastic component of concrete strain is:

$$\varepsilon_e(\tau_2) = \sigma_c(\tau_0)/E_c(\tau_0) + \Delta\sigma_c(\tau_1)/E_c(\tau_1) + \Delta\sigma_c(\tau_2)/E_c(\tau_2) = -291 \times 10^{-6};$$

the shrinkage component is $\varepsilon_{sh}(\tau_2) = -246 \times 10^{-6}$; and therefore the creep component of strain is:

$$\varepsilon_{cr}(\tau_2) = \varepsilon(\tau_2) - \varepsilon_e(\tau_2) - \varepsilon_{sh}(\tau_2) = -323 \times 10^{-6}.$$

At subsequent time instants

Similarly, the stress and strain components at the end of each subsequent time step are calculated and the results are tabulated below.

Instant j	*Time* τ_j *(days)*	*Concrete stress increment* $\Delta\sigma_c\,(\tau_j)$ *(MPa)*	*Concrete stress* $\sigma_c(\tau_j)$ *(MPa)*	*Steel stress* $\sigma_s(\tau_j)$ *(MPa)*	*Total strain* $\varepsilon(\tau_j)$ $(\times 10^{-6})$	*Elastic strain* $\varepsilon_e(\tau_j)$ $(\times 10^{-6})$	*Creep strain* $\varepsilon_{cr}(\tau_j)$ $(\times 10^{-6})$	*Shrinkage strain* $\varepsilon_{sh}(\tau_j)$ $(\times 10^{-6})$
0	14	–	−9.67	−72.3	−361	−361	0	0
1	24	1.17	−8.50	−131	−653	−319	−192	−142
2	44	0.83	−7.67	−172	−860	−291	−323	−246
3	84	0.63	−7.04	−204	−1018	−269	−424	−325
4	214	0.65	−6.39	−236	−1180	−248	−525	−407
5	514	0.34	−6.05	−253	−1264	−237	−571	−456
6	10,014	0.41	−5.64	−273	−1367	−225	−632	−510

The effect of increasing the number of time steps on the accuracy of the predictions using the SSM is illustrated below. In the following table, the results of further analyses are presented in which the time discretisation proposed in Eqs 4.12 has been adopted. Either 6 or 18 time steps have been considered between τ_0 and each of the previously considered time instants (τ_k in the first column of the table below). The results provided on each line of the table have been produced by separate time analyses using the SSM.

Time τ_k *(days)*	*No of time steps used*	*Concrete stress* $\sigma_c(\tau_k)$ *(MPa)*	*Steel stress (MPa)*	*Total strain* $\varepsilon(\tau_k)$ $(\times 10^{-6})$	*Elastic strain* $\varepsilon_e(\tau_k)$ $(\times 10^{-6})$	*Creep strain* $\varepsilon_{cr}(\tau_k)$ $(\times 10^{-6})$	*Shrinkage strain* $\varepsilon_{sh}(\tau_k)$ $(\times 10^{-6})$
14	1	−9.67	−72.3	−361	−361	0	0
24	6	−8.54	−129	−644	−320	−182	−142
	(18)	(−8.55)	(−128)	(−641)	(−320)	(−179)	(−142)
44	6	−7.68	−171	−857	−291	−320	−246
	(18)	(−7.72)	(−169)	(−847)	(−292)	(−309)	(−246)
84	6	−7.02	−205	−1023	−269	−430	−325
	(18)	(−7.09)	(−201)	(−1006)	(−270)	(−411)	(−325)
214	6	−6.36	−238	−1188	−247	−534	−407
	(18)	(−6.46)	(−233)	(−1164)	(−249)	(−508)	(−407)
514	6	−6.00	−255	−1277	−236	−586	−456
	(18)	(−6.10)	(−251)	(−1253)	(−238)	(−559)	(−456)
10,014	6	−5.65	−273	−1366	−225	−631	−510
	(18)	(−5.71)	(−270)	(−1349)	(−226)	(−613)	(−510)

Based on these results, increasing the number of time steps by a factor of three has produced changes in the final total strain and in the final concrete and steel stresses of less than 2 per cent. Considering the uncertainty associated with the estimates of the concrete properties, there is little to be gained, from a practical point of view, by considering more than about 6–10 time steps in most situations. The axial shortening of the column is calculated using Eq. 4.23. If the overall length of the column is 4000 mm, the column shortening at first loading and after 10,000 days are $e_L(\tau_0 = 14) = 1.44$ mm and $e_L(\tau_{18} = 10{,}014) = 5.40$ mm, respectively. The axial shortening of the column has increased with time by a factor of almost four.

4.3.3 Example application (SSM) – Approach 2

In approach 2, the concrete constitutive relationship is manipulated so that the concrete stress at the end of each time instant is obtained directly. If the rectangular approximation of the integral-type creep law is used, Eq. 4.13c may be rearranged as follows:

$$\varepsilon_j - \varepsilon_{sh,j} = J_{j,0}\sigma_{c,0} + \sum_{i=1}^{j} J_{j,i}\Delta\sigma_{c,i}$$

$$= J_{j,0}\sigma_{c,0} + J_{j,j}\left(\sigma_{c,j} - \sigma_{c,j-1}\right) + \sum_{i=1}^{j-1} J_{j,i}\left(\sigma_{c,i} - \sigma_{c,i-1}\right)$$

$$= J_{j,0}\sigma_{c,0} + J_{j,j}\sigma_{c,j} - J_{j,j}\sigma_{c,j-1} + \sum_{i=1}^{j-1} J_{j,i}\sigma_{c,i} - \sum_{i=1}^{j-1} J_{j,i}\sigma_{c,i-1}$$

$$= J_{j,0}\sigma_{c,0} + J_{j,j}\sigma_{c,j} - J_{j,j}\sigma_{c,j-1} + \sum_{i=1}^{j-1} J_{j,i}\sigma_{c,i} - \sum_{i=0}^{j-2} J_{j,i+1}\sigma_{c,i}$$

$$= J_{j,0}\sigma_{c,0} + J_{j,j}\sigma_{c,j} - J_{j,j}\sigma_{c,j-1} + J_{j,j-1}\sigma_{c,j-1} + \sum_{i=1}^{j-2}(J_{j,i} - J_{j,i+1})\,\sigma_{c,i} - J_{j,1}\sigma_{c,0}$$

$$= J_{j,0}\sigma_{c,0} + J_{j,j}\sigma_{c,j} + \sum_{i=1}^{j-1}\left(J_{j,i} - J_{j,i+1}\right)\sigma_{c,i} - J_{j,1}\sigma_{c,0}$$

$$= J_{j,j}\sigma_{c,j} + \left(J_{j,0} - J_{j,1}\right)\,\sigma_{c,0} + \sum_{i=1}^{j-1}\left(J_{j,i} - J_{j,i+1}\right)\,\sigma_{c,i}$$

and therefore:

$$\varepsilon_j - \varepsilon_{sh,j} = J_{j,j}\,\sigma_{c,j} + \sum_{i=0}^{j-1}\left(J_{j,i} - J_{j,i+1}\right)\,\sigma_{c,i} \tag{4.24}$$

Expressing Eq. 4.24 in terms of the stress at the end of the j-th time step, the constitutive relationship for the concrete at any time τ_j becomes:

$$\sigma_{c,j} = \frac{\varepsilon_j - \varepsilon_{sh,j}}{J_{j,j}} - \sum_{i=0}^{j-1}\frac{J_{j,i} - J_{j,i+1}}{J_{j,j}}\,\sigma_{c,i} \quad = E_{c,j}\left(\varepsilon_j - \varepsilon_{sh,j}\right) + \sum_{i=0}^{j-1} F_{e,j,i}\sigma_{c,i} \tag{4.25}$$

where $E_{c,j}$ is the instantaneous elastic modulus of concrete at τ_j; $F_{e,j,i}$ is the stress modification factor; and these terms are defined as

$$E_{c,j} = \frac{1}{J_{j,j}} \tag{4.26a}$$

and

$$F_{e,j,i} = \frac{J_{j,i+1} - J_{j,i}}{J_{j,j}} \tag{4.26b}$$

For completeness, the corresponding expressions for $E_{c,j}$ and $F_{e,j,i}$ to be used in Eq. 4.25 when adopting the trapezoidal approximation of the integral-type creep law are:

For $i = 0$:

$$E_{c,j} = \frac{1}{J_{j,j}} \tag{4.26c}$$

and

$$F_{e,j,i} = \frac{J_{j,1} - J_{j,0}}{J_{j,j} + J_{j,j-1}} \tag{4.26d}$$

For $i > 0$:

$$E_{c,j} = \frac{2}{J_{j,j} + J_{j,j-1}} \tag{4.26e}$$

and

$$F_{e,j,i} = \frac{J_{j,i+1} - J_{j,i-1}}{J_{j,j} + J_{j,j-1}} \tag{4.26f}$$

Noting that the steel and concrete strains are always identical, the equilibrium equation at the end of the j-th time step for the axially loaded column example can be expressed as:

$$\begin{aligned} P &= \sigma_{c,j} A_c + \sigma_{s,j} A_s \\ &= A_c E_{c,j}\,\varepsilon_j - A_c E_{c,j}\,\varepsilon_{sh,j} + \sum_{i=0}^{j-1} A_c F_{e,j,i} \sigma_{c,i} + A_s E_s \varepsilon_j \end{aligned} \tag{4.27}$$

where $\sigma_{s,j} = \sigma_s(\tau_j)$ and $\varepsilon_j (= \varepsilon_{s,j})$ is the strain in both the concrete and the steel at τ_j.

Solving for the concrete strain ε_j gives:

$$\varepsilon_j = \frac{P}{A_c E_{c,j} + A_s E_s} + \frac{A_c E_{c,j} \varepsilon_{sh,j}}{A_c E_{c,j} + A_s E_s} - \sum_{i=0}^{j-1} \frac{A_c F_{e,j,i}}{A_c E_{c,j} + A_s E_s} \sigma_{c,i} \tag{4.28}$$

and observing that

$$\frac{A_c E_{c,j}}{A_c E_{c,j} + A_s E_s} = \frac{1}{1 + \dfrac{E_s A_s}{E_{c,j} A_c}} = \frac{1}{1 + n_j \rho}$$

Eq. 4.28 can be rewritten as:

$$\varepsilon_j = \frac{1}{1 + n_j \rho} \left(\frac{P}{A_c E_{c,j}} + \varepsilon_{sh,j} - \sum_{i=0}^{j-1} \frac{F_{e,j,i}}{E_{c,j}} \sigma_{c,i} \right) \tag{4.29}$$

It is also noted that:

$$\begin{aligned} \frac{F_{e,j,i}}{E_{c,j}} &= \frac{J_{j,i+1} - J_{j,i}}{J_{j,j}} J_{j,j} = J_{j,i+1} - J_{j,i} = \frac{1}{E_e(\tau_j, \tau_{i+1})} - \frac{1}{E_e(\tau_j, \tau_i)} = \frac{1}{E_{e,j,i+1}} - \frac{1}{E_{e,j,i}} \\ &= \frac{n_{e,j,i+1}}{E_s} - \frac{n_{e,j,i}}{E_s} \end{aligned}$$

where $n_{e,j,i} = E_s / E_{e,j,i}$ and, for ease of notation, $E_{e,j,i} = E_e(\tau_j, \tau_i)$. Eq. 4.29 can therefore be simplified to:

$$\varepsilon_j = \frac{1}{1 + n_j \rho} \left(\frac{P}{A_c E_{c,j}} + \varepsilon_{sh,j} - \sum_{i=0}^{j-1} \frac{n_{e,j,i+1} - n_{e,j,i}}{E_s} \sigma_{c,i} \right) \tag{4.30}$$

Substituting Eq. 4.30 into Eq. 4.25 gives:

$$\sigma_{c,j} = \frac{1}{1+n_j\rho}\left(\frac{P}{A_c} + E_{c,j}\varepsilon_{sh,j} - E_{c,j}\sum_{i=0}^{j-1}\frac{n_{e,j,i+1}-n_{e,j,i}}{E_s}\sigma_{c,i}\right) - E_{c,j}\varepsilon_{sh,j} + \sum_{i=0}^{j-1} F_{e,j,i}\sigma_{c,i}$$

$$= \frac{1}{1+n_j\rho}\left(\frac{P}{A_c} - n_j\rho E_{c,j}\varepsilon_{sh,j} - E_{c,j}\sum_{i=0}^{j-1}\frac{n_{e,j,i+1}-n_{e,j,i}}{E_s}\sigma_{c,i} + \sum_{i=0}^{j-1} F_{e,j,i}\sigma_{c,i} + n_j\rho\sum_{i=0}^{j-1} F_{e,j,i}\sigma_{c,i}\right)$$

which can be simplified to:

$$\sigma_{c,j} = \frac{1}{1+n_j\rho}\left[\frac{P}{A_c} - E_s\rho\,\varepsilon_{sh,j} + \rho\sum_{i=0}^{j-1}\left(n_{e,j,i+1} - n_{e,j,i}\right)\sigma_{c,i}\right] \tag{4.31}$$

With the concrete stress $\sigma_{c,j}$ at time, τ_j, calculated from Eq. 4.31, the steel stress is obtained from the equilibrium equation:

$$\sigma_{s,j} = \frac{P - \sigma_{c,j}A_c}{A_s} \tag{4.32}$$

and the steel and concrete strains are obtained from Eqs 4.18b and 4.17, respectively.

Example 4.3

The column analysed in Example 4.2 is to be re-analysed here using approach 2. As before, $P = -1000$ kN; $A_c = 90{,}000$ mm^2; $A_s = 1800$ mm^2; $\rho = A_s/A_c = 0.02$; $E_s = 200$ GPa and the time varying concrete properties are:

j	0	1	2	3	4	5	6
τ_j (days)	14	24	44	84	214	514	10,014
$t-\tau_j$ (days)	0	10	30	70	200	500	10,000
$\varphi(\tau_j, \tau_0)$	0	0.53	0.98	1.38	1.83	2.10	2.39
$\varphi(\tau_j, \tau_1)$		0	0.72	1.18	1.63	1.89	2.15
$\varphi(\tau_j, \tau_2)$			0	0.90	1.44	1.69	1.94
$\varphi(\tau_j, \tau_3)$				0	1.22	1.51	1.75
$\varphi(\tau_j, \tau_4)$					0	1.26	1.54
$\varphi(\tau_j, \tau_5)$						0	1.38
$E_c(\tau_j)$ (MPa)	26,700	27,900	28,900	29,700	30,500	31,000	31,700
$\varepsilon_{sh}(\tau_j)\times 10^{-6}$	0	−142	−246	−325	−407	−456	−510

At first loading (τ_0):

As in Example 4.2, with $E_{c,0} = E_{e,0} = 26.7$ GPa; and $n_0 = E_s/E_{c,0} = 7.48$, the instantaneous stresses and strains immediately after first loading are: $\varepsilon_0 = -361 \times 10^{-6}$; $\varepsilon_{s,0} = -361 \times 10^{-6}$; $\sigma_{c,0} = -9.67$ MPa and $\sigma_{s,0} = -72.3$ MPa.

At time instant τ_1 = 24 days (time step 1 from τ_0 to τ_1):

For time interval, $j = 1$:
$E_{c,0} = 26.7$ GPa; $E_{c,1} = 27.9$ GPa; $n_1 = E_s/E_{c,1} = 7.17$; $\varphi_{1,0} = 0.53$; $E_{e,1,0} = E_{c,0}/(1+\varphi_{1,0}) = 17.48$ GPa; $n_{e,1,0} = E_s/E_{e,1,0} = 11.44$; $n_{e,1,1} = E_s/E_{e,1,1} = 7.17$ and $\varepsilon_{sh,1} = -142 \times 10^{-6}$.
From Eq. 4.31, the concrete stress at the end of the first time step ($j = 1$) is:

$$\sigma_{c,1} = \frac{1}{1+7.17 \times 0.02}\left[\frac{-1000 \times 10^3}{90{,}000} - 200{,}000 \times 0.02 \times (-142 \times 10^{-6})\right.$$

$$\left. +0.02 \times (7.17 - 11.44) \times (-9.67)\right] = -8.50 \text{ MPa.}$$

and from Eq. 4.32: $\sigma_{s,1} = \dfrac{-1000 \times 10^3 + 8.50 \times 90{,}000}{1800} = -131$ MPa;
Eq. 4.18b and Eq. 4.17 give: $\varepsilon_{s,1} = \dfrac{-131}{200{,}000} = -653 \times 10^{-6}$ and $\varepsilon_1 = -653 \times 10^{-6}$.
As calculated in Example 4.2, the elastic, creep and shrinkage components of the concrete strain at time τ_1 are $\varepsilon_{e,1} = -319 \times 10^{-6}$, $\varepsilon_{cr,1} = -192 \times 10^{-6}$, and $\varepsilon_{sh}(\tau_1) = -142 \times 10^{-6}$.

At time instant $\tau_2 = 44$ days (time step 2 from τ_1 to τ_2):

Time interval, $j = 2$:
$E_{c,0} = 26.7$ GPa; $E_{c,1} = 27.9$ GPa; $E_{c,2} = 28.9$ GPa; $n_2 = E_s/E_{c,2} = 6.92$; $\varphi_{2,0} = 0.98$; $E_{e,2,0} = E_{c,0}/(1+\varphi_{2,0}) = 13.53$ GPa; $n_{e,2,0} = E_s/E_{e,2,0} = 14.78$; $\varphi_{.2,1} = 0.72$; $E_{e,2,1} = E_{c,1}/(1+\varphi_{2,1}) = 16.25$ GPa; $n_{e,2,1} = E_s/E_{e,2,1} = 12.31$; and $\varepsilon_{sh,2} = -246 \times 10^{-6}$.
From Eq. 4.31, the concrete stress at the end of the first time step ($j = 1$) is:

$$\sigma_{c,2} = \frac{1}{1+6.92 \times 0.02}\left[\frac{-1000 \times 10^3}{90{,}000} - 200{,}000 \times 0.02 \times (-246 \times 10^{-6})\right.$$

$$\left. +0.02 \times [(12.31 - 14.78) \times (-9.67) + (6.92 - 12.31) \times (-8.50)]\right]$$

$$= -7.67 \text{ MPa}$$

and from Eq. 4.32: $\sigma_{s,1} = \dfrac{-1000 \times 10^3 + 7.67 \times 90{,}000}{1800} = -172$ MPa.

Eqs 4.18b and 4.17 give: $\varepsilon_{s,1} = \dfrac{-172}{200{,}000} = -860 \times 10^{-6}$ and $\varepsilon_1 = -860 \times 10^{-6}$.

As in Example 2.2, the elastic, creep and shrinkage components of concrete strain at τ_2 are $\varepsilon_{e,2} = -291 \times 10^{-6}$; $\varepsilon_{cr,2} = -323 \times 10^{-6}$, and $\varepsilon_{sh}(\tau_2) = -246 \times 10^{-6}$.

At subsequent time instants:

Similarly, the stress and strain components at the end of each subsequent time step are calculated and the results are tabulated below. It can be seen that the results obtained using the SSM with approaches 1 and 2 are identical.

Instant j	*Time* τ_j *(days)*	*Concrete stress* $\sigma_c(\tau_j)$ *(MPa)*	*Steel stress* $\sigma_s(\tau_j)$ *(MPa)*	*Total strain* $\varepsilon(\tau_j)$ $(\times 10^{-6})$	*Elastic strain* $\varepsilon_e(\tau_j)$ $(\times 10^{-6})$	*Creep strain* $\varepsilon_{cr}(\tau_j)$ $(\times 10^{-6})$	*Shrinkage strain* $\varepsilon_{sh}(\tau_j)$ $(\times 10^{-6})$
0	14	−9.67	−72.3	−361	−361	0	0
1	24	−8.50	−131	−653	−319	−192	−142
2	44	−7.67	−172	−860	−291	−323	−246
3	84	−7.04	−204	−1018	−269	−424	−325
4	214	−6.39	−236	−1180	−248	−525	−407
5	514	−6.05	−253	−1264	−237	−571	−456
6	10,014	−5.64	−273	−1367	−225	−632	−510

4.4 The age-adjusted effective modulus method (AEMM)

4.4.1 Formulation

A simple adjustment to the effective modulus method to account for the ageing of concrete was proposed by Trost (Ref. 6). Later the method was more rigorously formulated and further developed by Dilger and Neville (Ref. 7) and Bazant (Ref. 8). The method is sometimes called the *Trost–Bazant Method* (Ref. 9), but Bazant's more descriptive title, the *age-adjusted effective modulus method*, is preferred here.

Consider the two concrete stress histories and the corresponding creep-time curves shown in Fig. 4.6. In stress history (a), $\sigma_c(\tau_0)$ is applied at time τ_0 and subsequently held constant with time. In stress history (b), the stress $\sigma_c(t)$ is gradually applied, beginning at τ_0 and reaching a magnitude equal to $\sigma_c(\tau_0)$ at time τ_k. The creep strain at any time $t(> \tau_0)$ produced by the gradually applied stress $\sigma_c(t)$ is significantly smaller than that resulting from the stress $\sigma_c(\tau_0)$ abruptly applied at τ_0, as shown. This is due to ageing. The earlier a concrete specimen is loaded, the greater the final creep strain. A reduced creep coefficient can therefore be used to calculate creep strain if stress is gradually applied. Let this reduced creep coefficient be $\chi(t, \tau_0)\varphi(t, \tau_0)$, where the coefficient $\chi(t, \tau_0)$ is called the *ageing coefficient* (< 1.0).

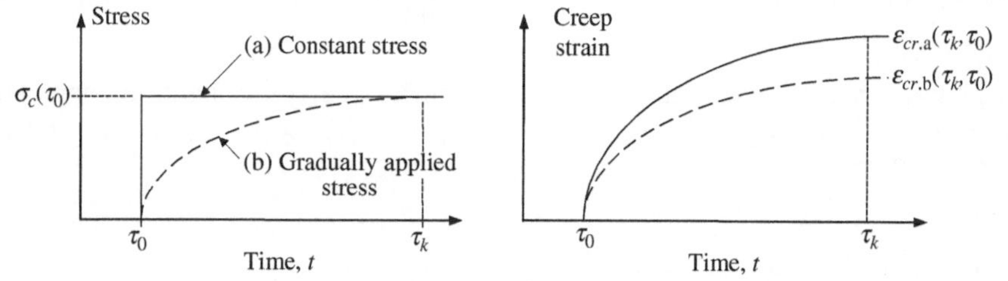

$$\varepsilon_{cr.a}(\tau_k, \tau_0) = \frac{\sigma_c(\tau_0)}{E_c(\tau_0)} \varphi(\tau_k, \tau_0)\,; \quad \varepsilon_{cr.b}(\tau_k, \tau_0) = \frac{\sigma_0(\tau_0)}{E_c(\tau_0)} \chi(\tau_k, \tau_0)\varphi(\tau_k, \tau_0)$$

Figure 4.6 Creep due to constant and variable stress histories.

The creep strain at time, t, due to a stress, $\sigma_c(t)$, that has been gradually applied over the time interval $t - \tau_0$, may be expressed as:

$$\varepsilon_{cr}(t) = \frac{\sigma_c(t)}{E_c(\tau_0)} \chi(t, \tau_0)\, \varphi(t, \tau_0) \tag{4.33}$$

The magnitude of the *ageing coefficient* $\chi(t, \tau_0)$ generally falls within the range 0.4 to 1.0 depending on the rate of application of the gradually applied stress in the period after τ_0.

Consider the gradually reducing stress history shown in Fig. 4.7. An initial compressive stress $\sigma_c(\tau_0)$ applied at time τ_0 is reduced with time due to the application of a gradually increasing tensile stress increment $\Delta\sigma_c(t)$. The change of stress may be due to a change of the external loads, or restraint to creep and shrinkage, or variations of temperature, or combinations of these, and is usually unknown at the beginning of an analysis.

Using the AEMM, the total strain at time t may be expressed as the sum of the strains produced by $\sigma_c(\tau_0)$ (instantaneous and creep), the strains produced by the

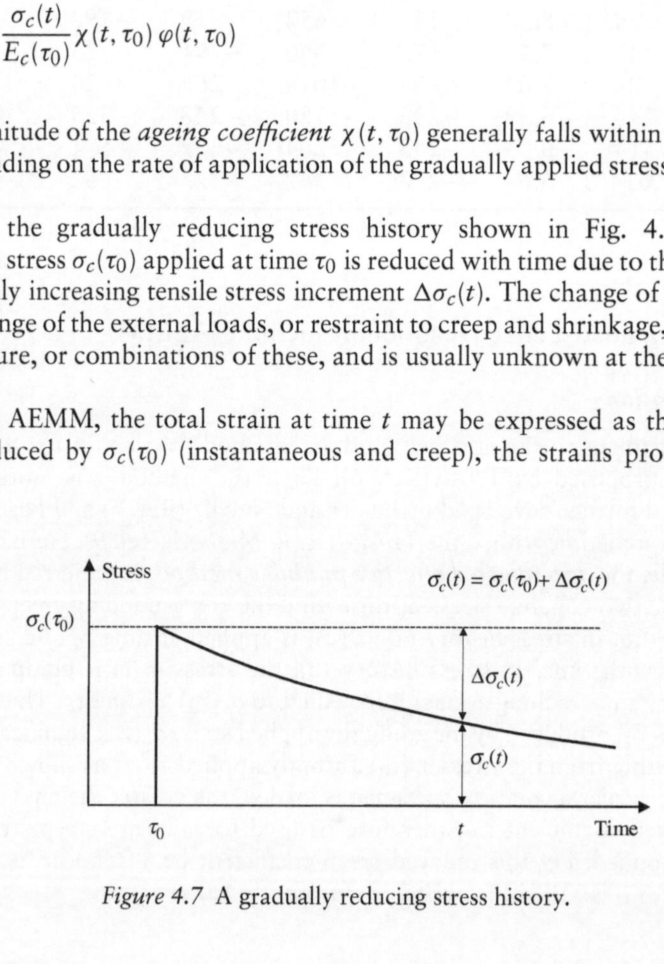

Figure 4.7 A gradually reducing stress history.

gradually applied stress increment $\Delta\sigma_c(t)$ (instantaneous and creep), and the shrinkage strain:

$$\varepsilon(t) = \frac{\sigma_c(\tau_0)}{E_c(\tau_0)}[1+\varphi(t,\tau_0)] + \frac{\Delta\sigma_c(t)}{E_c(\tau_0)}[1+\chi(t,\tau_0)\,\varphi(t,\tau_0)] + \varepsilon_{sh}(t)$$

$$= \frac{\sigma_c(\tau_0)}{E_e(t,\tau_0)} + \frac{\Delta\sigma_c(t)}{\overline{E}_e(t,\tau_0)} + \varepsilon_{sh}(t) \tag{4.34}$$

where $E_e(t,\tau_0)$ is the effective modulus of Eq. 4.3 and $\overline{E}_e(t,\tau_0)$ is the *age-adjusted effective modulus* given by:

$$\overline{E}_e(t,\tau_0) = \frac{E_c(\tau_0)}{1+\chi(t,\tau_0)\,\varphi(t,\tau_0)} \tag{4.35}$$

With the AEMM, two analyses need to be carried out: one at first loading (time τ_0) and one at time t after the period of sustained stress. Unlike the EMM, the final strain depends on the stress at first loading $\sigma_c(\tau_0)$ and the change in stress that occurs with time $\Delta\sigma_c(t)$.

The reinforced concrete column shown in Fig. 4.1 is to be analysed using the AEMM. As for the SSM, two approaches will also be considered here:

(i) Approach 1 obtains the solution for the concrete stress at time t, $\sigma_c(t)$, by calculating the change in stress that occurs over the entire period of sustained loading, $\Delta\sigma_c(t)$; and
(ii) Approach 2 calculated the stress at time t, $\sigma_c(t)$, directly.

Both approaches lead to the same solution and differ only in the manipulation of the governing equations.

4.4.2 Example application (AEMM) – Approach 1

The instantaneous analysis at time τ_0 (immediately after the application of the axial force P, and before any creep and shrinkage has occurred) is identical to that presented earlier and the instantaneous stresses and strains are:

$$\sigma_c(\tau_0) = \frac{P}{A_c(1+n_0\rho)};\ \sigma_s(\tau_0) = \frac{P-\sigma_c(\tau_0)A_c}{A_s};\ \varepsilon(\tau_0) = \frac{\sigma_c(\tau_0)}{E_c(\tau_0)} \text{ and } \varepsilon_s(\tau_0) = \frac{\sigma_s(\tau_0)}{E_s}$$

where $n_0 = E_s/E_c(\tau_0)$ = the modular ratio at time τ_0 and $\rho = A_s/A_c$ = the reinforcement ratio.

As described in Section 4.2.2, the equilibrium and compatibility requirements at any time t are specified by Eqs 4.4 and 4.5, respectively:

$$P = \sigma_c(t)A_c + \sigma_s(t)A_s \tag{4.4}$$

and

$$\varepsilon_s(t) = \varepsilon(t) \tag{4.5}$$

The constitutive relationship for concrete is described by Eq. 4.34 and, as before, the steel is linear-elastic:

$$\varepsilon(t) = \frac{\sigma_c(\tau_0)}{E_e(t,\tau_0)} + \frac{\Delta\sigma_c(t)}{\overline{E}_e(t,\tau_0)} + \varepsilon_{sh}(t) \tag{4.36a}$$

and

$$\varepsilon_s(t) = \frac{\sigma_s(t)}{E_s} \tag{4.36b}$$

where $\Delta\sigma_c(t)$ is the unknown time-dependent change of stress caused by creep and shrinkage. If $\sigma_c(\tau_0)$ is compressive, the stress increment $\Delta\sigma_c(t)$ is usually tensile.

Rearranging Eqs 4.36 in terms of time-varying stresses gives:

$$\Delta\sigma_c(t) = \overline{E}_e(t,\tau_0)\varepsilon(t) - \frac{\overline{E}_e(t,\tau_0)}{E_e(t,\tau_0)}\sigma_c(\tau_0) - \overline{E}_e(t,\tau_0)\varepsilon_{sh}(t) \tag{4.37a}$$

and

$$\sigma_s(t) = E_s\varepsilon_s(t) \tag{4.37b}$$

At any time, t, the concrete stress is given by:

$$\sigma_c(t) = \sigma_c(\tau_0) + \Delta\sigma_c(t) \tag{4.38}$$

and the governing equation of the problem is obtained by substituting Eqs 4.37, 4.38 and 4.5 into Eq. 4.4 and solving for the unknown total strain $\varepsilon(t)$:

$$\varepsilon\,(t) = \frac{1}{1+\bar{n}_{e,0}\rho}\left[\frac{P}{\overline{E}_e(t,\tau_0)A_c} + \sigma_c(\tau_0)\left(\frac{1}{E_e(t,\tau_0)} - \frac{1}{\overline{E}_e(t,\tau_0)}\right) + \varepsilon_{sh}(t)\right] \tag{4.39}$$

where $n_{e,0} = E_s/E_e(t,\tau_0)$, $\rho = A_s/A_c$, and $\bar{n}_{e,0} = E_s/\overline{E}_e(t,\tau_0)$ = the age-adjusted modular ratio.

By substituting Eq. 4.39 into Eq. 4.36a, the concrete stress increment $\Delta\sigma_c(t)$ can be expressed as:

$$\Delta\sigma_c(t) = \frac{1}{1+\bar{n}_{e,0}\rho}\left[\frac{P}{A_c} - \sigma_c(\tau_0)(1+n_{e,0}\rho) - \varepsilon_{sh}(t)\,E_s\,\rho\right] \tag{4.40}$$

With $\Delta\sigma_c(t)$ calculated from Eq. 4.40, the resultant steel and concrete stresses at time t are given by Eqs 4.37b and 4.38. Alternatively, these could be calculated from equilibrium considerations (Eq. 4.4).

4.4.3 Determination of the ageing coefficient

Numerical values for $\chi(t,\tau_0)$ can be obtained from the associated creep coefficient $\varphi(t,\tau_0)$ if the stress change $\Delta\sigma_c(t)$ is known. Like the creep coefficient, the ageing coefficient $\chi(t, \tau_0)$ depends on the age of the concrete at first loading τ_0, the duration of load $(t - \tau_0)$, the size and shape of the member, and so on.

For the gradually decreasing concrete stress history when a member is subjected to a constant sustained load, the ageing coefficient $\chi(t, \tau_0)$ may be obtained by rearranging and expanding Eq. 4.40. With shrinkage set to zero, $\chi(t, \tau_0)$ is given by:

$$\chi(t, \tau_0) = \frac{\sigma_c(\tau_0)}{\sigma_c(\tau_0) - \sigma_c(t)} - \frac{1}{\varphi(t, \tau_0)} \frac{1 + n\rho}{n\rho} \tag{4.41}$$

If the stress history of Fig. 4.7 is the result of pure relaxation (i.e. when the member is restrained and the strain $\varepsilon(t)$ remains constant with time at its initial value of $\sigma_c(\tau_0)/E_c(\tau_0)$), the ageing coefficient $\chi(t, \tau_0)$ may be obtained by rearranging the expanded form of Eq. 4.34:

$$\chi(t, \tau_0) = \frac{\sigma_c(\tau_0)}{\sigma_c(\tau_0) - \sigma_c(t)} - \frac{1}{\varphi(t, \tau_0)} \tag{4.42a}$$

Eq. 4.42a is commonly expressed in the literature in the following form:

$$\chi(t, \tau_0) = \frac{E_c(\tau_0)}{E_c(\tau_0) - R(t, \tau_0)} - \frac{1}{\varphi(t, \tau_0)} \tag{4.42b}$$

in which $R(t, \tau_0)$ represents the relaxation function defined as the stress at time t due to a unit strain applied at time τ_0 and kept constant throughout the period τ_0 to t. Values for $R(t, \tau_0)$ can be readily calculated by applying the SSM (Eq. 4.25) to the case of a constant unit strain history beginning at τ_0.

Using a step-by-step numerical analysis (Section 4.3.2) to establish the gradual loss of concrete stress with time ($\Delta\sigma_c(t) = \sigma_c(t) - \sigma_c(\tau_0)$), ageing coefficients $\chi(t, \tau_0)$ may be readily determined for the creep coefficients $\varphi(t, \tau_0)$ specified by Eq. 2.3 in Section 2.1.4. For the cross-section, concrete properties and loading details of Examples 4.1 and 4.2, the ageing coefficients associated with the gradual concrete stress change under a constant external load P first applied at age τ_0 are calculated using Eq. 4.41 and shown in Fig. 4.8. Also shown in Fig. 4.8 are the ageing coefficients calculated using Eq. 4.42a associated with the more rapid change in concrete stress caused by pure relaxation (i.e. when the total strain $\varepsilon(t)$ is held constant with time and the applied load P varies as a result of relaxation). Neville *et al.* (Ref. 9) and others have shown that ageing coefficients calculated in this way, when used in conjunction with the age-adjusted effective modulus method, provided close agreement with experimental data.

Fig. 4.9 shows the variation with time of the ageing coefficients obtained from the creep coefficients of Eq. 2.3 for a creep problem similar to that of Examples 4.1 and 4.2.

For concrete loaded in the first few weeks after casting (such as for the curves corresponding to $\tau_0 = 7$ and 14 days in Fig. 4.9), and for load durations exceeding about 100 days, the ageing coefficient is between 0.55 and 0.72 for normal strength concrete with a value of 0.65 being suitable for most practical situations where final deformations are required. Considering how uncertain are the predictions of $\varphi(t, \tau_0)$

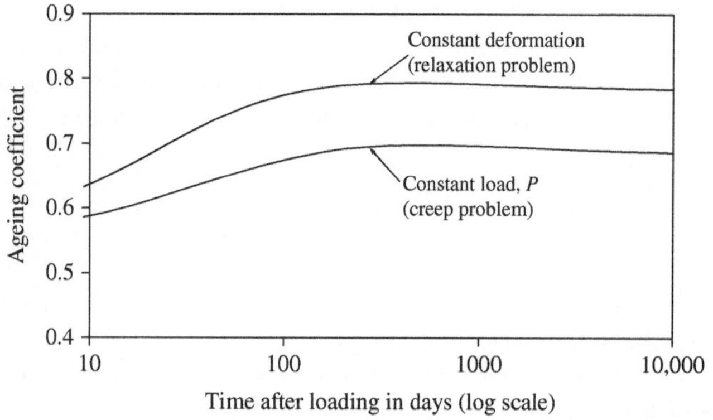

Figure 4.8 Ageing coefficient versus time for the concrete used in Example 4.2.

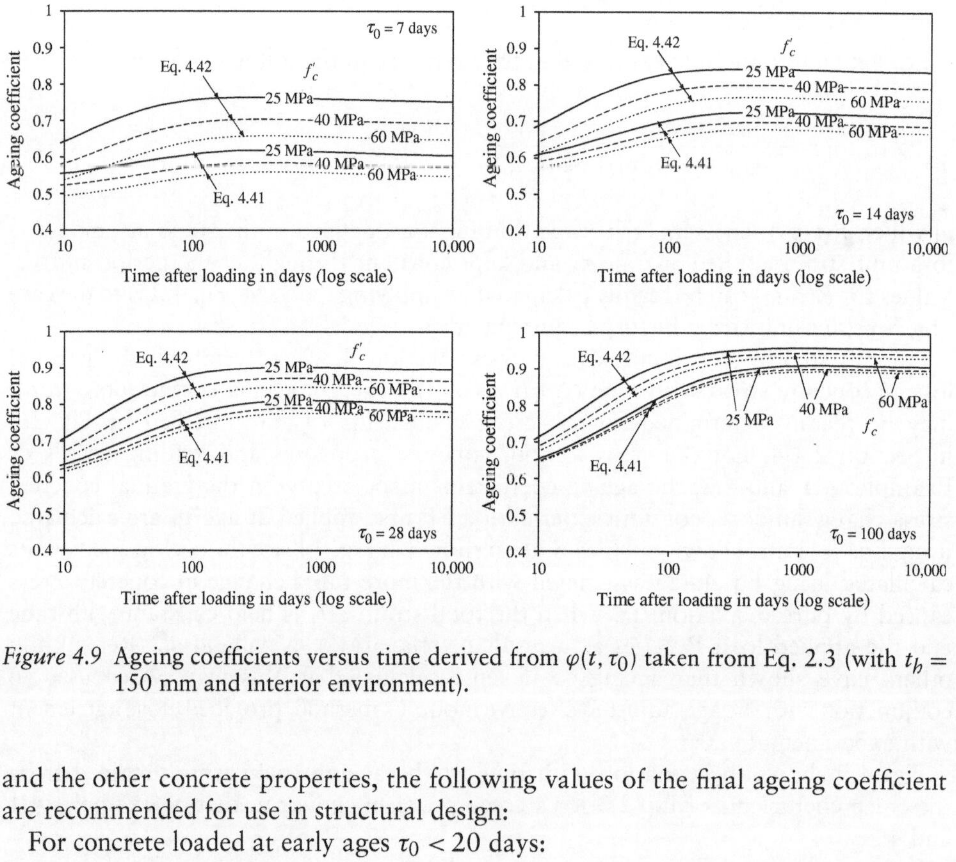

Figure 4.9 Ageing coefficients versus time derived from $\varphi(t, \tau_0)$ taken from Eq. 2.3 (with t_h = 150 mm and interior environment).

and the other concrete properties, the following values of the final ageing coefficient are recommended for use in structural design:

For concrete loaded at early ages $\tau_0 < 20$ days:

$$\chi(\infty, \tau_0) = \chi^*(\tau_0) = 0.65 \text{ for creep problems (constant load)} \tag{4.43a}$$

$$\chi(\infty, \tau_0) = \chi^*(\tau_0) = 0.80 \text{ for relaxation problems (constant deformation)} \tag{4.43b}$$

For concrete loaded at later ages $\tau_0 > 28$ days:

$$\chi(\infty, \tau_0) = \chi^*(\tau_0) = 0.75 \text{ for creep problems} \quad (4.43c)$$

$$\chi(\infty, \tau_0) = \chi^*(\tau_0) = 0.85 \text{ for relaxation problems} \quad (4.43d)$$

The use of Eq. 4.43 simplifies the age-adjusted effective modulus method and usually leads to good approximations of both material and structural behaviour.

The inadequacies of the effective modulus method are to a large extent overcome by the introduction of the ageing coefficient. Bazant (Ref. 8) pointed out that the method is theoretically exact for any problem in which strain varies proportionally with the creep coefficient (provided, of course, the ageing coefficient has been carefully calculated from the assumed creep coefficient). Extracting $\chi(t, \tau_0)$ from Fig. 4.9 is not particularly suitable for computer application nor is it generally necessary. The approximations of Eq. 4.43 are suitable for most applications.

Example 4.4

The symmetrically reinforced column analysed in Example 4.1 (see Fig. 4.1) under a constant sustained axial load P is re-analysed here using the AEMM. As before: $P = -1000$ kN, $A_c = 90{,}000$ mm^2; $A_s = 1800$ mm^2; $\rho = A_s/A_c = 0.02$ and $E_s = 200$ GPa.

The variation of the creep coefficient and shrinkage with time were given in Example 4.1 and, for convenience, are repeated in the following table. The ageing coefficients are taken from Fig. 4.9 and are also given below. All other data as given in Example 4.1.

$(t - \tau_0)$ in days	0	10	30	70	200	500	10,000 (∞)
$\varphi(t, \tau_0)$	0	0.53	0.98	1.38	1.83	2.10	2.39
$\chi(t, \tau_0)$	–	0.60	0.64	0.67	0.70	0.70	0.69
$\varepsilon_{sh}(t - \tau_0) \times 10^{-6}$	0	−142	−246	−325	−407	−456	−510

At first loading, $(t - \tau_0) = 0$:

$\varphi(t, \tau_0) = 0$ and $\varepsilon_{sh}(\tau_0) = 0$.

As in Example 4.1: with $E_c(\tau_0) = E_e(\tau_0) = 26.7$ GPa; and $n_0 = E_s/E_c(\tau_0) = 7.48$, the instantaneous stresses and strains immediately after first loading are: $\varepsilon(\tau_0) = -361 \times 10^{-6}$; $\varepsilon_s(\tau_0) = -361 \times 10^{-6}$; $\sigma_c(\tau_0) = -9.67$ MPa and $\sigma_s(\tau_0) = -72.3$ MPa.

At $(t - \tau_0) = 10$ days:

$\varphi(t, \tau_0) = 0.53$, $\chi(t, \tau_0) = 0.60$ and $\varepsilon_{sh}(t) = -142 \times 10^{-6}$.

$E_e(t, \tau_0) = E_c(\tau_0)/(1 + \varphi(t, \tau_0)) = 17.48$ GPa; and $n_{e,0} = E_s/E_e(t, \tau_0) = 11.44$;

$\bar{E}_e(t, \tau_0) = E_c(\tau_0)/(1 + \chi(t, \tau_0)\varphi(t, \tau_0)) = 20.27$ GPa; and $\bar{n}_{e,0} = E_s/\bar{E}_e(t, \tau_0) = 9.87$.

Eq. 4.40: $\Delta\sigma_c(t) = \frac{1}{1+9.87\times 0.02}\left[\frac{-1000\times 10^3}{90{,}000} + 9.67(1+11.44\times 0.02)\right.$

$\left. + 142\times 10^{-6}\times 200{,}000\times 0.02\right] = +1.12\text{ MPa};$

Eq. 4.38: $\sigma_c(t) = -9.67 + 1.12 = -8.55\text{ MPa};$

Eq. 4.4: $\sigma_s(t) = ((-1000\times 10^3 + 8.55\times 90{,}000)/1800 = -128.0\text{ MPa};$

Eq. 4.36b: $\varepsilon_s(t) = -128.0/200{,}000 = -640\times 10^{-6}$; and

Eq. 4.5: $\varepsilon(t) = -640\times 10^{-6}$.

The elastic and shrinkage components of the total concrete strain are $\varepsilon_e(t) = \sigma_c(t)/E_c(\tau_0) = -320\times 10^{-6}$ and $\varepsilon_{sh}(t) = -142\times 10^{-6}$; and therefore the creep component of strain is $\varepsilon_{cr}(t) = \varepsilon(t) - \varepsilon_e(t) - \varepsilon_{sh}(t) = -178\times 10^{-6}$.
The results of the analyses at first loading and throughout the period of sustained loading are provided in the following table.

Time t (days)	*Duration of load* $t-\tau_0$ *(days)*	*Concrete stress* $\sigma_c(t)$ *(MPa)*	*Steel stress* $\sigma_s(t)$ *(MPa)*	*Total strain* $\varepsilon(t)$ $(\times 10^{-6})$	*Elastic strain* $\varepsilon_e(t)$ $(\times 10^{-6})$	*Creep strain* $\varepsilon_{cr}(t)$ $(\times 10^{-6})$	*Shrinkage strain* $\varepsilon_{sh}(t)$ $(\times 10^{-6})$
14	0	−9.67	−72.3	−361	−361	0	0
24	10	−8.55	−128	−640	−320	−178	−142
44	30	−7.74	−169	−843	−289	−308	−246
84	70	−7.10	−200	−1002	−266	−411	−325
214	200	−6.48	−232	−1158	−242	−509	−407
514	500	−6.12	−249	−1247	−229	−563	−456
10,014	10,000	−5.73	−269	−1346	−214	−622	−510

By comparison, the variations of stresses and strains in this example if the approximation of Eq. 4.43 is adopted throughout (i.e. if $\chi(t,\tau_0)$ is equal to 0.65 for all values of t) are provided below. For practical purposes, the two sets of results are the same.

Time t (days)	*Duration of load* $t-\tau_0$ *(days)*	*Concrete stress* $\sigma_c(t)$ *(MPa)*	*Steel stress* $\sigma_s(t)$ *(MPa)*	*Total strain* $\varepsilon(t)$ $(\times 10^{-6})$	*Elastic strain* $\varepsilon_e(t)$ $(\times 10^{-6})$	*Creep strain* $\varepsilon_{cr}(t)$ $(\times 10^{-6})$	*Shrinkage strain* $\varepsilon_{sh}(t)$ $(\times 10^{-6})$
14	0	−9.67	−72.3	−361	−361	0	0
24	10	−8.55	−128	−639	−320	−177	−142
44	30	−7.74	−169	−843	−289	−307	−246
84	70	−7.10	−201	−1004	−265	−413	−325
214	200	−6.45	−233	−1166	−241	−518	−407
514	500	−6.08	−252	−1258	−227	−575	−456
10,014	10,000	−5.69	−271	−1355	−213	−632	−510

4.4.4 *Example application (AEMM) – Approach 2*

As discussed for the SSM, approach 2 involves expressing the constitutive relationship in terms of the initial stress $\sigma_c(\tau_0)$ and final stress $\sigma_c(t)$. Eq. 4.36a is re-expressed as:

$$\begin{aligned}\varepsilon(t) &= \frac{\sigma_c(\tau_0)}{E_e(t,\tau_0)} + \frac{\sigma_c(t)-\sigma_c(\tau_0)}{\overline{E}_e(t,\tau_0)} + \varepsilon_{sh}(t) \\ &= \frac{\sigma_c(\tau_0)\varphi(t,\tau_0)[1-\chi(t,\tau_0)]}{E_c(\tau_0)} + \frac{\sigma_c(t)[1+\chi(t,\tau_0)\varphi(t,\tau_0)]}{E_c(\tau_0)} + \varepsilon_{sh}(t)\end{aligned} \tag{4.44}$$

and rearranging gives:

$$\sigma_c(t) = \overline{E}_e(t,\tau_0)[\varepsilon(t) - \varepsilon_{sh}(t)] + \sigma_c(\tau_0)\overline{F}_{e,0} \tag{4.45}$$

where

$$\overline{F}_{e,0} = \varphi(t,\tau_0)\frac{[\chi(t,\tau_0)-1]}{[1+\chi(t,\tau_0)\varphi(t,\tau_0)]} \tag{4.46}$$

The steel stress is given by Eq. 4.37b:

$$\sigma_s(t) = E_s\varepsilon_s(t)$$

and, from Eqs 4.4 and 4.5, the equilibrium and compatibility requirements at time t are:

$$P = \sigma_c(t)A_c + \sigma_s(t)A_s$$

and

$$\varepsilon_s(t) = \varepsilon(t)$$

Solving for $\varepsilon(t)$ gives:

$$\varepsilon(t) = \frac{1}{1+\bar{n}_{e,0}\rho}\left[\frac{P}{A_c\overline{E}_e(t,\tau_0)} - \frac{\sigma_c(\tau_0)\overline{F}_{e,0}\,\bar{n}_{e,0}}{E_s} + \varepsilon_{sh}(t)\right] \tag{4.47}$$

By substituting Eqs 4.45, 4.37b and 4.5 into Eq. 4.4, the concrete stress at time t can be expressed as:

$$\sigma_c(t) = \frac{1}{1+\bar{n}_{e,0}\rho}\left[\frac{P}{A_c} + \sigma_c(\tau_0)\overline{F}_{e,0}\,\bar{n}_{e,0}\rho - \varepsilon_{sh}(t)\,E_s\,\rho\right] \tag{4.48}$$

where $\rho = A_s/A_c$, and $\bar{n}_{e,0} = E_s/\overline{E}_e(t,\tau_0)$ = the age-adjusted modular ratio.

The steel stress $\sigma_s(t)$ is determined by substituting $\varepsilon_s(t) = \varepsilon(t)$ into Eq. 4.37b or by enforcing the requirements of equilibrium (Eq. 4.4).

Example 4.5

Example 4.4 is repeated here using the AEMM approach 2. As before: $E_c(\tau_0) = E_e(\tau_0) = 26.7$ GPa; $\sigma_c(\tau_0) = -9.67$ MPa; $\sigma_s(\tau_0) = -72.3$ MPa; $\varepsilon_s(\tau_0) = -361 \times 10^{-6}$ and $\varepsilon(\tau_0) = -361 \times 10^{-6}$.

At $(t - \tau_0) = 10$ days:

$\varphi(t, \tau_0) = 0.53$, $\chi(t, \tau_0) = 0.60$ and $\varepsilon_{sh}(t) = -142 \times 10^{-6}$.
$\overline{E}_e(t, \tau_0) = E_c(\tau_0)/(1 + \chi(t, \tau_0)\varphi(t, \tau_0)) = 20.27$ GPa; and $\bar{n}_{e,0} = E_s/\overline{E}_e(t, \tau_0) = 9.87$.

From Eq. 4.46: $\overline{F}_{e,0} = 0.53 \dfrac{[0.60 - 1]}{[1 + 0.60 \times 0.53)]} = -0.160$

Eq. 4.48 gives:

$$\sigma_c(t) = \frac{1}{1 + 9.87 \times 0.02}\left[\frac{-1000 \times 10^3}{90{,}000} - 9.67 \times (-0.160) \times 9.87 \times 0.02 + 142 \times 10^{-6} \times 200{,}000 \times 0.02\right] = -8.55 \text{ MPa}$$

Eq. 4.4: $\sigma_s(t) = (-1000 \times 10^3 + 8.55 \times 90{,}000)/1800 = -128.0$ MPa;

Eq. 4.36b: $\varepsilon_s(t) = -128.0/200{,}000 = -640 \times 10^{-6}$; and

Eq. 4.5: $\varepsilon(t) = -640 \times 10^{-6}$.

The results of the analysis are identical to those obtained using approach 1.

4.5 The rate of creep method (RCM)

4.5.1 Formulation

The rate of creep method was originally proposed by Glanville (Ref. 10) and further developed by Whitney (Ref. 11) in the early 1930s to describe the time-varying behaviour of concrete. Dischinger (Ref. 12) first applied the method to the analysis of concrete structures.

The RCM is based on the assumption that the rate of change of creep with time, $d\varphi(t, \tau_0)/dt (\equiv \dot{\varphi}(t, \tau))$, is independent of the age at loading, τ. This means that creep curves for concrete loaded at different times are assumed to be parallel, as shown in Fig 4.10. This, of course, is not true. Better agreement with actual behaviour is obtained by assuming that creep curves for concrete loaded at different times are affine rather than parallel, that is, they are assumed to have the same shape. However, by assuming that the rate of creep is independent of the age at loading, only a single creep curve is required to calculate creep strains due to any stress history.

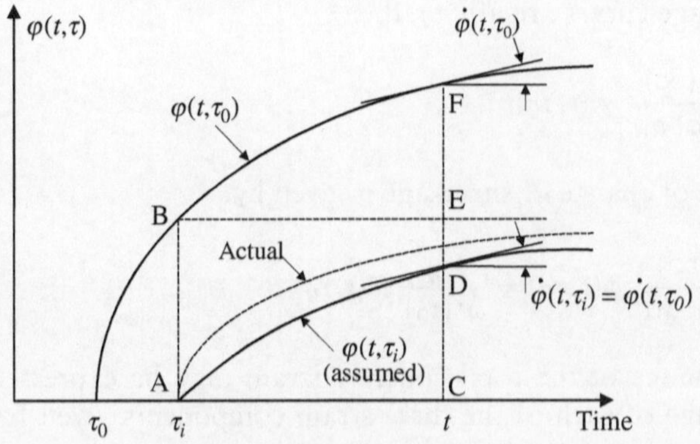

Figure 4.10 Parallel creep curves assumed in the rate of creep method.

Consider the parallel creep curves of Fig. 4.10. Since the ordinates EF and CD are assumed to be equal, CD = CF – AB or:

$$\varphi(t,\tau_i) = \varphi(t,\tau_0) - \varphi(\tau_i,\tau_0) \tag{4.49}$$

The creep–time curve caused by a stress applied at any age τ_i is thus completely defined by a single curve $\varphi(t,\tau_0)$, where τ_0 is usually taken to be the age at first loading.

The creep strain at time t caused by a constant stress $\sigma_c(\tau_i)$ applied at τ_i is therefore taken to be:

$$\varepsilon_{cr}(t,\tau_i) = \frac{\sigma_c(\tau_i)}{E_c(\tau_0)}\varphi(t,\tau_i) = \frac{\sigma_c(\tau_i)}{E_c(\tau_0)}\left[\varphi(t,\tau_0) - \varphi(\tau_i,\tau_0)\right] \tag{4.50}$$

and the change in creep strain between any two time instants t_1 and t_2 after the loading at τ_i is:

$$\Delta\varepsilon_{cr}(t,\tau_i) = \varepsilon_{cr}(t_2,\tau_i) - \varepsilon_{cr}(t_1,\tau_i) = \frac{\sigma_c(\tau_i)}{E_c(\tau_0)}\left[\varphi(t_2,\tau_0) - \varphi(t_1,\tau_0)\right] \tag{4.51}$$

If a continuously varying stress history is divided into small time intervals δt and the stress during each interval is considered to be constant, the change of creep strain during any time interval is given by Eq. 4.51. In the limit, as δt approaches zero, the rate of change of creep depends only on the current stress and the rate of change of the creep coefficient and is given by:

$$\frac{d\,\varepsilon_{cr}(t,\tau)}{d\,t} = \frac{\sigma_c(t)}{E_c(\tau_0)}\frac{d\,\varphi(t,\tau_0)}{d\,t} \quad \text{or} \quad \dot{\varepsilon}_{cr}(t,\tau) = \frac{\sigma_c(t)}{E_c(\tau_0)}\dot{\varphi}(t,\tau_0) \tag{4.52}$$

The rate of change of the instantaneous strain at any time depends on the rate of change of stress:

$$\frac{d\,\varepsilon_e(t)}{d\,t} = \frac{1}{E_c(\tau_0)}\frac{d\,\sigma_c(t)}{d\,t} \quad \text{or} \quad \dot{\varepsilon}_e(t) = \frac{\dot{\sigma}_c(t)}{E_c(\tau_0)} \tag{4.53}$$

where the elastic modulus of concrete is here assumed to be constant in time.

If it is further assumed that shrinkage develops at the same rate as creep (i.e. the creep and shrinkage curves are affine), then:

$$\varepsilon_{sh}(t) = \frac{\varepsilon_{sh}(\infty)}{\varphi(\infty, \tau_0)} \varphi(t, \tau_0) \tag{4.54}$$

and the time rate of change of shrinkage is given by:

$$\dot{\varepsilon}_{sh}(t) = \frac{\varepsilon_{sh}(\infty)}{\varphi(\infty, \tau_0)} \dot{\varphi}(t, \tau_0) = \frac{\varepsilon^*_{sh}}{\phi^*(\tau_0)} \dot{\varphi}(t, \tau_0) \tag{4.55}$$

The rate of change of the total concrete strain may be expressed as the sum of the rates of change of each of the three strain components given by Eqs 4.52, 4.53 and 4.55:

$$\dot{\varepsilon}(t, \tau) = \frac{\dot{\sigma}_c(t)}{E_c(\tau_0)} + \dot{\varphi}(t, \tau_0)\left[\frac{\sigma_c(t)}{E_c(\tau_0)} + \frac{\varepsilon^*_{sh}}{\varphi^*(\tau_0)}\right] \tag{4.56}$$

This first order differential equation is a constitutive relationship for concrete that can be readily included in structural analysis. The solution of the resulting differential equations is easily obtained for a variety of practical problems. Eq. 4.56 may be rewritten with the creep coefficient φ ($\equiv \varphi(t, \tau)$) as the independent variable:

$$\frac{\mathrm{d}\,\varepsilon(t, \tau)}{\mathrm{d}\,\varphi} = \frac{1}{E_c(\tau_0)} \frac{\mathrm{d}\,\sigma_c(t)}{\mathrm{d}\,\varphi} + \frac{\sigma_c(t)}{E_c(\tau_0)} + \frac{\varepsilon^*_{sh}}{\varphi^*(\tau_0)} \tag{4.57}$$

Eqs 4.56 and 4.57 were developed by Dischinger (Ref. 12) and this type of differential constitutive relationship for concrete is often referred to as *Dischinger's equation of state.*

4.5.2 Discussion

The rate of creep method, although mathematically attractive, has several deficiencies. The assumption of parallel creep curves causes a significant underestimation of creep in old concrete. As time increases, $\dot{\varphi}(t, \tau_0)$ approaches zero and, according to the RCM, so too does the creep strain for concrete loaded at times much later than τ_0. This is not so. Old concrete does creep, albeit much less than young concrete. The RCM therefore underestimates creep for an increasing stress history. This error may be large if significant increases of stress occur at times $\tau_i >> \tau_0$.

Eq. 4.52 also implies that when stress is removed, the rate of change of creep is zero, that is, creep recovery is not predicted. Creep strains are therefore overestimated for a decreasing stress history. However, for the many practical situations in which loads are applied on relatively young concrete and the stress does not vary too much with time, the RCM gives excellent results.

4.5.3 Example application (RCM)

The rate of creep analysis of the column shown in Fig. 4.1 follows the same general steps as the elastic analysis described in Section 4.2.2. However, since the constitutive relationship for concrete is in differential form (Eq. 4.56), it is convenient to also express the equilibrium and compatibility equations in this form. For clarity, the arguments t and τ_0 are omitted from the various functions in the following analysis.

Noting that the rate of change of the applied load P with time is zero, differentiating the equilibrium equation with respect to time gives:

$$\dot{P} = \dot{N}_c + \dot{N}_s = \dot{\sigma}_c A_c + \dot{\sigma}_s A_s = 0 \tag{4.58}$$

For compatibility, the rates of change of concrete and steel strains are identical, that is:

$$\dot{\varepsilon} = \dot{\varepsilon}_s \tag{4.59}$$

and the constitutive relationships for both the concrete and the steel are:

$$\dot{\varepsilon} = \frac{\dot{\sigma}_c}{E_c} + \dot{\varphi}\left[\frac{\sigma_c}{E_c} + \frac{\varepsilon^*_{sh}}{\varphi^*}\right] \quad \text{and} \quad \dot{\varepsilon}_s = \frac{\dot{\sigma}_s}{E_s} \tag{4.60}$$

where $\dot{\varphi} = \dot{\varphi}\,(t, \tau_0)$; $\varepsilon^*_{sh} = \varepsilon_{sh}(\infty)$; and $\varphi^* = \varphi\,(\infty, \tau_0)$. From Eq. 4.58:

$$\dot{\sigma}_s = -\,\dot{\sigma}_c \frac{A_c}{A_s} = -\frac{\dot{\sigma}_c}{\rho} \tag{4.61}$$

and substituting Eqs 4.60 and 4.61 into Eq. 4.59 gives:

$$-\frac{\dot{\sigma}_c}{E_s \rho} = \frac{\dot{\sigma}_c}{E_c} + \dot{\varphi}\left[\frac{\sigma_c}{E_c} + \frac{\varepsilon^*_{sh}}{\varphi^*}\right]$$

Multiplying by E_c and gathering like terms yields:

$$\frac{\dot{\sigma}_c}{\sigma_c + S} = \frac{-n\rho}{1 + n\rho}\,\dot{\varphi} \tag{4.62}$$

where:

$$S = \frac{\varepsilon^*_{sh}\, E_c}{\varphi^*} \tag{4.63}$$

Integrating both sides of Eq. 4.62 with respect to time gives:

$$\ln(\sigma_c + S)\,\big|_{t=\tau_0}^{t} = -\frac{n\rho}{1 + n\rho}\,\varphi\,\big|_{t=\tau_0}^{t} \quad \text{i.e.} \quad \ln\frac{\sigma_c(t) + S}{\sigma_c(\tau_0) + S} = -\frac{n\rho}{1 + n\rho}\,\varphi\,(t, \tau_0).$$

Solving for concrete stress gives:

$$\frac{\sigma_c(t) + S}{\sigma_c(\tau_0) + S} = e^{\frac{-n\rho}{1+n\rho}\,\varphi\,(t,\tau_0)}$$

and therefore:

$$\sigma_c(t) = (\sigma_c(\tau_0) + S)\, e^{\frac{-n\rho}{1+n\rho}\varphi(t,\tau_0)} - S \tag{4.64}$$

With the concrete stress obtained from Eq. 4.64, the steel stress may be found from Eq. 4.4 and the steel strain from Eq. 4.6b. Compatibility requires that the concrete strain is the same as the steel strain.

Example 4.6

The symmetrically reinforced column subjected to a constant sustained axial load P that was analysed in Example 4.1 (see Fig. 4.1) is to be re-analysed using the RCM. As before: $P = -1000$ kN, $A_c = 90{,}000$ mm^2; $A_s = 1800$ mm^2; $\rho = A_s/A_c = 0.02$; $E_c(\tau_0) = 26.7$ GPa; $E_s = 200$ GPa; $n = E_s/E_c(\tau_0) = 7.48$; and $\varepsilon^*_{sh} = -510 \times 10^{-6}$.

The variations of the creep coefficient with time is:

$(t-\tau_0)$ in days	0	10	30	70	200	500	10,000 (∞)
$\varphi(t,\tau_0)$	0	0.53	0.98	1.38	1.83	2.10	2.39

The constants in Eq. 4.64 are: $\dfrac{n\rho}{1+n\rho} = \dfrac{7.48 \times 0.02}{1+7.48 \times 0.02} = 0.130$;

and from Eq. 4.62: $S = \dfrac{-510 \times 10^{-6} \times 26{,}700}{2.39} = -5.71$ MPa.

At first loading, $(t - \tau_0) = 0$:

$\varphi(t,\tau_0) = 0$; $\varepsilon_{sh}(t) = 0$

$\sigma_c(\tau_0) = -9.67$ MPa; $\sigma_s(\tau_0) = -72.3$ MPa; $\varepsilon_s(\tau_0) = -361 \times 10^{-6}$ and $\varepsilon(\tau_0) = -361 \times 10^{-6}$.

At $(t - \tau_0) = 10$ days:

$\varphi(t,\tau_0) = 0.53$

Eq. 4.64: $\sigma_c(t) = (-9.67 - 5.71)e^{-0.130\times 0.53} + 5.71 = -8.65$ MPa;

Eq. 4.4: $\sigma_s(t) = \dfrac{-1000 \times 10^{-3} + 8.65 \times 90{,}000}{1800} = 123.3$ MPa;

Eq. 4.6b: $\varepsilon_s(t) = \dfrac{-123.3}{200{,}000} = -616 \times 10^{-6}$;

Eq. 4.5: $\varepsilon(t) = -616 \times 10^{-6}$.

The elastic and shrinkage components of the total concrete strain are $\varepsilon_e(t) = \sigma_c(t)/E_c(\tau_0) = -323 \times 10^{-6}$; from Eq. 4.54, the shrinkage component of strain is $\varepsilon_{sh}(t) = (0.53/2.39) \times (-510 \times 10^{-6}) = -113 \times 10^{-6}$ (since the creep and shrinkage curves are assumed to be affine); and therefore the creep component of strain is $\varepsilon_{cr}(t) = \varepsilon(t) - \varepsilon_e(t) - \varepsilon_{sh}(t) = -180 \times 10^{-6}$. It is noted that in this example using RCM, because of the assumption that the creep and shrinkage curves have the same shape, different levels of shrinkage from those specified in the previous examples occur during each time step.

The results of the analyses at first loading and throughout the period of sustained loading are as follows:

Time t (days)	*Duration of load* $t-\tau_0$ *(days)*	*Concrete stress* $\sigma_c(t)$ *(MPa)*	*Steel stress* $\sigma_s(t)$ *(MPa)*	*Total strain* $\varepsilon(t)$ $(\times 10^{-6})$	*Elastic concrete strain* $\varepsilon_e(t)$ $(\times 10^{-6})$	*Creep strain* $\varepsilon_{cr}(t)$ $(\times 10^{-6})$	*Shrinkage strain* $\varepsilon_{sh}(t)$ $(\times 10^{-6})$
14	0	−9.67	−72.3	−361	−361	0	0
24	10	−8.65	−123	−616	−323	−180	−113
44	30	−7.83	−164	−820	−293	−319	−209
84	70	−7.14	−198	−992	−267	−431	−294
214	200	−6.41	−235	−1175	−240	−544	−391
514	500	−5.99	−256	−1279	−224	−607	−448
10,014	10,000	−5.56	−278	−1388	−208	−670	−510

4.6 Comparison of methods of analysis

A comparison of the numerical results calculated using each method of analysis for the axially loaded column section considered in Examples 4.1–4.6 is provided in Table 4.1. The prediction of time-dependent redistribution of stresses and the gradual development of creep strain with time are compared in Figs 4.11 and 4.12, respectively.

A large redistribution of stresses is predicted by all of the methods of analysis. In this axially loaded column, the concrete stress decreases substantially with time and the steel stress increases dramatically. At initial loading, the concrete in the section carried 87 per cent of the total load. After 10,000 days under load, this has reduced to 59.3 per cent (according to the AEMM, for example). During this same time period, the steel stress increased from 72.3 MPa to 269 MPa. This is, in fact, typical of the redistribution of stresses that takes place with time in reinforced concrete structures.

The concrete and steel stresses obtained using the AEMM at any time t are almost identical with those predicted by the much more laborious SSM (calculated with 18 intervals). This is not surprising; since the ageing coefficients were determined from the concrete stress changes calculated using the SSM. However, if one assumes that the ageing coefficient is constant and equal to 0.65 at every value of t, as suggested by Eq. 4.43, the AEMM predicts stresses and total strains to within 1 per cent of these more refined values.

Table 4.1 Comparisons of calculated stresses and strains (Examples 4.1–4.6)

Method of analysis	*Duration of load (days)*	*Concrete stress* $\sigma_c(\tau_i)$ *(MPa)*	*Steel stress* $\sigma_s(\tau_i)$ *(MPa)*	*Total strain* $\varepsilon(\tau_i)$ $(\times 10^{-6})$	*Elastic strain* $\varepsilon_e(\tau_i)$ $(\times 10^{-6})$	*Creep strain* $\varepsilon_{cr}(\tau_i)$ $(\times 10^{-6})$	*Shrinkage strain* $\varepsilon_{sh}(\tau_i)$ $(\times 10^{-6})$
Effective modulus method (EMM)	0	−9.67	−72.3	−361	−361	0	0
	10	−8.58	−127	−633	−321	−170	−142
	30	−7.82	−165	−824	−292	−285	−246
	70	−7.23	−194	−969	−270	−374	−325
	200	−6.66	−222	−1112	−249	−456	−407
	500	−6.35	−238	−1191	−237	−497	−456
	10,000	−6.02	−255	−1273	−225	−538	−510
Age-adjusted effective modulus method (AEMM)	0	−9.67	−72.3	−361	−361	0	0
	10	−8.55	−128	−640	−320	−178	−142
	30	−7.74	−169	−843	−289	−308	−246
	70	−7.10	−200	−1002	−266	−411	−325
	200	−6.48	−232	−1158	−242	−509	−407
	500	−6.12	−249	−1247	−229	−563	−456
	10,000	−5.73	−269	−1346	−214	−622	−510
Step-by-step method (SSM)	0	−9.67	−72.3	−361	−361	0	0
	10	−8.55	−128	−641	−320	−179	−142
	30	−7.72	−169	−847	−292	−309	−246
	70	−7.09	−201	−1006	−270	−411	−325
	200	−6.46	−233	−1164	−249	−508	−407
	500	−6.10	−251	−1253	−238	−559	−456
	10,000	−5.71	−270	−1349	−226	−613	−510
Rate of creep method (RCM)	0	−9.67	−72.3	−361	−361	0	0
	10	−8.65	−123	−616	−323	−180	−113
	30	−7.83	−164	−820	−293	−319	−209
	70	−7.14	−198	−992	−267	−431	−294
	200	−6.41	−235	−1175	−240	−544	−391
	500	−5.99	−256	−1279	−224	−607	−448
	10,000	−5.56	−278	−1388	−208	−670	−510

Figure 4.12 confirms that the EMM underestimates creep strains for a decreasing stress history, predicting full creep recovery as the stress reduces. The EMM predicts the smallest creep strain at any time $t > \tau$ and the smallest eventual stress redistribution (see Fig. 4.11). By contrast, the RCM overestimates creep for this decreasing stress history because it does not allow for any creep recovery. The RCM predicts the largest creep strain and the largest stress redistribution. However, for this simple example, good results are obtained using the RCM at all times after loading with relatively little computational effort.

It is noted that the magnitudes of the stresses and deformations predicted by the SSM and the AEMM lie between the magnitudes predicted by the EMM (which is known to underestimate the effects of creep for a decreasing stress history) and the magnitudes predicted by the RCM (which is known to overestimate the effects of creep). Although the total strain predicted by the AEMM and the SSM are almost identical, the elastic strain is different. The AEMM assumes that the total change of stress is applied at τ_0 and so the elastic strain at any time t is taken as the concrete stress at that time divided

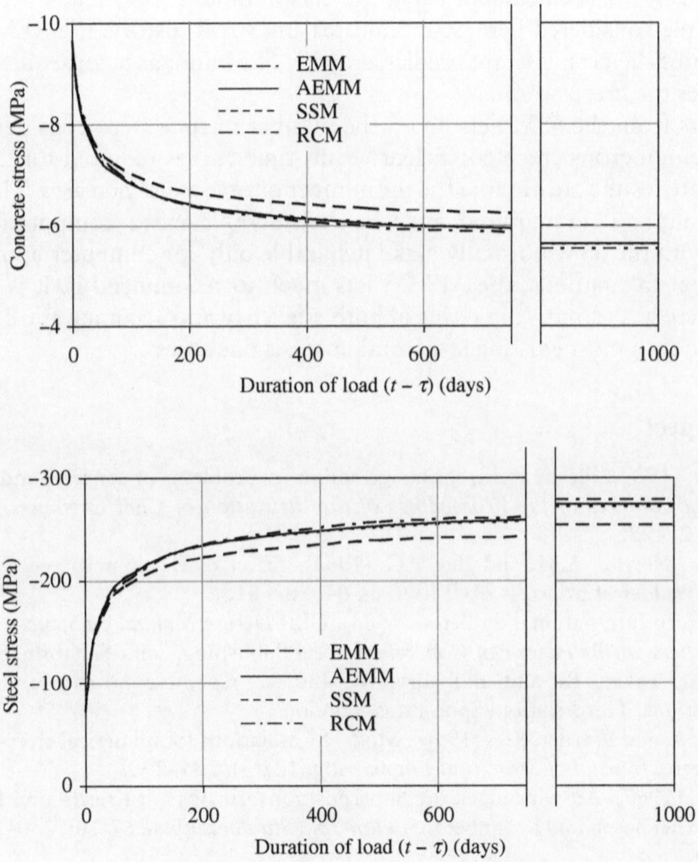

Figure 4.11 Stress redistributions in axially loaded column examples.

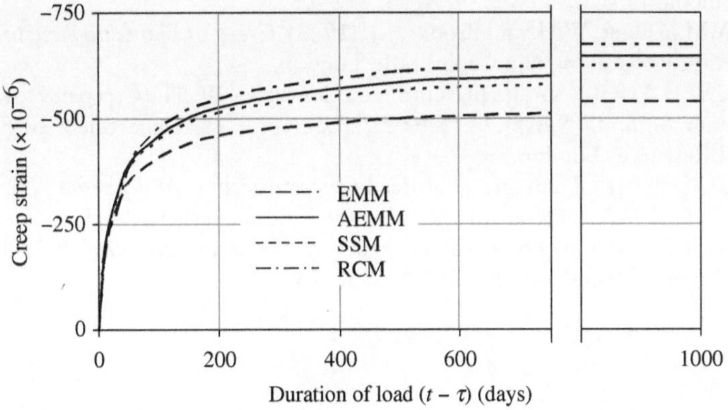

Figure 4.12 Creep strain versus time in the axially loaded column examples.

by the elastic modulus at τ_0. In the SSM, the elastic strain caused by a change of stress at any time $t(> \tau_0)$ is calculated using the elastic modulus at that time. Therefore, in the example considered here, with a decreasing stress history, the AEMM slightly underestimates the elastic strain predicted by the SSM and, as a consequence, slightly overestimates the creep strain.

The results from the SSM depend on the number of time steps considered and the shape of the numerous creep coefficient versus time curves required for the analysis. More accurate results are obtained as the number of time steps increases. The increased amount of input data required and the considerable extra computational effort associated with the SSM normally make it suitable only for computer applications.

For manual calculations, the AEMM has much to recommend it. It is usually the most satisfactory alternative in terms of both efficiency and accuracy and can be easily incorporated into most existing structural analysis packages.

4.7 References

1. Faber, O. (1927). Plastic yield, shrinkage and other problems of concrete and their effects on design. *Minutes of the Proceedings of the Institution of Civil Engineers*, 225, Part I, London, 27–23.
2. Ghali, A., Neville, A.M. and Jha, P.C. (1967). Effect of elastic and creep recoveries of concrete on loss of prestress. *ACI Journal*, 64, 802–810.
3. Comité Euro-International du Béton (1984). *CEB Design Manual on Structural Effects of Time-Dependent Behaviour of Concrete*. Georgi Publishing, Saint-Saphorin, Switzerland.
4. Ghali, A., Favre, R. and Eldbadry, M. (2002). *Concrete Structures: Stresses and Deformations*. Third edition, Spon Press, London.
5. Kawano, A. and Warner, R.F. (1996). Model formulations for numerical creep calculations for concrete. *Journal of Structural Engineering*, 122(3), 284–290.
6. Trost, H. (1967). Auswirkungen des Superpositionsprinzips auf Kriech- und Relaxations-Probleme bei Beton und Spannbeton. *Beton- und Stahlbetonbau*, 62(10), 230–238, No. 11, 261–269.
7. Dilger, W. and Neville, A.M. (1971). Method of creep analysis of structural members. *ACI SP 27-17*, 349–379.
8. Bazant, Z.P. (1972). Prediction of concrete creep effects using age-adjusted effective modulus method. *ACI Journal*, 69, April, 212–217.
9. Neville, A.M., Dilger, W.H. and Brooks, J.J. (1983). *Creep of Plain and Structural Concrete*. Construction Press (Longman Group Ltd), London.
10. Glanville, W.H. (1930). Studies in Reinforced Concrete – III, The Creep or Flow of Concrete under Load. *Building Research Technical Paper No. 12*, Department of Scientific and Industrial Research, London.
11. Whitney, C.S. (1932). Plain and reinforced concrete arches. *ACI Journal*, 28, 479–519.
12. Dischinger, F. (1937). Untersuchungen über die Knicksicherheit, die elastische Verformung und das Kriechen des Betons bei Bogenbrücken. *Der Bauingenieur*, 18(33/34), 487–529 and No. 35/36, 539–552 and No.39/40, 595–621.

5 Uncracked sections

Axial force and uniaxial bending

5.1 Introductory remarks

In this chapter, time analyses are presented for a variety of uncracked cross-sections containing at least one axis of symmetry and subjected to axial load and uniaxial bending. The concrete is assumed to be able to carry any applied tension and is therefore uncracked. Symmetrically reinforced and asymmetrically reinforced sections containing any number of levels of both non-prestressed and prestressed reinforcement are considered. Composite steel–concrete and composite concrete–concrete sections are also examined.

The proposed methods of analysis are based on the assumptions of the Euler–Bernoulli beam theory, in which plane sections are assumed to remain plane and perpendicular to the beam axis before, and after, both short- and long-term deformations. This means that the strain distributions on the cross-section, both immediately upon loading and after a prolonged period of sustained loading and shrinkage, are assumed to be linear.

5.2 Overview of cross-sectional analysis

Cross-sectional analysis is used extensively for modelling structural response at both service and ultimate load conditions. With this approach, the governing equilibrium equations describing the behaviour of any cross-section are expressed in terms of two unknowns that define the strain diagram. These unknowns can be identified by a single value of strain measured at the level of the reference axis ε_r and the slope of the strain diagram κ (which is, of course, the curvature of the cross-section).

Considering the cross-section shown in Fig. 5.1a, the section is symmetrical about the y-axis and the orthogonal x-axis is selected as the reference axis. If the cross-section is subjected to an axial force applied at the origin of the x- and y-axes and a bending moment applied about the x-axis, the strain diagram is shown in Fig. 5.1b and the strain at any depth y below the reference axis is given by:

$$\varepsilon = \varepsilon_r + y\kappa \tag{5.1}$$

The two unknowns of the problem, ε_r and κ, are then determined by enforcing horizontal and rotational equilibrium at the cross-section:

$$N_i = N_e \tag{5.2a}$$

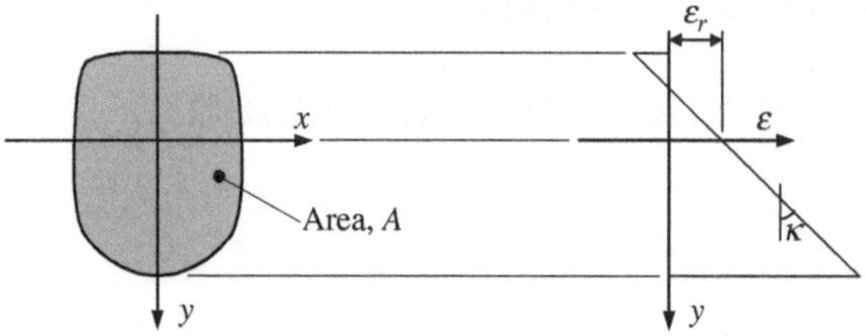

Figure 5.1 Generic cross-section.

and

$$M_i = M_e \tag{5.2b}$$

where N_e and M_e denote the external axial force and moment, respectively, at the section and N_i and M_i are the internal axial force and moment, respectively, given by:

$$N_i = \int_A \sigma \, dA \tag{5.3a}$$

and

$$M_i = \int_A y\sigma \, dA \tag{5.3b}$$

When the two unknowns (ε_r and κ) are calculated from the two equilibrium equations (Eqs 5.2) and the strain is determined using Eq. 5.1, the stresses in the concrete and steel may be obtained from the appropriate constitutive relationships. The internal actions are then readily determined from the stresses using Eqs 5.3.

This procedure forms the basis of both the short- and long-term analyses presented in the remainder of this chapter.

5.3 Short-term analysis of reinforced or prestressed concrete cross-sections

The short-term analysis of an uncracked reinforced or prestressed concrete cross-section at service loads is usually made assuming linear-elastic behaviour of the concrete (in compression and tension) and linear-elastic behaviour of the non-prestressed reinforcement and the prestressing steel. The procedure can be applied to typical cross-sections such as those shown in Fig. 5.2.

Without any loss of generality, it is assumed that the contribution of each reinforcing bar or prestressing tendon is lumped into layers according to its location, as shown in Fig. 5.3. The numbers of layers of non-prestressed and prestressed reinforcement are m_s and m_p, respectively. In Fig. 5.3b, $m_s = 3$ and $m_p = 2$. In particular, the properties of each layer of non-prestressed reinforcement are defined by its area, elastic modulus and location with respect to the arbitrarily chosen x-axis and are labelled as $A_{s(i)}$, $E_{s(i)}$ and

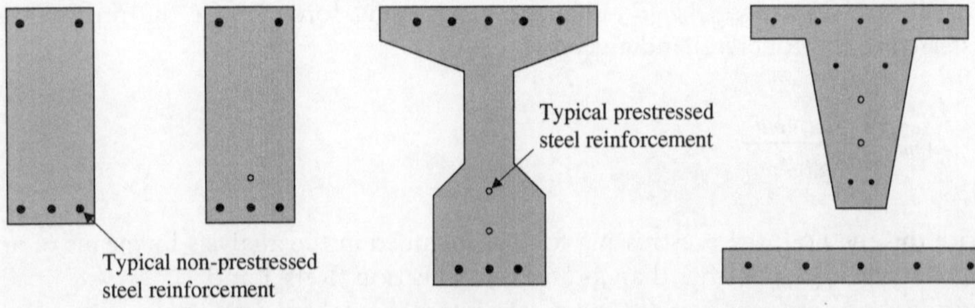

Figure 5.2 Typical reinforced and prestressed concrete sections.

$y_{s(i)}$, respectively. With this notation the subscript '*s*' stands for (non-prestressed) steel reinforcement and $i = 1, \ldots, m_s$. Similarly, $A_{p(i)}$, $E_{p(i)}$ and $y_{p(i)}$ represent, respectively, the area, elastic modulus and location of the prestressing steel with respect to the x-axis and $i = 1, \ldots, m_p$.

The geometric properties of the concrete part of the cross-section are A_c, B_c and I_c, which denote the area and the first and second moments of area of the concrete about the x-axis, respectively.

For the short-term or instantaneous analysis, assumed to take place at time τ_0, the linear-elastic stress-strain relationships of the concrete and the steel are:

$$\sigma_{c,0} = E_{c,0}\varepsilon_0 \tag{5.4a}$$

$$\sigma_{s(i),0} = E_{s(i)}\varepsilon_0 \tag{5.4b}$$

$$\sigma_{p(i),0} = E_{p(i)}\left(\varepsilon_0 + \varepsilon_{p(i),init}\right) \tag{5.4c}$$

in which $\sigma_{c,0}$, $\sigma_{s(i),0}$ and $\sigma_{p(i),0}$ represent the stresses in the concrete, in the i-th layer of non-prestressed reinforcement (with $i = 1, \ldots, m_s$) and in the i-th layer of prestressing steel (with $i = 1, \ldots, m_p$), respectively, immediately after first loading at time τ_0, while

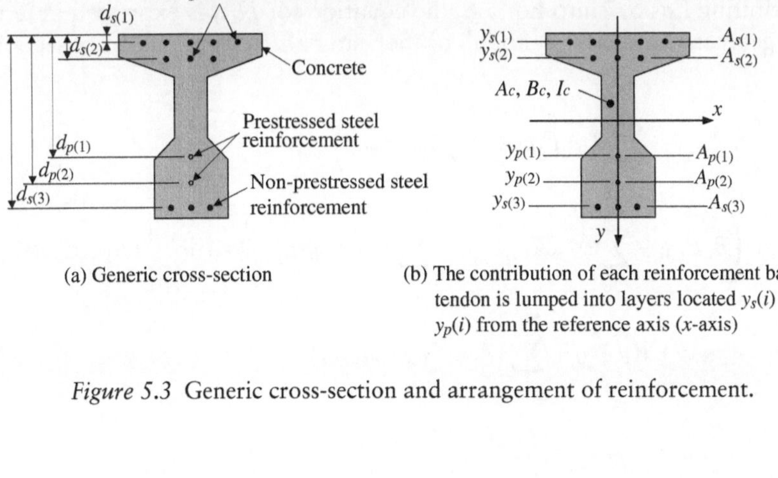

(a) Generic cross-section

(b) The contribution of each reinforcement bar or tendon is lumped into layers located $y_s(i)$ and $y_p(i)$ from the reference axis (x-axis)

Figure 5.3 Generic cross-section and arrangement of reinforcement.

$\varepsilon_{p(i),init}$ is the initial strain in the i-th layer of prestressing steel produced by the initial tensile prestressing force $P_{p(i),init}$ (i.e. the prestressing force before the prestress is transferred to the concrete) and given by:

$$\varepsilon_{p(i),init} = \frac{P_{p(i),init}}{A_{p(i)}E_{p(i)}} \tag{5.5}$$

With this method, the prestressing force is included in the analysis by means of an induced strain $\varepsilon_{p(i),init}$ rather than as an external action (Refs 1 and 2).

For the short-term analysis at time τ_0, the internal axial force and moment resisted by the cross-section about the reference axis are denoted $N_{i,0}$ and $M_{i,0}$, respectively. The internal axial force $N_{i,0}$ is the sum of the axial forces resisted by the component materials forming the cross-section and is given by:

$$N_{i,0} = N_{c,0} + N_{s,0} + N_{p,0} \tag{5.6}$$

where $N_{c,0}$, $N_{s,0}$ and $N_{p,0}$ represent the axial forces resisted by the concrete, the non-prestressed reinforcement and the prestressing steel, respectively, and are calculated from:

$$N_{c,0} = \int_{A_c} \sigma_{c,0}\,dA = \int_{A_c} E_{c,0}(\varepsilon_{r,0} + y\kappa_0)\,dA = A_c E_{c,0}\varepsilon_{r,0} + B_c E_{c,0}\kappa_0 \tag{5.7a}$$

$$N_{s,0} = \sum_{i=1}^{m_s}\left(A_{s(i)}E_{s(i)}\right)\left(\varepsilon_{r,0} + y_{s(i)}\kappa_0\right) = \sum_{i=1}^{m_s}\left(A_{s(i)}E_{s(i)}\right)\varepsilon_{r,0} + \sum_{i=1}^{m_s}\left(y_{s(i)}A_{s(i)}E_{s(i)}\right)\kappa_0 \tag{5.7b}$$

$$N_{p,0} = \sum_{i=1}^{m_p}\left(A_{p(i)}E_{p(i)}\right)\varepsilon_{r,0} + \sum_{i=1}^{m_p}\left(y_{p(i)}A_{p(i)}E_{p(i)}\right)\kappa_0 + \sum_{i=1}^{m_p}\left(A_{p(i)}E_{p(i)}\varepsilon_{p(i),init}\right) \tag{5.7c}$$

where the additional subscripts '0' used for the strain at the level of the reference axis ($\varepsilon_{r,0}$) and the curvature (κ_0) highlight that these are calculated at time τ_0 after the application of $N_{e,0}$ and $M_{e,0}$ and after the transfer of prestress.

By substituting Eqs 5.7 into Eq. 5.6, the equation for $N_{i,0}$ is expressed in terms of the actual geometry and elastic moduli of the materials forming the cross-section:

$$\begin{aligned} N_{i,0} &= \left(A_c E_{c,0} + \sum_{i=1}^{m_s} A_{s(i)}E_{s(i)} + \sum_{i=1}^{m_p} A_{p(i)}E_{p(i)}\right)\varepsilon_{r,0} \\ &\quad + \left(B_c E_{c,0} + \sum_{i=1}^{m_s} y_{s(i)}A_{s(i)}E_{s(i)} + \sum_{i=1}^{m_p} y_{p(i)}A_{p(i)}E_{p(i)}\right)\kappa_0 + \sum_{i=1}^{m_p}\left(A_{p(i)}E_{p(i)}\varepsilon_{p(i),init}\right) \\ &= R_{A,0}\varepsilon_{r,0} + R_{B,0}\kappa_0 + \sum_{i=1}^{m_p}\left(A_{p(i)}E_{p(i)}\varepsilon_{p(i),init}\right) \end{aligned} \tag{5.8}$$

in which $R_{A,0}$ and $R_{B,0}$ represent, respectively, the axial rigidity and the stiffness related to the first moment of area about the reference axis calculated at time τ_0 and are given by:

$$R_{A,0} = A_c E_{c,0} + \sum_{i=1}^{m_s} A_{s(i)} E_{s(i)} + \sum_{i=1}^{m_p} A_{p(i)} E_{p(i)} \tag{5.9a}$$

$$R_{B,0} = B_c E_{c,0} + \sum_{i=1}^{m_s} y_{s(i)} A_{s(i)} E_{s(i)} + \sum_{i=1}^{m_p} y_{p(i)} A_{p(i)} E_{p(i)} \tag{5.9b}$$

Similarly, the equation for $M_{i,0}$ may be expressed as:

$$\begin{aligned} M_{i,0} &= \left(B_c E_{c,0} + \sum_{i=1}^{m_s} y_{s(i)} A_{s(i)} E_{s(i)} + \sum_{i=1}^{m_p} y_{p(i)} A_{p(i)} E_{p(i)} \right) \varepsilon_{r,0} \\ &\quad + \left(I_c E_{c,0} + \sum_{i=1}^{m_s} y_{s(i)}^2 A_{s(i)} E_{s(i)} + \sum_{i=1}^{m_p} y_{p(i)}^2 A_{p(i)} E_{p(i)} \right) \kappa_0 \\ &\quad + \sum_{i=1}^{m_p} \left(y_{p(i)} A_{p(i)} E_{p(i)} \varepsilon_{p(i),init} \right) \\ &= R_{B,0}\, \varepsilon_{r,0} + R_{I,0}\, \kappa_0 + \sum_{i=1}^{m_p} \left(y_{p(i)} A_{p(i)} E_{p(i)} \varepsilon_{p(i),init} \right) \end{aligned} \tag{5.10}$$

where $R_{I,0}$ is the flexural rigidity at time τ_0 given by:

$$R_{I,0} = I_c E_{c,0} + \sum_{i=1}^{m_s} y_{s(i)}^2 A_{s(i)} E_{s(i)} + \sum_{i=1}^{m_p} y_{p(i)}^2 A_{p(i)} E_{p(i)} \tag{5.11}$$

Substituting the expressions for $N_{i,0}$ and $M_{i,0}$ (Eqs 5.8 and 5.10) into Eqs 5.2 produces the system of equilibrium equations that may be written in compact form as:

$$\mathbf{r}_{e,0} = \mathbf{D}_0 \boldsymbol{\varepsilon}_0 + \mathbf{f}_{p,init} \tag{5.12}$$

where

$$\mathbf{r}_{e,0} = \begin{bmatrix} N_{e,0} \\ M_{e,0} \end{bmatrix} \tag{5.13a}$$

$$\mathbf{D}_0 = \begin{bmatrix} R_{A,0} & R_{B,0} \\ R_{B,0} & R_{I,0} \end{bmatrix} \tag{5.13b}$$

$$\boldsymbol{\varepsilon}_0 = \begin{bmatrix} \varepsilon_{r,0} \\ \kappa_0 \end{bmatrix} \tag{5.13c}$$

$$\mathbf{f}_{p,init} = \sum_{i=1}^{m_p} \begin{bmatrix} A_{p(i)} E_{p(i)} \varepsilon_{p(i),init} \\ y_{p(i)} A_{p(i)} E_{p(i)} \varepsilon_{p(i),init} \end{bmatrix} \tag{5.13d}$$

The vector $\mathbf{r}_{e,0}$ is the vector of the external actions at first loading (at time τ_0), i.e. axial force $N_{e,0}$ and moment $M_{e,0}$; the matrix $\mathbf{D}_0$ contains the cross-sectional material and geometric properties calculated at τ_0; the strain vector $\boldsymbol{\varepsilon}_0$ contains the unknown independent variables describing the strain diagram at time τ_0 ($\varepsilon_{r,0}$ and κ_0); and the vector $\mathbf{f}_{p,init}$ contains the actions caused by the initial prestressing.

The vector $\boldsymbol{\varepsilon}_0$ is readily obtained by solving the equilibrium equations (Eq. 5.12) giving:

$$\boldsymbol{\varepsilon}_0 = \mathbf{D}_0^{-1}\left(\mathbf{r}_{e,0} - \mathbf{f}_{p,init}\right) = \mathbf{F}_0\left(\mathbf{r}_{e,0} - \mathbf{f}_{p,init}\right) \tag{5.14}$$

where

$$\mathbf{F}_0 = \frac{1}{R_{A,0}R_{I,0} - R_{B,0}^2}\begin{bmatrix} R_{I,0} & -R_{B,0} \\ -R_{B,0} & R_{A,0} \end{bmatrix} \tag{5.15}$$

The stress distribution related to the concrete and reinforcement can then be calculated from the constitutive equations (Eqs 5.4) re-expressed here as:

$$\sigma_{c,0} = E_{c,0}\varepsilon_0 = E_{c,0}[1 \quad y]\boldsymbol{\varepsilon}_0 \tag{5.16a}$$

$$\sigma_{s(i),0} = E_{s(i)}\varepsilon_0 = E_{s(i)}[1 \quad y_{s(i)}]\boldsymbol{\varepsilon}_0 \tag{5.16b}$$

$$\sigma_{p(i),0} = E_{p(i)}\left(\varepsilon_0 + \varepsilon_{p(i),init}\right) = E_{p(i)}[1 \quad y_{p(i)}]\,\boldsymbol{\varepsilon}_0 + E_{p(i)}\varepsilon_{p(i),init} \tag{5.16c}$$

where $\varepsilon_0 = \varepsilon_{r,0} + y\kappa_0 = [1 \quad y]\,\boldsymbol{\varepsilon}_0$.

Although this procedure is presented here assuming linear-elastic material properties, it is quite general and is also applicable to non-linear material behaviour, in which case the integrals of Eqs 5.3 might have to be evaluated numerically. However, when calculating the short-term response of uncracked reinforced and prestressed concrete cross-sections under typical in-service loads, material behaviour is essentially linear elastic.

In reinforced and prestressed concrete design, it is common to calculate the cross-sectional properties by transforming the section into equivalent areas of one of the constituent materials. For example, for the cross-section of Fig. 5.3a, the transformed concrete cross-section for the short-term analysis is shown in Fig. 5.4, with the area of each layer of bonded steel reinforcement and tendons ($A_{s(i)}$ and $A_{p(i)}$, respectively) transformed into equivalent areas of concrete ($n_{s(i),0}A_{s(i)}$ and $n_{p(i),0}A_{p(i)}$, respectively), where $n_{s(i),0} = E_{s(i)}/E_{c,0}$ is the modular ratio of the i-th layer of non-prestressed steel and $n_{p(i),0} = E_{p(i)}/E_{c,0}$ is the modular ratio of the i-th layer of prestressing steel.

For the transformed section of Fig. 5.4, the cross-sectional rigidities defined in Eqs 5.9 and 5.11 can be re-calculated as:

$$R_{A,0} = A_0E_{c,0} \tag{5.17a}$$

$$R_{B,0} = B_0E_{c,0} \tag{5.17b}$$

$$R_{I,0} = I_0E_{c,0} \tag{5.17c}$$

where A_0 is the area of the transformed concrete section, and B_0 and I_0 are the first and second moments of the transformed area about the reference x-axis at first loading.

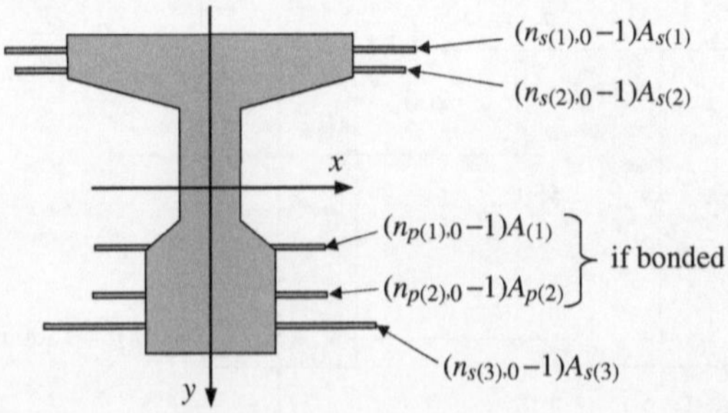

Figure 5.4 Transformed section with bonded reinforcement transformed into equivalent areas of concrete.

Substituting Eqs 5.17 into Eq. 5.15 enables $\mathbf{F}_0$ to be expressed in terms of the properties of the transformed concrete section as:

$$\mathbf{F}_0 = \frac{1}{E_{c,0}(A_0 I_0 - B_0^2)} \begin{bmatrix} I_0 & -B_0 \\ -B_0 & A_0 \end{bmatrix} \tag{5.18}$$

The two approaches proposed for the calculation of the cross-sectional rigidities, i.e. the one based on Eqs 5.9 and 5.11 and the one relying on the properties of the transformed section (Eqs 5.17), are equivalent. The procedure based on the transformed section (Eqs 5.17) is often preferred for the analysis of reinforced and prestressed concrete sections, while for composite steel–concrete and concrete–concrete cross-sections, the procedure based on Eqs 5.9 and 5.11 is generally more convenient. The use of both approaches is illustrated in the following examples.

Example 5.1

For the reinforced concrete section shown in Fig. 5.5, the strain and stress distributions are to be determined immediately after application of an axial force $N_{e,0} = -30$ kN and a sagging (positive) bending moment of $M_{e,0} = 50$ kNm, applied with respect to the x-axis located 200 mm below the top fibre of the cross-section. Both the concrete and reinforcement are assumed to be linear-elastic with $E_{c,0} = 25$ GPa and $E_s = 200$ GPa. The modular ratio for the reinforcing steel is therefore $n_{s,0} = n_{s(1),0} = n_{s(2),0} = E_s/E_{c,0} = 8$.
For this reinforced concrete cross-section, $\mathbf{f}_{p,init}$ is a nil vector and the vector of external actions is:

$$\mathbf{r}_{e,0} = \begin{bmatrix} -30 \times 10^3 \text{ N} \\ 50 \times 10^6 \text{ Nmm} \end{bmatrix}$$

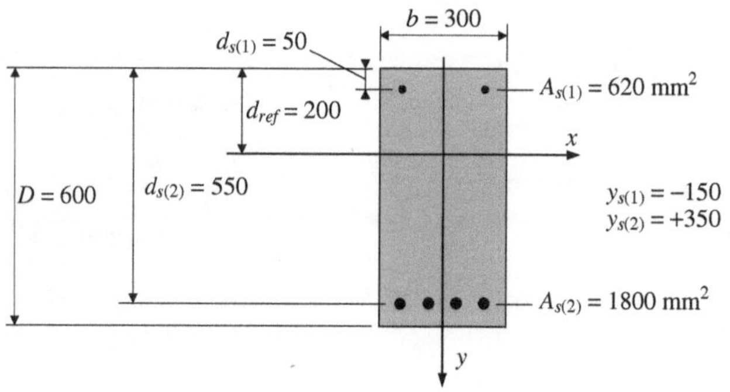

Figure 5.5 Reinforced concrete cross-section for Example 5.1 (dimensions in mm).

With the steel reinforcement transformed into equivalent areas of concrete, the relevant properties of the transformed cross-section for inclusion in the $\mathbf{F}_0$ matrix (Eq. 5.18) are calculated as follows:

$$A_0 = bD + (n_{s,0} - 1)[A_{s(1)} + A_{s(2)}] = 300 \times 600 + (8-1) \times [620 + 1800]$$
$$= 196{,}900\ \text{mm}^2$$

$$B_0 = bD\left(\frac{D}{2} - d_{ref}\right) + (n_{s,0} - 1)[A_{s(1)}y_{s(1)} + A_{s(2)}y_{s(2)}]$$
$$= 300 \times 600 \times (300 - 200) + (8-1) \times [620 \times (-150) + 1800 \times 350]$$
$$= 21{,}760 \times 10^3\ \text{mm}^3$$

$$I_0 = \frac{bD^3}{12} + bD\left(\frac{D}{2} - d_{ref}\right)^2 + (n_{s,0} - 1)[A_{s(1)}y_{s(1)}^2 + A_{s(2)}y_{s(2)}^2]$$
$$= \frac{300 \times 600^3}{12} + 300 \times 600\,(300 - 200)^2 + (8-1)$$
$$\times [620 \times (-150)^2 + 1800 \times 350^2] = 8841 \times 10^6\ \text{mm}^4$$

From Eq. 5.18:

$$\mathbf{F}_0 = \frac{1}{25{,}000 \times (196{,}900 \times 8841 \times 10^6 - (21{,}760 \times 10^3)^2)}$$
$$\times \begin{bmatrix} 8841 \times 10^6 & -21{,}760 \times 10^3 \\ -21{,}760 \times 10^3 & 196{,}900 \end{bmatrix}$$
$$= \begin{bmatrix} 278.9 \times 10^{-12}\ \text{N}^{-1} & -686.5 \times 10^{-15}\ \text{N}^{-1}\text{mm}^{-1} \\ -686.5 \times 10^{-15}\ \text{N}^{-1}\text{mm}^{-1} & 6.214 \times 10^{-15}\ \text{N}^{-1}\text{mm}^{-2} \end{bmatrix}$$

The strain vector $\boldsymbol{\varepsilon}_0$ containing the unknown strain variables is determined from Eq. 5.14:

$$\boldsymbol{\varepsilon}_0 = \begin{bmatrix} 278.9 \times 10^{-12} & -686.5 \times 10^{-15} \\ -686.5 \times 10^{-15} & 6.214 \times 10^{-15} \end{bmatrix} \times \left\{ \begin{bmatrix} -30 \times 10^3 \\ 50 \times 10^6 \end{bmatrix} - \begin{bmatrix} 0 \\ 0 \end{bmatrix} \right\}$$

$$= \begin{bmatrix} -42.7 \times 10^{-6} \\ 0.331 \times 10^{-6}\ \mathrm{mm}^{-1} \end{bmatrix}$$

The strain at the reference axis and the curvature are therefore:

$$\varepsilon_{r,0} = -42.7 \times 10^{-6} \text{ and } \kappa_0 = 0.331 \times 10^{-6}\ \mathrm{mm}^{-1}$$

and, from Eq. 5.1, the top ($y = -200$ mm) and bottom ($y = +400$ mm) fibre strains are:

$$\varepsilon_{0(top)} = \varepsilon_{r,0} - 200 \times \kappa_0 = (-42.7 - 200 \times 0.331) \times 10^{-6} = -108.9 \times 10^{-6};$$

and

$$\varepsilon_{0(btm)} = \varepsilon_{r,0} + 400 \times \kappa_0 = (-42.7 + 400 \times 0.331) \times 10^{-6} = +89.8 \times 10^{-6}.$$

The top and bottom fibre stresses in the concrete and the stresses in the two layers of reinforcement are obtained from Eqs 5.16:

$$\sigma_{c,0(top)} = E_{c,0}\begin{bmatrix} 1 & y_{c(top)} \end{bmatrix} \boldsymbol{\varepsilon}_0 = 25 \times 10^3 \times \begin{bmatrix} 1 & -200 \end{bmatrix} \begin{bmatrix} -42.7 \times 10^{-6} \\ 0.331 \times 10^{-6} \end{bmatrix}$$
$$= -2.72 \text{ MPa}$$

$$\sigma_{c,0(btm)} = E_{c,0}\begin{bmatrix} 1 & y_{c(btm)} \end{bmatrix} \boldsymbol{\varepsilon}_0 = 25 \times 10^3 \times \begin{bmatrix} 1 & 400 \end{bmatrix} \begin{bmatrix} -42.7 \times 10^{-6} \\ 0.331 \times 10^{-6} \end{bmatrix}$$
$$= 2.25 \text{ MPa}$$

$$\sigma_{s(1),0} = E_{s(1)}\begin{bmatrix} 1 & y_{s(1)} \end{bmatrix} \boldsymbol{\varepsilon}_0 = 200 \times 10^3 \times \begin{bmatrix} 1 & -150 \end{bmatrix} \begin{bmatrix} -42.7 \times 10^{-6} \\ 0.331 \times 10^{-6} \end{bmatrix}$$
$$= -18.5 \text{ MPa}$$

$$\sigma_{s(2),0} = E_{s(2)}\begin{bmatrix} 1 & y_{s(2)} \end{bmatrix} \boldsymbol{\varepsilon}_0 = 200 \times 10^3 \begin{bmatrix} 1 & 350 \end{bmatrix} \begin{bmatrix} -42.7 \times 10^{-6} \\ 0.331 \times 10^{-6} \end{bmatrix}$$
$$= 14.6 \text{ MPa}$$

The results are plotted in Fig. 5.6.

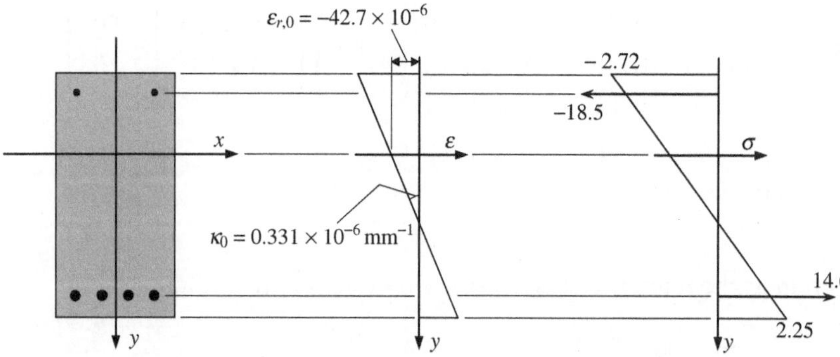

Figure 5.6 Strain and stress diagrams for Example 5.1.

The cross-sectional rigidities could have also been calculated using Eqs 5.9 and 5.11 without transforming the section as follows:

$$R_{A,0} = A_{gross}E_{c,0} + \sum_{i=1}^{m_s} A_{s(i)}(E_{s(i)} - E_{c,0}) = bDE_{c,0} + \left(A_{s(1)} + A_{s(2)}\right)(E_s - E_{c,0})$$

$$= 300 \times 600 \times 25{,}000 + (620 + 1800) \times (200{,}000 - 25{,}000)$$

$$= 4923 \times 10^6 \text{ N}$$

$$R_{B,0} = A_{gross}\left(\frac{D}{2} - d_{ref}\right)E_{c,0} + \sum_{i=1}^{m_s} y_{s(i)}A_{s(i)}(E_{s(i)} - E_{c,0})$$

$$= 300 \times 600 \times \left(\frac{600}{2} - 200\right) \times 25{,}000 + [620 \times (-150) + 1800 \times 350]$$

$$\times (200{,}000 - 25{,}000) = 543.9 \times 10^9 \text{Nmm}$$

$$R_{I,0} = \left[I_{gross} + A_{gross}\left(\frac{D}{2} - d_{ref}\right)^2\right]E_{c,0} + \sum_{i=1}^{m_s} y_{s(i)}^2 A_{s(i)}(E_{s(i)} - E_{c,0})$$

$$= \left[\frac{300 \times 600^3}{12} + 300 \times 600(300 - 200)^2\right] \times 25{,}000 + \left[620 \times (-150)^2\right.$$

$$\left.+1800 \times 350^2\right] \times (200{,}000 - 25{,}000) = 221.0 \times 10^{12} \text{ Nmm}^2$$

and $\mathbf{F}_0$ can then be obtained from Eq. 5.15 as:

$$\mathbf{F}_0 = \frac{1}{R_{A,0}R_{I,0} - R_{B,0}^2}\begin{bmatrix} R_{I,0} & -R_{B,0} \\ -R_{B,0} & R_{A,0} \end{bmatrix}$$

$$= \frac{1}{4923 \times 10^6 \times 221.0 \times 10^{12} - (543.9 \times 10^9)^2}$$

$$\times \begin{bmatrix} 221.0 \times 10^{12} & -543.9 \times 10^9 \\ -543.9 \times 10^9 & 4923 \times 10^6 \end{bmatrix}$$

$$= \begin{bmatrix} 278.9 \times 10^{-12}\,\text{N}^{-1} & -686.5 \times 10^{-15}\,\text{N}^{-1}\text{mm}^{-1} \\ -686.5 \times 10^{-15}\,\text{N}^{-1}\text{mm}^{-1} & 6.214 \times 10^{-15}\,\text{N}^{-1}\text{mm}^{-2} \end{bmatrix}$$

and this, of course, is identical to that calculated previously using Eq. 5.18.

Example 5.2

The instantaneous stress and strain distributions on the precast pretensioned concrete section shown in Fig. 5.7 are to be calculated. The cross-section is that of a Girder Type 3 from the AS5100.5 (Ref. 3) guidelines, with the area of the gross section $A_{gross} = 317 \times 10^3$ mm^2 and the second moment of area of the gross-section about the centroidal axis $I_{gross} = 49{,}900 \times 10^6$ mm^4. The centroid of the gross cross-sectional area is located 602 mm below its top fibre, i.e. $d_c = 602$ mm. The section is subjected to a compressive axial force $N_{e,0} = -100$ kN and a hogging moment of $M_{e,0} = -50$ kNm applied with respect to the reference x-axis, that is taken in this example to be 300 mm below the top fibre of the cross-section. All materials are linear-elastic with $E_{c,0} = 32$ GPa and $E_s = E_p = 200$ GPa. The modular ratios of the reinforcing steel and the prestressing steel are therefore $n_{s(i),0} = n_{p(i),0} = 6.25$. The prestressing forces applied to the top and bottom tendons ($A_{p(1)}$ and $A_{p(2)}$, respectively) prior to the transfer of prestress are $P_{p(1),init} = P_{p(2),init} = 1000$ kN.
The distances of the steel layers from the reference axis are $y_{s(1)} = -240$ mm, $y_{s(2)} = +790$ mm, $y_{p(1)} = +580$ mm and $y_{p(2)} = +710$ mm. From Eq. 5.5, the initial strains in the prestressing steel layers prior to the transfer of prestress to the concrete are:

$$\varepsilon_{p(1),init} = \varepsilon_{p(2),init} = \frac{1000 \times 10^3}{800 \times 200 \times 10^3} = 0.00625$$

The vector of external actions at first loading is

$$\mathbf{r}_{e,0} = \begin{bmatrix} -100 \times 10^3\ \text{N} \\ -50 \times 10^6\ \text{Nmm} \end{bmatrix}$$

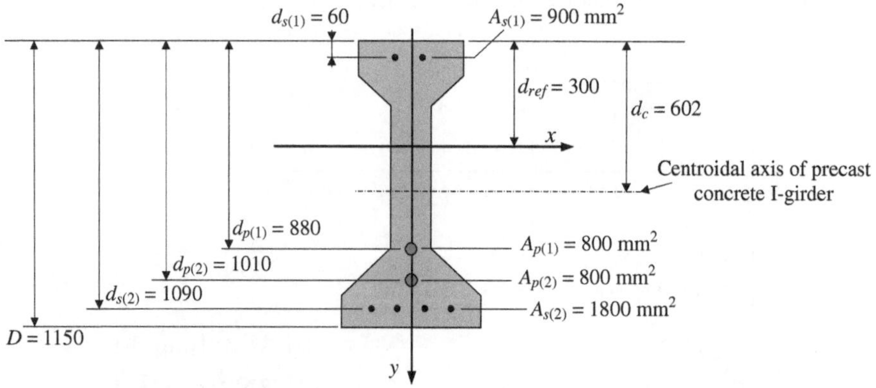

Figure 5.7 Precast prestressed concrete section for Example 5.2 (dimensions in mm).

and, from Eq. 5.13d, the vector of initial prestressing forces is:

$$\mathbf{f}_{p,init} = \sum_{i=1}^{2} \begin{bmatrix} A_{p(i)}E_{p(i)}\varepsilon_{p(i),init} \\ y_{p(i)}A_{p(i)}E_{p(i)}\varepsilon_{p(i),init} \end{bmatrix}$$

$$= \begin{bmatrix} 800 \times 200{,}000 \times 0.00625 \\ 580 \times 800 \times 200{,}000 \times 0.00625 \end{bmatrix} + \begin{bmatrix} 800 \times 200{,}000 \times 0.00625 \\ 710 \times 800 \times 200{,}000 \times 0.00625 \end{bmatrix}$$

$$= \begin{bmatrix} 2000 \times 10^3 \text{ N} \\ 1290 \times 10^6 \text{ Nmm} \end{bmatrix}$$

The relevant cross-sectional properties of the transformed section expressed in equivalent concrete areas are:

$$A_0 = A_{gross} + (n_{s,0} - 1)[A_{s(1)} + A_{s(2)}] + (n_{p,0} - 1)[A_{p(1)} + A_{p(2)}]$$
$$= 317 \times 10^3 + (6.25 - 1) \times [900 + 1800] + (6.25 - 1) \times [800 + 800]$$
$$= 339.6 \times 10^3 \text{ mm}^2$$

$$B_0 = A_{gross}(d_c - d_{ref}) + (n_{s,0} - 1)[A_{s(1)}y_{s(1)} + A_{s(2)}y_{s(2)}]$$
$$+ (n_{p,0} - 1)[A_{p(1)}y_{p(1)} + A_{p(2)}y_{p(2)}] = 317 \times 10^3 \times (602 - 300)$$
$$+ (6.25 - 1) \times [900 \times (-240) + 1800 \times 790] + (6.25 - 1)$$
$$\times [800 \times 580 + 800 \times 710] = 107.5 \times 10^6 \text{ mm}^3$$

$$I_0 = I_{gross} + A_{gross}(d_c - d_{ref})^2 + (n_{s,0} - 1)(A_{s(1)}y_{s(1)}^2 + A_{s(2)}y_{s(2)}^2)$$
$$+ (n_{p,0} - 1)(A_{p(1)}y_{p(1)}^2 + A_{p(2)}y_{p(2)}^2) = 49{,}900 \times 10^6 + 317 \times 10^3$$
$$\times (602 - 300)^2 + (6.25 - 1) \times [900 \times (-240)^2 + 1800 \times 790^2]$$
$$+ (6.25 - 1) \times [800 \times 580^2 + 800 \times 710^2] = 88{,}510 \times 10^6 \text{ mm}^4$$

From Eq. 5.18:

$$\mathbf{F}_0 = \frac{1}{32{,}000 \times (339{,}600 \times 88{,}510 \times 10^6 - (107.5 \times 10^6)^2)}$$

$$\times \begin{bmatrix} 88{,}510 \times 10^6 & -107.5 \times 10^6 \\ -107.5 \times 10^6 & 339{,}600 \end{bmatrix}$$

$$= \begin{bmatrix} 149.5 \times 10^{-12}\ \mathrm{N^{-1}} & -181.5 \times 10^{-15}\ \mathrm{N^{-1}mm^{-1}} \\ -181.5 \times 10^{-15}\ \mathrm{N^{-1}mm^{-1}} & 573.5 \times 10^{-18}\ \mathrm{N^{-1}mm^{-2}} \end{bmatrix}$$

The strain vector $\boldsymbol{\varepsilon}_0$ containing the unknown strain variables is determined from Eq. 5.14:

$$\boldsymbol{\varepsilon}_0 = \mathbf{F}_0 \left(\mathbf{r}_{e,0} - \mathbf{f}_{p,init}\right) = \begin{bmatrix} 149.5 \times 10^{-12} & -181.5 \times 10^{-15} \\ -181.5 \times 10^{-15} & 573.5 \times 10^{-18} \end{bmatrix}$$

$$\times \left\{ \begin{bmatrix} -100 \times 10^3 \\ -50 \times 10^6 \end{bmatrix} - \begin{bmatrix} 2000 \times 10^3 \\ 1290 \times 10^6 \end{bmatrix} \right\}$$

$$= \begin{bmatrix} -70.7 \times 10^{-6} \\ -0.387 \times 10^{-6}\ \mathrm{mm^{-1}} \end{bmatrix}$$

The strain at the reference axis and the curvature are therefore $\varepsilon_{r,0} = -70.7 \times 10^{-6}$ and $\kappa_0 = -0.387 \times 10^{-6}$ mm^{-1} and, from Eq. 5.1, the top ($y = -300$ mm) and bottom ($y = +850$ mm) fibre strains are:

$$\varepsilon_{0(top)} = \varepsilon_{r,0} - 300 \times \kappa_0 = (-70.7 - 300 \times (-0.387)) \times 10^{-6} = +45.5 \times 10^{-6}$$

and

$$\varepsilon_{0(btm)} = \varepsilon_{r,0} + 850 \times \kappa_0 = (-70.7 + 850 \times (-0.387)) \times 10^{-6} = -399.8 \times 10^{-6}$$

From Eq. 5.16a, the top and bottom fibre stresses in the concrete are:

$$\sigma_{c,0(top)} = E_{c,0}\varepsilon_{0(top)} = 32{,}000 \times 45.5 \times 10^{-6} = +1.45\ \mathrm{MPa}$$

$$\sigma_{c,0(btm)} = E_{c,0}\varepsilon_{0(btm)} = 32{,}000 \times (-399.8) \times 10^{-6} = -12.8\ \mathrm{MPa}$$

and, from Eq. 5.16b, the stresses in the two layers of non-prestressed reinforcement are:

$$\sigma_{s(1),0} = 200 \times 10^3 \times \begin{bmatrix} 1 & -240 \end{bmatrix} \begin{bmatrix} -70.7 \times 10^{-6} \\ -0.387 \times 10^{-6} \end{bmatrix} = 4.5\ \mathrm{MPa}$$

and

$$\sigma_{s(2),0} = 200 \times 10^3 \times \begin{bmatrix} 1 & 790 \end{bmatrix} \begin{bmatrix} -70.7 \times 10^{-6} \\ -0.387 \times 10^{-6} \end{bmatrix} = -75.3\ \mathrm{MPa}$$

The stresses in the two layers of prestressing steel are obtained from Eq. 5.16c:

$$\sigma_{p(1),0} = E_p \begin{bmatrix} 1 & y_{p(1)} \end{bmatrix} \boldsymbol{\varepsilon}_0 + E_p \varepsilon_{p(1),init}$$

$$= 200 \times 10^3 \times \begin{bmatrix} 1 & 580 \end{bmatrix} \begin{bmatrix} -70.7 \times 10^{-6} \\ -0.387 \times 10^{-6} \end{bmatrix} + 200{,}000 \times 0.00625$$

$$= 1191 \text{ MPa}$$

and

$$\sigma_{p(2),0} = E_p \begin{bmatrix} 1 & y_{p(2)} \end{bmatrix} \boldsymbol{\varepsilon}_0 + E_p \varepsilon_{p(2),init}$$

$$= 200 \times 10^3 \times \begin{bmatrix} 1 & 710 \end{bmatrix} \begin{bmatrix} -70.7 \times 10^{-6} \\ -0.387 \times 10^{-6} \end{bmatrix} + 200{,}000 \times 0.00625$$

$$= 1181 \text{ MPa.}$$

The stress and strain distributions are plotted in Fig. 5.8.

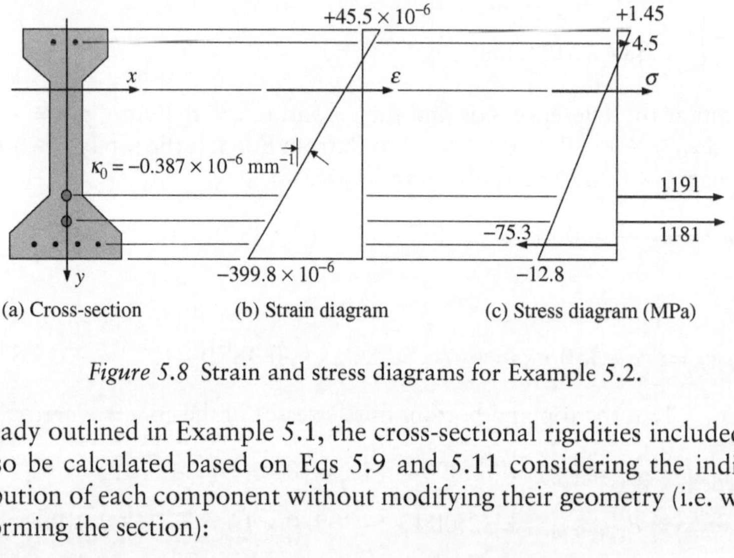

Figure 5.8 Strain and stress diagrams for Example 5.2.

As already outlined in Example 5.1, the cross-sectional rigidities included in $\mathbf{F}_0$ can also be calculated based on Eqs 5.9 and 5.11 considering the individual contribution of each component without modifying their geometry (i.e. without transforming the section):

$$R_{A,0} = A_{gross} E_{c,0} + \left(A_{s(1)} + A_{s(2)}\right)\left(E_s - E_{c,0}\right) + \left(A_{p(1)} + A_{p(2)}\right)\left(E_p - E_{c,0}\right)$$

$$= 10{,}866 \times 10^6 \text{ N}$$

$$R_{B,0} = A_{gross}\left(d_c - d_{ref}\right) E_{c,0} + \left(A_{s(1)} y_{s(1)} + A_{s(2)} y_{s(2)}\right)\left(E_s - E_{c,0}\right)$$

$$+ \left(A_{p(1)} y_{p(1)} + A_{p(2)} y_{p(2)}\right)\left(E_p - E_{c,0}\right) = 3439 \times 10^9 \text{ Nmm}$$

$$R_{I,0} = \left[I_{gross} + A_{gross}\left(d_c - d_{ref}\right)^2\right] E_{c,0} + \left(A_{s(1)} y_{s(1)}^2 + A_{s(2)} y_{s(2)}^2\right)\left(E_s - E_{c,0}\right)$$

$$+ \left(A_{p(1)} y_{p(1)}^2 + A_{p(2)} y_{p(2)}^2\right)\left(E_p - E_{c,0}\right) = 2832 \times 10^{12} \text{ Nmm}^2$$

Matrix $\mathbf{F}_0$ can then be determined as (Eq. 5.15):

$$\mathbf{F}_0 = \frac{1}{R_{A,0}R_{I,0} - R_{B,0}^2}\begin{bmatrix} R_{I,0} & -R_{B,0} \\ -R_{B,0} & R_{A,0} \end{bmatrix}$$

$$= \begin{bmatrix} 149.5 \times 10^{-12}\ \text{N}^{-1} & -181.5 \times 10^{-15}\ \text{N}^{-1}\text{mm}^{-1} \\ -181.5 \times 10^{-15}\ \text{N}^{-1}\text{mm}^{-1} & 573.5 \times 10^{-18}\ \text{N}^{-1}\text{mm}^{-2} \end{bmatrix}$$

5.4 Long-term analysis of reinforced or prestressed concrete cross-sections using the age-adjusted effective modulus method (AEMM)

Cross-sectional analysis using the age-adjusted effective modulus method (see Section 4.4) provides an effective tool for determining how stresses and strains vary with time due to creep and shrinkage of concrete and relaxation of the prestressing steel (if any). For this purpose, two instants in time are identified, as shown in Fig. 5.9. One time instant is the time at first loading, i.e. $t = \tau_0$, and one represents the instant in time at which stresses and strains need to be evaluated, i.e. $t = \tau_k$. It is usually convenient to measure time in days starting from the time when the concrete is poured.

During the time interval Δt_k $(= \tau_k - \tau_0)$, creep and shrinkage strains develop in the concrete and relaxation occurs in the prestressing steel. The gradual change of strain in the concrete with time causes changes of stress in the bonded reinforcement. In general, as the concrete shortens due to compressive creep and shrinkage, the reinforcement is compressed and there is a gradual increase in the compressive stress in the non-prestressed reinforcement and a gradual loss of prestress in any bonded prestressing tendons. To maintain equilibrium, the gradual change of force in the steel at each bonded reinforcement level is opposed by an equal and opposite restraining force on the concrete, as shown in Fig. 5.10. These gradually applied forces ($\Delta T_{c.s(i)}$ and $\Delta T_{c.p(i)}$) are usually tensile and, for a prestressed or partially-prestressed cross-section, tend to relieve the concrete of its initial compression.

The resultants of the creep and shrinkage-induced internal restraining forces on the concrete are an increment of axial force $\Delta N(\tau_k)$ and an increment of moment about the reference axis $\Delta M(\tau_k)$ given by:

$$\Delta N(\tau_k) = \sum_{i=1}^{m_s} \Delta T_{c.s(i)} + \sum_{i=1}^{m_p} \Delta T_{c.p(i)} \tag{5.19a}$$

and

$$\Delta M(\tau_k) = \sum_{i=1}^{m_s} \Delta T_{c.s(i)} y_{s(i)} + \sum_{i=1}^{m_p} \Delta T_{c.p(i)} y_{p(i)} \tag{5.19b}$$

Equal and opposite actions $-\Delta N(\tau_k)$ and $-\Delta M(\tau_k)$ are applied to the bonded steel parts of the cross-section.

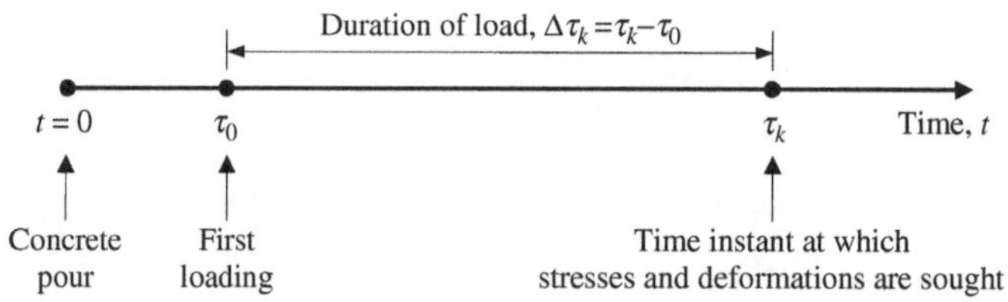

Figure 5.9 Relevant instants in time (AEMM).

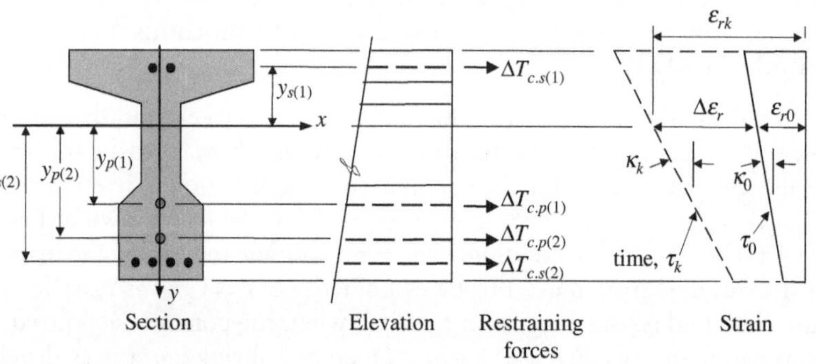

Figure 5.10 Time-dependent actions and deformations.

The strain at any depth y below the reference axis at time τ_k may be expressed in terms of the strain at the reference axis ε_{rk} and the curvature κ_k:

$$\varepsilon_k = \varepsilon_{r,k} + y\kappa_k \tag{5.20}$$

The magnitude of the change of strain that occurs with time $\Delta\varepsilon_k$ $(= \varepsilon_k - \varepsilon_0)$ is the sum of each of the following components:

(a) the free shrinkage strain $\varepsilon_{sh}(\tau_k)$ (which is usually considered to be uniform over the section);
(b) the unrestrained creep strain caused by the initial concrete stress $\sigma_{c,0}$ existing at the beginning of the time period, i.e. $\varepsilon_{cr,k} = \varphi(\tau_k, \tau_0)\sigma_{c,0}/E_{c,0}$; and
(c) the creep and elastic strain caused by $\Delta N(\tau_k)$ and $\Delta M(\tau_k)$ gradually applied to the concrete cross-section throughout the time period.

For the time analysis, the steel reinforcement and prestressing tendons (if any) are assumed to be linear-elastic (as for the short-term analysis) and the constitutive relationship for the concrete is that given by Eq. 4.45. The stress-strain relationships for each material at τ_0 and at τ_k are as follows.

At τ_0:

$$\sigma_{c,0} = E_{c,0}\varepsilon_0 \tag{5.21a}$$

$$\sigma_{s(i),0} = E_{s(i)}\varepsilon_0 \tag{5.21b}$$

$$\sigma_{p(i)} = E_{p(i)}\left(\varepsilon_0 + \varepsilon_{p(i),init}\right) \tag{5.21c}$$

At τ_k:

$$\sigma_{c,k} = \overline{E}_{e,k}\left(\varepsilon_k - \varepsilon_{sh,k}\right) + \overline{F}_{e,0}\sigma_{c,0} \tag{5.22a}$$

$$\sigma_{s(i),k} = E_{s(i)}\varepsilon_k \tag{5.22b}$$

$$\sigma_{p(i),k} = E_{p(i)}\left(\varepsilon_k + \varepsilon_{p(i),init} - \varepsilon_{p.rel(i),k}\right) \tag{5.22c}$$

where $\overline{F}_{e,0}$ is given by Eq. 4.46, $\overline{E}_{e,k}$ is the age-adjusted effective modulus at $t = \tau_k$ (Eq. 4.35), and $\varepsilon_{p.rel(i),k}$ is the tensile creep strain that has developed in the i-th prestressing tendon at time τ_k (often referred to as the *relaxation strain*) and may be calculated from:

$$\varepsilon_{p.rel(i),k} = \frac{P_{p(i),init}}{A_{p(i)}E_{p(i)}}\varphi_p(\tau_k, \sigma_{p(i),init}) = \varepsilon_{p(i),init}\ \varphi_p(\tau_k, \sigma_{p(i),init}) \tag{5.23a}$$

where $\varphi_p(\tau_k, \sigma_{p(i),init})$ is the creep coefficient for the prestressing steel at time τ_k due to an initial stress $\sigma_{p(i),init}$ in the i-th prestressing tendon just prior to transfer (as given in Table 2.7). Alternatively, $\varepsilon_{p.rel(i),k}$ may be calculated from the relaxation R given in Eq. 2.11 or included in Table 2.7 as:

$$\varepsilon_{p.rel(i),k} = \varepsilon_{p(i),init}\ R \tag{5.23b}$$

The governing equations describing the long-term behaviour of a cross-section are obtained by enforcing equilibrium at the cross-section at time τ_k following the approach already presented in the previous section for the instantaneous analysis at time τ_0 (Eqs 5.4–5.18), and by extending the approach outlined for axially loaded uncracked cross-sections in Sections 4.4.1 and 4.4.4. Restating the equilibrium equations (Eqs 5.2) at time τ_k gives:

$$\mathbf{r}_{e,k} = \mathbf{r}_{i,k} \tag{5.24}$$

where

$$\mathbf{r}_{e,k} = \begin{bmatrix} N_{e,k} \\ M_{e,k} \end{bmatrix} \tag{5.25a}$$

and

$$\mathbf{r}_{i,k} = \begin{bmatrix} N_{i,k} \\ M_{i,k} \end{bmatrix} \tag{5.25b}$$

$N_{i,k}$ and $M_{i,k}$ depict the internal axial force and moment resisted by the cross-section at time τ_k, while $N_{e,k}$ and $M_{e,k}$ represent the external applied loads at this time. As in Eq. 5.6, the axial force $N_{i,k}$ is the sum of the axial forces carried by the concrete, the reinforcement and the tendons:

$$N_{i,k} = N_{c,k} + N_{s,k} + N_{p,k} \tag{5.26}$$

Considering the time-dependent constitutive relationship for the concrete (Eq. 5.22a), the axial force resisted by the concrete at time τ_k can be expressed as:

$$N_{c,k} = \int_{A_c} \sigma_{c,k}\, \mathrm{d}A = \int_{A_c} \left[\overline{E}_{e,k}\left(\varepsilon_{r,k} + y\kappa_k - \varepsilon_{sh,k}\right) + \overline{F}_{e,0}\sigma_{c,0}\right]\mathrm{d}A$$

$$= A_c\overline{E}_{e,k}\varepsilon_{r,k} + B_c\overline{E}_{e,k}\kappa_k - A_c\overline{E}_{e,k}\varepsilon_{sh,k} + \overline{F}_{e,0}N_{c,0} \tag{5.27}$$

where $\varepsilon_{r,k}$ and κ_k are the strain at the level of the reference axis and the curvature at time τ_k, while $N_{c,0}$ is the axial force resisted by the concrete at time τ_0. For the time analysis, $N_{c,0}$ is assumed to be known having been determined during the instantaneous analysis from Eq. 5.7a.

Using the constitutive equations for the steel (Eqs 5.22b and c), the axial forces carried by the reinforcing bars and the prestressing steel at time τ_k are calculated as:

$$N_{s,k} = \sum_{i=1}^{m_s}\left(A_{s(i)}E_{s(i)}\right)\varepsilon_{r,k} + \sum_{i=1}^{m_s}\left(y_{s(i)}A_{s(i)}E_{s(i)}\right)\kappa_k \tag{5.28a}$$

$$N_{p,k} = \sum_{i=1}^{m_p}\left(A_{p(i)}E_{p(i)}\right)\varepsilon_{r,k} + \sum_{i=1}^{m_p}\left(y_{p(i)}A_{p(i)}E_{p(i)}\right)\kappa_k$$

$$+\sum_{i=1}^{m_p}\left[A_{p(i)}E_{p(i)}\left(\varepsilon_{p(i),init} - \varepsilon_{p.rel(i),k}\right)\right] \tag{5.28b}$$

By substituting Eqs 5.27 and 5.28 into Eq. 5.26, the internal axial force is given by:

$$N_{i,k} = \left(A_c\overline{E}_{e,k} + \sum_{i=1}^{m_s}A_{s(i)}E_{s(i)} + \sum_{i=1}^{m_p}A_{p(i)}E_{p(i)}\right)\varepsilon_{r,k} + \left(B_c\overline{E}_{e,k} + \sum_{i=1}^{m_s}y_{s(i)}A_{s(i)}E_{s(i)}\right.$$

$$\left. + \sum_{i=1}^{m_p}y_{p(i)}A_{p(i)}E_{p(i)}\right)\kappa_k - A_c\overline{E}_{e,k}\varepsilon_{sh,k} + \overline{F}_{e,0}N_{c,0}$$

$$+\sum_{i=1}^{m_p}\left[A_{p(i)}E_{p(i)}\left(\varepsilon_{p(i),init} - \varepsilon_{p.rel(i),k}\right)\right]$$

$$= R_{A,k}\varepsilon_{r,k} + R_{B,k}\kappa_k - A_c\overline{E}_{e,k}\varepsilon_{sh,k} + \overline{F}_{e,0}N_{c,0}$$

$$+\sum_{i=1}^{m_p}\left[A_{p(i)}E_{p(i)}\left(\varepsilon_{p(i),init} - \varepsilon_{p.rel(i),k}\right)\right] \tag{5.29}$$

where the axial rigidity and the stiffness related to the first moment of area calculated at time τ_k have been referred to as $R_{A,k}$ and $R_{B,k}$, respectively, and are given by:

$$R_{A,k} = A_c\overline{E}_{e,k} + \sum_{i=1}^{m_s}A_{s(i)}E_{s(i)} + \sum_{i=1}^{m_p}A_{p(i)}E_{p(i)} \tag{5.30a}$$

$$R_{B,k} = B_c\overline{E}_{e,k} + \sum_{i=1}^{m_s}y_{s(i)}A_{s(i)}E_{s(i)} + \sum_{i=1}^{m_p}y_{p(i)}A_{p(i)}E_{p(i)} \tag{5.30b}$$

In a similar manner, the internal moment $M_{i,k}$ resisted by the cross-section at time τ_k can be expressed as:

$$\begin{aligned} M_{i,k} = &\left(B_c\overline{E}_{e,k} + \sum_{i=1}^{m_s} y_{s(i)}A_{s(i)}E_{s(i)} + \sum_{i=1}^{m_p} y_{p(i)}A_{p(i)}E_{p(i)} \right) \varepsilon_{r,k} \\ &+ \left(I_c\overline{E}_{e,k} + \sum_{i=1}^{m_s} y_{s(i)}^2 A_{s(i)}E_{s(i)} + \sum_{i=1}^{m_p} y_{p(i)}^2 A_{p(i)}E_{p(i)} \right) \kappa_k - B_c\overline{E}_{e,k}\varepsilon_{sh,k} \\ &+ \overline{F}_{e,0}M_{c,0} + \sum_{i=1}^{m_p} \left[y_{p(i)}A_{p(i)}E_{p(i)} \left(\varepsilon_{p(i),init} - \varepsilon_{p.rel(i),k} \right) \right] \\ = &R_{B,k}\varepsilon_{r,k} + R_{I,k}\kappa_k - B_c\overline{E}_{e,k}\varepsilon_{sh,k} + \overline{F}_{e,0}M_{c,0} \\ &+ \sum_{i=1}^{m_p} \left[y_{p(i)}A_{p(i)}E_{p(i)} \left(\varepsilon_{p(i),init} - \varepsilon_{p.rel(i),k} \right) \right] \end{aligned} \tag{5.31}$$

where the flexural rigidity of the cross-section calculated at time τ_k, referred to as $R_{I,k}$ in Eq. 5.31, is given by:

$$R_{I,k} = I_c\overline{E}_{e,k} + \sum_{i=1}^{m_s} y_{s(i)}^2 A_{s(i)}E_{s(i)} + \sum_{i=1}^{m_p} y_{p(i)}^2 A_{p(i)}E_{p(i)} \tag{5.32}$$

and $M_{c,0}$ is the moment resisted by the concrete component at time τ_0 and from the instantaneous analysis:

$$M_{c,0} = \int_{A_c} y\sigma_{c,0}\, \mathrm{d}A = \int_{A_c} yE_{c,0}\left(\varepsilon_{r,0} + y\kappa_0 \right) \mathrm{d}A = B_cE_{c,0}\varepsilon_{r,0} + I_cE_{c,0}\kappa_0 \tag{5.33}$$

After substituting Eqs 5.29 and 5.31 into Eq. 5.24, the equilibrium equations at time τ_k may be written in compact form as:

$$\mathbf{r}_{e,k} = \mathbf{D}_k\boldsymbol{\varepsilon}_k + \mathbf{f}_{cr,k} - \mathbf{f}_{sh,k} + \mathbf{f}_{p,init} - \mathbf{f}_{p.rel,k} \tag{5.34}$$

where

$$\boldsymbol{\varepsilon}_k = \begin{bmatrix} \varepsilon_{r,k} \\ \kappa_k \end{bmatrix} \tag{5.35a}$$

and

$$\mathbf{D}_k = \begin{bmatrix} R_{A,k} & R_{B,k} \\ R_{B,k} & R_{I,k} \end{bmatrix} \tag{5.35b}$$

The vector $\mathbf{f}_{cr,k}$ represents the creep effect produced by the stress $\sigma_{c,0}$ resisted by the concrete at time τ_0 and is given by:

$$\mathbf{f}_{cr,k} = \overline{F}_{e,0}\begin{bmatrix} N_{c,0} \\ M_{c,0} \end{bmatrix} = \overline{F}_{e,0}E_{c,0}\begin{bmatrix} A_c\varepsilon_{r,0} + B_c\kappa_0 \\ B_c\varepsilon_{r,0} + I_c\kappa_0 \end{bmatrix} \tag{5.36}$$

and $\overline{F}_{e,0}$ is given in Eq. 4.46. The vector $\mathbf{f}_{sh,k}$ accounts for the uniform shrinkage strain that develops in the concrete over the time period and is given by:

$$\mathbf{f}_{sh,k} = \begin{bmatrix} A_c \\ B_c \end{bmatrix} \overline{E}_{e,k} \varepsilon_{sh,k} \tag{5.37}$$

The vector $\mathbf{f}_{p,init}$ in Eq. 5.34 accounts for the initial prestress and the vector $\mathbf{f}_{p.rel,k}$ accounts for the resultant actions caused by the loss of prestress in the tendon due to relaxation. These are given by:

$$\mathbf{f}_{p,init} = \sum_{i=1}^{m_p} \begin{bmatrix} A_{p(i)} E_{p(i)} \varepsilon_{p(i),init} \\ y_{p(i)} A_{p(i)} E_{p(i)} \varepsilon_{p(i),init} \end{bmatrix} \tag{5.38a}$$

and

$$\mathbf{f}_{p.rel,k} = \mathbf{f}_{p,init}\, \varphi_p(\tau_k, \sigma_{p(i),init}) \tag{5.38b}$$

Eq. 5.34 can be solved for ε_k as:

$$\begin{aligned} \boldsymbol{\varepsilon}_k &= \mathbf{D}_k^{-1} \left(\mathbf{r}_{e,k} - \mathbf{f}_{cr,k} + \mathbf{f}_{sh,k} - \mathbf{f}_{p,init} + \mathbf{f}_{p.rel,k} \right) \\ &= \mathbf{F}_k \left(\mathbf{r}_{e,k} - \mathbf{f}_{cr,k} + \mathbf{f}_{sh,k} - \mathbf{f}_{p,init} + \mathbf{f}_{p.rel,k} \right) \end{aligned} \tag{5.39}$$

where

$$\mathbf{F}_k = \frac{1}{R_{A,k} R_{I,k} - R_{B,k}^2} \begin{bmatrix} R_{I,k} & -R_{B,k} \\ -R_{B,k} & R_{A,k} \end{bmatrix} \tag{5.40}$$

The stress distribution at time τ_k can then be calculated as follows:

$$\sigma_{c,k} = \overline{E}_{e,k} \left(\varepsilon_k - \varepsilon_{sh,k} \right) + \overline{F}_{e,0} \sigma_{c,0} = \overline{E}_{e,k} \left\{ \begin{bmatrix} 1 & y \end{bmatrix} \boldsymbol{\varepsilon}_k - \varepsilon_{sh,k} \right\} + \overline{F}_{e,0} \sigma_{c,0} \tag{5.41a}$$

$$\sigma_{s,k(i)} = E_{s(i)} \varepsilon_k = E_{s(i)} \begin{bmatrix} 1 & y_{s(i)} \end{bmatrix} \boldsymbol{\varepsilon}_k \tag{5.41b}$$

$$\begin{aligned} \sigma_{p(i),k} &= E_{p(i)} \left(\varepsilon_k + \varepsilon_{p(i),init} - \varepsilon_{p.rel(i),k} \right) \\ &= E_{p(i)} \begin{bmatrix} 1 & y_{p(i)} \end{bmatrix} \boldsymbol{\varepsilon}_k + E_{p(i)} \varepsilon_{p(i),init} - E_{p(i)} \varepsilon_{p.rel(i),k} \end{aligned} \tag{5.41c}$$

where at any point y from the reference axis $\varepsilon_k = \varepsilon_{r,k} + y\kappa_k = \begin{bmatrix} 1 & y \end{bmatrix} \boldsymbol{\varepsilon}_k$.

The cross-sectional rigidities $R_{A,k}$, $R_{B,k}$ and $R_{I,k}$, required for the solution at time τ_k can also be calculated from the properties of the age-adjusted transformed section, obtained by transforming the bonded steel areas (reinforcement and tendons) into equivalent areas of the aged concrete at time τ_k, as follows:

$$R_{A,k} = \overline{A}_k \overline{E}_{e,k} \tag{5.42a}$$

$$R_{B,k} = \overline{B}_k \overline{E}_{e,k} \tag{5.42b}$$

$$R_{I,k} = \overline{I}_k \overline{E}_{e,k} \tag{5.42c}$$

where $\overline{E}_{e,k}$ is the age-adjusted effective modulus, $\overline{A}_k$ is the area of the age-adjusted transformed section, and $\overline{B}_k$ and $\overline{I}_k$ are the first and second moments of the area of the age-adjusted transformed section about the reference axis. For the determination of $\overline{A}_k$, $\overline{B}_k$ and $\overline{I}_k$, the areas of the bonded steel are transformed into equivalent areas of concrete by multiplying by the age-adjusted modular ratio $\bar{n}_{es(i),k}(= E_{s(i)}/\overline{E}_{e,k})$ or $\bar{n}_{ep(i),k} = (E_{p(i)}/\overline{E}_{e,k})$, as appropriate.

Based on Eqs 5.42, the expression for $\mathbf{F}_k$ (Eq. 5.40) can be re-written as:

$$\mathbf{F}_k = \frac{1}{\overline{E}_{e,k}(\overline{A}_k\overline{I}_k - \overline{B}_k^2)}\begin{bmatrix} \overline{I}_k & -\overline{B}_k \\ -\overline{B}_k & \overline{A}_k \end{bmatrix} \tag{5.43}$$

The calculation of the time-dependent stresses and deformations using the above procedure is illustrated in Examples 5.3 and 5.4.

Example 5.3

For the reinforced concrete section shown in Fig. 5.5, the strain and stress distributions at τ_0 immediately after the application of an axial force $N_{e,0} = -30$ kN and a bending moment of $M_{e,0} = 50$ kNm were calculated in Example 5.1. If the applied loads remain constant during the time interval τ_0 to τ_k, the strain and stress distributions at time τ_k are to be determined using the age-adjusted effective modulus method. As in Example 5.1, $E_{c,0} = 25$ GPa; $E_s = 200$ GPa; and $n_{s,0} = E_s/E_{c,0} = 8$. Take $\varphi\ (\tau_k, \tau_0) = 2.5$; $\chi(\tau_k, \tau_0) = 0.65$; $\varepsilon_{sh}(\tau_k) = -600 \times 10^{-6}$ and assume the steel reinforcement is linear elastic.

From Example 5.1: At τ_0: $\varepsilon_{r,0} = -42.7 \times 10^{-6}$ and $\kappa_0 = 0.331 \times 10^{-6}$ mm^{-1}.

From Eqs 4.35 and 4.46:

$$\overline{E}_{e,k} = \frac{E_{c,0}}{1 + \chi(\tau_k, \tau_0)\varphi(\tau_k, \tau_0)} = \frac{25{,}000}{1 + 0.65 \times 2.5} = 9524 \text{ MPa}$$

and therefore $\bar{n}_{es,k} = 21.0$;

$$\overline{F}_{e,0} = \frac{\varphi(\tau_k, \tau_0)[\chi(\tau_k, \tau_0) - 1]}{1 + \chi(\tau_k, \tau_0)\varphi(\tau_k, \tau_0)} = \frac{2.5 \times (0.65 - 1.0)}{1.0 + 0.65 \times 2.5} = -0.333$$

The properties of the concrete part of the cross-section are:

$$A_c = bD - A_{s(1)} - A_{s(2)} = 300 \times 600 - 620 - 1800 = 177{,}600 \text{ mm}^2$$

$$B_c = bD(D/2 - d_{ref}) - A_{s(1)}y_{s(1)} - A_{s(2)}y_{s(2)} = 300 \times 600 \times (300 - 200)$$
$$- 620 \times (-150) - 1800 \times 350 = 17.46 \times 10^6 \text{mm}^3$$

$$I_c = \frac{bD^3}{12} + bD\left(\frac{D}{2} - d_{ref}\right)^2 - A_{s(1)}y_{s(1)}^2 - A_{s(2)}y_{s(2)}^2 = \frac{300 \times 600^3}{12}$$
$$+ 300 \times 600 \times (300 - 200)^2 - 620 \times (-150)^2 - 1800 \times 350^2$$
$$= 6966 \times 10^6 \text{ mm}^4$$

and the properties of the age-adjusted transformed cross-section are:

$$\overline{A}_k = bD + (\bar{n}_{es,k} - 1)[A_{s(1)} + A_{s(2)}] = 300 \times 600 + (21 - 1) \times [620 + 1800]$$
$$= 228{,}400 \text{ mm}^2$$

$$\overline{B}_k = bD\left(\frac{D}{2} - d_{ref}\right) + (\bar{n}_{es,k} - 1)[A_{s(1)}y_{s(1)} + A_{s(2)}y_{s(2)}] = 300 \times 600$$
$$\times (300 - 200) + (21 - 1) \times [620 \times (-150) + 1800 \times 350]$$
$$= 28.74 \times 10^6 \text{ mm}^3$$

$$\bar{I}_k = \frac{bD^3}{12} + bD\left(\frac{D}{2} - d_{ref}\right)^2 + (\bar{n}_{es,k} - 1)[A_{s(1)}y_{s(1)}^2 + A_{s(2)}y_{s(2)}^2]$$
$$= \frac{300 \times 600^3}{12} + 300 \times 600 \times (300 - 200)^2 + 20 \times [620 \times (-150)^2$$
$$+ 1800 \times 350^2] = 11{,}890 \times 10^6 \text{ mm}^4$$

From Eq. 5.36:

$$\mathbf{f}_{cr.k} = -0.333 \times 25{,}000 \times \begin{bmatrix} 177{,}600 \times (-42.7 \times 10^{-6}) + 17.46 \times 10^6 \\ \times 0.331 \times 10^{-6} \\ 17.46 \times 10^6 \times (-42.7 \times 10^{-6}) + 6966 \times 10^6 \\ \times 0.331 \times 10^{-6} \end{bmatrix}$$

$$= \begin{bmatrix} +14.9 \times 10^3 \text{ N} \\ -13.0 \times 10^6 \text{ Nmm} \end{bmatrix}$$

and from Eq. 5.37:

$$\mathbf{f}_{sh,k} = \begin{bmatrix} 177{,}600 \times 9524 \times (-600 \times 10^{-6}) \\ 17.46 \times 10^6 \times 9524 \times (-600 \times 10^{-6}) \end{bmatrix} = \begin{bmatrix} -1015 \times 10^3 \text{ N} \\ -99.79 \times 10^6 \text{ Nmm} \end{bmatrix}$$

Eq. 5.43 gives:

$$\mathbf{F}_k = \frac{1}{9524 \times (228{,}400 \times 11{,}890 \times 10^6 - (28.74 \times 10^6)^2)}$$
$$\times \begin{bmatrix} 11{,}890 \times 10^6 & -28.74 \times 10^6 \\ -28.74 \times 10^6 & 228{,}400 \end{bmatrix}$$
$$= \begin{bmatrix} 660.7 \times 10^{-12} \text{ N}^{-1} & -1.597 \times 10^{-12} \text{ N}^{-1}\text{mm}^{-1} \\ -1.597 \times 10^{-12} \text{ N}^{-1}\text{mm}^{-1} & 12.69 \times 10^{-15} \text{ N}^{-1}\text{mm}^{-2} \end{bmatrix}$$

The strain $\boldsymbol{\varepsilon}_k$ at time τ_k is determined using Eq. 5.39:

$$\boldsymbol{\varepsilon}_k = \mathbf{F}_k(\mathbf{r}_{e,k} - \mathbf{f}_{cr,k} + \mathbf{f}_{sh,k})$$

$$= \begin{bmatrix} 660.7 \times 10^{-12} & -1.597 \times 10^{-12} \\ -1.597 \times 10^{-12} & 12.69 \times 10^{-15} \end{bmatrix} \begin{bmatrix} (-30 - 14.9 - 1015) \times 10^3 \\ (50 + 13.0 - 99.79) \times 10^6 \end{bmatrix}$$

$$= \begin{bmatrix} -641.4 \times 10^{-6} \\ +1.226 \times 10^{-6}\ \text{mm}^{-1} \end{bmatrix}$$

That is, the strain at the reference axis and the curvature at time τ_k are $\varepsilon_{r,k} = -641.4 \times 10^{-6}$ and $\kappa_k = +1.226 \times 10^{-6}\ \text{mm}^{-1}$, respectively, and from Eq. 5.1, the top ($y = -200$ mm) and bottom ($y = +400$ mm) fibre strains are:

$$\varepsilon_{k(top)} = \varepsilon_{r,k} + (-200) \times \kappa_k = (-641.4 - 200 \times 1.226) \times 10^{-6}$$
$$= -886.5 \times 10^{-6}$$

and

$$\varepsilon_{k(btm)} = \varepsilon_{r,k} + 400 \times \kappa_k = (-641.4 + 400 \times 1.226) \times 10^{-6} = -151.1 \times 10^{-6}$$

The concrete stress distribution at time τ_k is calculated using Eq. 5.41a:

$$\sigma_{c,k(top)} = 9524 \left\{ \begin{bmatrix} 1 & -200 \end{bmatrix} \begin{bmatrix} -641.4 \times 10^{-6} \\ +1.226 \times 10^{-6} \end{bmatrix} - (-600 \times 10^{-6}) \right\}$$
$$+ (-0.333)(-2.27) = -1.82\ \text{MPa}$$

$$\sigma_{c,k(btm)} = 9524 \left\{ \begin{bmatrix} 1 & 400 \end{bmatrix} \begin{bmatrix} -641.4 \times 10^{-6} \\ +1.226 \times 10^{-6} \end{bmatrix} - (-600 \times 10^{-6}) \right\}$$
$$+ (-0.333)(2.25) = 3.52\ \text{MPa}$$

and, from Eq. 5.42b, the stresses in the reinforcement are:

$$\sigma_{s(1),k} = E_{s(1)} \begin{bmatrix} 1 & y_{s(1)} \end{bmatrix} \boldsymbol{\varepsilon}_k = 200 \times 10^3 \begin{bmatrix} 1 & -150 \end{bmatrix} \begin{bmatrix} -641.4 \times 10^{-6} \\ +1.226 \times 10^{-6} \end{bmatrix}$$
$$= -165\ \text{MPa}$$

and

$$\sigma_{s(2),k} = E_{s(2)} \begin{bmatrix} 1 & y_{s(2)} \end{bmatrix} \boldsymbol{\varepsilon}_k = 200 \times 10^3 \begin{bmatrix} 1 & 350 \end{bmatrix} \begin{bmatrix} -641.4 \times 10^{-6} \\ +1.226 \times 10^{-6} \end{bmatrix}$$
$$= -42.5\ \text{MPa}$$

The stress and strain distributions at τ_o (from Example 5.1) and at τ_k are shown in Fig. 5.11.

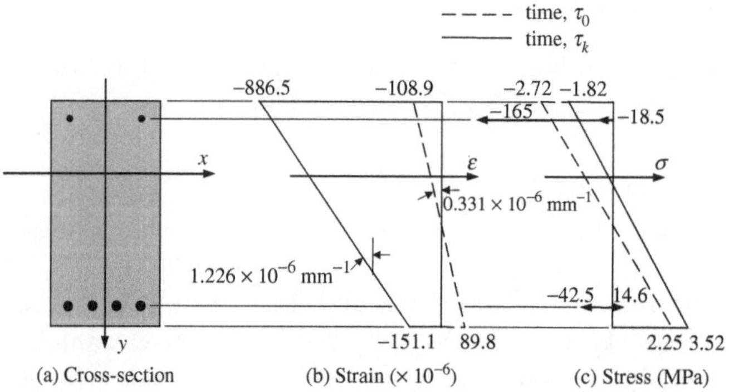

Figure 5.11 Strain and stress diagrams for Example 5.3 (all units in mm, MPa).

The cross-sectional rigidities could also have been calculated based on Eqs 5.30 and 5.32 without modifying the geometry of the section as:

$$R_{A,k} = A_c\overline{E}_{e,k} + \sum_{i=1}^{m_s} A_{s(i)}E_{s(i)} + \sum_{i=1}^{m_p} A_{p(i)}E_{p(i)} = A_cE_{c0} + \left(A_{s(1)} + A_{s(2)}\right)E_s$$

$$= 2175 \times 10^6 \text{ N};$$

$$R_{B,k} = B_c\overline{E}_{e,k} + \sum_{i=1}^{m_s} y_{s(i)}A_{s(i)}E_{s(i)} + \sum_{i=1}^{m_p} y_{p(i)}A_{p(i)}E_{p(i)}$$

$$= B_c\overline{E}_{e,k} + \left(A_{s(1)}y_{s(1)} + A_{s(2)}y_{s(2)}\right)E_s = 273.7 \times 10^9 \text{ Nmm; and}$$

$$R_{I,k} = I_c\overline{E}_{e,k} + \sum_{i=1}^{m_s} y_{s(i)}^2A_{s(i)}E_{s(i)} + \sum_{i=1}^{m_p} y_{p(i)}^2A_{p(i)}E_{p(i)}$$

$$= I_c\overline{E}_{e,k} + \left(A_{s(1)}y_{s(1)}^2 + A_{s(2)}y_{s(2)}^2\right)E_s = 113.2 \times 10^{12} \text{ Nmm}^2.$$

$\mathbf{F}_k$ is then determined using Eq. 5.40 (and is identical to that calculated earlier using Eq. 5.43):

$$\mathbf{F}_k = \frac{1}{R_{A,k}R_{I,k} - R_{B,k}^2}\begin{bmatrix} R_{I,k} & -R_{B,k} \\ -R_{B,k} & R_{A,k} \end{bmatrix}$$

$$= \frac{1}{2175 \times 10^6 \times 113.2 \times 10^{12} - (273.7 \times 10^9)^2}$$

$$\times \begin{bmatrix} 113.2 \times 10^{12} & -273.7 \times 10^9 \\ -273.7 \times 10^9 & 2175 \times 10^6 \end{bmatrix}$$

$$= \begin{bmatrix} 660.7 \times 10^{-12} \text{ N}^{-1} & -1.597 \times 10^{-12} \text{ N}^{-1}\text{mm}^{-1} \\ -1.597 \times 10^{-12} \text{ N}^{-1}\text{mm}^{-1} & 12.69 \times 10^{-15} \text{ N}^{-1}\text{mm}^{-2} \end{bmatrix}$$

Example 5.4

A time-dependent analysis of the prestressed concrete section of Example 5.2 (see Figure 5.7) is to be undertaken using the age-adjusted effective modulus method. The strain and stress distributions at τ_0 immediately after the application of an axial force $N_{e,0} = -100$ kN and a bending moment of $M_{e,0} = -50$ kNm were calculated in Example 5.2. If the applied loads remain constant during the time interval τ_0 to τ_k, the strain and stress distributions at time τ_k are to be determined.

As in Example 5.2: $E_{c,0} = 32$ GPa; $E_s = E_p = 200$ GPa; and therefore $n_{s(i),0} = n_{p(i),0} = 6.25$. Take $\varphi(\tau_k, \tau_0) = 2.0$; $\chi(\tau_k, \tau_0) = 0.65$; $\varepsilon_{sh}(\tau_k) = -400 \times 10^{-6}$ and, for the prestressing steel, $\varphi_p(\tau_k, \tau_0, \sigma_{p(i),init}) = 0.03$. Assume the steel reinforcement is linear elastic.

From Example 5.2 at τ_0: $\varepsilon_{r,0} = -70.7 \times 10^{-6}$ and $\kappa_0 = -0.387 \times 10^{-6}$ mm^{-1}.

From Eqs 4.35 and 4.46:

$$\overline{E}_{e,k} = \frac{32{,}000}{1 + 0.65 \times 2.0} = 13{,}910 \text{ MPa and therefore } \bar{n}_{es,k} = \bar{n}_{ep,k} = 14.37$$

$$\overline{F}_{e,0} = \frac{2.0 \times (0.65 - 1.0)}{1.0 + 0.65 \times 2.0} = -0.304$$

The properties of the concrete part of the cross-section are:

$$A_c = A_{gross} - A_{s(1)} - A_{s(2)} - A_{p(1)} - A_{p(2)} = 312{,}700 \text{ mm}^2$$

$$B_c = A_{gross}(d_c - d_{ref}) - [A_{s(1)}y_{s(1)} + A_{s(2)}y_{s(2)}] - [A_{p(1)}y_{p(1)} + A_{p(2)}y_{p(2)}]$$
$$= 93.49 \times 10^6 \text{ mm}^3$$

$$I_c = I_{gross} + A_{gross}(d_c - d_{ref})^2 - [A_{s(1)}y_{s(1)}^2 + A_{s(2)}y_{s(2)}^2] - [A_{p(1)}y_{p(1)}^2$$
$$+ A_{p(1)}y_{p(1)}^2] = 76{,}960 \times 10^6 \text{ mm}^4$$

and the properties of the age-adjusted transformed section in equivalent concrete areas are:

$$\overline{A}_k = A_{gross} + (\bar{n}_{es,k} - 1)[A_{s(1)} + A_{s(2)}] + (\bar{n}_{ep,k} - 1)[A_{p(1)} + A_{p(2)}]$$
$$= 374.5 \times 10^3 \text{ mm}^2$$

$$\overline{B}_k = A_{gross}(d_c - d_{ref}) + (\bar{n}_{es,k} - 1)[A_{s(1)}y_{s(1)} + A_{s(2)}y_{s(2)}]$$
$$+ (\bar{n}_{ep,k} - 1)[A_{p(1)}y_{p(1)} + A_{p(2)}y_{p(2)}] = 125.7 \times 10^6 \text{ mm}^3$$

$$\bar{I}_k = I_{gross} + A_{gross}(d_c - d_{ref})^2 + (\bar{n}_{es,k} - 1)(A_{s(1)}y_{s(1)}^2 + A_{s(2)}y_{s(2)}^2)$$
$$+ (\bar{n}_{ep,k} - 1)(A_{p(1)}y_{p(1)}^2 + A_{p(2)}y_{p(2)}^2) = 103.5 \times 10^9 \text{ mm}^4$$

Eqs 5.36 and 5.37 give:

$$\mathbf{f}_{cr.k} = \begin{bmatrix} +567.9 \times 10^3 \text{ N} \\ +354.6 \times 10^6 \text{ Nmm} \end{bmatrix}$$

and

$$\mathbf{f}_{sh,k} = \begin{bmatrix} -1740 \times 10^3 \text{ N} \\ -520 \times 10^6 \text{ Nmm} \end{bmatrix}$$

and from Eq. 5.38a (and Example 5.2) and Eq. 5.38b:

$$\mathbf{f}_{p,init} = \begin{bmatrix} 2000 \times 10^3 \text{ N} \\ 1290 \times 10^6 \text{ Nmm} \end{bmatrix}$$

and

$$\mathbf{f}_{p.rel,k} = \begin{bmatrix} 2000 \times 10^3 \times 0.03 \\ 1290 \times 10^6 \times 0.03 \end{bmatrix} = \begin{bmatrix} 60 \times 10^3 \text{ N} \\ 38.7 \times 10^6 \text{ Nmm} \end{bmatrix}$$

Using Eq. 5.43:

$$\mathbf{F}_k = \begin{bmatrix} 323.8 \times 10^{-12}\text{N}^{-1} & -393.1 \times 10^{-15}\text{N}^{-1}\text{mm}^{-1} \\ -393.1 \times 10^{-15}\text{N}^{-1}\text{mm}^{-1} & 1.171 \times 10^{-15}\text{N}^{-1}\text{mm}^{-2} \end{bmatrix}$$

and the strain vector $\boldsymbol{\varepsilon}_k$ at time τ_k is determined using Eq. 5.39:

$$\boldsymbol{\varepsilon}_k = \begin{bmatrix} 323.8 \times 10^{-12} & -393.1 \times 10^{-15} \\ -393.1 \times 10^{-15} & 1.171 \times 10^{-15} \end{bmatrix} \times \begin{bmatrix} (-100 - 567.9 - 1740 - 2000 + 60) \times 10^3 \\ (-50 - 354.6 - 520 - 1290 + 38.7) \times 10^6 \end{bmatrix} = \begin{bmatrix} -552.5 \times 10^{-6} \\ -0.840 \times 10^{-6} \text{ mm}^{-1} \end{bmatrix}$$

The strain at the reference axis and the curvature at time τ_k are therefore $\varepsilon_{r,k} = -552.5 \times 10^{-6}$ and $\kappa_k = -0.840 \times 10^{-6}$ mm^{-1}, respectively. From Eq. 5.1, the top ($y = -300$ mm) and bottom ($y = +850$ mm) fibre strains are:

$$\varepsilon_{k(top)} = \varepsilon_{r,k} + (-300) \times \kappa_k = -300.5 \times 10^{-6}$$

and

$$\varepsilon_{k(btm)} = \varepsilon_{r,k} + 850 \times \kappa_k = -1266 \times 10^{-6}$$

The concrete stress distribution at time τ_k is calculated using Eq. 5.41a:

$$\sigma_{c,k(top)} = 13{,}913\left\{[1 \;\; -300]\begin{bmatrix}-552.5\times10^{-6}\\-0.840\times10^{-6}\end{bmatrix} - (-400\times10^{-6})\right\}$$

$$+(-0.304)(1.45) = 0.94 \text{ MPa}$$

$$\sigma_{c,k(btm)} = 13{,}913\left\{[1 \;\; 850]\begin{bmatrix}-552.5\times10^{-6}\\-0.840\times10^{-6}\end{bmatrix} - (-400\times10^{-6})\right\}$$

$$+(-0.304)(-12.8) = -8.16 \text{ MPa}$$

and, from Eq. 5.41b, the stresses in the non-prestressed reinforcement are:

$$\sigma_{s(1),k} = E_{s(1)}\left[1 \;\; y_{s(1)}\right]\boldsymbol{\varepsilon}_k = 200\times10^3\;[1 \;\; -240]\begin{bmatrix}-552.5\times10^{-6}\\-0.840\times10^{-6}\end{bmatrix}$$

$$= -70.2 \text{ MPa}$$

and

$$\sigma_{s(2),k} = E_{s(2)}\left[1 \;\; y_{s(2)}\right]\boldsymbol{\varepsilon}_k = 200\times10^3\;[1 \;\; 790]\begin{bmatrix}-552.5\times10^{-6}\\-0.840\times10^{-6}\end{bmatrix}$$

$$= -243.3 \text{ MPa}$$

From Example 5.2, the initial strains in the prestress before transfer are $\varepsilon_{p(1),init} = \varepsilon_{p(2),init} = 0.00625$ and the relaxation strains are therefore $\varepsilon_{p.rel(1),k} = \varepsilon_{p.rel(2),k} = 0.00625\times0.03 = 187.5\times10^{-6}$. The stresses in the prestressing tendons at τ_k are obtained from Eq. 5.41c:

$$\sigma_{p(1),k} = 200{,}000\times\left[[1 \;\; 580]\begin{bmatrix}-552.5\times10^{-6}\\-0.840\times10^{-6}\end{bmatrix} + 0.00625 - 0.0001875\right]$$

$$= 1004 \text{ MPa}$$

and

$$\sigma_{p(2),k} = 200{,}000\times\left[[1 \;\; 710]\begin{bmatrix}-552.5\times10^{-6}\\-0.840\times10^{-6}\end{bmatrix} + 0.00625 - 0.0001875\right]$$

$$= 982 \text{ MPa}$$

The changes in stress in each prestressing tendon between first loading and time τ_k are therefore $\Delta\sigma_{p(1),k} = \sigma_{p(1),0} - \sigma_{p(1),k} = 1191 - 1004 = 187$ MPa and $\Delta\sigma_{p(2),k} = \sigma_{p(2),0} - \sigma_{p(2),k} = 1181 - 982 = 199$ MPa and represent the time-dependent loss of prestress in each tendon, with $\Delta\sigma_{p(1),k}$ being 15.7 per cent of the initial stress in $A_{p(1)}$ immediately after transfer and $\Delta\sigma_{p(2),k}$ being 16.8 per cent of the initial stress in $A_{p(2)}$.

The stress and strain distributions at τ_0 (from Example 5.2) and at τ_k are shown in Fig. 5.12.

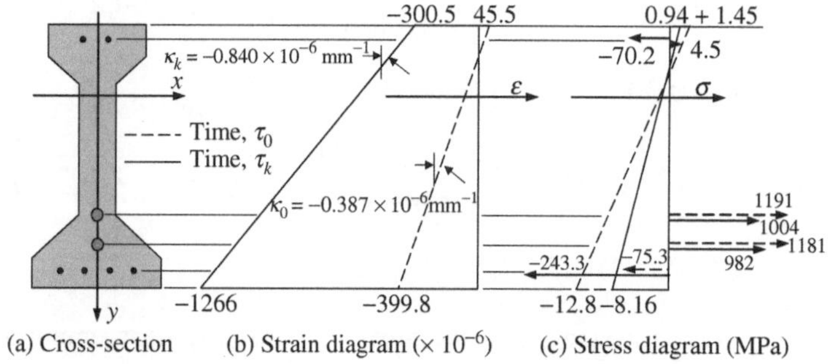

Figure 5.12 Strain and stress distributions for Example 5.4.

The cross-sectional rigidities calculated using Eqs 5.30 and 5.32 are:

$$R_{A,k} = A_c\overline{E}_{e,k} + \left(A_{s(1)} + A_{s(2)}\right)E_s + \left(A_{p(1)} + A_{p(2)}\right)E_p = 5210 \times 10^6 \text{ N}$$

$$R_{B,k} = B_c\overline{E}_{e,k} + \left(A_{s(1)}y_{s(1)} + A_{s(2)}y_{s(2)}\right)E_s$$
$$+ \left(A_{p(1)}y_{p(1)} + A_{p(2)}y_{p(2)}\right)E_p = 1748 \times 10^9 \text{ Nmm}$$

$$R_{I,k} = I_c\overline{E}_{e,k} + \left(A_{s(1)}y_{s(1)}^2 + A_{s(2)}y_{s(2)}^2\right)E_s$$
$$+ \left(A_{p(1)}y_{p(1)}^2 + A_{p(2)}y_{p(2)}^2\right)E_p = 1440 \times 10^{12} \text{ Nmm}^2$$

and the matrix $\mathbf{F}_k$ could have been determined conveniently using Eq. 5.41.

5.5 Long-term analysis of reinforced and prestressed cross-sections using the step-by-step procedure

The age-adjusted effective modulus method outlined in the previous section is a practical means to evaluate the time-dependent deformations and stresses on a reinforced or prestressed concrete section. It involves a single time interval, with stresses and deformations calculated at the beginning and end of the time interval, i.e. at τ_0 and τ_k as shown in Fig. 5.9. The level of accuracy of the results depends on the type of stress and/or deformation history undergone by the concrete.

For a more accurate representation of concrete behaviour, a more refined method may be required, such as the step-by-step method (see Section 4.3) or the Dirichlet Series approximation (Ref. 4–5). The step-by-step procedure involves the time domain being discretised into a number of instants τ_j (with $j = 0, \ldots, k$), as shown in Fig. 4.3, with the latter instant in time τ_k being the one at which the structural response is sought. Structural response is calculated at each time instant, in turn, with the solution at τ_j relying on the solutions obtained at the previous time instants. To achieve this, the concrete stresses calculated at each instant in time are stored for use in the subsequent analyses.

The steel reinforcement and prestressing tendons (if any) are again assumed to be linear-elastic (as for the short-term analysis) and the constitutive relationships for the steel at any time τ_j are:

$$\sigma_{s(i),j} = E_{s(i)}\varepsilon_j \tag{5.44a}$$

$$\sigma_{p(i),j} = E_{p(i)}\left(\varepsilon_j + \varepsilon_{p(i),init} - \varepsilon_{p.rel(i),j}\right) \tag{5.44b}$$

Following the approach outlined in Section 4.3.3, the constitutive relationship for the concrete at τ_j is given by Eq. 4.25 reproduced below:

$$\sigma_{c,j} = E_{c,j}\left(\varepsilon_j - \varepsilon_{sh,j}\right) + \sum_{i=0}^{j-1} F_{e,j,i}\sigma_{c,i} \tag{4.25}$$

and if the rectangular approximation of the integral-type creep law is adopted, the instantaneous elastic modulus of concrete $E_{c,j}$ and the term $F_{e,j,i}$ are given by Eqs 4.26a and b (reproduced below for ease of reference):

$$E_{c,j} = \frac{1}{J_{j,j}} \tag{4.26a}$$

and

$$F_{e,j,i} = \frac{J_{j,i+1} - J_{j,i}}{J_{j,j}} \tag{4.26b}$$

If the trapezoidal approximation is used for the numerical integration, the expressions for $E_{c,j}$ and $F_{e,j,i}$ are given in Eqs 4.26c–f. For ease of notation, the rectangular rule is considered in this chapter and used in the subsequent examples.

With the proposed approach, $k+1$ analyses must be carried out in order to yield the structural response at any time τ_k. For each time instant τ_j (with $j = 0, \ldots, k$), the equations utilised in the cross-sectional analysis are obtained from considerations of equilibrium:

$$\mathbf{r}_{i,j} = \mathbf{r}_{e,j} \tag{5.45}$$

where $\mathbf{r}_{i,j}$ and $\mathbf{r}_{e,j}$ are, respectively, the internal cross-sectional stress resultants and the external actions calculated at time τ_j, as follows:

$$\mathbf{r}_{i,j} = \begin{bmatrix} N_{i,j} \\ M_{i,j} \end{bmatrix} \tag{5.46a}$$

and

$$\mathbf{r}_{e,j} = \begin{bmatrix} N_{e,j} \\ M_{e,j} \end{bmatrix} \tag{5.46b}$$

where $N_{i,j}$ and $M_{i,j}$ depict the internal axial force and moment resisted by the cross-section at time τ_j, while $N_{e,j}$ and $M_{e,j}$ represent the external applied actions.

As already presented for the AEMM, the internal actions can be calculated by combining the contributions of the concrete, reinforcing steel and tendons. For example, the internal axial force resisted by the cross-section at τ_j can be expressed as:

$$N_{i,j} = N_{c,j} + N_{s,j} + N_{p,j} \tag{5.47}$$

The axial force in the concrete can be represented based on its constitutive relationship (Eq. 4.25) as:

$$\begin{aligned} N_{c,j} &= \int_{A_c} \sigma_{c,j}\, dA = \int_{A_c} \left[E_{c,j}\left(\varepsilon_j - \varepsilon_{sh,j}\right) + \sum_{i=0}^{j-1} F_{e,j,i}\sigma_{c,i} \right] dA \\ &= A_c E_{c,j}\varepsilon_{r,j} + B_c E_{c,j}\kappa_j - A_c E_{c,j}\varepsilon_{sh,j} + \sum_{i=0}^{j-1} F_{e,j,i} N_{c,i} \end{aligned} \tag{5.48}$$

where the axial forces $N_{c,i}$ resisted by the concrete at times τ_i (with $i = 0, \ldots, j-1$) are known (i.e. they have been calculated in previous time steps) at the beginning of the analysis at τ_j.

The axial forces in the reinforcement and tendons can be evaluated from the constitutive relations (Eqs 5.44) as:

$$N_{s,j} = \sum_{i=1}^{m_s} \left(A_{s(i)}E_{s(i)}\right)\varepsilon_{r,j} + \sum_{i=1}^{m_s} \left(y_{s(i)}A_{s(i)}E_{s(i)}\right)\kappa_j \tag{5.49a}$$

$$\begin{aligned} N_{p,j} &= \sum_{i=1}^{m_p} \left(A_{p(i)}E_{p(i)}\right)\varepsilon_{r,j} + \sum_{i=1}^{m_p} \left(y_{p(i)}A_{p(i)}E_{p(i)}\right)\kappa_j \\ &\quad + \sum_{i=1}^{m_p} \left[A_{p(i)}E_{p(i)}\left(\varepsilon_{p(i),init} - \varepsilon_{p.rel(i),j}\right)\right] \end{aligned} \tag{5.49b}$$

Substituting Eqs 5.48 and 5.49 into Eq. 5.47 produces:

$$\begin{aligned} N_{i,j} &= \left(A_c E_{c,j} + \sum_{i=1}^{m_s} A_{s(i)}E_{s(i)} + \sum_{i=1}^{m_p} A_{p(i)}E_{p(i)} \right)\varepsilon_{r,j} + \left(B_c E_{c,j} + \sum_{i=1}^{m_s} y_{s(i)}A_{s(i)}E_{s(i)} \right. \\ &\quad \left. + \sum_{i=1}^{m_p} y_{p(i)}A_{p(i)}E_{p(i)} \right)\kappa_j - A_c E_{c,j}\varepsilon_{sh,j} + \sum_{i=0}^{j-1} F_{e,j,i} N_{c,i} \\ &\quad + \sum_{i=1}^{m_p} \left[A_{p(i)}E_{p(i)}\left(\varepsilon_{p(i),init} - \varepsilon_{p.rel(i),j}\right)\right] \\ &= R_{A,j}\varepsilon_{r,j} + R_{B,j}\kappa_j - A_c E_{c,j}\varepsilon_{sh,j} + \sum_{i=0}^{j-1} F_{e,j,i} N_{c,i} \\ &\quad + \sum_{i=1}^{m_p} \left[A_{p(i)}E_{p(i)}\left(\varepsilon_{p(i),init} - \varepsilon_{p.rel(i),j}\right)\right] \end{aligned} \tag{5.50}$$

and following the same procedure, the internal moment $M_{i,j}$ at τ_j can be determined as:

$$M_{i,j} = \left(B_c E_{c,j} + \sum_{i=1}^{m_s} y_{s(i)} A_{s(i)} E_{s(i)} + \sum_{i=1}^{m_p} y_{p(i)} A_{p(i)} E_{p(i)} \right) \varepsilon_{r,j}$$

$$+ \left(I_c E_{c,j} + \sum_{i=1}^{m_s} y_{s(i)}^2 A_{s(i)} E_{s(i)} + \sum_{i=1}^{m_p} y_{p(i)}^2 A_{p(i)} E_{p(i)} \right) \kappa_j - B_c E_{c,j} \varepsilon_{sh,j}$$

$$+ \sum_{i=0}^{j-1} F_{e,j,i} M_{c,i} + \sum_{i=1}^{m_p} \left[y_{p(i)} A_{p(i)} E_{p(i)} \left(\varepsilon_{p(i),init} - \varepsilon_{p.rel(i),j} \right) \right]$$

$$= R_{B,j} \varepsilon_{r,j} + R_{I,j} \kappa_j - B_c E_{c,j} \varepsilon_{sh,j} + \sum_{i=0}^{j-1} F_{e,j,i} M_{c,i}$$

$$+ \sum_{i=1}^{m_p} \left[y_{p(i)} A_{p(i)} E_{p(i)} \left(\varepsilon_{p(i),init} - \varepsilon_{p.rel(i),j} \right) \right] \tag{5.51}$$

The cross-sectional rigidities $R_{A,j}$, $R_{B,j}$ and $R_{I,j}$ at time τ_j are defined as:

$$R_{A,j} = A_c E_{c,j} + \sum_{i=1}^{m_s} A_{s(i)} E_{s(i)} + \sum_{i=1}^{m_p} A_{p(i)} E_{p(i)} \tag{5.52a}$$

$$R_{B,j} = B_c E_{c,j} + \sum_{i=1}^{m_s} y_{s(i)} A_{s(i)} E_{s(i)} + \sum_{i=1}^{m_p} y_{p(i)} A_{p(i)} E_{p(i)} \tag{5.52b}$$

$$R_{I,j} = I_c E_{c,j} + \sum_{i=1}^{m_s} y_{s(i)}^2 A_{s(i)} E_{s(i)} + \sum_{i=1}^{m_p} y_{p(i)}^2 A_{p(i)} E_{p(i)} \tag{5.52c}$$

In the calculation of $M_{i,j}$ in Eq. 5.51, the moment $M_{c,i}$ resisted by the concrete component at time τ_i is calculated based on the solutions at each of the previous time instants:

$$M_{c,i} = \int_{A_c} y \sigma_{c,i} \, \mathrm{d}A = \int_{A_c} \left[y E_{c,i} \left(\varepsilon_i - \varepsilon_{sh,i} \right) + y \sum_{n=0}^{i-1} F_{e,i,n} \sigma_{c,n} \right] \mathrm{d}A$$

$$= B_c E_{c,i} \varepsilon_{r,i} + I_c E_{c,i} \kappa_i - B_c E_{c,i} \varepsilon_{sh,i} + \sum_{n=0}^{i-1} F_{e,i,n} M_{c,n} \tag{5.53}$$

Following the approach outlined previously, the equilibrium equations at time τ_j can then be written as:

$$\mathbf{r}_{e,j} = \mathbf{D}_j \boldsymbol{\varepsilon}_j + \mathbf{f}_{cr,j} - \mathbf{f}_{sh,j} + \mathbf{f}_{p,init} - \mathbf{f}_{p.rel,j} \tag{5.54}$$

where

$$\boldsymbol{\varepsilon}_j = \begin{bmatrix} \varepsilon_{r,j} \\ \kappa_j \end{bmatrix} \tag{5.55a}$$

$$\mathbf{D}_j = \begin{bmatrix} R_{A,j} & R_{B,j} \\ R_{B,j} & R_{I,j} \end{bmatrix} \tag{5.55b}$$

$$\mathbf{f}_{cr,j} = \sum_{i=0}^{j-1} F_{e,j,i} \mathbf{r}_{c,i} \tag{5.55c}$$

$$\mathbf{f}_{sh,j} = \begin{bmatrix} A_c \\ B_c \end{bmatrix} E_{c,j} \varepsilon_{sh,j} \tag{5.55d}$$

$$\mathbf{f}_{p,init} = \sum_{i=1}^{m_p} \begin{bmatrix} A_{p(i)} E_{p(i)} \varepsilon_{p(i),init} \\ y_{p(i)} A_{p(i)} E_{p(i)} \varepsilon_{p(i),init} \end{bmatrix} \tag{5.55e}$$

$$\mathbf{f}_{p.rel,j} = \mathbf{f}_{p,init}\, \varphi_p(\tau_j, \sigma_{p(i),init}) \tag{5.55f}$$

To simplify the notation, the internal actions resisted by the concrete at a previous time instant τ_i (calculated from Eqs 5.48 and 5.53) are collected in vector $\mathbf{r}_{c,i}$ and expressed as:

$$\mathbf{r}_{c,i} = \begin{bmatrix} N_{c,i} \\ M_{c,i} \end{bmatrix} = \mathbf{D}_{c,i} \boldsymbol{\varepsilon}_i + \sum_{n=0}^{i-1} F_{e,i,n} \mathbf{r}_{c,n} - \begin{bmatrix} A_c E_{c,i} \\ B_c E_{c,i} \end{bmatrix} \varepsilon_{sh,i} = \mathbf{D}_{c,i} \boldsymbol{\varepsilon}_i + \mathbf{f}_{cr,i} - \mathbf{f}_{sh,i} \tag{5.56}$$

with

$$\mathbf{D}_{c,i} = \begin{bmatrix} A_c & B_c \\ B_c & I_c \end{bmatrix} E_{c,i} \tag{5.57}$$

The strain vector $\boldsymbol{\varepsilon}_j$ at time τ_j is obtained by solving Eq. 5.54 and, as in the analyses presented earlier:

$$\boldsymbol{\varepsilon}_j = \mathbf{D}_j^{-1} \left(\mathbf{r}_{e,j} - \mathbf{f}_{cr,j} + \mathbf{f}_{sh,j} - \mathbf{f}_{p,init} + \mathbf{f}_{p.rel,j} \right) = \mathbf{F}_j \left(\mathbf{r}_{e,j} - \mathbf{f}_{cr,j} + \mathbf{f}_{sh,j} - \mathbf{f}_{p,init} + \mathbf{f}_{p.rel,j} \right) \tag{5.58}$$

in which

$$\mathbf{F}_j = \frac{1}{R_{A,j} R_{I,j} - R_{B,j}^2} \begin{bmatrix} R_{I,j} & -R_{B,j} \\ -R_{B,j} & R_{A,j} \end{bmatrix} \tag{5.59}$$

Finally, the stress distributions at time, τ_j, can be determined from the constitutive relationships (Eqs 5.44 and 4.25):

$$\sigma_{c,j} = E_{c,j} \varepsilon_j - E_{c,j} \varepsilon_{sh,j} + \sum_{i=0}^{j-1} F_{e,j,i} \sigma_{c,i} = E_{c,j} \left\{ \begin{bmatrix} 1 & y \end{bmatrix} \boldsymbol{\varepsilon}_j - \varepsilon_{sh,j} \right\} + \sum_{i=0}^{j-1} F_{e,j,i} \sigma_{c,i} \tag{5.60a}$$

$$\sigma_{s(i),j} = E_{s(i)}\varepsilon_j = E_{s(i)}\begin{bmatrix}1 & y_{p(i)}\end{bmatrix}\boldsymbol{\varepsilon}_j \tag{5.60b}$$

$$\begin{aligned}\sigma_{p(i),j} &= E_{p(i)}\left(\varepsilon_j + \varepsilon_{p(i),init} - \varepsilon_{p.rel(i),j}\right)\\ &= E_{p(i)}\begin{bmatrix}1 & y_{p(i)}\end{bmatrix}\boldsymbol{\varepsilon}_j + E_{p(i)}\varepsilon_{p(i),init} - E_{p(i)}\varepsilon_{p.rel(i),j}\end{aligned} \tag{5.60c}$$

where $\varepsilon_j = \varepsilon_{r,j} + y\kappa_j = \begin{bmatrix}1 & y\end{bmatrix}\boldsymbol{\varepsilon}_j$.

The cross-sectional rigidities at time τ_j, $R_{A,j}$, $R_{B,j}$ and $R_{I,j}$, can also be calculated from the properties of the transformed cross-section as follows:

$$R_{A,j} = A_j E_{e,j} \tag{5.61a}$$

$$R_{B,j} = B_j E_{e,j} \tag{5.61b}$$

$$R_{I,j} = I_j E_{e,j} \tag{5.61c}$$

where A_j, B_j and I_j are the area and the first and second moments of the area, respectively, of the transformed section about the reference axis. For the transformed section, the area of the reinforcement and bonded tendons are usually transformed into equivalent areas of concrete, based on $E_{c,j}$ using modular ratios of $n_{s(i),j}$ $(= E_{s(i)}/E_{c,j})$ for the bonded steel reinforcement and $n_{p(i),j}$ $(= E_{p(i)}/E_{c,j})$ for the bonded tendons. The matrix $\mathbf{F}_0$ in Eq. 5.58 can then be calculated as:

$$\mathbf{F}_j = \frac{1}{E_{e,j}(A_j I_j - B_j^2)}\begin{bmatrix} I_j & -B_j \\ -B_j & A_j \end{bmatrix} \tag{5.62}$$

In the following two examples, Examples 5.3 and 5.4 are re-analysed using the step-by-step method.

Example 5.5

For the reinforced concrete section shown in Fig. 5.5 (and analysed previously in Examples 5.1 and 5.3), determine the stress distribution at 30,000 days after first loading (i.e. approx. 80 years) using the step-by-step procedure. The external loading is applied at time $\tau_0 = 28$ days and kept constant over time (i.e. $N_{e,j} = -30$ kN and $M_{e,j} = 50$ kNm). A very coarse time discretisation is adopted in this example to illustrate all the steps involved in the solution process. The time period is subdivided into two intervals, τ_0 to τ_1 and τ_1 to τ_2, where $\tau_0 = 28$ days, $\tau_1 = 100$ days and $\tau_2 = 30{,}000$ days. In structural design, a finer discretisation is usually adopted to obtain more accurate results. For convenience in this example, shrinkage is assumed to begin at the time of first loading (i.e. at τ_0). The reinforcement is assumed to be linear-elastic.

The relevant properties of the concrete are as follows: $E_{c,0} = 25$ GPa; $E_{c,1} = 28$ GPa; $E_{c,2} = 30$ GPa; $\varphi(\tau_0, \tau_0) = 0$; $\varphi(\tau_1, \tau_0) = 1.5$; $\varphi(\tau_2, \tau_0) = 2.5$; $\varphi(\tau_1, \tau_1) = 0$; $\varphi(\tau_2, \tau_1) = 2.0$; $\varepsilon_{sh}(\tau_0) = \varepsilon_{sh,0} = 0$; $\varepsilon_{sh}(\tau_1) = \varepsilon_{sh,1} = -300 \times 10^{-6}$; and $\varepsilon_{sh}(\tau_2) = \varepsilon_{sh,2} = -600 \times 10^{-6}$.

Usually when using the step-by-step method (SSM), it is convenient to calculate the material properties for the concrete at the beginning of the analysis. From Eqs 1.10 and 4.26a and b:

$$J(\tau_0,\tau_0)=\frac{1}{E_{c,0}}=\frac{1}{25{,}000}=4\times 10^{-5}\ \text{MPa}^{-1}$$

$$J(\tau_1,\tau_1)=\frac{1}{E_{c,1}}=\frac{1}{28{,}000}=3.571\times 10^{-5}\ \text{MPa}^{-1}$$

$$J(\tau_2,\tau_2)=\frac{1}{E_{c,2}}=\frac{1}{30{,}000}=3.333\times 10^{-5}\ \text{MPa}^{-1}$$

$$J(\tau_1,\tau_0)=\frac{1+\varphi(\tau_1,\tau_0)}{E_{c,0}}=\frac{1+1.5}{25{,}000}=10\times 10^{-5}\ \text{MPa}^{-1}$$

$$J(\tau_2,\tau_0)=\frac{1+\varphi(\tau_2,\tau_0)}{E_{c,0}}=\frac{1+2.5}{25{,}000}=14\times 10^{-5}\ \text{MPa}^{-1}$$

$$J(\tau_2,\tau_1)=\frac{1+\varphi(\tau_2,\tau_1)}{E_{c,1}}=\frac{1+2.0}{28{,}000}=10.71\times 10^{-5}\ \text{MPa}^{-1}$$

$$F_{e,1,0}=\frac{J(\tau_1,\tau_1)-J(\tau_1,\tau_0)}{J(\tau_1,\tau_1)}=-1.8$$

$$F_{e,2,0}=\frac{J(\tau_2,\tau_1)-J(\tau_2,\tau_0)}{J(\tau_2,\tau_2)}=-0.986$$

$$F_{e,2,1}=\frac{J(\tau_2,\tau_2)-J(\tau_2,\tau_1)}{J(\tau_2,\tau_2)}=-2.214$$

Considering that the externally applied actions remain constant with time and the problem involves no prestressing tendons (for $j = 0, 1$ and 2):

$$\mathbf{r}_{e,j}=\begin{bmatrix}-30\times 10^3\ \text{N}\\ 50\times 10^6\ \text{Nmm}\end{bmatrix}$$

$$\mathbf{f}_{p,init}=\begin{bmatrix}0\ \text{N}\\ 0\ \text{Nmm}\end{bmatrix}$$

$$\mathbf{f}_{p.rel,j}=\begin{bmatrix}0\ \text{N}\\ 0\ \text{Nmm}\end{bmatrix}$$

Short-term analysis ($\tau_0 = 28$ days)

From Example 5.1: $\varepsilon_{r,0} = -42.7\times 10^{-6}$ and $\kappa_0 = 0.331\times 10^{-6}\ \text{mm}^{-1}$.
From Example 5.3: $A_c = 177.6\times 10^3\ \text{mm}^2$, $B_c = 17.46\times 10^6\ \text{mm}^3$ and $I_c = 6966\times 10^6\ \text{mm}^4$.

The stress resultants resisted by the concrete at τ_0 are determined from Eq. 5.56:

$$\mathbf{r}_{c,0} = \mathbf{D}_{c,0}\boldsymbol{\varepsilon}_0 + \mathbf{f}_{cr,0} - \mathbf{f}_{sh,0}$$

$$= 25{,}000 \times \begin{bmatrix} 177.6 \times 10^3 & 17.46 \times 10^6 \\ 17.46 \times 10^6 & 6966 \times 10^6 \end{bmatrix} \begin{bmatrix} -42.7 \times 10^{-6} \\ 0.331 \times 10^{-6} \end{bmatrix} + \begin{bmatrix} 0 \\ 0 \end{bmatrix} - \begin{bmatrix} 0 \\ 0 \end{bmatrix}$$

$$= \begin{bmatrix} -44.9 \times 10^3 \text{ N} \\ 39.1 \times 10^6 \text{ Nmm} \end{bmatrix}$$

Time analysis – SSM ($\tau_1 = 100$ days) – time step 1:

Considering $E_{c,1}$ for the concrete component, the cross-sectional rigidities at time τ_1 can be determined from Eqs 5.52 as:

$$R_{A,1} = 5456 \times 10^6 \text{ N}$$

$$R_{B,1} = 596.4 \times 10^9 \text{ Nmm}$$

$$R_{I,1} = 241.9 \times 10^{12} \text{ Nmm}^2$$

from which $\mathbf{D}_1$ and $\mathbf{F}_1$ can be determined (Eqs 5.55b and 5.59) as:

$$\mathbf{D}_1 = \begin{bmatrix} R_{A,1} & R_{B,1} \\ R_{B,1} & R_{I,1} \end{bmatrix} = \begin{bmatrix} 5456 \times 10^6 \text{ N} & 596.4 \times 10^9 \text{ Nmm} \\ 596.4 \times 10^9 \text{ Nmm} & 241.9 \times 10^{12} \text{ Nmm}^2 \end{bmatrix}$$

and

$$\mathbf{F}_1 = \frac{1}{R_{A,1}R_{I,1} - R_{B,1}^2} \begin{bmatrix} R_{I,1} & -R_{B,1} \\ -R_{B,1} & R_{A,1} \end{bmatrix}$$

$$= \begin{bmatrix} 250.8 \times 10^{-12} \text{ N}^{-1} & -618.4 \times 10^{-15} \text{ N}^{-1}\text{mm}^{-1} \\ -618.4 \times 10^{-15} \text{ N}^{-1}\text{mm}^{-1} & 5.658 \times 10^{-15} \text{ N}^{-1}\text{mm}^{-2} \end{bmatrix}$$

Creep effects due to the initial concrete stresses and shrinkage effects are obtained from Eq. 5.55c and d:

$$\mathbf{f}_{cr,1} = F_{e,1,0}\mathbf{r}_{c,0} = -1.8 \times \begin{bmatrix} -44.9 \times 10^3 \\ 39.1 \times 10^6 \end{bmatrix} = \begin{bmatrix} 80.8 \times 10^3 \text{ N} \\ -70.3 \times 10^6 \text{ Nmm} \end{bmatrix}$$

and

$$\mathbf{f}_{sh,1} = \begin{bmatrix} A_c \\ B_c \end{bmatrix} E_{c,1}\varepsilon_{sh,1} = \begin{bmatrix} 177.6 \times 10^3 \\ 17.46 \times 10^6 \end{bmatrix} \times 28{,}000 \times \left(-300 \times 10^{-6}\right)$$

$$= \begin{bmatrix} -1492 \times 10^3 \text{ N} \\ -146.7 \times 10^6 \text{ Nmm} \end{bmatrix}$$

The strain distribution at time τ_1 becomes (Eq. 5.58):

$$\boldsymbol{\varepsilon}_1 = \mathbf{F}_1 \left(\mathbf{r}_{e,1} - \mathbf{f}_{cr,1} + \mathbf{f}_{sh,1} \right)$$

$$= \begin{bmatrix} 250.8 \times 10^{-12} & -618.4 \times 10^{-15} \\ -618.4 \times 10^{-15} & 5.658 \times 10^{-15} \end{bmatrix}$$

$$\times \left(\begin{bmatrix} -30 \times 10^3 \\ 50 \times 10^6 \end{bmatrix} - \begin{bmatrix} 80.8 \times 10^3 \\ -70.3 \times 10^6 \end{bmatrix} + \begin{bmatrix} -1492 \times 10^3 \\ -146.7 \times 10^6 \end{bmatrix} \right)$$

$$= \begin{bmatrix} -385.7 \times 10^{-6} \\ 0.841 \times 10^{-6} \text{ mm}^{-1} \end{bmatrix}$$

Therefore, $\varepsilon_{r,1} = -385.7 \times 10^{-6}$ and $\kappa_1 = 0.841 \times 10^{-6}$ mm^{-1} and the strains in the top and bottom fibres of the cross-sections are (Eq. 5.1):

$$\varepsilon_{1(top)} = \varepsilon_{r,1} + (-200) \times \kappa_1 = -554.0 \times 10^{-6}$$

and

$$\varepsilon_{1(btm)} = \varepsilon_{r,1} + 400 \times \kappa_1 = -49.0 \times 10^{-6}$$

The top and bottom fibre concrete stresses at time τ_1 are (Eq. 5.60):

$$\sigma_{c,1(top)} = E_{c,1} \left\{ \begin{bmatrix} 1 & y_{c,(top)} \end{bmatrix} \boldsymbol{\varepsilon}_1 - \varepsilon_{sh,1} \right\} + F_{e,1,0}\sigma_{c,0(top)} = -2.21 \text{ MPa}$$

$$\sigma_{c,1(btm)} = E_{c,1} \left\{ \begin{bmatrix} 1 & y_{c,(btm)} \end{bmatrix} \boldsymbol{\varepsilon}_1 - \varepsilon_{sh,1} \right\} + F_{e,1,0}\sigma_{c,0(btm)} = 2.98 \text{ MPa}$$

and the stresses in the reinforcement are:

$$\sigma_{s(1),1} = E_{s(1)} \begin{bmatrix} 1 & y_{s(1)} \end{bmatrix} \boldsymbol{\varepsilon}_1 = -102.4 \text{ MPa}$$

$$\sigma_{s(2),1} = E_{s(2)} \begin{bmatrix} 1 & y_{s(2)} \end{bmatrix} \boldsymbol{\varepsilon}_1 = -18.2 \text{ MPa}$$

Based on these results the axial force and moment resisted by the concrete component at time τ_1 are (Eq. 5.56):

$$\mathbf{r}_{c,1} = \mathbf{D}_{c,1}\boldsymbol{\varepsilon}_1 + \mathbf{f}_{cr,1} - \mathbf{f}_{sh,1} = E_{c,1} \begin{bmatrix} A_c & B_c \\ B_c & I_c \end{bmatrix} \begin{bmatrix} \varepsilon_{r1} \\ \kappa_1 \end{bmatrix} + \mathbf{f}_{cr,1} - \mathbf{f}_{sh,1}$$

$$= 28{,}000 \times \begin{bmatrix} 177.6 \times 10^3 & 17.46 \times 10^6 \\ 17.46 \times 10^6 & 6966 \times 10^6 \end{bmatrix} \begin{bmatrix} -385.7 \times 10^{-6} \\ 0.841 \times 10^{-6} \end{bmatrix}$$

$$+ \begin{bmatrix} 80.8 \times 10^3 \\ -70.3 \times 10^6 \end{bmatrix} - \begin{bmatrix} -1492 \times 10^3 \\ -146.7 \times 10^6 \end{bmatrix}$$

$$= \begin{bmatrix} 66.3 \times 10^3 \text{ N} \\ 51.9 \times 10^6 \text{ Nmm} \end{bmatrix}$$

Time analysis – SSM ($\tau_2 = 30{,}000$ days) – time step 2

The cross-sectional rigidities at time τ_2 are (Eqs 5.52):

$$R_{A,2} = 5811 \times 10^6 \text{ N};$$

$$R_{B,2} = 631.3 \times 10^9 \text{ Nmm};$$

$$R_{I,2} = 255.8 \times 10^{12} \text{ Nmm}^2.$$

and from Eqs 5.55b and 5.59:

$$\mathbf{D}_2 = \begin{bmatrix} 5811 \times 10^6 \text{ N} & 631.3 \times 10^9 \text{ Nmm} \\ 631.3 \times 10^9 \text{ Nmm} & 255.8 \times 10^{12} \text{ Nmm}^2 \end{bmatrix}$$

$$\mathbf{F}_2 = \begin{bmatrix} 235.1 \times 10^{-12} \text{ N}^{-1} & -580.0 \times 10^{-15} \text{ N}^{-1}\text{mm}^{-1} \\ -580.0 \times 10^{-15} \text{ N}^{-1}\text{mm}^{-1} & 5.339 \times 10^{-15} \text{ N}^{-1}\text{mm}^{-2} \end{bmatrix}$$

The creep effects due to the concrete stresses in the previous time steps, at τ_0 and τ_1, and the shrinkage effects are obtained from Eqs 5.55c and d:

$$\mathbf{f}_{cr,2} = F_{e,2,0}\mathbf{r}_{c,0} + F_{e,2,1}\mathbf{r}_{c,1} = -0.986 \times \begin{bmatrix} -44.9 \times 10^3 \\ 39.1 \times 10^6 \end{bmatrix}$$

$$-\,2.214 \times \begin{bmatrix} 66.3 \times 10^3 \\ 51.9 \times 10^6 \end{bmatrix} = \begin{bmatrix} -102.5 \times 10^3 \text{ N} \\ -153.5 \times 10^6 \text{ Nmm} \end{bmatrix}$$

$$\mathbf{f}_{sh,2} = \begin{bmatrix} A_c \\ B_c \end{bmatrix} E_{c,2}\varepsilon_{sh,2} = \begin{bmatrix} 177.6 \times 10^3 \\ 17.46 \times 10^6 \end{bmatrix} \times 30{,}000 \times \left(-600 \times 10^{-6}\right)$$

$$= \begin{bmatrix} -3196 \times 10^3 \text{ N} \\ -314.3 \times 10^6 \text{ Nmm} \end{bmatrix}$$

The strain distribution at time τ_2 can be calculated as (Eq. 5.58):

$$\boldsymbol{\varepsilon}_2 = \mathbf{F}_2\left(\mathbf{r}_{e,2} - \mathbf{f}_{cr,2} + \mathbf{f}_{sh,2}\right) = \begin{bmatrix} -670.1 \times 10^{-6} \\ 1.220 \times 10^{-6} \text{ mm}^{-1} \end{bmatrix}$$

With $\varepsilon_{r,2} = -670.1 \times 10^{-6}$ and $\kappa_2 = 1.220 \times 10^{-6}$ mm^{-1}, the strains in the top and bottom fibres of the cross-sections are (Eq. 5.1):

$$\varepsilon_{2(top)} = \varepsilon_{r,2} + (-200) \times \kappa_2 = -914.2 \times 10^{-6}$$

$$\varepsilon_{2(btm)} = \varepsilon_{r,2} + 400 \times \kappa_2 = -181.9 \times 10^{-6}$$

The top and bottom fibre concrete stresses at τ_2 are (Eq. 5.60):

$$\sigma_{c,2(top)} = E_{c,2}\left\{\begin{bmatrix}1 & y_{c,(top)}\end{bmatrix}\boldsymbol{\varepsilon}_2 - \varepsilon_{sh,2}\right\} + F_{e,2,0}\sigma_{c,0(top)} + F_{e,2,1}\sigma_{c,1(top)}$$

$$= 30{,}000\left\{\begin{bmatrix}1 & -200\end{bmatrix}\begin{bmatrix}-670.1\times10^{-6}\\ 1.220\times10^{-6}\end{bmatrix} - \left(-600\times10^{-6}\right)\right\}$$

$$-0.986\times(-2.72) - 2.214\times(-2.21) = -1.85\text{ MPa}$$

$$\sigma_{c,2(btm)} = E_{c,2}\left\{\begin{bmatrix}1 & y_{c,(btm)}\end{bmatrix}\boldsymbol{\varepsilon}_2 - \varepsilon_{sh,2}\right\} + F_{e,2,0}\sigma_{c,0(btm)} + F_{e,2,1}\sigma_{c,1(btm)}$$

$$= 3.72\text{ MPa}$$

The final stresses in the reinforcement are:

$$\sigma_{s(1),2} = E_{s(1)}\begin{bmatrix}1 & y_{s(1)}\end{bmatrix}\boldsymbol{\varepsilon}_2 = -170.6\text{ MPa}$$

$$\sigma_{s(2),2} = E_{s(2)}\begin{bmatrix}1 & y_{s(2)}\end{bmatrix}\boldsymbol{\varepsilon}_2 = -48.6\text{ MPa}$$

The stress and strain distributions at τ_0 and at τ_2 are shown in Fig. 5.13.

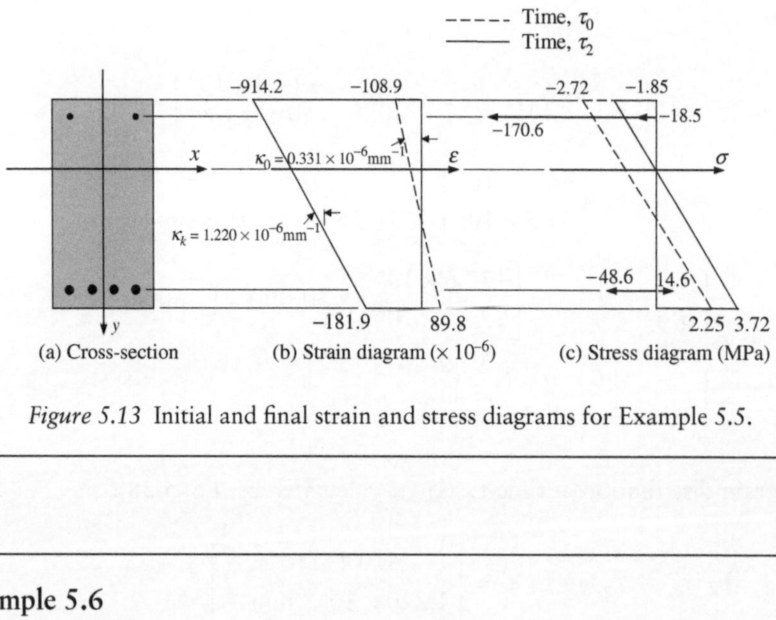

Figure 5.13 Initial and final strain and stress diagrams for Example 5.5.

Example 5.6

For the prestressed cross-section shown in Fig. 5.7 and analysed in Examples 5.2 and 5.4, determine the stress distribution at 30,000 days using the step-by-step procedure. The sustained external actions, $N_{e,j} = -100$ kN and $M_{e,j} = -50$ kNm are applied at time $\tau_0 = 28$ days and kept constant with time. As for Example 5.5, the time domain is discretised into two time steps, from τ_0 to τ_1 and from τ_1 to τ_2, with $\tau_0 = 28$ days, $\tau_1 = 100$ days and $\tau_2 = 30{,}000$ days. The reinforcement is assumed to be linear-elastic and the relevant properties of

the concrete are: $E_{c,0} = 32$ GPa; $E_{c,1} = 36$ GPa; $E_{c,2} = 40$ GPa; $\varphi(\tau_0, \tau_0) = 0$; $\varphi(\tau_1, \tau_0) = 1.0$; $\varphi(\tau_2, \tau_0) = 2.0$; $\varphi(\tau_1, \tau_1) = 0$; $\varphi(\tau_2, \tau_1) = 1.5$; $\varepsilon_{sh}(\tau_0) = \varepsilon_{sh,0} = 0$; $\varepsilon_{sh}(\tau_1) = \varepsilon_{sh,1} = -200 \times 10^{-6}$; and $\varepsilon_{sh}(\tau_2) = \varepsilon_{sh,2} = -400 \times 10^{-6}$.
The initial strains in the prestressing tendons prior to transfer is $\varepsilon_{p(1),init} = \varepsilon_{p(2),init} = 0.00625$ and the creep coefficient for the prestressing steel is $\varphi_p(\tau_1, \sigma_{p,init}) = 0.02$ and $\varphi_p(\tau_2, \sigma_{p,init}) = 0.03$.
The relevant creep functions for concrete are (Eqs 1.10 and 4.26a and b):

$$J(\tau_0, \tau_0) = 3.125 \times 10^{-5} \text{ MPa}^{-1}$$

$$J(\tau_1, \tau_1) = 2.777 \times 10^{-5} \text{ MPa}^{-1}$$

$$J(\tau_2, \tau_2) = 2.5 \times 10^{-5} \text{ MPa}^{-1}$$

$$J(\tau_1, \tau_0) = 6.25 \times 10^{-5} \text{ MPa}^{-1}$$

$$J(\tau_2, \tau_0) = 9.375 \times 10^{-5} \text{ MPa}^{-1}$$

$$J(\tau_2, \tau_1) = 6.944 \times 10^{-5} \text{ MPa}^{-1}$$

$$F_{e,1,0} = -1.25$$

$$F_{e,2,0} = -0.972$$

$$F_{e,2,1} = -1.777$$

Considering that the externally applied actions remain constant with time, for $j = 0, 1$ and 2:

$$\mathbf{r}_{e,j} = \begin{bmatrix} -100 \times 10^3 \text{ N} \\ -50 \times 10^6 \text{ Nmm} \end{bmatrix}$$

and

$$\mathbf{f}_{p,init} = \begin{bmatrix} 2000 \times 10^3 \text{ N} \\ 1290 \times 10^6 \text{ Nmm} \end{bmatrix}$$

Short-term analysis ($\tau_0 = 28$ days)

From Example 5.2: $\varepsilon_{r,0} = -70.7 \times 10^{-6}$ and $\kappa_0 = -0.387 \times 10^{-6}$ mm^{-1}.
From Example 5.4: $A_c = 312.7 \times 10^3$ mm^2; $B_c = 93.49 \times 10^6$ mm^3; $I_c = 76{,}960 \times 10^6$ mm^4.
From Eq. 5.56, the stress resultants resisted by the concrete at τ_0 are:

$$\mathbf{r}_{c,0} = \mathbf{D}_{c,0}\boldsymbol{\varepsilon}_0 + \mathbf{f}_{cr,0} - \mathbf{f}_{sh,0} = E_{c,0}\begin{bmatrix} A_c & B_c \\ B_c & I_c \end{bmatrix}\begin{bmatrix} \varepsilon_{r,0} \\ \kappa_0 \end{bmatrix} + \begin{bmatrix} 0 \\ 0 \end{bmatrix} - \begin{bmatrix} 0 \\ 0 \end{bmatrix}$$

$$= \begin{bmatrix} -1865 \times 10^3 \text{ N} \\ -1165 \times 10^6 \text{ Nmm} \end{bmatrix}$$

Time analysis – SSM ($\tau_1 = 100$ days) – time step 1

From Eqs 5.52, the cross-sectional rigidities at time τ_1 are:

$$R_{A,1} = 12{,}120 \times 10^6 \text{ N}$$

$$R_{B,1} = 3813 \times 10^9 \text{ Nmm}$$

$$R_{I,1} = 3140 \times 10^{12} \text{ Nmm}^2$$

and from Eq. 5.59:

$$\mathbf{F}_1 = \begin{bmatrix} 133.6 \times 10^{-12} \text{ N}^{-1}\text{mm}^{-2} & -162.2 \times 10^{-15} \text{ N}^{-1}\text{mm}^{-1} \\ -162.2 \times 10^{-15} \text{ N}^{-1}\text{mm}^{-1} & 515.4 \times 10^{-18} \text{ N}^{-1} \end{bmatrix}$$

The effects of creep due to the concrete stresses calculated at τ_0 and the effects of shrinkage are represented as:

$$\mathbf{f}_{cr,1} = F_{e,1,0}\mathbf{r}_{c,0} = -1.25 \times \begin{bmatrix} -1865 \times 10^3 \\ -1165 \times 10^6 \end{bmatrix} = \begin{bmatrix} 2332 \times 10^3 \text{ N} \\ 1456 \times 10^6 \text{ Nmm} \end{bmatrix}$$

$$\mathbf{f}_{sh,1} = \begin{bmatrix} A_c \\ B_c \end{bmatrix} E_{c,1}\varepsilon_{sh,1} = \begin{bmatrix} 312.7 \times 10^3 \\ 93.46 \times 10^6 \end{bmatrix} \times 36{,}000 \times \left(-200 \times 10^{-6}\right)$$

$$= \begin{bmatrix} -2251 \times 10^3 \text{ N} \\ -673.2 \times 10^6 \text{ Nmm} \end{bmatrix}$$

and the relaxation of the prestressing tendons is:

$$\mathbf{f}_{p.rel,1} = \begin{bmatrix} A_{p(1)}E_{p(1)}\varepsilon_{p(1),init} \\ y_{p(1)}A_{p(1)}E_{p(1)}\varepsilon_{p(1),init} \end{bmatrix} \varphi_p(\tau_1, \sigma_{p(1),init})$$

$$+ \begin{bmatrix} A_{p(2)}E_{p(2)}\varepsilon_{p(2),init} \\ y_{p(2)}A_{p(2)}E_{p(2)}\varepsilon_{p(2),init} \end{bmatrix} \varphi_p(\tau_1, \sigma_{p(2),init})$$

$$= \begin{bmatrix} 800 \times 200{,}000 \times 0.00625 \\ 580 \times 800 \times 200{,}000 \times 0.00625 \end{bmatrix} \times 0.02$$

$$+ \begin{bmatrix} 800 \times 200{,}000 \times 0.00625 \\ 710 \times 800 \times 200{,}000 \times 0.00625 \end{bmatrix} \times 0.02$$

$$= \begin{bmatrix} 40 \times 10^3 \text{ N} \\ 25.8 \times 10^6 \text{ Nmm} \end{bmatrix}$$

The strain distribution at time τ_1 becomes:

$$\boldsymbol{\varepsilon}_1 = \mathbf{F}_1\left(\mathbf{r}_{e,1} - \mathbf{f}_{cr,1} + \mathbf{f}_{sh,1} - \mathbf{f}_{p,init} + \mathbf{f}_{p.rel,1}\right)$$

$$= \begin{bmatrix} 133.6 \times 10^{-12} & -162.2 \times 10^{-15} \\ -162.2 \times 10^{-15} & 515.4 \times 10^{-18} \end{bmatrix}$$

$$\times \left(\begin{bmatrix} -100 \times 10^3 \\ -50 \times 10^6 \end{bmatrix} - \begin{bmatrix} 2332 \times 10^3 \\ 1456 \times 10^6 \end{bmatrix} + \begin{bmatrix} -2251 \times 10^3 \\ -673.2 \times 10^6 \end{bmatrix} \right.$$

$$\left. - \begin{bmatrix} 2000 \times 10^3 \\ 1290 \times 10^6 \end{bmatrix} + \begin{bmatrix} 40 \times 10^3 \\ 25.8 \times 10^6 \end{bmatrix} \right) = \begin{bmatrix} -328.8 \times 10^{-6} \\ -0.697 \times 10^{-6} \text{ mm}^{-1} \end{bmatrix}$$

With $\varepsilon_{r,1} = -328.8 \times 10^{-6}$ and $\kappa_1 = -0.697 \times 10^{-6}$ mm^{-1}, the strains in the top and bottom fibres of the cross-sections are (Eq. 5.1):

$$\varepsilon_{1(top)} = \varepsilon_{r,1} + (-300) \times \kappa_1 = -119.6 \times 10^{-6}$$

and

$$\varepsilon_{1(btm)} = \varepsilon_{r,1} + 850 \times \kappa_1 = -921.6 \times 10^{-6}.$$

The stresses in the top and bottom concrete fibres at time τ_1 are (Eqs 5.60):

$$\sigma_{c,1(top)} = E_{c,1}\left\{\begin{bmatrix} 1 & y_{c,(top)} \end{bmatrix} \boldsymbol{\varepsilon}_1 - \varepsilon_{sh,1}\right\} + F_{e,1,0}\sigma_{c,0(top)} = 1.07 \text{ MPa}$$

$$\sigma_{c,1(btm)} = E_{c,1}\left\{\begin{bmatrix} 1 & y_{c,(btm)} \end{bmatrix} \boldsymbol{\varepsilon}_1 - \varepsilon_{sh,1}\right\} + F_{e,1,0}\sigma_{c,0(btm)} = -9.98 \text{ MPa}$$

and the stress in the non-prestressed reinforcement is:

$$\sigma_{s(1),1} = E_{s(1)}\begin{bmatrix} 1 & y_{s(1)} \end{bmatrix} \boldsymbol{\varepsilon}_1 = 200 \times 10^3 \begin{bmatrix} 1 & -240 \end{bmatrix} \begin{bmatrix} -328.8 \times 10^{-6} \\ -0.697 \times 10^{-6} \end{bmatrix}$$

$$= -32.3 \text{ MPa}$$

$$\sigma_{s(2),1} = E_{s(2)}\begin{bmatrix} 1 & y_{s(2)} \end{bmatrix} \boldsymbol{\varepsilon}_1 = 200 \times 10^3 \begin{bmatrix} 1 & 790 \end{bmatrix} \begin{bmatrix} -328.8 \times 10^{-6} \\ -0.697 \times 10^{-6} \end{bmatrix}$$

$$= -176.0 \text{ MPa}$$

The stress in the prestressing steel is:

$$\sigma_{p(1),1} = E_{p(1)}\begin{bmatrix} 1 & y_{p(1)} \end{bmatrix} \boldsymbol{\varepsilon}_1 + E_{p(1)}\left(\varepsilon_{p(1),init} - \varepsilon_{p.rel(1),1}\right)$$

$$= 200 \times 10^3 \left(\begin{bmatrix} 1 & 580 \end{bmatrix} \begin{bmatrix} -328.8 \times 10^{-6} \\ -0.697 \times 10^{-6} \end{bmatrix} \right)$$

$$+ 200 \times 10^3 \times 0.00625 \times (1 - 0.02) = 1078 \text{ MPa}$$

$$\sigma_{p(2),1} = E_{p(2)}\begin{bmatrix} 1 & y_{p(2)} \end{bmatrix} \boldsymbol{\varepsilon}_1 + E_{p(2)}\left(\varepsilon_{p(2),init} - \varepsilon_{p.rel(2),1}\right) = 1060 \text{ MPa}$$

The axial force and moment resisted by the concrete component at time τ_1 are (Eq. 5.56):

$$\mathbf{r}_{c,1} = \mathbf{D}_{c,1}\boldsymbol{\varepsilon}_1 + \mathbf{f}_{cr,1} - \mathbf{f}_{sh,1} = E_{c,1}\begin{bmatrix} A_c & B_c \\ B_c & I_c \end{bmatrix}\begin{bmatrix} \varepsilon_{r1} \\ \kappa_1 \end{bmatrix} + \mathbf{f}_{cr,1} - \mathbf{f}_{sh,1}$$

$$= \begin{bmatrix} -1465 \times 10^3 \text{ N} \\ -909.3 \times 10^6 \text{ Nmm} \end{bmatrix}$$

Time analysis – SSM ($\tau_2 = 30{,}000$ days) – time step 2

The cross-sectional rigidities at time τ_2 are (Eqs 5.52):

$$R_{A,2} = 13{,}370 \times 10^6 \text{ N}$$

$$R_{B,2} = 4187 \times 10^9 \text{ Nmm}$$

$$R_{I,2} = 3448 \times 10^{12} \text{ Nmm}^2$$

$$\mathbf{F}_2 = \begin{bmatrix} 120.7 \times 10^{-12} \text{ N}^{-1}\text{mm}^{-2} & -146.6 \times 10^{-15} \text{ N}^{-1} \text{ mm}^{-1} \\ -146.6 \times 10^{-15} \text{ N}^{-1}\text{mm}^{-1} & 468.1 \times 10^{-18} \text{ N}^{-1} \end{bmatrix}$$

Creep effects due to the concrete stresses in previous time steps, i.e. at τ_0 and τ_1, are:

$$\mathbf{f}_{cr,2} = F_{e,2,0}\mathbf{r}_{c,0} + F_{e,2,1}\mathbf{r}_{c,1}$$

$$= -0.972 \times \begin{bmatrix} -1865 \times 10^3 \\ -1165 \times 10^6 \end{bmatrix} - 1.777 \times \begin{bmatrix} -1465 \times 10^3 \\ -909.3 \times 10^6 \end{bmatrix}$$

$$= \begin{bmatrix} 4418 \times 10^3 \text{ N} \\ 2749 \times 10^6 \text{ Nmm} \end{bmatrix}$$

and shrinkage effects are:

$$\mathbf{f}_{sh,2} = \begin{bmatrix} A_c \\ B_c \end{bmatrix} E_{c,2}\varepsilon_{sh,2} = \begin{bmatrix} 312.7 \times 10^3 \\ 93.49 \times 10^6 \end{bmatrix} \times 40{,}000 \times \left(-400 \times 10^{-6}\right)$$

$$= \begin{bmatrix} -5003 \times 10^3 \text{ N} \\ -1496 \times 10^6 \text{ Nmm} \end{bmatrix}$$

The relaxation of the prestressing tendons is:

$$\mathbf{f}_{p.rel,2} = \begin{bmatrix} A_{p(1)}E_{p(1)}\varepsilon_{p(1),init} \\ y_{p(1)}A_{p(1)}E_{p(1)}\varepsilon_{p(1),init} \end{bmatrix} \varphi_p(\tau_2, \sigma_{p(1),init})$$

$$+ \begin{bmatrix} A_{p(2)}E_{p(2)}\varepsilon_{p(2),init} \\ y_{p(2)}A_{p(2)}E_{p(2)}\varepsilon_{p(2),init} \end{bmatrix} \varphi_p(\tau_2, \sigma_{p(2),init}) = \begin{bmatrix} 60 \times 10^3 \text{ N} \\ 38.7 \times 10^6 \text{ Nmm} \end{bmatrix}$$

The strain distribution at time τ_2 is:

$$\boldsymbol{\varepsilon}_2 = \mathbf{F}_2\left(\mathbf{r}_{e,2} - \mathbf{f}_{cr,2} + \mathbf{f}_{sh,2} - \mathbf{f}_{p,init} + \mathbf{f}_{p.rel,2}\right) = \begin{bmatrix} -570.6 \times 10^{-6} \\ -0.915 \times 10^{-6}\ \text{mm}^{-1} \end{bmatrix}$$

With $\varepsilon_{r,2} = -570.6 \times 10^{-6}$ and $\kappa_2 = -0.915 \times 10^{-6}\ \text{mm}^{-1}$, the strains in the top and bottom fibres of the cross-sections are (Eq. 5.1):

$$\varepsilon_{2(top)} = \varepsilon_{r,2} + (-300) \times \kappa_2 = -295.8 \times 10^{-6}$$

and

$$\varepsilon_{2(btm)} = \varepsilon_{r,2} + 850 \times \kappa_2 = -1349 \times 10^{-6}$$

The stresses in the top and bottom concrete fibres at time τ_1 are (Eqs 5.60):

$$\sigma_{c,2(top)} = E_{c,2}\left\{\begin{bmatrix} 1 & y_{c,(top)} \end{bmatrix}\boldsymbol{\varepsilon}_2 - \varepsilon_{sh,2}\right\} + F_{e,2,0}\sigma_{c,0(top)} + F_{e,2,1}\sigma_{c,1(top)}$$

$$= 40{,}000\left\{\begin{bmatrix} 1 & -300 \end{bmatrix}\begin{bmatrix} -570.6 \times 10^{-6} \\ -0.915 \times 10^{-6} \end{bmatrix} - \left(-400 \times 10^{-6}\right)\right\}$$

$$- 0.972 \times 1.45 - 1.777 \times 1.07 = 0.84\ \text{MPa}$$

$$\sigma_{c,2(btm)} = E_{c,2}\left\{\begin{bmatrix} 1 & y_{c,(btm)} \end{bmatrix}\boldsymbol{\varepsilon}_2 - \varepsilon_{sh,2}\right\} + F_{e,2,0}\sigma_{c,0(btm)} + F_{e,2,1}\sigma_{c,1(btm)}$$

$$= -7.77\ \text{MPa}$$

and the stresses in the non-prestressed reinforcement are:

$$\sigma_{s(1),2} = E_{s(1)}\begin{bmatrix} 1 & y_{s(1)} \end{bmatrix}\boldsymbol{\varepsilon}_2 = 200 \times 10^3\begin{bmatrix} 1 & -240 \end{bmatrix}\begin{bmatrix} -570.6 \times 10^{-6} \\ -0.915 \times 10^{-6} \end{bmatrix}$$

$$= -70.1\ \text{MPa}$$

$$\sigma_{s(2),2} = E_{s(2)}\begin{bmatrix} 1 & y_{s(2)} \end{bmatrix}\boldsymbol{\varepsilon}_2 = 200 \times 10^3\begin{bmatrix} 1 & 790 \end{bmatrix}\begin{bmatrix} -570.6 \times 10^{-6} \\ -0.915 \times 10^{-6} \end{bmatrix}$$

$$= -258.8\ \text{MPa}$$

The final stresses in the prestressing tendons are:

$$\sigma_{p(1),2} = E_{p(1)}\begin{bmatrix} 1 & y_{p(1)} \end{bmatrix}\boldsymbol{\varepsilon}_2 + E_{p(1)}\left(\varepsilon_{p(1),init} - \varepsilon_{p.rel(1),2}\right) = 992.1\ \text{MPa}$$

$$\sigma_{p(2),2} = E_{p(2)}\begin{bmatrix} 1 & y_{p(2)} \end{bmatrix}\boldsymbol{\varepsilon}_2 + E_{p(2)}\left(\varepsilon_{p(2),init} - \varepsilon_{p.rel(2),2}\right) = 968.3\ \text{MPa}$$

The stress and strain distributions at τ_0 and at τ_2 are shown in Fig. 5.14.

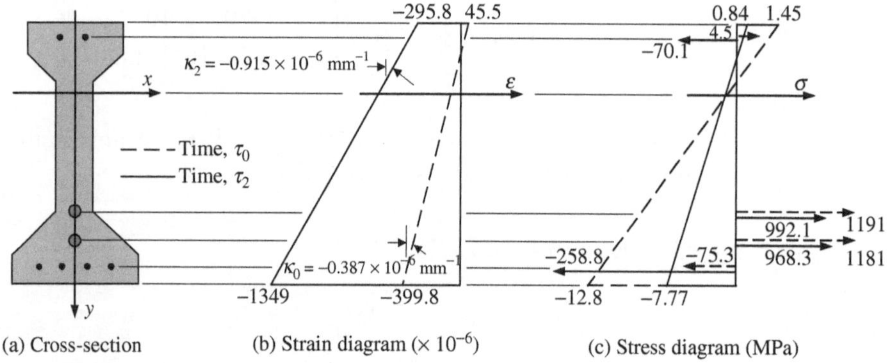

Figure 5.14 Initial and final strain and stress diagrams for Example 5.6.

5.6 Composite steel-concrete cross-sections

The analysis of composite steel-concrete cross-sections, such as those shown in Fig. 5.15, is presented here. Each cross-section has a single axis of symmetry and may be subjected to any combination of axial force and bending moment. In addition to the rolled or fabricated steel section or profiled decking, the concrete deck may contain layers of steel reinforcement.

Mechanical shear connections are usually used to ensure that the steel-concrete interface is capable of carrying the imposed horizontal shear, and that the steel and concrete portions act compositely. For most practical steel-concrete composite sections, little slip occurs between the steel and the concrete under normal in-service conditions, with the magnitude of any slip depending on the rigidity of the interface connection (Refs 6–11). In the following, a perfect bond between the steel and the concrete is assumed.

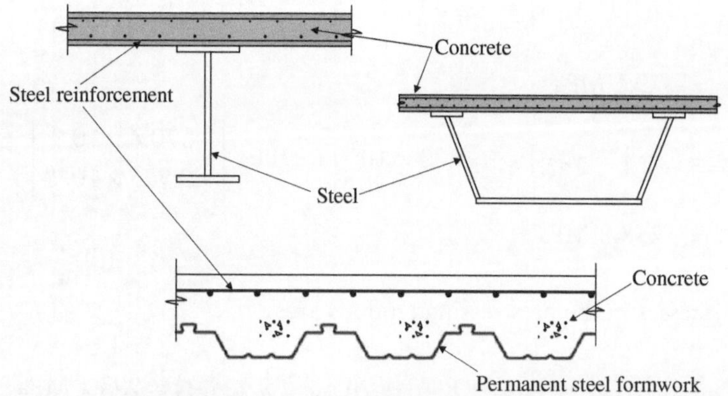

Figure 5.15 Composite steel-concrete cross-sections.

The procedures for the short- and long-term analyses of the composite sections shown in Fig. 5.15 are similar to those presented for reinforced and prestressed concrete sections in Sections 5.3–5.5, except that the flexural rigidity of the steel section (or its second moment of area) must be included, as well as its cross-sectional area. At any time instant τ_k, the internal axial force and moment resisted by the composite cross-section, $N_{i,k}$ and $M_{i,k}$, respectively, may be expressed as:

$$N_{i,k} = N_{c,k} + N_{s,k} + N_{ss,k} \tag{5.63a}$$

$$M_{i,k} = M_{c,k} + M_{s,k} + M_{ss,k} \tag{5.63b}$$

where $N_{c,k}$, $N_{s,k}$ and $N_{ss,k}$ are the resultant axial forces in the concrete, the non-prestressed reinforcement and the steel part of the cross-section, respectively, and $M_{c,k}$, $M_{s,k}$ and $M_{ss,k}$ are the corresponding moments about the reference axis. Since the rolled or fabricated steel (or permanent steel decking) does not creep or shrink, $N_{ss,k}$ and $M_{ss,k}$ may be defined at any time τ_k as:

$$N_{ss,k} = \int_{A_{ss}} \sigma_{ss,k}\,\mathrm{d}A = \int_{A_{ss}} E_{ss}(\varepsilon_{r,k} + y\kappa_k)\,\mathrm{d}A = A_{ss}E_{ss}\varepsilon_{r,k} + B_s E_{ss}\kappa_k \tag{5.64a}$$

$$M_{ss,k} = \int_{A_{ss}} y\sigma_{ss,k}\,\mathrm{d}A = \int_{A_{ss}} yE_{ss}(\varepsilon_{r,k} + y\kappa_k)\mathrm{d}\,A = B_{ss}E_{ss}\varepsilon_{r,k} + I_s E_{ss}\kappa_k \tag{5.64b}$$

where A_{ss} is the cross-sectional area of the rolled or fabricated steel section, and B_{ss} and I_{ss} are the first and second moments of area of the steel section calculated with respect to the adopted reference axis. The constitutive relationship for the rolled or fabricated steel is assumed to be linear-elastic as follows:

$$\sigma_{ss,k} = E_{ss}\varepsilon_k = E_{ss}[1 \quad y]\boldsymbol{\varepsilon}_k \tag{5.65}$$

The advantage of this representation is that it describes both the instantaneous analysis (at τ_0) and the time-dependent analysis (at τ_k). At $k = 0$, Eqs 5.63 can be substituted into Eqs 5.2 to produce the governing equilibrium equations for a short-term analysis (as previously presented in Eq. 5.12). Using the AEMM for the time analysis (at τ_k), Eqs 5.63 can be substituted into Eq. 5.24 to obtain the governing equations (previously presented in Eqs 5.34). The solution using the SSM is obtained by replacing Eqs 5.47 and 5.51 with Eqs 5.63 (and with time τ_j used instead of τ_k).

The only differences between the analyses of the composite steel-concrete cross-section and the analyses presented in Sections 5.3–5.5 are in the calculation of the cross-sectional rigidities, $R_{A,k}$, $R_{B,k}$ and $R_{I,k}$. The solution processes, involving the determination of the unknown strain and stress distributions at τ_k, are unchanged. For a composite steel-concrete cross-section, using the AEMM, the cross-sectional rigidities are:

$$R_{A,k} = A_c\overline{E}_{e,k} + A_{ss}E_{ss} + \sum_{i=1}^{m_s} A_{s(i)}E_{s(i)} \tag{5.66a}$$

$$R_{B,j} = B_c\overline{E}_{e,j} + B_{ss}E_{ss} + \sum_{i=1}^{m_s} y_{s(i)}A_{s(i)}E_{s(i)} \tag{5.66b}$$

$$R_{I,j} = I_c\overline{E}_{e,j} + I_{ss}E_{ss} + \sum_{i=1}^{m_s} y_{s(i)}^2A_{s(i)}E_{s(i)} \tag{5.66c}$$

where for the short-term analysis at τ_0, $\overline{E}_{e,k} = E_{c,0}$. When using the SSM, the concrete modulus to be used in Eq. 5.66 is the one associated with the time step under consideration.

As already discussed for reinforced and prestressed concrete sections, it is also common to transform the steel parts of the cross-section into equivalent areas of concrete, to calculate the rigidities of the transformed concrete section, and to use these in the analysis instead of the rigidities calculated using Eq. 5.66.

Example 5.7

The short-term and final long-term behaviour of the composite cross-section shown in Fig. 5.16 is to be calculated. The cross-section is subjected to a sustained bending moment $M_e = 500$ kNm. The axial force N_e in this example is zero. The time-dependent behaviour of the concrete is to be modelled using the AEMM and the relevant material properties are:

$$E_{c,0} = 25{,}000 \text{ MPa}$$

$$E_{ss} = E_{s(1)} = 200{,}000 \text{ MPa}$$

$$\varphi_0^* = \varphi(\infty, \tau_0) = 2.5$$

$$\chi_0^* = \chi(\infty, \tau_0) = 0.65$$

$$\varepsilon_{sh}^* = \varepsilon_{sh}(\infty) = -600 \times 10^{-6}$$

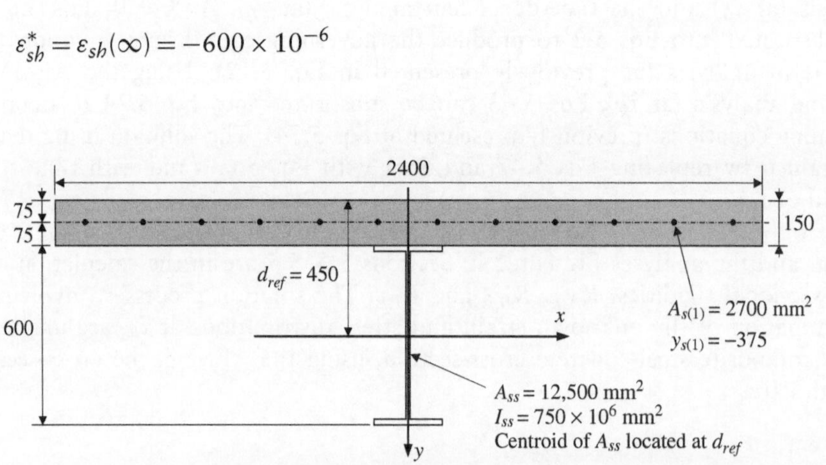

Figure 5.16 Cross-sectional details for Example 5.7 (all dimensions in mm).

The external actions remain constant with time and are given by (with $j = 0, k$)

$$\mathbf{r}_{e,j} = \begin{bmatrix} N_{e,j} \\ M_{e,j} \end{bmatrix} = \begin{bmatrix} 0 \text{ N} \\ 500 \times 10^6 \text{ Nmm} \end{bmatrix}$$

Short-term analysis (at time τ_0)

The geometric properties of the concrete slab are:

$$A_c = bD - A_{s(1)} = 2400 \times 150 - 2700 = 357{,}300 \text{ mm}^2$$

$$B_c = bD(D/2 - d_{ref}) - A_{s(1)}y_{s(1)} = 2400 \times 150 \times (75 - 450) - 2700 \times (-375)$$
$$= -134.0 \times 10^6 \text{ mm}^3$$

$$I_c = \frac{bD^3}{12} + bD\left(\frac{D}{2} - d_{ref}\right)^2 - A_{s(1)}y_{s(1)}^2 = \frac{2400 \times 150^3}{12}$$
$$+ 2400 \times 150 \times (75 - 450)^2 - 2700 \times (-375)^2 = 50{,}920 \times 10^6 \text{ mm}^4$$

The properties of the steel beam are already provided with respect to d_{ref}:

$A_{ss} = 12{,}500 \text{ mm}^2$;
$B_{ss} = 0 \text{ mm}^3$ (as the centroid of the steel beam is located on the reference axis); and
$I_{ss} = 750 \times 10^6 \text{ mm}^4$.

The cross-sectional rigidities are calculated using Eqs 5.66:

$$R_{A,0} = A_c E_{c,0} + A_{ss}E_{ss} + \sum_{i=1}^{m_s} A_{s(i)}E_{s(i)} = A_c E_{c,0} + A_{ss}E_{ss} + A_{s(1)}E_{s(1)}$$
$$= 11{,}970 \times 10^6 \text{ N}$$

$$R_{B,0} = B_c E_{c,0} + B_{ss}E_{ss} + \sum_{i=1}^{m_s} y_{s(i)}A_{s(i)}E_{s(i)} = B_c E_{c,0} + B_{ss}E_{ss} + y_{s(1)}A_{s(1)}E_{s(1)}$$
$$= -3552 \times 10^9 \text{ Nmm}$$

$$R_{I,0} = I_c E_{c,0} + I_{ss}E_{ss} + \sum_{i=1}^{m_s} y_{s(i)}^2 A_{s(i)}E_{s(i)} = I_c E_{c,0} + I_{ss}E_{ss} + y_{s(1)}^2 A_{s(1)}E_{s(1)}$$
$$= 1499 \times 10^{12} \text{ Nmm}^2$$

From Eq. 5.15:

$$\mathbf{F}_0 = \frac{1}{R_{A,0}R_{I,0} - R_{B,0}^2} \begin{bmatrix} R_{I,0} & -R_{B,0} \\ -R_{B,0} & R_{A,0} \end{bmatrix}$$
$$= \begin{bmatrix} 2.81 \times 10^{-10} \text{ N}^{-1} & 6.66 \times 10^{-13} \text{ N}^{-1}\text{mm}^{-1} \\ 6.66 \times 10^{-13} \text{ N}^{-1}\text{mm}^{-1} & 2.24 \times 10^{-15} \text{ N}^{-1}\text{mm}^{-2} \end{bmatrix}$$

At time τ_0, before any creep or shrinkage has occurred, Eq. 5.14 reduces to:

$$\boldsymbol{\varepsilon}_0 = \mathbf{F}_0 \mathbf{r}_{e,0} = \begin{bmatrix} 2.81 \times 10^{-10} & 6.66 \times 10^{-13} \\ 6.66 \times 10^{-13} & 2.24 \times 10^{-15} \end{bmatrix} \begin{bmatrix} 0 \\ 500 \times 10^6 \end{bmatrix}$$

$$= \begin{bmatrix} 333.3 \times 10^{-6} \\ 1.123 \times 10^{-6} \text{ mm}^{-1} \end{bmatrix}$$

With $\varepsilon_{r,0} = 333.3 \times 10^{-6}$ and $\kappa_0 = 1.123 \times 10^{-6}$ mm^{-1}, the strains at the top and bottom fibres of the cross-sections are (Eq. 5.1):

$$\varepsilon_{0(top)} = \varepsilon_{r,0} + (-450) \times \kappa_0 = -172.2 \times 10^{-6}$$

and

$$\varepsilon_{0(btm)} = \varepsilon_{r,0} + 300 \times \kappa_0 = 670.4 \times 10^{-6}$$

The concrete stresses at the top and bottom of the slab at time τ_0 are (Eq. 5.16a):

$$\sigma_{c,0(top)} = E_{c,0} \begin{bmatrix} 1 & y_{c,(top)} \end{bmatrix} \boldsymbol{\varepsilon}_0 = 25{,}000 \begin{bmatrix} 1 & -450 \end{bmatrix} \begin{bmatrix} 333.3 \times 10^{-6} \\ 1.123 \times 10^{-6} \end{bmatrix}$$

$$= -4.31 \text{ MPa}$$

$$\sigma_{c,0(btm)} = E_{c,0} \begin{bmatrix} 1 & y_{c,(btm)} \end{bmatrix} \boldsymbol{\varepsilon}_0 = 25{,}000 \begin{bmatrix} 1 & -300 \end{bmatrix} \begin{bmatrix} 333.3 \times 10^{-6} \\ 1.123 \times 10^{-6} \end{bmatrix}$$

$$= -0.09 \text{ MPa}$$

The stress in the steel reinforcement is (Eq. 5.16b):

$$\sigma_{s(1),0} = E_{s(1)} \begin{bmatrix} 1 & y_{s(1)} \end{bmatrix} \boldsymbol{\varepsilon}_0 = 200 \times 10^3 \begin{bmatrix} 1 & -375 \end{bmatrix} \begin{bmatrix} 333.3 \times 10^{-6} \\ 1.123 \times 10^{-6} \end{bmatrix}$$

$$= -17.6 \text{ MPa}$$

The stresses at the top and bottom fibres of the steel I-beam are (Eq. 5.65):

$$\sigma_{ss,0(top)} = E_{ss} \left\{ \begin{bmatrix} 1 & y_{ss,(top)} \end{bmatrix} \boldsymbol{\varepsilon}_0 \right\} = 200 \times 10^3 \left\{ \begin{bmatrix} 1 & -300 \end{bmatrix} \begin{bmatrix} 333.3 \times 10^{-6} \\ 1.123 \times 10^{-6} \end{bmatrix} \right\}$$

$$= -0.74 \text{ MPa}$$

$$\sigma_{ss,0(btm)} = E_{ss} \left\{ \begin{bmatrix} 1 & y_{ss,(btm)} \end{bmatrix} \boldsymbol{\varepsilon}_0 \right\} = 200 \times 10^3 \left\{ \begin{bmatrix} 1 & +300 \end{bmatrix} \begin{bmatrix} 333.3 \times 10^{-6} \\ 1.123 \times 10^{-6} \end{bmatrix} \right\}$$

$$= 134 \text{ MPa}$$

The stress resultants resisted by the concrete at τ_0 can be determined from:

$$\mathbf{r}_{c,0} = \begin{bmatrix} N_{c,0} \\ M_{c,0} \end{bmatrix} = \mathbf{D}_{c,0}\boldsymbol{\varepsilon}_0 = E_{c,0}\begin{bmatrix} A_c & B_c \\ B_c & I_c \end{bmatrix}\begin{bmatrix} \varepsilon_{r,0} \\ \kappa_0 \end{bmatrix} = \begin{bmatrix} -785.8 \times 10^3 \text{ N} \\ 313.6 \times 10^6 \text{ Nmm} \end{bmatrix}$$

Time analysis – AEMM (at time $\tau_k = \infty$)

From Eqs 4.35 and 4.46:

$$\overline{E}_{e,k} = \frac{25{,}000}{1 + 0.65 \times 2.5} = 9524 \text{ MPa}$$

$$\overline{F}_{e,0} = \frac{2.5 \times (0.65 - 1.0)}{1.0 + 0.65 \times 2.5} = -0.333$$

The cross-sectional rigidities at time τ_k are (Eq. 5.66):

$$R_{A,k} = A_c\overline{E}_{e,k} + A_{ss}E_{ss} + A_{s(1)}E_{s(1)} = 6443 \times 10^6 \text{ N}$$

$$R_{B,k} = B_c\overline{E}_{e,k} + B_{ss}E_{ss} + y_{s(1)}A_{s(1)}E_{s(1)} = -1479 \times 10^9 \text{ Nmm}$$

$$R_{I,k} = I_c\overline{E}_{e,k} + I_{ss}E_{ss} + y_{s(1)}^2A_{s(1)}E_{s(1)} = 710.9 \times 10^{12} \text{ Nmm}^2$$

$\mathbf{F}_k$ is then obtained using Eq. 5.40:

$$\mathbf{F}_k = \frac{1}{R_{A,k}R_{I,k} - R_{B,k}^2}\begin{bmatrix} R_{I,k} & -R_{B,k} \\ -R_{B,k} & R_{A,k} \end{bmatrix}$$

$$= \begin{bmatrix} 2.96 \times 10^{-10} \text{ N}^{-1} & 6.17 \times 10^{-13} \text{ N}^{-1}\text{mm}^{-1} \\ 6.17 \times 10^{-13} \text{ N}^{-1}\text{mm}^{-1} & 2.69 \times 10^{-15} \text{ N}^{-1}\text{mm}^{-2} \end{bmatrix}$$

The creep effects at τ_k due to the concrete stresses calculated at τ_0 are calculated using Eq. 5.36:

$$\mathbf{f}_{cr,k} = \overline{F}_{e,0}\mathbf{r}_{c,0} = -0.333 \times \begin{bmatrix} -785.8 \times 10^3 \\ 313.6 \times 10^6 \end{bmatrix} = \begin{bmatrix} 261.9 \times 10^3 \text{ N} \\ -104.6 \times 10^6 \text{ Nmm} \end{bmatrix}$$

and the shrinkage effects are given by Eq. 5.37:

$$\mathbf{f}_{sh,k} = \begin{bmatrix} A_c \\ B_c \end{bmatrix}\overline{E}_{e,k}\varepsilon_{sh,k} = \begin{bmatrix} 357{,}300 \times 9524 \\ -134 \times 10^6 \times 9524 \end{bmatrix}\left(-600 \times 10^{-6}\right)$$

$$= \begin{bmatrix} -2042 \times 10^3 \text{ N} \\ 765.6 \times 10^6 \text{ Nmm} \end{bmatrix}$$

From Eq. 5.39, the final strain distribution at time τ_k is:

$$\boldsymbol{\varepsilon}_k = \mathbf{F}_k\left(\mathbf{r}_{e,k} - \mathbf{f}_{cr,k} + \mathbf{f}_{sh,k}\right) = \begin{bmatrix} 162.2 \times 10^{-6} \\ 2.265 \times 10^{-6}\ \text{mm}^{-1} \end{bmatrix}$$

With $\varepsilon_{r,k} = 162.2 \times 10^{-6}$ and $\kappa_k = 2.265 \times 10^{-6}\ \text{mm}^{-1}$, the strains in the top and bottom fibres of the cross-sections are (Eq. 5.1):

$$\varepsilon_{k(top)} = \varepsilon_{r,k} + (-450) \times \kappa_k = -856.9 \times 10^{-6}$$

$$\varepsilon_{k(btm)} = \varepsilon_{r,k} + 300 \times \kappa_k = 841.6 \times 10^{-6}$$

The final concrete stresses at the top and bottom of the slab at time τ_k are (Eq. 5.41a):

$$\sigma_{c,k(top)} = \overline{E}_{e,k}\ \left\{\left[1 \quad y_{c,(top)}\right]\boldsymbol{\varepsilon}_k - \varepsilon_{sh,k}\right\} + \overline{F}_{e,0}\sigma_{c,0}$$

$$= 9524\left\{\left[1 \quad -450\right]\begin{bmatrix} 162.2 \times 10^{-6} \\ 2.265 \times 10^{-6} \end{bmatrix} - \left(-600 \times 10^{-6}\right)\right\}$$

$$-\,0.333 \times (-4.31) = -1.01\ \text{MPa}$$

$$\sigma_{c,k(btm)} = \overline{E}_{e,k}\ \left\{\left[1 \quad y_{c,(btm)}\right]\boldsymbol{\varepsilon}_k - \varepsilon_{sh,k}\right\} + \overline{F}_{e,0}\sigma_{c,0} = 0.82\ \text{MPa}$$

The final stress in the steel reinforcement is (Eq. 5.41b):

$$\sigma_{s(1),k} = E_{s(1)}\left[1 \quad y_{s(1)}\right]\ \boldsymbol{\varepsilon}_k = 200 \times 10^3\ \left[1 \quad -375\right]\begin{bmatrix} 162.2 \times 10^{-6} \\ 2.265 \times 10^{-6} \end{bmatrix}$$

$$= -137.4\,\text{MPa}$$

and the final stresses at the top and bottom fibres of the steel I-beam are (Eq. 5.65):

$$\sigma_{ss,k(top)} = E_{ss}\left[1 \quad y_{ss,(top)}\right]\ \boldsymbol{\varepsilon}_k = 200 \times 10^3\ \left[1 \quad -300\right]\begin{bmatrix} 162.2 \times 10^{-6} \\ 2.265 \times 10^{-6} \end{bmatrix}$$

$$= -103.4\,\text{MPa}$$

$$\sigma_{ss,k(btm)} = E_{ss}\left[1 \quad y_{ss,(btm)}\right]\ \boldsymbol{\varepsilon}_k = 200 \times 10^3\ \left[1 \quad 300\right]\begin{bmatrix} 162.2 \times 10^{-6} \\ 2.265 \times 10^{-6} \end{bmatrix}$$

$$= 168.3\ \text{MPa}$$

The final stress and strain distributions are illustrated in Fig. 5.17.

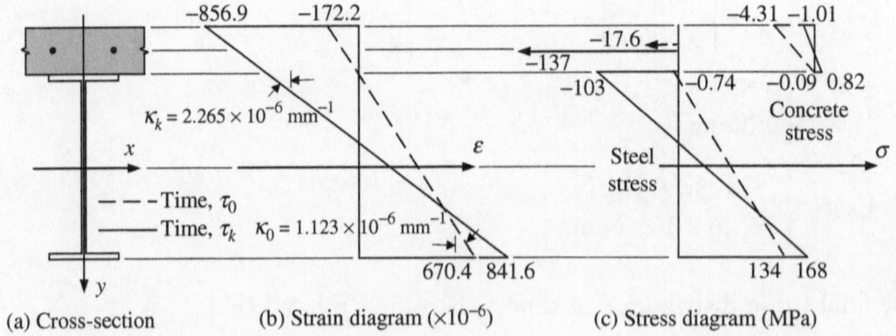

Figure 5.17 Stress and strain distributions due to $M_e = 500$ kNm (Example 5.7).

As can be seen in Fig. 5.17, the effect of creep and shrinkage in the concrete portions of a composite section in positive bending is to substantially increase positive curvature with time. In this example, the initial curvature has approximately doubled (from $\kappa_0 = 1.123 \times 10^{-6}$ mm^{-1} to $\kappa_k = 2.265 \times 10^{-6}$ mm^{-1}). In addition, the concrete deck has been relieved of much of its initial compression, with tensile stress developing in the bottom concrete fibres. The steel I-section therefore carries significantly more of the external moment after a period of sustained loading. Both the tensile stresses in the bottom fibres of the steel I-section and the compressive stresses in the top fibres have increased substantially with time.

Example 5.8

In order to examine the effects of shrinkage on the behaviour of a composite section, the cross-section shown in Fig. 5.16 (and analysed in Example 5.7) is here re-analysed for the case when $M_e = 0$ and $N_e = 0$. As in Example 5.7, the final shrinkage is -600×10^{-6}. This unloaded situation is quite common in practice, for example near the supports of a simply-supported member or at the points of contraflexure in a continuous composite girder.

Obviously, in the short-term, the section is unloaded, with zero deformation ($\varepsilon_{r,0} = 0$ and $\kappa_0 = 0$) and no external actions. As time increases, the concrete deck shrinks while being restrained by the steel I-section and by the bonded steel reinforcement. As for Example 5.7: $\overline{E}_{e,k} = 9524$ MPa; $\overline{F}_{e,0} = -0.333$; $A_c = 357{,}300$ mm^2; $B_c = -134 \times 10^6$ mm^3; $I_c = 50{,}920 \times 10^6$ mm^4; $R_{A,k} = 6443 \times 10^6$ N; $R_{B,k} = -1479 \times 10^9$ Nmm; $R_{I,k} = 710.9 \times 10^{12}$ Nmm^2; and

$$\mathbf{F}_k = \begin{bmatrix} 2.96 \times 10^{-10}\ \text{N}^{-1} & 6.17 \times 10^{-13}\ \text{N}^{-1}\text{mm}^{-1} \\ 6.17 \times 10^{-13}\ \text{N}^{-1}\text{mm}^{-1} & 2.69 \times 10^{-15}\ \text{N}^{-1}\text{mm}^{-2} \end{bmatrix}$$

With no external loads and from Eqs 5.36 and 5.37:

$$\mathbf{r}_{e,0} = \mathbf{r}_{e,k} = \begin{bmatrix} 0\text{ N} \\ 0\text{ Nmm} \end{bmatrix}$$

$$\mathbf{f}_{cr,k} = \begin{bmatrix} 0\text{ N} \\ 0\text{ Nmm} \end{bmatrix}$$

$$\mathbf{f}_{sh,k} = \begin{bmatrix} -2042 \times 10^3\text{ N} \\ 765.6 \times 10^6\text{ Nmm} \end{bmatrix}$$

The final strain distribution at time τ_k becomes (Eq. 5.39):

$$\boldsymbol{\varepsilon}_k = \mathbf{F}_k \left(\mathbf{r}_{e,k} - \mathbf{f}_{cr,k} + \mathbf{f}_{sh,k}\right) = \begin{bmatrix} -133.4 \times 10^{-6}\text{ N} \\ +0.799 \times 10^{-6}\text{ Nmm} \end{bmatrix}$$

With $\varepsilon_{r,k} = -133.4 \times 10^{-6}$ and $\kappa_k = +0.799 \times 10^{-6}\text{ mm}^{-1}$, the final strain at the top of the slab ($y = -450$ mm) and at the bottom of the steel I-section ($y = 300$ mm) at time τ_k are:

$$\varepsilon_{k(top)} = \varepsilon_{r,k} + (-450) \times \kappa_k = -493.2 \times 10^{-6}$$

$$\varepsilon_{k(btm)} = \varepsilon_{r,k} + 300 \times \kappa_k = +106.5 \times 10^{-6}$$

The final concrete stresses at the top and bottom of the slab at time τ_k are (Eq. 5.41a):

$$\sigma_{c,k(top)} = 9524 \left\{ \begin{bmatrix} 1 & -450 \end{bmatrix} \begin{bmatrix} -133.4 \times 10^{-6} \\ 0.799 \times 10^{-6} \end{bmatrix} - (-600 \times 10^{-6}) \right\}$$

$$= +1.02\text{ MPa}$$

$$\sigma_{c,k(btm)} = 9524 \left\{ \begin{bmatrix} 1 & -300 \end{bmatrix} \begin{bmatrix} -133.4 \times 10^{-6} \\ 0.799 \times 10^{-6} \end{bmatrix} - (-600 \times 10^{-6}) \right\}$$

$$= +2.16\text{ MPa}$$

and the final stress in the steel reinforcement is (Eq. 5.41b):

$$\sigma_{s(1),k} = 200{,}000\,[-133.4 - 375 \times 0.799] \times 10^{-6} = -86.6\text{ MPa}$$

The final stresses at the top and bottom fibres of the steel I-beam are (Eq. 5.65):

$$\sigma_{ss,k(top)} = 200{,}000\,[-133.4 - 300 \times 0.799] \times 10^{-6} = -74.6\text{ MPa}$$

$$\sigma_{ss,k(btm)} = 200{,}000\,[-133.4 + 300 \times 0.799] \times 10^{-6} = +21.3\text{ MPa}$$

The final stress and strain distributions are illustrated in Fig 5.18. The shrinkage-induced final curvature ($+0.799 \times 10^{-6}$ mm^{-1}) is almost 70 per cent of the change in curvature with time in Example 5.7 (where $\Delta\kappa = \kappa_k - \kappa_0 = 1.142 \times 10^{-6}$ mm^{-1}). In addition, the tensile stress induced in the bottom fibre of the concrete due to the restraint to shrinkage may be sufficient to cause cracking in the concrete.

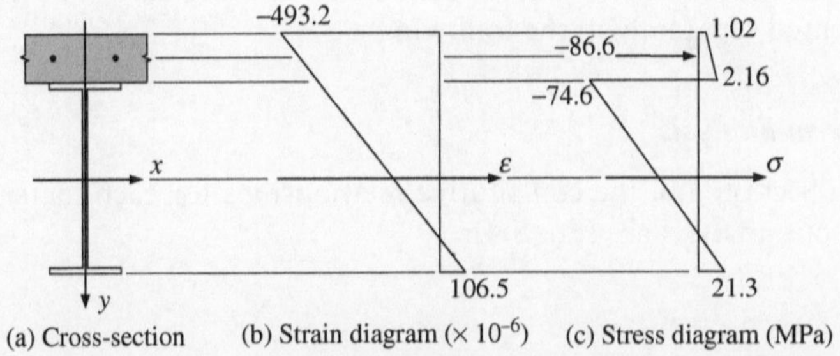

Figure 5.18 Shrinkage-induced stress and strain distributions at τ_k for Example 5.8.

5.7 Composite concrete-concrete cross-sections

The analysis of a cross-section consisting of different reinforced or prestressed concrete elements acting compositely together and subjected to a sustained bending moment and axial force is presented here. A typical cross-section is shown in Fig 5.19. This particular cross-section is made up of two concrete elements, a precast pretensioned I-section (Element 1) and an *in-situ* slab deck (Element 2). Each concrete element may have different creep and shrinkage characteristics, as well as different elastic moduli. In the subsequent analyses, perfect bond is assumed between each concrete element, as well as between the concrete and the bonded steel reinforcement.

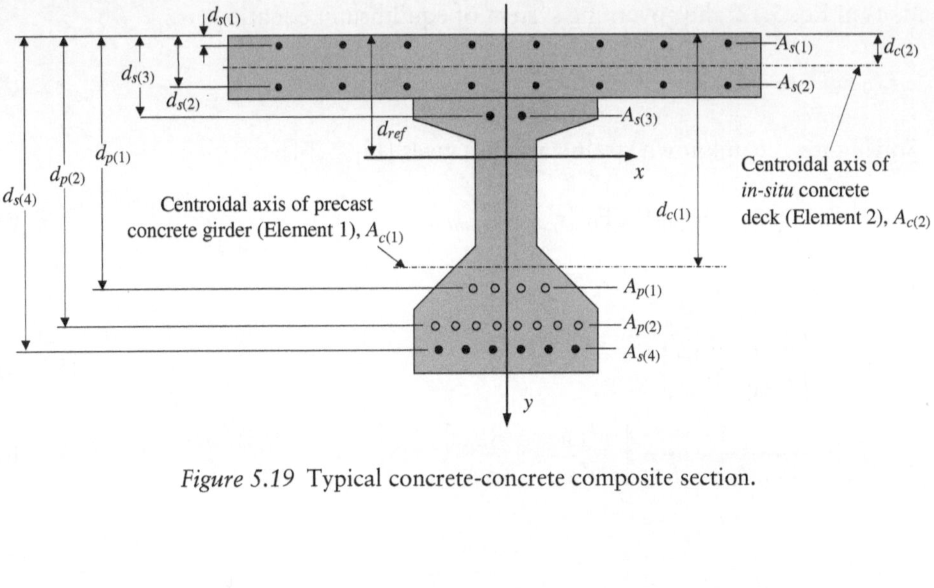

Figure 5.19 Typical concrete-concrete composite section.

The general procedure introduced in Sections 5.3–5.5 for reinforced and prestressed concrete cross-sections is readily extended to include composite sections comprising two or more concrete components. Consider a composite section formed with m_c different concrete components. For the i-th concrete component (with $i = 1, \ldots, m_c$), the area is $A_{c(i)}$, and the first and second moments of the area about the reference axis are $B_{c(i)}$ and $I_{c(i)}$, respectively.

For clarity and completeness, the short-term analysis (assuming linear-elastic material behaviour) and the time-dependent analyses using both the AEMM and the SSM are presented separately in the following.

5.7.1 *Short-term analysis*

As outlined in Section 5.3, the constitutive relationships for each material for use in the instantaneous analysis are (Eq. 5.4):

$$\sigma_{c(i),0} = E_{c(i),0}\varepsilon_0 \tag{5.67a}$$

$$\sigma_{s(i),0} = E_{s(i)}\varepsilon_0 \tag{5.67b}$$

$$\sigma_{p(i),0} = E_{p(i)}\left(\varepsilon_0 + \varepsilon_{p(i),init}\right) \tag{5.67c}$$

Similar to Eq. 5.7a, the internal actions carried by the i-th concrete element (for inclusion in the equilibrium equations) can be expressed as:

$$N_{c(i),0} = \int_{A_{c(i)}} \sigma_{c(i),0}\, dA = \int_{A_{c(i)}} E_{c(i),0}\left(\varepsilon_{r,0} + y\kappa_0\right) dA$$
$$= A_{c(i)}E_{c(i),0}\varepsilon_{r,0} + B_{c(i)}E_{c(i),0}\kappa_0 \tag{5.68a}$$

$$M_{c(i),0} = \int_{A_{c(i)}} y\sigma_{c(i),0}\, dA = \int_{A_{c(i)}} E_{c(i),0}\,y\left(\varepsilon_{r,0} + y\kappa_0\right) dA$$
$$= B_{c(i)}E_{c(i),0}\varepsilon_{r,0} + I_{c(i)}E_{c(i),0}\kappa_0 \tag{5.68b}$$

and, as in Eq. 5.12, the governing system of equilibrium equations is:

$$\mathbf{r}_{e,0} = \mathbf{D}_0\boldsymbol{\varepsilon}_0 + \mathbf{f}_{p,init} \tag{5.69}$$

Solving for the unknown strain variables gives (Eq. 5.14):

$$\boldsymbol{\varepsilon}_0 = \mathbf{D}_0^{-1}\left(\mathbf{r}_{e,0} - \mathbf{f}_{p,init}\right) = \mathbf{F}_0\left(\mathbf{r}_{e,0} - \mathbf{f}_{p,init}\right) \tag{5.70}$$

where

$$\mathbf{D}_0 = \begin{bmatrix} R_{A,0} & R_{B,0} \\ R_{B,0} & R_{I,0} \end{bmatrix} \tag{5.71a}$$

$$\mathbf{F}_0 = \frac{1}{R_{A,0}R_{I,0} - R_{B0}^2}\begin{bmatrix} R_{I,0} & -R_{B,0} \\ -R_{B,0} & R_{A,0} \end{bmatrix} \tag{5.71b}$$

$$\boldsymbol{\varepsilon}_0 = \begin{bmatrix} \varepsilon_{r,0} \\ \kappa_0 \end{bmatrix} \tag{5.71c}$$

$$\mathbf{r}_{e,0} = \begin{bmatrix} N_{e,0} \\ M_{e,0} \end{bmatrix} \tag{5.71d}$$

$$\mathbf{f}_{p,init} = \sum_{i=1}^{m_p} \begin{bmatrix} A_{p(i)}E_{p(i)}\varepsilon_{p(i),init} \\ y_{p(i)}A_{p(i)}E_{p(i)}\varepsilon_{p(i),init} \end{bmatrix} \tag{5.71e}$$

The cross-sectional rigidities forming the $\mathbf{D}_0$ matrix in Eq. 5.71a are:

$$R_{A,0} = \sum_{i=1}^{m_c} A_{c(i)}E_{c(i),0} + \sum_{i=1}^{m_s} A_{s(i)}E_{s(i)} + \sum_{i=1}^{m_p} A_{p(i)}E_{p(i)} = \sum_{i=1}^{m_c} A_{c(i)}E_{c(i),0} + R_{A,s} + R_{A,p} \tag{5.72a}$$

$$R_{B,0} = \sum_{i=1}^{m_c} B_{c(i)}E_{c(i),0} + \sum_{i=1}^{m_s} y_{s(i)}A_{s(i)}E_{s(i)} + \sum_{i=1}^{m_p} y_{p(i)}A_{p(i)}E_{p(i)}$$
$$= \sum_{i=1}^{m_c} B_{c(i)}E_{c(i),0} + R_{B,s} + R_{B,p} \tag{5.72b}$$

$$R_{I,0} = \sum_{i=1}^{m_c} I_{c(i)}E_{c(i),0} + \sum_{i=1}^{m_s} y^2_{s(i)}A_{s(i)}E_{s(i)} + \sum_{i=1}^{m_p} y^2_{p(i)}A_{p(i)}E_{p(i)}$$
$$= \sum_{i=1}^{m_c} I_{c(i)}E_{c(i),0} + R_{I,s} + R_{I,p} \tag{5.72c}$$

where for convenience the following notation is introduced for the rigidities of the reinforcement and tendons:

$$R_{A,s} = \sum_{i=1}^{m_s} A_{s(i)}E_{s(i)} \tag{5.73a}$$

$$R_{B,s} = \sum_{i=1}^{m_s} y_{s(i)}A_{s(i)}E_{s(i)} \tag{5.73b}$$

$$R_{I,s} = \sum_{i=1}^{m_s} y^2_{s(i)}A_{s(i)}E_{s(i)} \tag{5.73c}$$

$$R_{A,p} = \sum_{i=1}^{m_p} A_{p(i)}E_{p(i)} \tag{5.74a}$$

$$R_{B,p} = \sum_{i=1}^{m_p} y_{p(i)}A_{p(i)}E_{p(i)} \tag{5.74b}$$

$$R_{I,p} = \sum_{i=1}^{m_p} y^2_{p(i)}A_{s(i)}E_{p(i)} \tag{5.74c}$$

The stress distribution is calculated from Eq. 5.16:

$$\sigma_{c(i),0} = E_{c(i),0}\varepsilon_0 = E_{c(i),0}[1 \quad y]\boldsymbol{\varepsilon}_0 \tag{5.75a}$$

$$\sigma_{s(i),0} = E_{s(i)}\varepsilon_0 = E_{s(i)}[1 \quad y_{s(i)}]\boldsymbol{\varepsilon}_0 \tag{5.75b}$$

$$\sigma_{p(i),0} = E_{p(i)}\left(\varepsilon_0 + \varepsilon_{p(i),init}\right) = E_{p(i)}[1 \quad y_{p(i)}]\boldsymbol{\varepsilon}_0 + E_{p(i)}\varepsilon_{p(i),init} \tag{5.75c}$$

where $\varepsilon_0 = \varepsilon_{r,0} + y\kappa_0 = [1 \quad y]\boldsymbol{\varepsilon}_0$.

5.7.2 Time analysis – AEMM

For the analysis of stresses and deformations on a composite concrete-concrete cross-section at time τ_k after a period of sustained loading, the AEMM may be used, as outlined in Section 5.4. The stress–strain relationships for each concrete element and for each layer of reinforcement and tendons at τ_k are as follows (Eq. 5.22):

$$\sigma_{c(i),k} = \overline{E}_{e(i),k}\left(\varepsilon_k - \varepsilon_{sh(i),k}\right) + \overline{F}_{e(i),0}\sigma_{c(i),0} \tag{5.76a}$$

$$\sigma_{s(i),k} = E_{s(i)}\varepsilon_k \tag{5.76b}$$

$$\sigma_{p(i),k} = E_{p(i)}\left(\varepsilon_k + \varepsilon_{p(i),init} - \varepsilon_{p.rel(i),k}\right) \tag{5.76c}$$

In this case the contribution of the i-th concrete component to the internal axial force and moment can be determined as (similar to Eq. 5.27):

$$\begin{aligned} N_{c(i),k} &= \int_{A_{c(i)}} \sigma_{c(i),k}\,\mathrm{d}A = \int_{A_{c(i)}} \left[\overline{E}_{e(i),k}\left(\varepsilon_{r,k} + y\kappa_k - \varepsilon_{sh(i),k}\right) + \overline{F}_{e(i),0}\sigma_{c(i),0}\right]\mathrm{d}A \\ &= A_{c(i)}\overline{E}_{e(i),k}\varepsilon_{r,k} + B_{c(i)}\overline{E}_{e(i),k}\kappa_k - A_{c(i)}\overline{E}_{e(i),k}\varepsilon_{sh(i),k} + \overline{F}_{e(i),0}N_{c(i),0} \end{aligned} \tag{5.77a}$$

$$\begin{aligned} M_{c(i),k} &= \int_{A_{c(i)}} y\sigma_{c(i),k}\,\mathrm{d}A = \int_{A_{c(i)}} y\left[\overline{E}_{e(i),k}\left(\varepsilon_{r,k} + y\kappa_k - \varepsilon_{sh(i),k}\right) + \overline{F}_{e(i),0}\sigma_{c(i),0}\right]\mathrm{d}A \\ &= B_{c(i)}\overline{E}_{e(i),k}\varepsilon_{r,k} + I_{c(i)}\overline{E}_{e(i),k}\kappa_k - B_{c(i)}\overline{E}_{e(i),k}\varepsilon_{sh(i),k} + \overline{F}_{e(i),0}M_{c(i),0} \end{aligned} \tag{5.77b}$$

The equilibrium equations are (Eq. 5.34):

$$\mathbf{r}_{e,k} = \mathbf{D}_k\boldsymbol{\varepsilon}_k + \mathbf{f}_{cr,k} - \mathbf{f}_{sh,k} + \mathbf{f}_{p,init} - \mathbf{f}_{p.rel,k} \tag{5.78}$$

and solving gives the strain at time τ_k (Eq. 5.39):

$$\begin{aligned} \boldsymbol{\varepsilon}_k &= \mathbf{D}_k^{-1}\left(\mathbf{r}_{e,k} - \mathbf{f}_{cr,k} + \mathbf{f}_{sh,k} - \mathbf{f}_{p,init} + \mathbf{f}_{p.rel,k}\right) \\ &= \mathbf{F}_k\left(\mathbf{r}_{e,k} - \mathbf{f}_{cr,k} + \mathbf{f}_{sh,k} - \mathbf{f}_{p,init} + \mathbf{f}_{p.rel,k}\right) \end{aligned} \tag{5.79}$$

where

$$\boldsymbol{\varepsilon}_k = \begin{bmatrix} \varepsilon_{r,k} \\ \kappa_k \end{bmatrix} \tag{5.80a}$$

$$\mathbf{r}_{e,k} = \begin{bmatrix} N_{e,k} \\ M_{e,k} \end{bmatrix} \tag{5.80b}$$

$$\mathbf{D}_k = \begin{bmatrix} R_{A,k} & R_{B,k} \\ R_{B,k} & R_{I,k} \end{bmatrix} \tag{5.80c}$$

$$\mathbf{F}_k = \frac{1}{R_{A,k}R_{I,k} - R_{B,k}^2} \begin{bmatrix} R_{I,k} & -R_{B,k} \\ -R_{B,k} & R_{A,k} \end{bmatrix} \tag{5.80d}$$

and the cross-sectional rigidities at τ_k are:

$$R_{A,k} = \sum_{i=1}^{m_c} A_{c(i)}\overline{E}_{e(i),k} + \sum_{i=1}^{m_s} A_{s(i)}E_{s(i)} + \sum_{i=1}^{m_p} A_{p(i)}E_{p(i)} = \sum_{i=1}^{m_c} A_{c(i)}\overline{E}_{e(i),k} + R_{A,s} + R_{A,p} \tag{5.81a}$$

$$R_{B,k} = \sum_{i=1}^{m_c} B_{c(i)}\overline{E}_{e(i),k} + \sum_{i=1}^{m_s} y_{s(i)}A_{s(i)}E_{s(i)} + \sum_{i=1}^{m_p} y_{p(i)}A_{p(i)}E_{p(i)}$$
$$= \sum_{i=1}^{m_c} B_{c(i)}\overline{E}_{e(i),k} + R_{B,s} + R_{B,p} \tag{5.81b}$$

$$R_{I,k} = \sum_{i=1}^{m_c} I_{c(i)}\overline{E}_{e(i),k} + \sum_{i=1}^{m_s} y_{s(i)}^2 A_{s(i)}E_{s(i)} + \sum_{i=1}^{m_p} y_{p(i)}^2 A_{p(i)}E_{p(i)}$$
$$= \sum_{i=1}^{m_c} I_{c(i)}\overline{E}_{e(i),k} + R_{I,s} + R_{I,p} \tag{5.81c}$$

In Eq. 5.78, the effects of creep and shrinkage of the concrete elements are included by (see Eqs 5.36 and 5.37):

$$\mathbf{f}_{cr,k} = \sum_{i=1}^{m_c} \overline{F}_{e(i),0} \begin{bmatrix} N_{c(i),0} \\ M_{c(i),0} \end{bmatrix} = \sum_{i=1}^{m_c} \overline{F}_{e(i),0} E_{c(i),0} \begin{bmatrix} A_{c(i)}\varepsilon_{r,0} + B_{c(i)}\kappa_0 \\ B_{c(i)}\varepsilon_{r,0} + I_{c(i)}\kappa_0 \end{bmatrix} \tag{5.82}$$

$$\mathbf{f}_{sh,k} = \sum_{i=1}^{m_c} \begin{bmatrix} A_{c(i)} \\ B_{c(i)} \end{bmatrix} \overline{E}_{e(i),k}\varepsilon_{sh(i),k} \tag{5.83}$$

The initial strain in the prestressing steel and relaxation are accounted for using (Eq. 5.38):

$$\mathbf{f}_{p,init} = \sum_{i=1}^{m_p} \begin{bmatrix} A_{p(i)} E_{p(i)} \varepsilon_{p(i),init} \\ y_{p(i)} A_{p(i)} E_{p(i)} \varepsilon_{p(i),init} \end{bmatrix} \tag{5.84a}$$

$$\mathbf{f}_{p.rel,k} = \sum_{i=1}^{m_p} \begin{bmatrix} A_{p(i)} E_{p(i)} \varepsilon_{p(i),init} \varphi_p(\tau_k, \sigma_{p(i),init}) \\ y_{p(i)} A_{p(i)} E_{p(i)} \varepsilon_{p(i),init} \varphi_p(\tau_k, \sigma_{p(i),init}) \end{bmatrix} \tag{5.84b}$$

The stress distributions at time τ_k in each concrete element and in the reinforcement and tendons are (Eq. 5.41):

$$\begin{aligned} \sigma_{c(i),k} &= \overline{E}_{e(i),k} \left(\varepsilon_k - \varepsilon_{sh(i),k} \right) + \overline{F}_{e(i),0} \sigma_{c(i),0} \\ &= \overline{E}_{e(i),k} \left\{ \begin{bmatrix} 1 & y \end{bmatrix} \boldsymbol{\varepsilon}_k - \varepsilon_{sh(i),k} \right\} + \overline{F}_{e(i),0} \sigma_{c(i),0} \end{aligned} \tag{5.85a}$$

$$\sigma_{s(i),k} = E_{s(i)} \varepsilon_k = E_{s(i)} \begin{bmatrix} 1 & y_{s(i)} \end{bmatrix} \boldsymbol{\varepsilon}_k \tag{5.85b}$$

$$\begin{aligned} \sigma_{p(i),k} &= E_{p(i)} \left(\varepsilon_k + \varepsilon_{p(i),init} - \varepsilon_{p.rel(i),k} \right) \\ &= E_{p(i)} \begin{bmatrix} 1 & y_{p(i)} \end{bmatrix} \boldsymbol{\varepsilon}_k + E_{p(i)} \varepsilon_{p(i),init} - E_{p(i)} \varepsilon_{p.rel(i),k} \end{aligned} \tag{5.85c}$$

where $\varepsilon_k = \varepsilon_{r,k} + y\kappa_k = \begin{bmatrix} 1 & y \end{bmatrix} \boldsymbol{\varepsilon}_k$.

Example 5.9

The composite cross-section shown in Fig. 5.20 forms part of a bridge deck and consists of an *in-situ* reinforced concrete deck and a precast post-tensioned I-section (with bonded tendons). At time τ_0 with both concrete elements acting compositely, the road surface and other superimposed dead load are placed on the bridge, thereby introducing an increment of external moment $M_e = 400$ kNm at the section shown. Both the immediate and final long-term changes of stress caused by M_e are to be calculated.

For the precast I-girder:
$E_{c(1)}(\tau_0) = 32{,}000$ MPa; $\varphi_{(1)}(\infty, \tau_0) = 2.0$; $\chi_{(1)}(\infty, \tau_0) = 0.65$.
For the concrete deck:
$E_{c(2)}(\tau_0) = 25{,}000$ MPa; $\varphi_{(2)}(\infty, \tau_0) = 2.5$; $\chi_{(2)}(\infty, \tau_0) = 0.65$.
For the steel reinforcement and tendons:
$E_s = E_p = 200{,}000$ MPa.

In this example, only the increments of stress and strain caused by M_e are to be calculated. In the actual bridge girder, these must be added to the increments of stress and strain caused by the prestress, self-weight, live load, shrinkage of the concrete, relaxation of the prestressing steel and, if applicable, temperature changes.

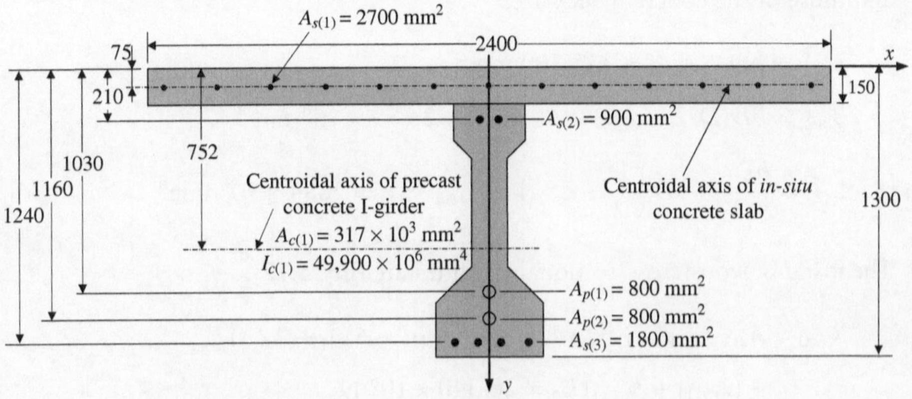

Figure 5.20 Details of the cross-section for Example 5.9.

In the following solution, the top surface of the *in-situ* slab is selected as the reference axis, as shown in Fig. 5.20, (i.e. $d_{ref} = 0$ mm) and its distances from the steel layers are: $y_{s(1)} = 75$ mm, $y_{s(2)} = 210$ mm, $y_{s(3)} = 1240$ mm, $y_{p(1)} = 1030$ mm and $y_{p(2)} = 1160$ mm.
The applied loads are:

$$\mathbf{r}_{e,0} = \begin{bmatrix} 0 \text{ N} \\ 400 \times 10^6 \text{ Nmm} \end{bmatrix}$$

As no prestressing forces and shrinkage effects are considered in this problem, the vectors $\mathbf{f}_{p,init}$, $\mathbf{f}_{p.rel,0}$, $\mathbf{f}_{p.rel,k}$ and $\mathbf{f}_{sh,k}$ are all nil:

$$\mathbf{f}_{p,init} = \begin{bmatrix} 0 \text{ N} \\ 0 \text{ Nmm} \end{bmatrix}$$

$$\mathbf{f}_{p.rel,0} = \mathbf{f}_{p.rel,k} = \begin{bmatrix} 0 \text{ N} \\ 0 \text{ Nmm} \end{bmatrix}$$

$$\mathbf{f}_{sh,k} = \begin{bmatrix} 0 \text{ N} \\ 0 \text{ Nmm} \end{bmatrix}$$

Short-term analysis (at time τ_0)

The geometric properties of the precast girder cross-section are calculated as:

$$A_{c(1)} = A_{gross} - A_{s(2)} - A_{s(3)} - A_{p(1)} - A_{p(2)} = 312{,}700 \text{ mm}^2$$

$$B_{c(1)} = A_{gross}(d_{c(1)} - d_{ref}) - (A_{s(2)}y_{s(2)} + A_{s(3)}y_{s(3)}) - (A_{p(1)}y_{p(1)} + A_{p(2)}y_{p(2)})$$
$$= 234.2 \times 10^6 \text{ mm}^3$$

$$I_{c(1)} = I_{gross} + A_{gross}(d_{c(1)} - d_{ref})^2 - \left(A_{s(2)}y_{s(2)}^2 + A_{s(3)}y_{s(3)}^2\right)$$
$$- \left(A_{p(1)}y_{p(1)}^2 + A_{p(2)}y_{p(2)}^2\right) = 224.4 \times 10^9 \text{ mm}^4$$

and those of the concrete deck are:

$$A_{c(2)} = bD - A_{s(1)} = 357{,}300 \text{ mm}^2$$

$$B_{c(2)} = bD(D/2 - d_{ref}) - A_{s(1)}y_{s(1)} = 26.79 \times 10^6 \text{ mm}^3$$

$$I_{c(2)} = \frac{bD^3}{12} + bD\left(\frac{D}{2} - d_{ref}\right)^2 - A_{s(1)}y_{s(1)}^2 = 2684 \times 10^6 \text{ mm}^4$$

The instantaneous cross-sectional rigidities are (Eq. 5.72):

$$R_{A,0} = A_{c(1)}E_{c(1),0} + A_{c(2)}E_{c(2),0} + \left(A_{s(1)} + A_{s(2)} + A_{s(3)}\right)E_s + \left(A_{p(1)} + A_{p(2)}\right)E_p = 20{,}340 \times 10^6 \text{ N}$$

$$R_{B,0} = B_{c(1)}E_{c(1),0} + B_{c(2)}E_{c(2),0} + \left(A_{s(1)}y_{s(1)} + A_{s(2)}y_{s(2)} + A_{s(3)}y_{s(3)}\right)E_s + \left(A_{p(1)}y_{p(1)} + A_{p(2)}y_{p(2)}\right)E_p = 9039 \times 10^9 \text{ Nmm}$$

$$R_{I,0} = I_{c(1)}E_{c(1),0} + I_{c(2)}E_{c(2),0} + \left(A_{s(1)}y_{s(1)}^2 + A_{s(2)}y_{s(2)}^2 + A_{s(3)}y_{s(3)}^2\right)E_s + \left(A_{p(1)}y_{p(1)}^2 + A_{p(2)}y_{p(2)}^2\right)E_p = 8198 \times 10^{12} \text{ Nmm}^2$$

The matrix $\mathbf{F}_0$ can then be determined as (Eq. 5.71b):

$$\mathbf{F}_0 = \frac{1}{R_{A,0}R_{I,0} - R_{B,0}^2}\begin{bmatrix} R_{I,0} & -R_{B,0} \\ -R_{B,0} & R_{A,0} \end{bmatrix} = \begin{bmatrix} 96.42 \times 10^{-12} \text{ N}^{-1} & -106.3 \times 10^{-15} \text{ N}^{-1}\text{mm}^{-1} \\ -106.3 \times 10^{-15} \text{ N}^{-1}\text{mm}^{-1} & 239.2 \times 10^{-18} \text{ N}^{-1}\text{mm}^{-2} \end{bmatrix}$$

The strain vector $\boldsymbol{\varepsilon}_0$ containing the unknown strain variables is determined from Eq. 5.70:

$$\boldsymbol{\varepsilon}_0 = \mathbf{F}_0\left(\mathbf{r}_{e,0} - \mathbf{f}_{p,init}\right) = \begin{bmatrix} 96.42 \times 10^{-12} & -106.3 \times 10^{-15} \\ -106.3 \times 10^{-15} & 239.2 \times 10^{-18} \end{bmatrix} \times \left\{\begin{bmatrix} 0 \\ 400 \times 10^6 \end{bmatrix} - \begin{bmatrix} 0 \\ 0 \end{bmatrix}\right\} = \begin{bmatrix} -42.5 \times 10^{-6} \\ 0.095 \times 10^{-6} \text{ mm}^{-1} \end{bmatrix}$$

The increment of strain at the reference axis and the increment of curvature caused by the imposed moment $M_e = 400$ kNm are therefore $\varepsilon_{r,0} = -42.5 \times 10^{-6}$ and $\kappa_0 = 0.095 \times 10^{-6}$ mm^{-1} and, from Eq. 5.1, the increments of top ($y = 0$ mm) and bottom ($y = 1300$ mm) fibre strains are:

$$\varepsilon_{0(top)} = \varepsilon_{r,0} + 0 \times \kappa_0 = -42.5 \times 10^{-6} \text{ and}$$

$$\varepsilon_{0(btm)} = \varepsilon_{r,0} + 1300 \times \kappa = 81.8 \times 10^{-6}$$

From Eq. 5.75a, the changes in the top and bottom fibre stresses on the precast I-section are:

$$\sigma_{c(1),0(top)} = E_{c(1),0}\varepsilon_{c(1),0(top)} = 32{,}000 \times (-42.5 + 150 \times 0.095) \times 10^{-6}$$
$$= -0.90 \text{ MPa}$$

$$\sigma_{c(1),0(btm)} = E_{c(1),0}\varepsilon_{c(1),0(btm)} = 32{,}000 \times (-42.5 + 1300 \times 0.095) \times 10^{-6}$$
$$= +2.62 \text{ MPa}$$

and on the concrete deck:

$$\sigma_{c(2),0(top)} = E_{c(2),0}\varepsilon_{c(2),0(top)} = 25{,}000 \times (-42.5 + 0 \times 0.095) \times 10^{-6}$$
$$= -1.06 \text{ MPa}$$

$$\sigma_{c(2),0(btm)} = E_{c(2),0}\varepsilon_{c(2),0(btm)} = 25{,}000 \times (-42.5 + 150 \times 0.095) \times 10^{-6}$$
$$= -0.70 \text{ MPa}$$

The increments of stress in the non-prestressed reinforcement are obtained from Eq. 5.75b as:

$$\sigma_{s(1),0} = 200 \times 10^3 \times [1 \quad 75] \begin{bmatrix} -42.5 \times 10^{-6} \\ 0.095 \times 10^{-6} \end{bmatrix} = -7.1 \text{ MPa}$$

$$\sigma_{s(2),0} = 200 \times 10^3 \times [1 \quad 210] \begin{bmatrix} -42.5 \times 10^{-6} \\ 0.095 \times 10^{-6} \end{bmatrix} = -4.5 \text{ MPa}$$

$$\sigma_{s(3),0} = 200 \times 10^3 [1 \quad 1240] \begin{bmatrix} -42.5 \times 10^{-6} \\ 0.095 \times 10^{-6} \end{bmatrix} = 15.2 \text{ MPa}$$

and in the prestressing steel (Eq. 5.75c)

$$\sigma_{p(1),0} = E_p [1 \quad y_{p(1)}] \varepsilon_0 = 200 \times 10^3 \times [1 \quad 1030] \begin{bmatrix} -42.5 \times 10^{-6} \\ 0.095 \times 10^{-6} \end{bmatrix}$$
$$= 11.2 \text{ MPa}$$

$$\sigma_{p(2),0} = E_p [1 \quad y_{p(2)}] \varepsilon_0 = 200 \times 10^3 \times [1 \quad 1160] \begin{bmatrix} -42.5 \times 10^{-6} \\ 0.095 \times 10^{-6} \end{bmatrix}$$
$$= 13.7 \text{ MPa}$$

Long-term analysis (at time $\tau_k = \infty$)

For the long-term analysis at time τ_k the age-adjusted material properties for the two concrete elements are obtained from Eqs 4.35 and 4.46:

$$\bar{E}_{e(1),k} = \frac{E_{c(1),0}}{1 + \chi_{(1)}(\tau_k, \tau_0)\varphi_{(1)}(\tau_k, \tau_0)} = \frac{32{,}000}{1 + 0.65 \times 2.0} = 13{,}910 \text{ MPa}$$

$$\overline{E}_{e(2),k} = \frac{E_{c(2),0}}{1+\chi_{(2)}(\tau_k,\tau_0)\varphi_{(2)}(\tau_k,\tau_0)} = \frac{25{,}000}{1+0.65\times 2.5} = 9523 \text{ MPa}$$

$$\overline{F}_{e(1),0} = \frac{\varphi_{(1)}(\tau_k,\tau_0)\left[\chi_{(1)}(\tau_k,\tau_0)-1\right]}{1+\chi_{(1)}(\tau_k,\tau_0)\varphi_{(1)}(\tau_k,\tau_0)} = \frac{2.0\times(0.65-1)}{1+0.65\times 2.0} = -0.304$$

$$\overline{F}_{e(2),0} = \frac{\varphi_{(2)}(\tau_k,\tau_0)\left[\chi_{(2)}(\tau_k,\tau_0)-1\right]}{1+\chi_{(2)}(\tau_k,\tau_0)\varphi_{(2)}(\tau_k,\tau_0)} = \frac{2.5\times(0.65-1)}{1+0.65\times 2.5} = -0.333$$

Creep effects are accounted for using Eq. 5.82:

$$\mathbf{f}_{cr,k} = \overline{F}_{e(1),0}E_{c(1),0}\begin{bmatrix} A_{c(1)}\varepsilon_{r,0}+B_{c(1)}\kappa_0 \\ B_{c(1)}\varepsilon_{r,0}+I_{c(1)}\kappa_0 \end{bmatrix} + \overline{F}_{e(2),0}E_{c(2),0}\begin{bmatrix} A_{c(2)}\varepsilon_{r,0}+B_{c(2)}\kappa_0 \\ B_{c(2)}\varepsilon_{r,0}+I_{c(2)}\kappa_0 \end{bmatrix}$$

$$= \begin{bmatrix} 16.5\times 10^3 \text{ N} \\ -104.8\times 10^6 \text{ Nmm} \end{bmatrix}$$

The cross-sectional rigidities at time τ_k are calculated from Eq. 5.81:

$$R_{A,k} = A_{c(1)}\overline{E}_{e(1),k} + A_{c(2)}\overline{E}_{e(2),k} + \left(A_{s(1)}+A_{s(2)}+A_{s(3)}\right)E_s + \left(A_{p(1)}+A_{p(2)}\right)E_p = 9153\times 10^6 \text{ N}$$

$$R_{B,k} = B_{c(1)}\overline{E}_{e(1),k} + B_{c(2)}\overline{E}_{e(2),k} + \left(A_{s(1)}y_{s(1)}+A_{s(2)}y_{s(2)}+A_{s(3)}y_{s(3)}\right)E_s + \left(A_{p(1)}y_{p(1)}+A_{p(2)}y_{p(2)}\right)E_p = 4389\times 10^9 \text{ Nmm}$$

$$R_{I,k} = I_{c(1)}\overline{E}_{e(1),k} + I_{c(2)}\overline{E}_{e(2),k} + \left(A_{s(1)}y_{s(1)}^2+A_{s(2)}y_{s(2)}^2+A_{s(3)}y_{s(3)}^2\right)E_s + \left(A_{p(1)}y_{p(1)}^2+A_{p(2)}y_{p(2)}^2\right)E_p = 4097\times 10^{12} \text{ Nmm}^2$$

Matrix $\mathbf{F}_k$ is then determined as (Eq. 5.80d):

$$\mathbf{F}_k = \begin{bmatrix} 224.6\times 10^{-12} \text{ N}^{-1} & -240.5\times 10^{-15} \text{ N}^{-1}\text{mm}^{-1} \\ -240.5\times 10^{-15} \text{ N}^{-1}\text{mm}^{-1} & 501.7\times 10^{-18} \text{ N}^{-1}\text{mm}^{-2} \end{bmatrix}$$

The strain vector $\boldsymbol{\varepsilon}_k$ at time τ_k caused by the sustained moment M_e is obtained from Eq. 5.79:

$$\boldsymbol{\varepsilon}_k = \mathbf{F}_k\left(\mathbf{r}_{e,k} - \mathbf{f}_{cr,k}\right) = \begin{bmatrix} -125.1\times 10^{-6} \text{ N} \\ 0.257\times 10^{-6} \text{ Nmm} \end{bmatrix}$$

The final increments of strains at the top and bottom fibres of the cross-section are (Eq. 5.1):

$$\varepsilon_{k(top)} = \varepsilon_{r,k} + 0\times\kappa_k = -125.1\times 10^{-6}$$

$$\varepsilon_{k(btm)} = \varepsilon_{r,k} + 1300\times\kappa_k = 209.2\times 10^{-6}$$

The increments of stress at time τ_k in the top and bottom fibres of the precast I-section are determined using Eq. 5.85a:

$$\sigma_{c(1),k(top)} = \overline{E}_{e(1),k}\left(\varepsilon_{c(1),k(top)} - \varepsilon_{sh(1),k}\right) + \overline{F}_{e(1),0}\sigma_{c(1),0(top)}$$
$$= 13{,}913 \times (-125.1 + 150 \times 0.257) \times 10^{-6} - 0.304 \times (-0.90)$$
$$= -0.93 \text{ MPa}$$

and

$$\sigma_{c(1),k(btm)} = \overline{E}_{e(1),k}\left(\varepsilon_{c(1),k(btm)} - \varepsilon_{sh(1),k}\right) + \overline{F}_{e(1),0}\sigma_{c(1),0(btm)} = 2.11 \text{ MPa}$$

Similarly, for the concrete deck:

$$\sigma_{c(2),k(top)} = \overline{E}_{e(2),k}\left(\varepsilon_{c(2),k(top)} - \varepsilon_{sh(2),k}\right) + \overline{F}_{e(2),0}\sigma_{c(2),0(top)}$$
$$= 9523 \times (-125.1 + 0 \times 0.257) \times 10^{-6} - 0.333 \times (-1.06)$$
$$= -0.84 \text{ MPa}$$
$$\sigma_{c(2),k(btm)} = \overline{E}_{e(2),k}\left(\varepsilon_{c(2),k(btm)} - \varepsilon_{sh(2),k}\right) + \overline{F}_{e(2),0}\sigma_{c(2),0(btm)} = -0.59 \text{ MPa}$$

The final increments of stress in the non-prestressed reinforcement are obtained from Eq. 5.85b:

$$\sigma_{s(1),k} = 200 \times 10^3 \times [1 \quad 75]\begin{bmatrix} -125.1 \times 10^{-6} \\ 0.257 \times 10^{-6} \end{bmatrix} = -21.2 \text{ MPa}$$

$$\sigma_{s(2),k} = 200 \times 10^3 \times [1 \quad 210]\begin{bmatrix} -125.1 \times 10^{-6} \\ 0.257 \times 10^{-6} \end{bmatrix} = -14.2 \text{ MPa}$$

$$\sigma_{s(3),k} = 200 \times 10^3 [1 \quad 1240]\begin{bmatrix} -125.1 \times 10^{-6} \\ 0.257 \times 10^{-6} \end{bmatrix} = +38.8 \text{ MPa}$$

and for the prestressing steel, Eq. 5.85c gives:

$$\sigma_{p(1),k} = E_p [1 \quad y_{p(1)}] \boldsymbol{\varepsilon}_k + E_p \varepsilon_{pinit(1)}$$
$$= 200 \times 10^3 \times [1 \quad 1030]\begin{bmatrix} -125.1 \times 10^{-6} \\ 0.257 \times 10^{-6} \end{bmatrix} + 0 = 27.9 \text{ MPa}$$

and

$$\sigma_{p(2),k} = E_p [1 \quad y_{p(2)}] \boldsymbol{\varepsilon}_k + E_p \varepsilon_{pinit(2)}$$
$$= 200 \times 10^3 \times [1 \quad 1160]\begin{bmatrix} -125.1 \times 10^{-6} \\ 0.257 \times 10^{-6} \end{bmatrix} + 0 = 34.6 \text{ MPa}$$

The stress and strain distributions at times τ_0 and at τ_k are shown in Fig. 5.21.

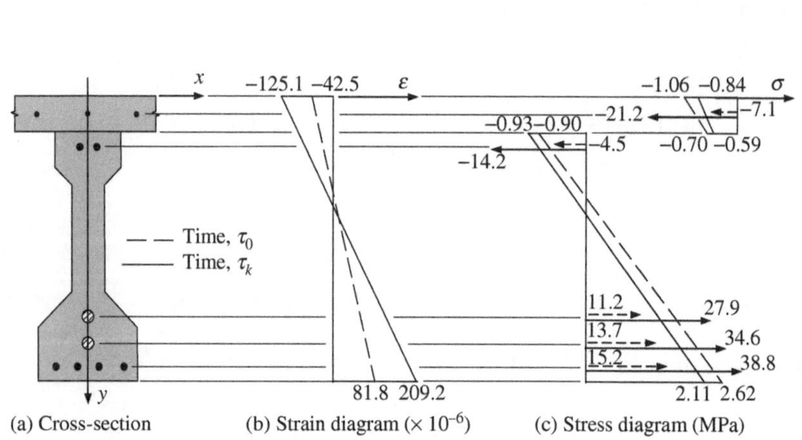

Figure 5.21 Strain and stress distributions for Example 5.9.

5.7.3 Time analysis – SSM

The SSM outlined in Section 5.5 is extended in the following to analyse a composite cross-section with m_c different concrete elements. Similar to Eqs 4.25 and 5.44, the constitutive relationships for the different materials at time τ_j are:

$$\sigma_{c(i),j} = E_{c(i),j}\left(\varepsilon_j - \varepsilon_{sh(i),j} - \varepsilon_{c(i),set-1}\right) + \sum_{n=0}^{j-1} F_{e(i),j,n}\sigma_{c(i),n} \tag{5.86a}$$

$$\sigma_{s(i),j} = E_{s(i)}\left(\varepsilon_j - \varepsilon_{s(i),set-1}\right) \tag{5.86b}$$

$$\sigma_{p(i),j} = E_{p(i)}\left(\varepsilon_j + \varepsilon_{p(i),init} - \varepsilon_{p.rel(i),j} - \varepsilon_{p(i),set-1}\right) \tag{5.86c}$$

where the additional strain terms $\varepsilon_{c(i),set-1}$, $\varepsilon_{s(i),set-1}$, $\varepsilon_{p(i),set-1}$ have been included to account for variations in the cross-section when different parts of the cross-section are cast at different times, such as occurs, for example, in sequential casting. The term *set* refers to the time step at which a particular concrete element sets and begins to contribute to the structural stiffness. The inclusion of the strains $\varepsilon_{c(i),set-1}$, $\varepsilon_{s(i),set-1}$, $\varepsilon_{p(i),set-1}$ in Eq. 5.86 ensures that all new components are initially unloaded. Similar treatment can be given to the grouting of post-tensioning ducts at some time after the transfer of prestress, as illustrated subsequently in Example 5.10.

In the formulation adopted here (and illustrated in Example 5.10), two steps are considered in the solution whenever a new concrete element is added to the cross-section. The first step involves the calculation of the cross-sectional response just before the element sets, at τ_{set-1}, and the second step involves the calculation of the response just after the element starts to contribute to the structural stiffness, at τ_{set}. Obviously, for situations not related to sequential casting, i.e. where all concrete elements are in place from the beginning of the analysis, the terms $\varepsilon_{c(i),set-1}$, $\varepsilon_{s(i),set-1}$ and $\varepsilon_{p(i),set-1}$ in Eq. 5.86 are all set to zero.

Similarly to Eq. 5.48, the axial force and moment resisted by the i-th concrete element are:

$$N_{c(i),j} = \int_{A_{c(i)}} \sigma_{c(i),j}\, dA = \int_{A_{c(i)}} \left[E_{c(i),j}\left(\varepsilon_j - \varepsilon_{sh(i),j} - \varepsilon_{c(i),set-1}\right) + \sum_{n=0}^{j-1} F_{e(i),j,n}\sigma_{c(i),n} \right] dA$$

$$= A_{c(i)}E_{c(i),j}\left(\varepsilon_{r,j} - \varepsilon_{r,set-1}\right) + B_{c(i)}E_{c(i),j}\left(\kappa_j - \kappa_{set-1}\right) - A_{c(i)}E_{c(i),j}\varepsilon_{sh(i),j}$$

$$+ \sum_{n=0}^{j-1} F_{e(i),j,n} N_{c(i),n} \tag{5.87a}$$

$$M_{c(i),j} = \int_{A_{c(i)}} y\sigma_{c(i),j}\, dA = \int_{A_{c(i)}} y \left[E_{c(i),j}\left(\varepsilon_j - \varepsilon_{sh(i),j} - \varepsilon_{c(i),set-1}\right) + \sum_{n=0}^{j-1} F_{e(i),j,n}\sigma_{c(i),n} \right] dA$$

$$= B_{c(i)}E_{c(i),j}\left(\varepsilon_{r,j} - \varepsilon_{r,set-1}\right) + I_{c(i)}E_{c(i),j}\left(\kappa_j - \kappa_{set-1}\right) - B_{c(i)}E_{c(i),j}\varepsilon_{sh(i),j}$$

$$+ \sum_{n=0}^{j-1} F_{e(i),j,n} M_{c(i),n} \tag{5.87b}$$

in which $\varepsilon_{r,set-1}$ and κ_{set-1} define the strain diagram immediately before the i-th concrete element sets at τ_{set-1}.

As for the earlier analyses (Eq. 5.54), the governing equilibrium equations are:

$$\mathbf{r}_{e,j} = \mathbf{D}_j\boldsymbol{\varepsilon}_j + \mathbf{f}_{cr,j} - \mathbf{f}_{sh,j} + \mathbf{f}_{p,init} - \mathbf{f}_{p.rel,j} - \mathbf{f}_{set,j} \tag{5.88}$$

and solving for the unknown strain distribution at τ_j gives:

$$\boldsymbol{\varepsilon}_j = \mathbf{D}_j^{-1}\left(\mathbf{r}_{e,j} - \mathbf{f}_{cr,j} + \mathbf{f}_{sh,j} - \mathbf{f}_{p,init} + \mathbf{f}_{p.rel,j} + \mathbf{f}_{set,j}\right)$$

$$= \mathbf{F}_j\left(\mathbf{r}_{e,j} - \mathbf{f}_{cr,j} + \mathbf{f}_{sh,j} - \mathbf{f}_{p,init} + \mathbf{f}_{p.rel,j} + \mathbf{f}_{set,j}\right) \tag{5.89}$$

where (as before):

$$\boldsymbol{\varepsilon}_j = \begin{bmatrix} \varepsilon_{r,j} \\ \kappa_j \end{bmatrix} \tag{5.90a}$$

$$\mathbf{D}_j = \begin{bmatrix} R_{A,j} & R_{B,j} \\ R_{B,j} & R_{I,j} \end{bmatrix} \tag{5.90b}$$

$$\mathbf{r}_{e,j} = \begin{bmatrix} N_{e,j} \\ M_{e,j} \end{bmatrix} \tag{5.90c}$$

$$\mathbf{F}_j = \frac{1}{R_{A,j}R_{I,j} - R_{B,j}^2} \begin{bmatrix} R_{I,j} & -R_{B,j} \\ -R_{B,j} & R_{A,j} \end{bmatrix} \tag{5.90d}$$

$$\mathbf{f}_{p,init} = \sum_{i=1}^{m_p} \begin{bmatrix} A_{p(i)}E_{p(i)}\varepsilon_{p(i),init} \\ y_{p(i)}A_{p(i)}E_{p(i)}\varepsilon_{p(i),init} \end{bmatrix} \tag{5.90e}$$

$$\mathbf{f}_{p.rel,j} = \sum_{i=1}^{m_p} \begin{bmatrix} A_{p(i)} E_{p(i)} \varepsilon_{p(i),init} \varphi_p(\tau_j, \sigma_{p(i),init}) \\ y_{p(i)} A_{p(i)} E_{p(i)} \varepsilon_{p(i),init} \varphi_p(\tau_j, \sigma_{p(i),init}) \end{bmatrix} \tag{5.90f}$$

$$\mathbf{f}_{cr,j} = \sum_{i=1}^{m_c} \sum_{n=0}^{j-1} F_{e(i),j,n} \mathbf{r}_{c(i),n} \tag{5.90g}$$

$$\mathbf{f}_{sh,j} = \sum_{i=1}^{m_c} \begin{bmatrix} A_{c(i)} \\ B_{c(i)} \end{bmatrix} E_{c(i),j} \varepsilon_{sh(i),j} \tag{5.90h}$$

while $\mathbf{f}_{set.j}$ is defined as:

$$\mathbf{f}_{set,j} = \sum_{i=1}^{m_c} \mathbf{f}_{c(i).set,j} + \sum_{i=1}^{m_s} \mathbf{f}_{s(i).set,j} + \sum_{i=1}^{m_p} \mathbf{f}_{p(i).set,j} \tag{5.90i}$$

and

$$\mathbf{f}_{c(i).set,j} = \begin{bmatrix} A_{c(i)} E_{c(i),j} \varepsilon_{r,set-1} + B_{c(i)} E_{c(i),j} \kappa_{set-1} \\ B_{c(i)} E_{c(i),j} \varepsilon_{r,set-1} + I_{c(i)} E_{c(i),j} \kappa_{set-1} \end{bmatrix} \tag{5.90j}$$

$$\mathbf{f}_{s(i).set,j} = \begin{bmatrix} A_{s(i)} E_{s(i)} \varepsilon_{r,set-1} + B_{s(i)} E_{s(i)} \kappa_{set-1} \\ B_{s(i)} E_{s(i)} \varepsilon_{r,set-1} + I_{s(i)} E_{s(i)} \kappa_{set-1} \end{bmatrix} \tag{5.90k}$$

$$\mathbf{f}_{p(i).set,j} = \begin{bmatrix} A_{p(i)} E_{p(i)} \varepsilon_{r,set-1} + B_{p(i)} E_{p(i)} \kappa_{set-1} \\ B_{p(i)} E_{p(i)} \varepsilon_{r,set-1} + I_{p(i)} E_{p(i)} \kappa_{set-1} \end{bmatrix} \tag{5.90l}$$

The i-th component of the cross-section is included in the calculation of $\mathbf{f}_{set,j}$ only when the instant in time τ_j is greater than or equal to τ_{set} of the i-th component.

The axial force and moment resisted by the concrete at any time τ_i (required in Eq. 5.90g) is determined as:

$$\mathbf{r}_{c(i),n} = \begin{bmatrix} N_{c(i),n} \\ M_{c(i),n} \end{bmatrix} = \mathbf{D}_{c(i),n} (\boldsymbol{\varepsilon}_n - \boldsymbol{\varepsilon}_{set-1}) + \sum_{l=0}^{n-1} F_{e(i),n,l} \mathbf{r}_{c(i),l} - \begin{bmatrix} A_{c(i)} \\ B_{c(i)} \end{bmatrix} E_{c(i),n} \varepsilon_{sh(i),n} \tag{5.91}$$

where $\boldsymbol{\varepsilon}_{set-1}$ is the vector containing $\varepsilon_{r,set-1}$ and κ_{set-1} and:

$$\mathbf{D}_{c(i),n} = \begin{bmatrix} A_{c(i)} & B_{c(i)} \\ B_{c(i)} & I_{c(i)} \end{bmatrix} E_{c(i),n} \tag{5.92}$$

The term $\mathbf{D}_{c(i),n}$ in Eq. 5.91 includes the properties of the i-th concrete element introduced at τ_{set} while $\boldsymbol{\varepsilon}_{set-1}$ defines the strain diagram at time τ_{set-1}, i.e. immediately prior to the setting of the concrete in the i-th element.

The cross-sectional rigidities to be used at τ_j are:

$$R_{A,j} = \sum_{i=1}^{m_c} A_{c(i)}E_{c(i),j} + \sum_{i=1}^{m_s} A_{s(i)}E_{s(i)} + \sum_{i=1}^{m_p} A_{p(i)}E_{p(i)} = \sum_{i=1}^{m_c} A_{c(i)}E_{c(i),j} + R_{A,s} + R_{A,p} \tag{5.93a}$$

$$R_{B,j} = \sum_{i=1}^{m_c} B_{c(i)}E_{c(i),j} + \sum_{i=1}^{m_s} y_{s(i)}A_{s(i)}E_{s(i)} + \sum_{i=1}^{m_p} y_{p(i)}A_{p(i)}E_{p(i)}$$

$$= \sum_{i=1}^{m_c} B_{c(i)}E_{c(i),j} + R_{B,s} + R_{B,p} \tag{5.93b}$$

$$R_{I,j} = \sum_{i=1}^{m_c} I_{c(i)}E_{c(i),j} + \sum_{i=1}^{m_s} y_{s(i)}^2 A_{s(i)}E_{s(i)} + \sum_{i=1}^{m_p} y_{p(i)}^2 A_{p(i)}E_{p(i)}$$

$$= \sum_{i=1}^{m_c} I_{c(i)}E_{c(i),j} + R_{I,s} + R_{I,p} \tag{5.93c}$$

and finally, the stress distributions at τ_j can be determined from

$$\sigma_{c(i),j} = E_{c(i),j}\left(\varepsilon_j - \varepsilon_{sh(i),j} - \varepsilon_{c(i),set-1}\right) + \sum_{n=0}^{j-1} F_{e(i),j,n}\sigma_{c(i),n}$$

$$= E_{c(i),j}\left\{\begin{bmatrix}1 & y\end{bmatrix}\left(\boldsymbol{\varepsilon}_j - \boldsymbol{\varepsilon}_{set-1}\right) - \varepsilon_{sh(i),j}\right\} + \sum_{n=0}^{j-1} F_{e(i),j,n}\sigma_{c(i),n} \tag{5.94a}$$

$$\sigma_{s(i),j} = E_{s(i)}\left(\varepsilon_j - \varepsilon_{s(i),set-1}\right) = E_{s(i)}\begin{bmatrix}1 & y_{s(i)}\end{bmatrix}\left(\boldsymbol{\varepsilon}_j - \boldsymbol{\varepsilon}_{set-1}\right) \tag{5.94b}$$

$$\sigma_{p(i),j} = E_{p(i)}\left(\varepsilon_j + \varepsilon_{p(i),init} - \varepsilon_{p.rel(i),j} - \varepsilon_{p(i),set-1}\right)$$

$$= E_{p(i)}\begin{bmatrix}1 & y_{p(i)}\end{bmatrix}\left(\boldsymbol{\varepsilon}_j - \boldsymbol{\varepsilon}_{set-1}\right) + E_{p(i)}\varepsilon_{p(i),init} - E_{p(i)}\varepsilon_{p.rel(i),j} \tag{5.94c}$$

where $\varepsilon_j = \varepsilon_{r,j} + y\kappa_j = \begin{bmatrix}1 & y\end{bmatrix}\boldsymbol{\varepsilon}_j$.

Example 5.10

The bridge girder with cross-section shown in Fig. 5.20 is subjected to the following load history. The I-section is cast and moist cured for 7 days. At age 7 days, the member is post-tensioned with an initial prestressing force of 2000 kN ($P_{p,init} = 1000$ kN in each tendon). The diameter of each prestressing duct is 60 mm. The moment at the cross-section due to the self-weight of the girder is $M_7 = 550$ kNm and is first applied to the cross-section during the post-tensioning operation at age 7 days. Shrinkage of the concrete in the I-girder also begins to develop at this time. At age 40 days, the slab deck is cast and cured and the moment caused by the weight of the wet concrete, $M_{40} = 620$ kNm, is first applied to the precast I-section. The post-tensioning ducts are also grouted at this time (i.e. at age 40 days).

Composite action gradually begins to develop as soon as the concrete in the deck sets and the grout within the ducts hardens. Full composite action may not be achieved for several days. However, it is here assumed that the deck and the I-section act compositely and that the prestressing tendons are bonded to the concrete at all times after age 40 days. Shrinkage of the deck is assumed to also begin at age 40 days. At $\tau = 60$ days, the road surface is placed and other superimposed dead loads are applied to the bridge, thereby introducing an additional moment of $M_{60} = 400$ kNm. Thereafter, the loads are assumed to remain constant in time.

The stress and strain distributions on the cross-section are to be calculated at the following times:

(i) immediately after the application of the prestress at age 7 days;
(ii) before and after the slab deck is cast at age 40 days;
(iii) after the deck and I-section act compositely at age 40 days;
(iv) before and after the road surface is placed at age 60 days;
(v) at age 30,000 days (i.e. about 80 years) assuming the applied actions acting at age 60 days remain constant thereafter.

Based on this loading history the following steps 0 to 6 are considered in the worked example:

Step	*Age (in days)*	*Description of event*
0	7	Prestressing force applied to the section. Application of $M_7 = 550$ kNm (short-term analysis).
1	40 –	Time step introduced to determine the effects of creep, shrinkage and relaxation in the I-girder from age 7 days to age 40 days under the action introduced at 7 days and held constant thereafter (time analysis).
2	40 +	The deck is poured introducing an additional moment of $M_{40} = 620$ kNm to be resisted by the I-section alone (short-term analysis)
3	40 ++	Change in cross-section due to grouting of ducts and deck acting compositely with the I-section (from this point in time).
4	60 –	Time step introduced to evaluate the time-dependent behaviour of the composite section under constant actions between ages 40 and 60 days (time analysis).
5	60 +	Road surface added causing an additional moment of $M_{60} = 400$ kNm to be applied to the composite section (short-term analysis).
6	30,000	Time step introduced to evaluate the final long-term behaviour assuming loads remain constant after age 60 days.

The material properties are:

For all reinforcing steel and tendons:

$$E_s = E_p = 200 \text{ GPa}$$
$$\varphi_p(7, \sigma_{p,init}) = 0$$
$$\varphi_p(40, \sigma_{p,init}) = 0.015$$
$$\varphi_p(60, \sigma_{p,init}) = 0.02$$
$$\varphi_p(30{,}000, \sigma_{p,init}) = 0.03$$

For the precast I-section (Concrete element 1):

$$E_{c(1),0} = 25 \text{ GPa}$$
$$E_{c(1),1} = E_{c(1),2} = E_{c(1),3} = 32 \text{ GPa}$$
$$E_{c(1),4} = E_{c(1),5} = 34 \text{ GPa}$$
$$E_{c(1),6} = 38 \text{ GPa}$$
$$\varphi_{(1)}(7,7) = 0$$
$$\varphi_{(1)}(40,7) = 0.8$$
$$\varphi_{(1)}(60,7) = 1.0$$
$$\varphi_{(1)}(30{,}000,7) = 2.5$$
$$\varphi_{(1)}(40,40) = 0$$
$$\varphi_{(1)}(60,40) = 0.8$$
$$\varphi_{(1)}(30{,}000,40) = 1.8$$
$$\varphi_{(1)}(60,60) = 0$$
$$\varphi_{(1)}(30{,}000,60) = 1.6$$
$$\varepsilon_{sh(1)}(7) = 0$$
$$\varepsilon_{sh(1)}(40) = -100 \times 10^{-6}$$
$$\varepsilon_{sh(1)}(60) = -150 \times 10^{-6}$$
$$\varepsilon_{sh(1)}(30{,}000) = -400 \times 10^{-6}$$

For the concrete deck (Concrete element 2):

$$E_{c(2),3} = 18 \text{ GPa}$$
$$E_{c(2),4} = E_{c(2),5} = 25 \text{ GPa}$$
$$E_{c(2),6} = 30 \text{ GPa}$$

$$\varphi_{(2)}(40,40) = 0$$

$$\varphi_{(2)}(60,40) = 2.0$$

$$\varphi_{(2)}(30{,}000,40) = 3.5$$

$$\varphi_{(2)}(60,60) = 0$$

$$\varphi_{(2)}(30{,}000,60) = 2.8$$

$$\varepsilon_{sh(2)}(40) = 0$$

$$\varepsilon_{sh(2)}(60) = -200 \times 10^{-6}$$

$$\varepsilon_{sh(1)}(30{,}000) = -600 \times 10^{-6}$$

As for Example 5.9, the top fibre of the concrete deck is selected as the reference axis, i.e. $d_{ref} = 0$ mm, and the distances of the steel and prestressing layers from the reference axis are $y_{s(1)} = 75$ mm, $y_{s(2)} = 210$ mm, $y_{s(3)} = 1240$ mm, $y_{p(1)} = 1030$ mm and $y_{p(2)} = 1160$ mm.

The external loads applied at each step are:

$$\mathbf{r}_{e,0} = \mathbf{r}_{e,1} = \begin{bmatrix} 0 \text{ N} \\ 550 \times 10^{6} \text{ Nmm} \end{bmatrix}$$

$$\mathbf{r}_{e,2} = \mathbf{r}_{e,3} = \mathbf{r}_{e,4} = \begin{bmatrix} 0 \text{ N} \\ 1170 \times 10^{6} \text{ Nmm} \end{bmatrix}$$

$$\mathbf{r}_{e,5} = \mathbf{r}_{e,6} = \begin{bmatrix} 0 \text{ N} \\ 1570 \times 10^{6} \text{ Nmm} \end{bmatrix}$$

The concrete material coefficients to be used in the various analyses are as follows:

For the precast I-section (Element 1):

$$J_{c(1)}(7,7) = 4 \times 10^{-5} \text{ MPa}^{-1}$$

$$J_{c(1)}(40,7) = 7.2 \times 10^{-5} \text{ MPa}^{-1}$$

$$J_{c(1)}(60,7) = 8 \times 10^{-5} \text{ MPa}^{-1}$$

$$J_{c(1)}(30{,}000,7) = 14 \times 10^{-5} \text{ MPa}^{-1}$$

$$J_{c(1)}(40,40) = 3.125 \times 10^{-5} \text{ MPa}^{-1}$$

$$J_{c(1)}(60,40) = 5.625 \times 10^{-5} \text{ MPa}^{-1}$$

$$J_{c(1)}(30{,}000,40) = 8.75 \times 10^{-5} \text{ MPa}^{-1}$$

$$J_{c(1)}(60,60) = 2.941 \times 10^{-5} \text{ MPa}^{-1}$$

$$J_{c(1)}(30{,}000,60) = 7.647 \times 10^{-5} \text{ MPa}^{-1}$$

$$J_{c(1)}(30{,}000, 30{,}000) = 2.631 \times 10^{-5}\ \text{MPa}^{-1}$$

$$F_{e(1),1,0} = F_{e(1),2,0} = F_{e(1),3,0} = -1.304$$

$$F_{e(1),4,0} = F_{e(1),5,0} = -0.807$$

$$F_{e(1),6,0} = -1.995$$

$$F_{e(1),4,3} = F_{e(1),5,3} = -0.912$$

$$F_{e(1),6,3} = -0.419$$

$$F_{e(1),6,5} = -1.906$$

$$F_{e(1),i,j} = 0 \text{ if } \tau_i = \tau_j$$

$$\varepsilon_{sh(1)}(7) = 0$$

$$\varepsilon_{sh(1)}(40) = \varepsilon_{sh(1),1} = \varepsilon_{sh(1),2} = \varepsilon_{sh(1),3} = -100 \times 10^{-6}$$

$$\varepsilon_{sh(1)}(60) = \varepsilon_{sh(1),4} = \varepsilon_{sh(1),5} = -150 \times 10^{-6}$$

$$\varepsilon_{sh(1)}(30{,}000) = \varepsilon_{sh(1),6} = -400 \times 10^{-6}$$

For the concrete deck (Element 2):

$$J_{c(2)}(40, 40) = 5.555 \times 10^{-5}\ \text{MPa}^{-1}$$

$$J_{c(2)}(60, 40) = 16.66 \times 10^{-5}\ \text{MPa}^{-1}$$

$$J_{c(2)}(30{,}000, 40) = 25 \times 10^{-5}\ \text{MPa}^{-1}$$

$$J_{c(2)}(60, 60) = 4.0 \times 10^{-5}\ \text{MPa}^{-1}$$

$$J_{c(2)}(30{,}000, 60) = 15.2 \times 10^{-5}\ \text{MPa}^{-1}$$

$$J_{c(2)}(30{,}000, 30{,}000) = 3.333 \times 10^{-5}\ \text{MPa}^{-1}$$

$$F_{e(2),4,3} = F_{e(2),5,3} = -3.166$$

$$F_{e(2),6,3} = -2.94$$

$$F_{e(2),6,5} = -3.56$$

$$F_{e(2),i,j} = 0 \text{ if } \tau_i = \tau_j$$

$$\varepsilon_{sh(2)}(40) = \varepsilon_{sh(2),1} = \varepsilon_{sh(2),2} = \varepsilon_{sh(2),3} = 0$$

$$\varepsilon_{sh(2)}(60) = \varepsilon_{sh(2),4} = \varepsilon_{sh(2),5} = -200 \times 10^{-6}$$

$$\varepsilon_{sh(2)}(30{,}000) = \varepsilon_{sh(2),6} = -600 \times 10^{-6}$$

The initial prestress is included in the analysis using:

$$\mathbf{f}_{p,init} = \begin{bmatrix} 2000 \times 10^3\ \text{N} \\ 2190 \times 10^6\ \text{Nmm} \end{bmatrix}$$

The concrete area of the I-section contributing to the cross-sectional rigidity changes when the prestressing ducts are grouted. For clarity, the concrete properties of the I-section before grouting are referred to as $A_{c(1a)}$, $B_{c(1a)}$ and $I_{c(1a)}$, and after grouting as $A_{c(1b)}$, $B_{c(1b)}$ and $I_{c(1b)}$.
The procedures required to analyse the I-section in the first three steps (before the concrete deck is cast) are identical to those illustrated in Examples 5.2 and 5.6 and, for this reason, only the results of the analyses for Steps 1, 2 and 3 are presented here.

Step 0: Short-term analysis at age $\tau = 7$ days

Concrete properties: $A_{c(1a)} = 308.6 \times 10^3$ mm^2; $B_{c(1a)} = 229.7 \times 10^6$ mm^3; and $I_{c(1a)} = 219.5 \times 10^9$ mm^4.
Cross-sectional rigidities: $R_{A,0} = 8256 \times 10^6$ N; $R_{B,0} = 6228 \times 10^9$ Nmm; and $R_{I,0} = 6050 \times 10^{12}$ Nmm^2.
The matrix $\mathbf{F}_0$:

$$\mathbf{F}_0 = \begin{bmatrix} 542.2 \times 10^{-12}\ \text{N}^{-1} & -558.2 \times 10^{-15}\ \text{N}^{-1}\text{mm}^{-1} \\ -558.2 \times 10^{-15}\ \text{N}^{-1}\text{mm}^{-1} & 739.9 \times 10^{-18}\ \text{N}^{-1}\text{mm}^{-2} \end{bmatrix}$$

Strain at reference axis and curvature:
$\varepsilon_{r,0} = -169.0 \times 10^{-6}$ and $\kappa_0 = -0.097 \times 10^{-6}$ mm^{-1}.
Strains in the top and bottom fibres of the I-section:
$\varepsilon_{0(top)} = -183.6 \times 10^{-6}$ and $\varepsilon_{0(btm)} = -295.2 \times 10^{-6}$.
Concrete stresses: $\sigma_{c(1),0(top)} = -4.60$ MPa and $\sigma_{c(1),0(btm)} = -7.38$ MPa.
Reinforcement stresses: $\sigma_{s(2),0} = -37.9$ MPa and $\sigma_{s(3),0} = -57.9$ MPa.
Stresses in tendons: $\sigma_{p(1),0} = 1250$ MPa and $\sigma_{p(2),0} = 1250$ MPa.
The concrete stress resultants at $\tau_0 = 7$ days are:

$$\mathbf{r}_{c(1),0} = \begin{bmatrix} -1862 \times 10^3\ \text{N} \\ -1504 \times 10^6\ \text{Nmm} \end{bmatrix}.$$

Step 1: Time analysis – SSM at age $\tau_1 = 40$ days

Creep and shrinkage of the concrete as well as relaxation of the prestressing steel in the I-section are accounted for with:

$$\mathbf{f}_{cr,1} = \begin{bmatrix} 2427 \times 10^3\ \text{N} \\ 1961 \times 10^6\ \text{Nmm} \end{bmatrix}$$

$$\mathbf{f}_{sh,1} = \begin{bmatrix} -987.7 \times 10^3\ \text{N} \\ -735.3 \times 10^6\ \text{Nmm} \end{bmatrix}$$

$$\mathbf{f}_{p.rel,1} = \begin{bmatrix} 30 \times 10^3\ \text{N} \\ 32.85 \times 10^6\ \text{Nmm} \end{bmatrix}$$

Cross-sectional rigidities: $R_{A,1} = 10416 \times 10^6$ N; $R_{B,1} = 7837 \times 10^9$ Nmm; and $R_{I,1} = 7587 \times 10^{12}$ Nmm^2.

The matrix $\mathbf{F}_1 = \begin{bmatrix} 430.7 \times 10^{-12}\ \text{N}^{-1} & -444.8 \times 10^{-15}\ \text{N}^{-1}\text{mm}^{-1} \\ -444.8 \times 10^{-15}\ \text{N}^{-1}\text{mm}^{-1} & 591.3 \times 10^{-18}\ \text{N}^{-1}\text{mm}^{-2} \end{bmatrix}$.

Strain at reference axis and curvature:
$\varepsilon_{r,1} = -405.1 \times 10^{-6}$ and $\kappa_1 = -0.148 \times 10^{-6}$ mm^{-1}.
Strains in the top and bottom fibres of the I-section:
$\varepsilon_{1(top)} = -427.4 \times 10^{-6}$ and $\varepsilon_{1(btm)} = -598.5 \times 10^{-6}$.
Concrete stresses: $\sigma_{c(1),1(top)} = -4.49$ MPa and $\sigma_{c(1),0(btm)} = -6.33$ MPa.
Reinforcement stresses: $\sigma_{s(2),1} = -87.3$ MPa and $\sigma_{s(3),1} = -117.9$ MPa.
Stresses in tendons: $\sigma_{p(1),1} = 1231$ MPa and $\sigma_{p(2),1} = 1231$ MPa.
The concrete stress resultants at $\tau_1 = 40$ days are:

$$\mathbf{r}_{c(1),1} = \begin{bmatrix} -1679 \times 10^3\ \text{N} \\ -1327 \times 10^6\ \text{Nmm} \end{bmatrix}.$$

Step 2: Short-term analysis at age $\tau = 40$ days

This step is identical to the previous one with the exception that the applied external load has increased due to the pour of the concrete deck.
Strain at reference axis and curvature:
$\varepsilon_{r,2} = -680.9 \times 10^{-6}$ and $\kappa_2 = 0.217 \times 10^{-6}$ mm^{-1}.
Strains in the top and bottom fibres of the I-section:
$\varepsilon_{2(top)} = -648.2 \times 10^{-6}$ and $\varepsilon_{2(btm)} = -397.7 \times 10^{-6}$.
Concrete stresses: $\sigma_{c(1),2(top)} = -11.56$ MPa and $\sigma_{c(1),2(btm)} = 0.09$ MPa.
Reinforcement stresses: $\sigma_{s(2),2} = -127.0$ MPa and $\sigma_{s(3),2} = -82.1$ MPa.
Stresses in tendons: $\sigma_{p(1),2} = 1231$ MPa and $\sigma_{p(2),2} = 1231$ MPa.
The concrete stress resultants at $\tau = 40$ days after the wet concrete is poured are:

$$\mathbf{r}_{c(1),2} = \begin{bmatrix} -1708 \times 10^3\ \text{N} \\ -779.8 \times 10^6\ \text{Nmm} \end{bmatrix}.$$

Step 3: Grouting of tendons and commencement of composite action at age $\tau_1 = 40$ days

During this step, the cross-sectional properties change due to the grouting of the tendons and due to the deck beginning to act compositely with the I-section. This change in cross-section will be captured in $\mathbf{f}_{set,j}$ to ensure that the added components are unloaded at the time of setting.
The cross-sectional properties of both I-section (element 1) and the concrete deck (element 2) are:

$$A_{c(1b)} = 312.7 \times 10^3\ \text{mm}^2$$

$$B_{c(1b)} = 234.2 \times 10^6\ \text{mm}^3$$

$$I_{c(1b)} = 224.4 \times 10^9\ \text{mm}^4$$

$$A_{c(2)} = 357.3 \times 10^3 \text{ mm}^2$$

$$B_{c(2)} = 26.79 \times 10^6 \text{ mm}^3$$

$$I_{c(2)} = 2685 \times 10^6 \text{ mm}^4$$

The cross-sectional rigidities are:

$$R_{A,3} = 17{,}840 \times 10^6 \text{ N}$$

$$R_{B,3} = 8852 \times 10^9 \text{ Nmm; and}$$

$$R_{I,3} = 8179 \times 10^{12} \text{ Nmm}^2$$

Matrix $\mathbf{F}_3$ is now:

$$\mathbf{F}_3 = \begin{bmatrix} 121.1 \times 10^{-12} \text{ N}^{-1} & -131.0 \times 10^{-15} \text{ N}^{-1}\text{mm}^{-1} \\ -131.0 \times 10^{-15} \text{ N}^{-1} & 264.1 \times 10^{-18} \text{ N}^{-1}\text{mm}^{-1} \end{bmatrix}$$

The time-dependent effects of creep and shrinkage of the concrete and the relaxation response, that is, $\mathbf{f}_{cr,3}$, $\mathbf{f}_{sh,3}$ and $\mathbf{f}_{p.rel,3}$, are identical to those of the previous two steps as they all correspond to age 40 days.

The vectors defining the unloaded condition for the grout, prestressing and concrete deck at the time of setting are determined below.

For prestressing cables and grout:

$$A_{grout} = A_{duct(1)} + A_{duct(2)} - A_{pre(1)} - A_{pre(2)} = 2827 + 2827 - 800 - 800 = 4054 \text{ mm}^2$$

$$B_{grout} = B_{duct(1)} + B_{duct(2)} - B_{pre(1)} - B_{pre(2)} = (2827 - 800) \times 1030 + (2827 - 800) \times 1160 = 4439.1 \times 10^3 \text{ mm}^3$$

$$I_{grout} = I_{duct(1)} + I_{duct(2)} - I_{pre(1)} - I_{pre(2)} = (2827 - 800) \times 1030^2 + (2827 - 800) \times 1160^2 = 4877.9 \times 10^6 \text{ mm}^4$$

$$\mathbf{f}_{c(1).set,3} = \begin{bmatrix} A_{grout} E_{c(1),3} \varepsilon_{r,2} + B_{grout} E_{c(1),3} \kappa_2 \\ B_{grout} E_{c(1),3} \varepsilon_{r,2} + I_{grout} E_{c(1),3} \kappa_2 \end{bmatrix}$$

$$= \begin{bmatrix} 4054 \times 32{,}000 \times (-680.9 \times 10^{-6}) + 4439.1 \times 10^3 \\ \times 32{,}000 \times (0.217 \times 10^{-6}) \\ 4439.1 \times 10^3 \times 32{,}000 \times (-680.9 \times 10^{-6}) + 4877.9 \times 10^6 \\ \times 32{,}000 \times (0.217 \times 10^{-6}) \end{bmatrix}$$

$$= \begin{bmatrix} -57.4 \times 10^3 \text{ N} \\ -62.7 \times 10^6 \text{ Nmm} \end{bmatrix}$$

$$\mathbf{f}_{p(1).set,3} = \begin{bmatrix} A_{p(1)}E_{p(1)}\varepsilon_{r,2} + B_{p(1)}E_{p(1)}\kappa_2 \\ B_{p(1)}E_{p(1)}\varepsilon_{r,2} + I_{p(1)}E_{p(1)}\kappa_2 \end{bmatrix}$$

$$= \begin{bmatrix} 800 \times 200 \times 10^3 \times (-680.9 \times 10^{-6}) + 800 \times 1030 \\ \times 200 \times 10^3 \times (0.217 \times 10^{-6}) \\ 800 \times 1030 \times 200 \times 10^3 \times (-680.9 \times 10^{-6}) + 800 \times 1030^2 \\ \times 200 \times 10^3 \times (0.217 \times 10^{-6}) \end{bmatrix}$$

$$= \begin{bmatrix} -73.0 \times 10^3 \text{ N} \\ -75.2 \times 10^6 \text{ Nmm} \end{bmatrix}$$

$$\mathbf{f}_{p(2).set,3} = \begin{bmatrix} A_{p(2)}E_{p(2)}\varepsilon_{r,2} + B_{p(2)}E_{p(2)}\kappa_2 \\ B_{p(2)}E_{p(2)}\varepsilon_{r,2} + I_{p(2)}E_{p(2)}\kappa_2 \end{bmatrix}$$

$$= \begin{bmatrix} 800 \times 200 \times 10^3 \times (-680.9 \times 10^{-6}) + 800 \times 1160 \\ \times 200 \times 10^3 \times (0.217 \times 10^{-6}) \\ 800 \times 1160 \times 200 \times 10^3 \times (-680.9 \times 10^{-6}) + 800 \times 1160^2 \\ \times 200 \times 10^3 \times (0.217 \times 10^{-6}) \end{bmatrix}$$

$$= \begin{bmatrix} -68.5 \times 10^3 \text{ N} \\ -79.5 \times 10^6 \text{ Nmm} \end{bmatrix}$$

For the concrete deck:

$$\mathbf{f}_{c(2),set,3} = \begin{bmatrix} A_{c(2)}E_{c(2),3}\varepsilon_{r,2} + B_{c(2)}E_{c(2),3}\kappa_2 \\ B_{c(2)}E_{c(2),3}\varepsilon_{r,2} + I_{c(2)}E_{c(2),3}\kappa_2 \end{bmatrix}$$

$$= \begin{bmatrix} 357{,}300 \times 18{,}000 \times (-680.9 \times 10^{-6}) + 26.79 \times 10^6 \\ \times 18{,}000 \times (0.217 \times 10^{-6}) \\ 26.79 \times 10^6 \times 18{,}000 \times (-680.9 \times 10^{-6}) + 2685 \times 10^6 \\ \times 18{,}000 \times (0.217 \times 10^{-6}) \end{bmatrix}$$

$$= \begin{bmatrix} -4274 \times 10^3 \text{ N} \\ -317.9 \times 10^6 \text{ Nmm} \end{bmatrix}$$

$$\mathbf{f}_{s(1).set,3} = \begin{bmatrix} A_{s(1)}E_{s(1)}\varepsilon_{r,2} + B_{s(1)}E_{s(1)}\kappa_2 \\ B_{s(1)}E_{s(1)}\varepsilon_{r,2} + I_{s(1)}E_{s(1)}\kappa_2 \end{bmatrix}$$

$$= \begin{bmatrix} 2700 \times 200 \times 10^3 \times (-680.9 \times 10^{-6}) + 202.5 \times 10^3 \\ \times 200 \times 10^3 \times (0.217 \times 10^{-6}) \\ 202.5 \times 10^3 \times 200 \times 10^3 \times (-680.9 \times 10^{-6}) + 15.19 \times 10^6 \\ \times 200 \times 10^3 \times (0.217 \times 10^{-6}) \end{bmatrix}$$

$$= \begin{bmatrix} -358.8 \times 10^3 \text{ N} \\ -26.9 \times 10^6 \text{ Nmm} \end{bmatrix}$$

Combining the above contributions related to concrete setting:

$$\mathbf{f}_{set,3} = \mathbf{f}_{c(1).set,3} + \mathbf{f}_{p(1).set,3} + \mathbf{f}_{p(2).set,3} + \mathbf{f}_{c(2).set,3} + \mathbf{f}_{s(1).set,3}$$

$$= \begin{bmatrix} -4832 \times 10^3 \text{ N} \\ -562.2 \times 10^6 \text{ Nmm} \end{bmatrix}$$

The strain distribution at time τ_3 can then be determined as:

$$\boldsymbol{\varepsilon}_3 = \mathbf{F}_3 \left(\mathbf{r}_{e,3} - \mathbf{f}_{cr,3} + \mathbf{f}_{sh,3} - \mathbf{f}_{p,init} + \mathbf{f}_{p.rel,3} + \mathbf{f}_{set,3} \right)$$

$$= \begin{bmatrix} 121.1 \times 10^{-12} & -131.0 \times 10^{-15} \\ -131.0 \times 10^{-15} & 264.1 \times 10^{-18} \end{bmatrix} \times \left(\begin{bmatrix} 0 \\ 1170 \times 10^6 \end{bmatrix} \right.$$

$$- \begin{bmatrix} 2427 \times 10^3 \\ 1961 \times 10^6 \end{bmatrix} + \begin{bmatrix} -987.7 \times 10^3 \\ -735.3 \times 10^6 \end{bmatrix} - \begin{bmatrix} 2000 \times 10^3 \\ 2190 \times 10^6 \end{bmatrix} + \begin{bmatrix} 30 \times 10^3 \\ 32.85 \times 10^6 \end{bmatrix}$$

$$\left. + \begin{bmatrix} -4832 \times 10^3 \\ -562.2 \times 10^6 \end{bmatrix} \right) = \begin{bmatrix} -680.9 \times 10^{-6} \\ 0.217 \times 10^{-6} \text{ mm}^{-1} \end{bmatrix}$$

As expected there is no change in deformation between steps 3 and 4 as the new cross-sectional components are kept unloaded. The strain diagram for the precast I-section at step 3 is identical to the one already calculated for step 2:

$$\varepsilon_{3(top)} = \varepsilon_{2(top)} = -648.2 \times 10^{-6}$$

$$\varepsilon_{3(btm)} = \varepsilon_{2(btm)} = -397.7 \times 10^{-6}$$

Stresses at step 3 are also identical to those obtained at Step 2 as shown below:

For the precast I-section:

$$\sigma_{c(1),3(top)} = E_{c(1),3} \left\{ \begin{bmatrix} 1 & y_{c(1),(top)} \end{bmatrix} \boldsymbol{\varepsilon}_3 - \varepsilon_{sh(1),3} \right\} + F_{e(1),3,0} \sigma_{c(1),0(top)}$$

$$= 32{,}000 \left\{ \begin{bmatrix} 1 & 150 \end{bmatrix} \begin{bmatrix} -680.9 \times 10^{-6} \\ 0.217 \times 10^{-6} \end{bmatrix} - \left(-100 \times 10^{-6} \right) \right\}$$

$$- 1.304 \times (-4.60) = -11.56 \text{ MPa}$$

$$\sigma_{c(1),3(btm)} = E_{c(1),3} \left\{ \begin{bmatrix} 1 & y_{c(1),(btm)} \end{bmatrix} \boldsymbol{\varepsilon}_3 - \varepsilon_{sh(1),3} \right\} + F_{e(1),3,0} \sigma_{c(1),0(btm)}$$

$$= 0.09 \text{ MPa}$$

For the concrete deck:

$$\sigma_{c(2),3(top)} = E_{c(2),3} \left\{ \begin{bmatrix} 1 & y_{c(2),(top)} \end{bmatrix} \boldsymbol{\varepsilon}_3 - \varepsilon_{sh(2),3} \right\} - E_{c(2),3} \begin{bmatrix} 1 & y_{c(2),(top)} \end{bmatrix} \boldsymbol{\varepsilon}_2$$

$$= 18{,}000 \left\{ \begin{bmatrix} 1 & 0 \end{bmatrix} \begin{bmatrix} -680.9 \times 10^{-6} \\ 0.217 \times 10^{-6} \end{bmatrix} - 0 \right\}$$

$$- 18{,}000 \left\{ \begin{bmatrix} 1 & 0 \end{bmatrix} \begin{bmatrix} -680.9 \times 10^{-6} \\ 0.217 \times 10^{-6} \end{bmatrix} \right\} = 0.0 \text{ MPa}$$

$$\sigma_{c(2),3(btm)} = E_{c(2),3}\left\{\begin{bmatrix}1 & y_{c(2),(btm)}\end{bmatrix}\boldsymbol{\varepsilon}_3 - \varepsilon_{sh(2),3}\right\} - E_{c(2),3}\begin{bmatrix}1 & y_{c(2),(btm)}\end{bmatrix}\boldsymbol{\varepsilon}_2$$
$$= 0.0 \text{ MPa}$$

For the non-prestressed reinforcement:

$$\sigma_{s(1),3} = 200 \times 10^3 \times \begin{bmatrix}1 & 75\end{bmatrix}\left(\begin{bmatrix}-680.9 \times 10^{-6} \\ 0.217 \times 10^{-6}\end{bmatrix} - \begin{bmatrix}-680.9 \times 10^{-6} \\ 0.217 \times 10^{-6}\end{bmatrix}\right)$$
$$= 0.0 \text{ MPa}$$

$$\sigma_{s(2),3} = 200 \times 10^3 \times \begin{bmatrix}1 & 210\end{bmatrix}\begin{bmatrix}-680.9 \times 10^{-6} \\ 0.217 \times 10^{-6}\end{bmatrix}$$
$$= -127.0 \text{ MPa}$$

$$\sigma_{s(3),3} = 200 \times 10^3 \begin{bmatrix}1 & 1240\end{bmatrix}\begin{bmatrix}-680.9 \times 10^{-6} \\ 0.217 \times 10^{-6}\end{bmatrix}$$
$$= -82.1 \text{ MPa}$$

For the prestressing steel:

$$\sigma_{p(1),3} = E_{p(1)}\begin{bmatrix}1 & y_{p(1)}\end{bmatrix}(\boldsymbol{\varepsilon}_3 - \boldsymbol{\varepsilon}_2) + E_{p(1)}(\varepsilon_{p(1),init} - \varepsilon_{p.rel(1),3}) = 1231 \text{ MPa}$$
$$\sigma_{p(2),3} = E_{p(2)}\begin{bmatrix}1 & y_{p(2)}\end{bmatrix}(\boldsymbol{\varepsilon}_3 - \boldsymbol{\varepsilon}_2) + E_{p(2)}(\varepsilon_{p(2),init} - \varepsilon_{p.rel(2),3}) = 1231 \text{ MPa}$$

The concrete stress resultants at step 3, $\tau = 40$ days after the wet concrete has set, are:

$$\mathbf{r}_{c(1),3} = \mathbf{D}_{c(1),3}\boldsymbol{\varepsilon}_3 + \mathbf{f}_{cr(1),3} - \mathbf{f}_{sh(1),3} - \mathbf{f}_{set(1),3} = \begin{bmatrix}-1708 \times 10^3 \text{ N} \\ -779.8 \times 10^6 \text{ Nmm}\end{bmatrix}$$

$$\mathbf{r}_{c(2),3} = \mathbf{D}_{c(2),3}\boldsymbol{\varepsilon}_3 + \mathbf{f}_{cr(2),3} - \mathbf{f}_{sh(2),3} - \mathbf{f}_{set(2),3} = \begin{bmatrix}0 \text{ N} \\ 0 \text{ Nmm}\end{bmatrix}$$

Step 4: Time analysis SSM at age 60 days

In step 4, the stresses and strains due to the previous loading history are evaluated at 60 days.
The following cross-sectional properties are calculated for step 4:

$R_{A,4} = 20{,}960 \times 10^6$ N;

$R_{B,4} = 9508 \times 10^9$ Nmm;

$R_{I,4} = 8647 \times 10^{12}$ Nmm^2.

The matrix $\mathbf{F}_4$ is:

$$\mathbf{F}_4 = \begin{bmatrix} 95.15 \times 10^{-12}\ \text{N}^{-1} & -104.6 \times 10^{-15}\ \text{N}^{-1}\text{mm}^{-1} \\ -104.6 \times 10^{-15}\ \text{N}^{-1}\ \text{mm}^{-1} & 230.7 \times 10^{-18}\ \text{N}^{-1}\ \text{mm}^{-2} \end{bmatrix}$$

The effects of creep and shrinkage in this time period are:

$$\mathbf{f}_{cr,4} = F_{e(1),4,0}\mathbf{r}_{c(1),0} + F_{e(1),4,3}\mathbf{r}_{c(1),3}$$

$$= -0.807 \times \begin{bmatrix} -1862 \times 10^3 \\ -1504 \times 10^6 \end{bmatrix} - 0.912 \times \begin{bmatrix} -1708 \times 10^3 \\ -779.8 \times 10^6 \end{bmatrix}$$

$$= \begin{bmatrix} 3061 \times 10^3\ \text{N} \\ 1925 \times 10^6\ \text{Nmm} \end{bmatrix}$$

$$\mathbf{f}_{sh,4} = \begin{bmatrix} A_{c(1)} \\ B_{c(1)} \end{bmatrix} E_{c(1),4}\varepsilon_{sh(1),4} + \begin{bmatrix} A_{c(2)} \\ B_{c(2)} \end{bmatrix} E_{c(2),4}\varepsilon_{sh(2),4}$$

$$= \begin{bmatrix} 312{,}700 \\ 234.2 \times 10^6 \end{bmatrix} \times 34{,}000 \times \left(-150 \times 10^{-6}\right) + \begin{bmatrix} 357{,}300 \\ 26.79 \times 10^6 \end{bmatrix}$$

$$\times 25{,}000 \times \left(-200 \times 10^{-6}\right)$$

$$= \begin{bmatrix} -1595 \times 10^3 \\ -1195 \times 10^6 \end{bmatrix} + \begin{bmatrix} -1786 \times 10^3 \\ -133.9 \times 10^6 \end{bmatrix} = \begin{bmatrix} -3381 \times 10^3\ \text{N} \\ -1328 \times 10^6\ \text{Nmm} \end{bmatrix}$$

Note that the grout is assumed to be shrinking at the same rate as the concrete used for the precast I-section.
The effects of relaxation of the steel tendon are calculated from Eq. 5.90f:

$$\mathbf{f}_{p.rel,4} = \begin{bmatrix} A_{p(1)}E_{p(1)}\varepsilon_{p(1),init}\varphi_p(60, \sigma_{p(1),init}) \\ y_{p(1)}A_{p(1)}E_{p(1)}\varepsilon_{p(1),init}\varphi_p(60, \sigma_{p(1),init}) \end{bmatrix}$$

$$+ \begin{bmatrix} A_{p(2)}E_{p(2)}\varepsilon_{p(2),init}\varphi_p(60, \sigma_{p(2),init}) \\ y_{p(2)}A_{p(2)}E_{p(2)}\varepsilon_{p(2),init}\varphi_p(60, \sigma_{p(2),init}) \end{bmatrix} = \begin{bmatrix} 40 \times 10^3\ \text{N} \\ 43.8 \times 10^6\ \text{Nmm} \end{bmatrix}$$

The terms defining $\mathbf{f}_{set,4}$ are calculated based on the corresponding values already obtained at time τ_3 and adjusted for the change in elastic moduli that has taken place:

$$\mathbf{f}_{c(1).set,4} = \mathbf{f}_{c(1).set,3}\frac{E_{c(1),4}}{E_{c(1),3}} = \begin{bmatrix} -57.4 \times 10^3 \\ -62.7 \times 10^6 \end{bmatrix} \frac{34{,}000}{32{,}000} = \begin{bmatrix} -61.0 \times 10^3\ \text{N} \\ -66.6 \times 10^6\ \text{Nmm} \end{bmatrix}$$

$$\mathbf{f}_{p(1).set,4} = \mathbf{f}_{p(1).set,3} = \begin{bmatrix} -73.0 \times 10^3\ \text{N} \\ -75.2 \times 10^6\ \text{Nmm} \end{bmatrix}$$

$$\mathbf{f}_{p(2).set,4} = \mathbf{f}_{p(2).set,3} = \begin{bmatrix} -68.5 \times 10^3 \text{ N} \\ -79.5 \times 10^6 \text{ Nmm} \end{bmatrix}$$

$$\mathbf{f}_{c(2).set,4} = \mathbf{f}_{c(2).set,3} \frac{E_{c(2),4}}{E_{c(2),3}} = \begin{bmatrix} -4274 \times 10^3 \\ -317.9 \times 10^6 \end{bmatrix} \frac{25{,}000}{18{,}000}$$

$$= \begin{bmatrix} -5936 \times 10^3 \text{ N} \\ -441.5 \times 10^6 \text{ Nmm} \end{bmatrix}$$

$$\mathbf{f}_{s(1).set,4} = \mathbf{f}_{s(1).set,3} = \begin{bmatrix} -358.8 \times 10^3 \text{ N} \\ -26.9 \times 10^6 \text{ Nmm} \end{bmatrix}$$

$$\mathbf{f}_{set,4} = \mathbf{f}_{c(1).set,4} + \mathbf{f}_{p(1).set,4} + \mathbf{f}_{p(2).set,4} + \mathbf{f}_{c(2).set,4} + \mathbf{f}_{s(1).set,4}$$

$$= \begin{bmatrix} -6498 \times 10^3 \text{ N} \\ -689.8 \times 10^6 \text{ Nmm} \end{bmatrix}$$

The strain distribution at time τ_4 becomes:

$$\boldsymbol{\varepsilon}_4 = \mathbf{F}_4 \left(\mathbf{r}_{e,4} - \mathbf{f}_{cr,4} + \mathbf{f}_{sh,4} - \mathbf{f}_{p,init} + \mathbf{f}_{p.rel,4} + \mathbf{f}_{set,4} \right)$$

$$= \begin{bmatrix} 95.15 \times 10^{-12} & -104.6 \times 10^{-15} \\ -104.6 \times 10^{-15} & 230.7 \times 10^{-18} \end{bmatrix} \times \left(\begin{bmatrix} 0 \\ 1170 \times 10^6 \end{bmatrix} \right.$$

$$- \begin{bmatrix} 3061 \times 10^3 \\ 1925 \times 10^6 \end{bmatrix} + \begin{bmatrix} -3381 \times 10^3 \\ -1328 \times 10^6 \end{bmatrix} - \begin{bmatrix} 2000 \times 10^3 \\ 2190 \times 10^6 \end{bmatrix} + \begin{bmatrix} 40 \times 10^3 \\ 43.8 \times 10^6 \end{bmatrix}$$

$$\left. + \begin{bmatrix} -6498 \times 10^3 \\ -689.8 \times 10^6 \end{bmatrix} \right) = \begin{bmatrix} -903.0 \times 10^{-6} \\ 0.424 \times 10^{-6} \text{ mm}^{-1} \end{bmatrix}$$

At age 60 days, $\varepsilon_{r,4} = -903.0 \times 10^{-6}$ and $\kappa_4 = 0.424 \times 10^{-6}$ mm^{-1} with strains in the top and bottom of the I-section equal to:

$$\varepsilon_{4(top)} = \varepsilon_{r,4} + 150 \times \kappa_4 = -839.4 \times 10^{-6}$$

and

$$\varepsilon_{4(btm)} = \varepsilon_{r,4} + 1300 \times \kappa_4 = -351.9 \times 10^{-6}$$

Stresses at the end of step 4 are calculated below:

For the precast I-section:

$$\sigma_{c(1),4(top)} = E_{c(1),4} \left\{ \begin{bmatrix} 1 & y_{c(1),(top)} \end{bmatrix} \boldsymbol{\varepsilon}_4 - \varepsilon_{sh(1),4} \right\} + F_{e(1),4,0} \sigma_{c(1),0(top)}$$

$$+ F_{e(1),4,3} \sigma_{c(1),3(top)}$$

$$= 34{,}000 \left\{ \begin{bmatrix} 1 & 150 \end{bmatrix} \begin{bmatrix} -903.0 \times 10^{-6} \\ 0.424 \times 10^{-6} \end{bmatrix} - \left(-150 \times 10^{-6} \right) \right\}$$

$$- 0.807 \times (-4.60) - 0.912 \times (-11.56) = -9.19 \text{ MPa}$$

and

$$\sigma_{c(1),4(btm)} = E_{c(1),4}\left\{\begin{bmatrix}1 & y_{c(1),(btm)}\end{bmatrix}\boldsymbol{\varepsilon}_4 - \varepsilon_{sh(1),4}\right\} + F_{e(1),4,0}\sigma_{c(1),0(btm)}$$
$$+ F_{e(1),4,3}\sigma_{c(1),3(btm)} = -0.99 \text{ MPa}$$

For the concrete deck:

$$\sigma_{c(2),4(top)} = E_{c(2),4}\left\{\begin{bmatrix}1 & y_{c(2),(top)}\end{bmatrix}(\boldsymbol{\varepsilon}_4 - \boldsymbol{\varepsilon}_2) + F_{e(2),4,3}\sigma_{c(1),3(top)} - \varepsilon_{sh(2),4}\right\}$$
$$= 25{,}000\left\{[1 \quad 0]\left(\begin{bmatrix}-903.0\times10^{-6}\\0.424\times10^{-6}\end{bmatrix} - \begin{bmatrix}-680.9\times10^{-6}\\0.217\times10^{-6}\end{bmatrix}\right) + 0\right.$$
$$\left. -\left(-200\times10^{-6}\right)\right\} = -0.55 \text{ MPa}$$

$$\sigma_{c(2),4(btm)} = E_{c(2),4}\left\{\begin{bmatrix}1 & y_{c(2),(btm)}\end{bmatrix}(\boldsymbol{\varepsilon}_4 - \boldsymbol{\varepsilon}_2) + F_{e(2),4,3}\sigma_{c(1),3(btm)} - \varepsilon_{sh(2),4}\right\}$$
$$= 25{,}000\left\{[1 \quad 150]\left(\begin{bmatrix}-903.0\times10^{-6}\\0.424\times10^{-6}\end{bmatrix} - \begin{bmatrix}-680.9\times10^{-6}\\0.217\times10^{-6}\end{bmatrix}\right) + 0\right.$$
$$\left. -\left(-200\times10^{-6}\right)\right\} = 0.22 \text{ MPa}$$

For the non-prestressed reinforcement:

$$\sigma_{s(1),4} = 200\times10^3\times[1 \quad 75]\left(\begin{bmatrix}-903.0\times10^{-6}\\0.424\times10^{-6}\end{bmatrix} - \begin{bmatrix}-680.9\times10^{-6}\\0.217\times10^{-6}\end{bmatrix}\right)$$
$$= -41.3 \text{ MPa}$$

$$\sigma_{s(2),4} = 200\times10^3\times[1 \quad 210]\begin{bmatrix}-903.0\times10^{-6}\\0.424\times10^{-6}\end{bmatrix} = -162.8 \text{ MPa}$$

$$\sigma_{s(3),4} = 200\times10^3[1 \quad 1240]\begin{bmatrix}-903.0\times10^{-6}\\0.424\times10^{-6}\end{bmatrix} = -75.5 \text{ MPa}$$

For the prestressing steel:

$$\sigma_{p(1),4} = E_{p(1)}\begin{bmatrix}1 & y_{p(1)}\end{bmatrix}(\boldsymbol{\varepsilon}_4 - \boldsymbol{\varepsilon}_2) + E_{p(1)}\left(\varepsilon_{p(1),init} - \varepsilon_{p.rel(1),4}\right)$$
$$= 200\times10^3\times[1 \quad 1030]\left(\begin{bmatrix}-903.0\times10^{-6}\\0.424\times10^{-6}\end{bmatrix} - \begin{bmatrix}-680.9\times10^{-6}\\0.217\times10^{-6}\end{bmatrix}\right)$$
$$+ 200{,}000\times0.00625\times(1-0.02) = 1223 \text{ MPa}$$

$$\sigma_{p(2),4} = E_{p(2)}\begin{bmatrix}1 & y_{p(2)}\end{bmatrix}(\boldsymbol{\varepsilon}_4 - \boldsymbol{\varepsilon}_2) + E_{p(2)}\left(\varepsilon_{p(2).init} - \varepsilon_{p.rel(2).4}\right)$$

$$= 200\times 10^3 \times \begin{bmatrix}1 & 1160\end{bmatrix}\left(\begin{bmatrix}-903.0\times 10^{-6}\\ 0.424\times 10^{-6}\end{bmatrix} - \begin{bmatrix}-680.9\times 10^{-6}\\ 0.217\times 10^{-6}\end{bmatrix}\right)$$

$$+ 200{,}000\times 0.00625\times(1-0.02) = 1228\ \text{MPa}$$

The concrete stress resultants at step 4 at $\tau = 60$ days are:

$$\mathbf{r}_{c(1),4} = \mathbf{D}_{c(1),4}\boldsymbol{\varepsilon}_4 + \mathbf{f}_{cr(1),4} - \mathbf{f}_{sh(1),4} - \mathbf{f}_{c(1).set,4}$$

$$= 34{,}000\times\begin{bmatrix}312{,}700 & 234.2\times 10^6\\ 234.2\times 10^6 & 224.4\times 10^9\end{bmatrix}\begin{bmatrix}-903.0\times 10^{-6}\\ 0.424\times 10^{-6}\end{bmatrix}$$

$$+\begin{bmatrix}3061\times 10^3\\ 1925\times 10^6\end{bmatrix} - \begin{bmatrix}-1595\times 10^3\\ -1195\times 10^6\end{bmatrix} - \begin{bmatrix}-61.0\times 10^3\\ -66.6\times 10^6\end{bmatrix}$$

$$= \begin{bmatrix}-1507\times 10^3\ \text{N}\\ -769.1\times 10^6\ \text{Nmm}\end{bmatrix}$$

$$\mathbf{r}_{c(2),4} = \mathbf{D}_{c(2),3}\boldsymbol{\varepsilon}_4 + \mathbf{f}_{cr(2),4} - \mathbf{f}_{sh(2),4} - \mathbf{f}_{c(2).set,4}$$

$$= 25{,}000\times\begin{bmatrix}357{,}300 & 26.79\times 10^6\\ 26.79\times 10^6 & 2685\times 10^6\end{bmatrix}\begin{bmatrix}-903.0\times 10^{-6}\\ 0.424\times 10^{-6}\end{bmatrix} + \begin{bmatrix}0\\ 0\end{bmatrix}$$

$$-\begin{bmatrix}-1786\times 10^3\\ -133.9\times 10^6\end{bmatrix} - \begin{bmatrix}-5936\times 10^3\\ -441.5\times 10^6\end{bmatrix} = \begin{bmatrix}-59.7\times 10^3\ \text{N}\\ -0.999\times 10^6\ \text{Nmm}\end{bmatrix}$$

Step 5: Time analysis SSM at age 60 days

The cross-sectional properties, the creep and shrinkage of the concrete and relaxation of the prestressing tendons are identical to the previous step. The only change for step 5 is an additional external moment applied to the composite section. Based on the results calculated for step 4, the strain variables can be calculated by modifying vector $\mathbf{r}_{e,5}$ to account for the change in external moment as:

$$\boldsymbol{\varepsilon}_5 = \mathbf{F}_5\left(\mathbf{r}_{e,5} - \mathbf{f}_{cr,5} + \mathbf{f}_{sh,5} - \mathbf{f}_{p,init} + \mathbf{f}_{p.rel,5} + \mathbf{f}_{set,5}\right)$$

$$= \begin{bmatrix}95.15\times 10^{-12} & -104.6\times 10^{-15}\\ -104.6\times 10^{-15} & 230.7\times 10^{-18}\end{bmatrix}\times\left(\begin{bmatrix}0\\ 1570\times 10^6\end{bmatrix}\right.$$

$$-\begin{bmatrix}3061\times 10^3\\ 1925\times 10^6\end{bmatrix} + \begin{bmatrix}-3381\times 10^3\\ -1328\times 10^6\end{bmatrix} - \begin{bmatrix}2000\times 10^3\\ 2190\times 10^6\end{bmatrix} + \begin{bmatrix}40\times 10^3\\ 43.8\times 10^6\end{bmatrix}$$

$$\left.+\begin{bmatrix}-6498\times 10^3\\ -689.8\times 10^6\end{bmatrix}\right) = \begin{bmatrix}-944.9\times 10^{-6}\\ 0.516\times 10^{-6}\ \text{mm}^{-1}\end{bmatrix}$$

At age 60 days, after the application of the road deck and other imposed dead loads, $\varepsilon_{r,5} = -944.9 \times 10^{-6}$ and $\kappa_5 = 0.516 \times 10^{-6}$ mm^{-1} with strains in the top and bottom of the I-section equal to:

$$\varepsilon_{5(top)} = \varepsilon_{r,5} + 150 \times \kappa_5 = -867.4 \times 10^{-6}$$

and

$$\varepsilon_{5(btm)} = \varepsilon_{r,5} + 1300 \times \kappa_5 = -273.8 \times 10^{-6}$$

Stresses at the end of step 5 are calculated below:

For the precast I-section:

$$\sigma_{c(1),5(top)} = E_{c(1),5}\left\{\begin{bmatrix}1 & y_{c(1),(top)}\end{bmatrix}\boldsymbol{\varepsilon}_5 - \varepsilon_{sh(1),5}\right\} + F_{e(1),5,0}\sigma_{c(1),0(top)} + F_{e(1),5,3}\sigma_{c(1),3(top)}$$

$$= 34{,}000\left\{[1 \quad 150]\begin{bmatrix}-944.9 \times 10^{-6} \\ 0.516 \times 10^{-6}\end{bmatrix} - \left(-150 \times 10^{-6}\right)\right\} - 0.807 \times (-4.60) - 0.912 \times (-11.56) = -10.14 \text{ MPa}$$

and

$$\sigma_{c(1),5(btm)} = E_{c(1),5}\left\{\begin{bmatrix}1 & y_{c(1),(btm)}\end{bmatrix}\boldsymbol{\varepsilon}_5 - \varepsilon_{sh(1),5}\right\} + F_{e(1),5,0}\sigma_{c(1),0(btm)} + F_{e(1),5,3}\sigma_{c(1),3(btm)} = 1.66 \text{ MPa}$$

For the concrete deck:

$$\sigma_{c(2),5(top)} = E_{c(2),5}\left\{\begin{bmatrix}1 & y_{c(2),(top)}\end{bmatrix}(\boldsymbol{\varepsilon}_5 - \boldsymbol{\varepsilon}_2) + F_{e(2),5,3}\sigma_{c(1),3(top)} - \varepsilon_{sh(2),5}\right\}$$

$$= 25{,}000\left\{[1 \quad 0]\left(\begin{bmatrix}-944.9 \times 10^{-6} \\ 0.516 \times 10^{-6}\end{bmatrix} - \begin{bmatrix}-680.9 \times 10^{-6} \\ 0.217 \times 10^{-6}\end{bmatrix}\right) + 0 - \left(-200 \times 10^{-6}\right)\right\} = -1.60 \text{ MPa}$$

and

$$\sigma_{c(2),5(btm)} = E_{c(2),5}\left\{\begin{bmatrix}1 & y_{c(2),(btm)}\end{bmatrix}(\boldsymbol{\varepsilon}_5 - \boldsymbol{\varepsilon}_2) + F_{e(2),5,3}\sigma_{c(1),3(btm)} - \varepsilon_{sh(2),5}\right\} = -0.48 \text{ MPa}$$

For the non-prestressed reinforcement:

$$\sigma_{s(1),5} = 200 \times 10^3 \times [1 \quad 75]\left(\begin{bmatrix}-944.9 \times 10^{-6} \\ 0.516 \times 10^{-6}\end{bmatrix} - \begin{bmatrix}-680.9 \times 10^{-6} \\ 0.217 \times 10^{-6}\end{bmatrix}\right) = -48.3 \text{ MPa}$$

$$\sigma_{s(2),5} = 200 \times 10^3 \times \begin{bmatrix} 1 & 210 \end{bmatrix} \begin{bmatrix} -944.9 \times 10^{-6} \\ 0.516 \times 10^{-6} \end{bmatrix} = -167.3 \text{ MPa}$$

$$\sigma_{s(3),5} = 200 \times 10^3 \begin{bmatrix} 1 & 1240 \end{bmatrix} \begin{bmatrix} -944.9 \times 10^{-6} \\ 0.516 \times 10^{-6} \end{bmatrix} = -60.9 \text{ MPa}$$

For the prestressing steel:

$$\sigma_{p(1),5} = E_{p(1)} \begin{bmatrix} 1 & y_{p(1)} \end{bmatrix} (\boldsymbol{\varepsilon}_5 - \boldsymbol{\varepsilon}_2) + E_{p(1)} \left(\varepsilon_{p(1).init} - \varepsilon_{p.rel(1).5}\right)$$

$$= 200 \times 10^3 \times \begin{bmatrix} 1 & 1030 \end{bmatrix} \left(\begin{bmatrix} -944.9 \times 10^{-6} \\ 0.516 \times 10^{-6} \end{bmatrix} - \begin{bmatrix} -680.9 \times 10^{-6} \\ 0.217 \times 10^{-6} \end{bmatrix} \right)$$

$$+ 200{,}000 \times 0.00625 \times (1 - 0.02) = 1234 \text{ MPa}$$

$$\sigma_{p(2),5} = E_{p(2)} \begin{bmatrix} 1 & y_{p(2)} \end{bmatrix} (\boldsymbol{\varepsilon}_5 - \boldsymbol{\varepsilon}_2) + E_{p(2)} \left(\varepsilon_{p(2).init} - \varepsilon_{p.rel(2).5}\right) = 1241 \text{ MPa}$$

The concrete stress resultants at step 5 at τ = 60 days are:

$$\mathbf{r}_{c(1),5} = \mathbf{D}_{c(1),5} \boldsymbol{\varepsilon}_5 + \mathbf{f}_{cr(1),5} - \mathbf{f}_{sh(1),5} - \mathbf{f}_{c(1).set,5}$$

$$= 34{,}000 \times \begin{bmatrix} 312{,}700 & 234.2 \times 10^6 \\ 234.2 \times 10^6 & 224.4 \times 10^9 \end{bmatrix} \begin{bmatrix} -944.9 \times 10^{-6} \\ 0.516 \times 10^{-6} \end{bmatrix}$$

$$+ \begin{bmatrix} 3061 \times 10^3 \\ 1925 \times 10^6 \end{bmatrix} - \begin{bmatrix} -1595 \times 10^3 \\ -1195 \times 10^6 \end{bmatrix} - \begin{bmatrix} -61.0 \times 10^3 \\ -66.6 \times 10^6 \end{bmatrix}$$

$$= \begin{bmatrix} -1217 \times 10^3 \text{ N} \\ -398.3 \times 10^6 \text{ Nmm} \end{bmatrix}$$

$$\mathbf{r}_{c(2),5} = \mathbf{D}_{c(2),5} \boldsymbol{\varepsilon}_5 + \mathbf{f}_{cr(2),5} - \mathbf{f}_{sh(2),5} - \mathbf{f}_{c(2).set,5}$$

$$= 25{,}000 \times \begin{bmatrix} 357{,}300 & 26.79 \times 10^6 \\ 26.79 \times 10^6 & 2685 \times 10^6 \end{bmatrix} \begin{bmatrix} -944.9 \times 10^{-6} \\ 0.516 \times 10^{-6} \end{bmatrix} + \begin{bmatrix} 0 \\ 0 \end{bmatrix}$$

$$- \begin{bmatrix} -1786 \times 10^3 \\ -133.9 \times 10^6 \end{bmatrix} - \begin{bmatrix} -5936 \times 10^3 \\ -441.5 \times 10^6 \end{bmatrix} = \begin{bmatrix} -371.7 \times 10^3 \text{ N} \\ -22.8 \times 10^6 \text{ Nmm} \end{bmatrix}$$

Step 6: Time analysis – SSM at age 30,000 days

The cross-sectional rigidities for step 6 are:

$R_{A,6} = 24{,}000 \times 10^6$ N;

$R_{B,6} = 10{,}580 \times 10^9$ Nmm;

$R_{I,6} = 9558 \times 10^{12}$ Nmm^2.

The matrix $\mathbf{F}_6$ is:

$$\mathbf{F}_6 = \begin{bmatrix} 81.35 \times 10^{-12}\ \text{N}^{-1} & -90.03 \times 10^{-15}\ \text{N}^{-1}\text{mm}^{-1} \\ -90.03 \times 10^{-15}\ \text{N}^{-1}\text{mm}^{-1} & 204.2 \times 10^{-18}\ \text{N}^{-1}\text{mm}^{-2} \end{bmatrix}$$

The effects of creep and shrinkage in this time period are:

$$\mathbf{f}_{cr,6} = F_{e(1),6,0}\mathbf{r}_{c(1),0} + F_{e(1),6,3}\mathbf{r}_{c(1),3} + F_{e(1),6,5}\mathbf{r}_{c(1),5} + F_{e(2),6,5}\mathbf{r}_{c(2),5}$$

$$= -1.995 \times \begin{bmatrix} -1862 \times 10^3 \\ -1504 \times 10^6 \end{bmatrix} - 0.419 \times \begin{bmatrix} -1708 \times 10^3 \\ -779.8 \times 10^6 \end{bmatrix}$$

$$- 1.906 \times \begin{bmatrix} -1217 \times 10^3 \\ -398.3 \times 10^6 \end{bmatrix} - 3.56 \times \begin{bmatrix} -371.7 \times 10^3 \\ -22.8 \times 10^6 \end{bmatrix}$$

$$= \begin{bmatrix} 8074 \times 10^3\ \text{N} \\ 4167 \times 10^6\ \text{Nmm} \end{bmatrix}$$

$$\mathbf{f}_{sh,6} = \begin{bmatrix} A_{c(1)} \\ B_{c(1)} \end{bmatrix} E_{c(1),6}\varepsilon_{sh(1),6} + \begin{bmatrix} A_{c(2)} \\ B_{c(2)} \end{bmatrix} E_{c(2),6}\varepsilon_{sh(2),6}$$

$$= \begin{bmatrix} 312{,}700 \\ 234.2 \times 10^6 \end{bmatrix} \times 38{,}000 \times \left(-400 \times 10^{-6}\right) + \begin{bmatrix} 357{,}300 \\ 26.79 \times 10^6 \end{bmatrix}$$

$$\times 30{,}000 \times \left(-600 \times 10^{-6}\right)$$

$$= \begin{bmatrix} -11{,}180 \times 10^3\ \text{N} \\ -4042 \times 10^6\ \text{Nmm} \end{bmatrix}$$

Relaxation of the steel tendon is accounted for using:

$$\mathbf{f}_{p.rel,6} = \sum_{i=1}^{2} \begin{bmatrix} A_{p(i)}E_{p(i)}\varepsilon_{p(i),init}\varphi_p(30{,}000, \sigma_{p(i),init}) \\ y_{p(i)}A_{p(i)}E_{p(i)}\varepsilon_{p(i),init}\varphi_p(30{,}000, \sigma_{p(i),init}) \end{bmatrix}$$

$$= \left(\begin{bmatrix} 800 \times 200{,}000 \\ 1030 \times 800 \times 200{,}000 \end{bmatrix} + \begin{bmatrix} 800 \times 200{,}000 \\ 1160 \times 800 \times 200{,}000 \end{bmatrix} \right)$$

$$\times 0.00625 \times 0.030 = \begin{bmatrix} 60.0 \times 10^3\ \text{N} \\ 65.7 \times 10^6\ \text{Nmm} \end{bmatrix}$$

The terms defining $\mathbf{f}_{set,6}$ are calculated using the same procedure adopted at time τ_4:

$$\mathbf{f}_{c(1).set,6} = \mathbf{f}_{c(1).set,3}\frac{E_{c(1),6}}{E_{c(1),3}} = \begin{bmatrix} -57.4 \times 10^3 \\ -62.7 \times 10^6 \end{bmatrix} \frac{38{,}000}{32{,}000} = \begin{bmatrix} -68.1 \times 10^3\ \text{N} \\ -74.5 \times 10^6\ \text{Nmm} \end{bmatrix}$$

$$\mathbf{f}_{p(1).set,6} = \mathbf{f}_{p(1).set,3} = \begin{bmatrix} -73.0 \times 10^3 \text{ N} \\ -75.2 \times 10^6 \text{ Nmm} \end{bmatrix}$$

$$\mathbf{f}_{p(2).set,6} = \mathbf{f}_{p(2).set,3} = \begin{bmatrix} -68.5 \times 10^3 \text{ N} \\ -79.5 \times 10^6 \text{ Nmm} \end{bmatrix}$$

$$\mathbf{f}_{c(2).set,6} = \mathbf{f}_{c(2).set,3} \frac{E_{c(2),6}}{E_{c(2),3}} = \begin{bmatrix} -4274 \times 10^3 \\ -317.9 \times 10^6 \end{bmatrix} \frac{30{,}000}{18{,}000}$$

$$= \begin{bmatrix} -7123 \times 10^3 \text{ N} \\ -529.8 \times 10^6 \text{ Nmm} \end{bmatrix}$$

$$\mathbf{f}_{s(1).set,6} = \mathbf{f}_{s(1).set,3} = \begin{bmatrix} -358.8 \times 10^3 \text{ N} \\ -26.9 \times 10^6 \text{ Nmm} \end{bmatrix}$$

$$\mathbf{f}_{set,6} = \mathbf{f}_{c(1).set,6} + \mathbf{f}_{p(1).set,6} + \mathbf{f}_{p(2).set,6} + \mathbf{f}_{c(2).set,6} + \mathbf{f}_{s(1).set,6}$$

$$= \begin{bmatrix} -7692 \times 10^3 \text{ N} \\ -785.9 \times 10^6 \text{ Nmm} \end{bmatrix}$$

The strain distribution at time τ_6 becomes:

$$\boldsymbol{\varepsilon}_6 = \mathbf{F}_6 \left(\mathbf{r}_{e,6} - \mathbf{f}_{cr,6} + \mathbf{f}_{sh,6} - \mathbf{f}_{p,init} + \mathbf{f}_{p.rel,6} + \mathbf{f}_{set,6} \right)$$

$$= \begin{bmatrix} 81.35 \times 10^{-12} & -90.03 \times 10^{-15} \\ -90.03 \times 10^{-15} & 204.2 \times 10^{-18} \end{bmatrix} \times \left(\begin{bmatrix} 0 \\ 1570 \times 10^6 \end{bmatrix} \right.$$

$$- \begin{bmatrix} 8074 \times 10^3 \\ 4167 \times 10^6 \end{bmatrix} + \begin{bmatrix} -11{,}180 \times 10^3 \\ -4042 \times 10^6 \end{bmatrix} - \begin{bmatrix} 2000 \times 10^3 \\ 2190 \times 10^6 \end{bmatrix} + \begin{bmatrix} 60.0 \times 10^3 \\ 65.7 \times 10^6 \end{bmatrix}$$

$$\left. + \begin{bmatrix} -7692 \times 10^3 \\ -785.9 \times 10^6 \end{bmatrix} \right) = \begin{bmatrix} -1490 \times 10^{-6} \\ 0.650 \times 10^{-6} \text{ mm}^{-1} \end{bmatrix}$$

At age 30,000 days, $\varepsilon_{r,6} = -1490 \times 10^{-6}$ and $\kappa_6 = 0.650 \times 10^{-6}$ mm^{-1} with strains in the top and bottom of the I-section equal to $\varepsilon_{6(top)} = -1393 \times 10^{-6}$ and $\varepsilon_{6(btm)} = -644.8 \times 10^{-6}$.
The strains in the top and bottom of the concrete deck are calculated as follows (measured from the time of setting):

$$\varepsilon_{6,(top)} = \begin{bmatrix} 1 & 0 \end{bmatrix} (\boldsymbol{\varepsilon}_6 - \boldsymbol{\varepsilon}_2) = -809.5 \times 10^{-6}$$

$$\varepsilon_{6,(btm)} = \begin{bmatrix} 1 & 150 \end{bmatrix} (\boldsymbol{\varepsilon}_6 - \boldsymbol{\varepsilon}_2) = -744.6 \times 10^{-6}$$

Final stresses at time 30,000 days are calculated below:

For the precast I-section: $\sigma_{c(1),6(top)} = -4.40$ MPa; $\sigma_{c(1),6(btm)} = 2.21$ MPa
For the concrete deck: $\sigma_{c(2),6(top)} = -0.59$ MPa; $\sigma_{c(2),6(btm)} = -2.63$ MPa

For the non-prestressed reinforcement:
$\sigma_{s(1),6} = -155.4$ MPa; $\sigma_{s(2),6} = -270.8$ MPa; $\sigma_{s(3),6} = -136.8$ MPa
For the prestressing steel: $\sigma_{p(1),6} = 1140$ MPa; $\sigma_{p(2),6} = 1151$ MPa.

The strains and stresses at times τ_0 and τ_6 are plotted in Fig. 5.22.

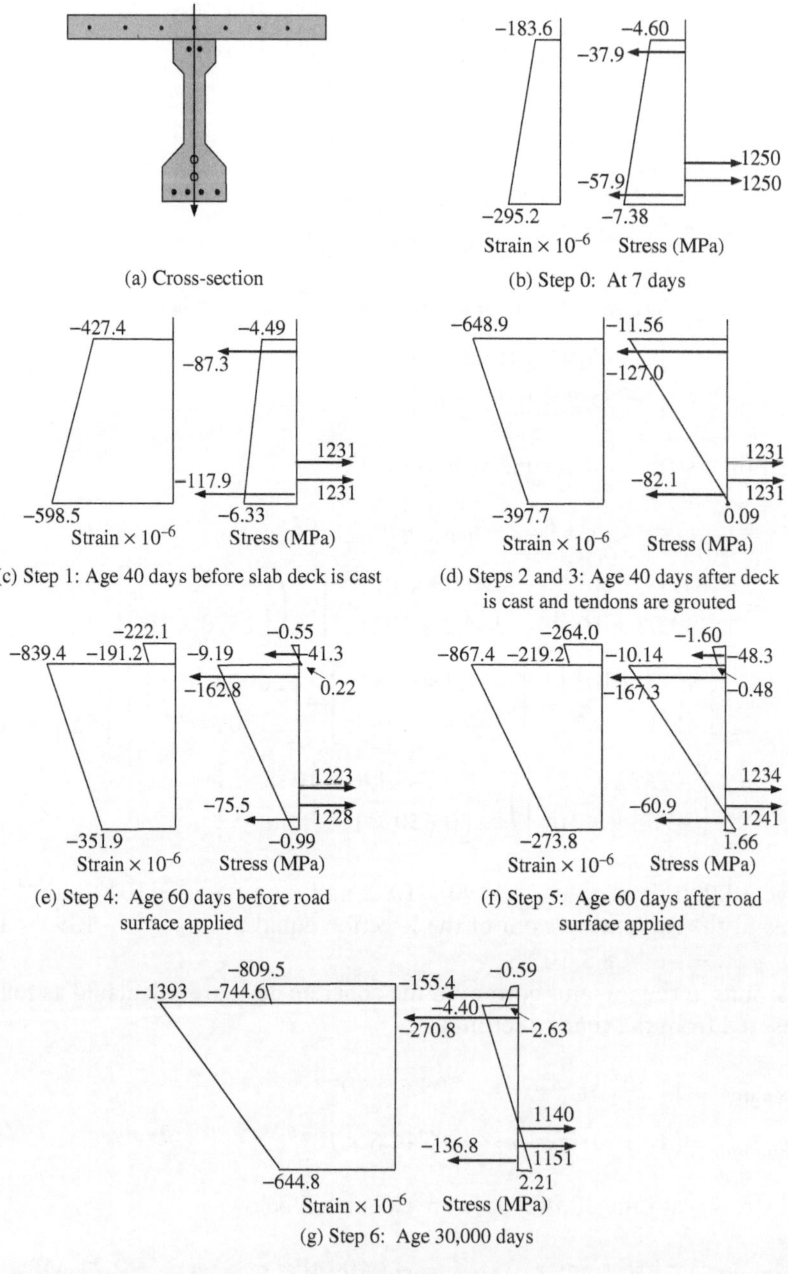

Figure 5.22 Strain and stress distributions for Example 5.10.

5.8 References

1. Gilbert, R.I. (1988). *Time Effects in Concrete Structures*. Elsevier Science Publishers, Amsterdam.
2. Ghali, A., Favre, R. and Eldbadry, M. (2002). *Concrete Structures: Stresses and Deformations*. Third edition, Spon Press, London.
3. Standards Australia (2004). *Australian Standard for Bridge Design – Part 5: Concrete (AS5100.5–2004)*. Sydney, Australia.
4. Bazant, Z.P. and Wu, S.T. (1973). Dirichlet series creep function for aging concrete. *Journal of the Engineering Mechanics Division, ASCE*, 99(2), 367–387.
5. Bazant, Z.P. and Wu, S.T. (1974). Rate-type creep law of aging concrete based on Maxwell chain. *Materials and Structures (RILEM)*, 7(37), 45–60.
6. Bradford, M.A. and Gilbert, R.I. (1992). Composite beams with partial interaction under sustained loads. *Journal of Structural Engineering, ASCE*, 118(7), 1871–1883.
7. Tarantino, A.M. and Dezi, L. (1992). Creep effects in composite beams with flexible shear connectors. *Journal of Structural Engineering, ASCE*, 118(8), 2063–2081.
8. Dezi, L. and Tarantino, A.M. (1993). Creep in composite continuous beams-I: theoretical treatment. *Journal of Structural Engineering, ASCE*, 119(7), 2095–2111.
9. Gilbert, R.I. and Bradford, M.A. (1995). Time-dependent behaviour of continuous composite beams at service loads. *Journal of Structural Engineering, ASCE*, 121(2), 319–327.
10. Ranzi, G. and Bradford, M.A. (2006). Analytical solutions for the time-dependent behaviour of composite beams with partial interaction. *International Journal of Solids and Structures*, 43, 3770–3793.
11. Ranzi, G. and Bradford, M.A. (2009). Analysis of composite beams with partial interaction using the direct stiffness approach for time effects. *International Journal for Numerical Methods in Engineering*, 78, 564–586.

6 Uncracked sections

Axial force and biaxial bending

6.1 Introduction

Methods of analysis to evaluate the short- and long-term response of concrete cross-sections subjected to axial force and biaxial bending are presented in this chapter. These methods are particularly useful for the analysis of reinforced, prestressed or composite steel-concrete columns in building structures. The concrete is assumed to remain uncracked and be able to carry any applied tension. The proposed formulation relies on the Euler–Bernoulli beam assumptions which require plane sections to remain plane and perpendicular to the member axis before and after deformation.

As for the case of axial loading and uniaxial bending (Chapter 5), the governing system of equations describing the structural response is expressed as a function of the unknown variables required to define the strain diagram. For the case of biaxial bending, the strain diagram is defined in terms of three variables, specifically the strain measured at the origin of the adopted orthogonal co-ordinate system ε_r and the curvatures calculated with respect to the two orthogonal x- and y-axes, κ_x and κ_y, respectively (as illustrated in Fig. 6.1). The external actions applied to the cross-section may include an axial force applied at the origin N_e and bending moments M_{xe} and M_{ye} applied with respect to the x- and y-axes, respectively. As shown in Fig. 6.1b, the flexural action is equivalent to an external moment M_e applied at an angle θ_M. The strain at an arbitrary point on the cross-section can be calculated from:

$$\varepsilon = \varepsilon_r + y\kappa_x - x\kappa_y \tag{6.1}$$

The negative sign placed before the third term in Eq. 6.1 is a consequence of the sign convention adopted for the applied moments (as illustrated in Fig. 6.1b).

The three variables defining the strain diagrams, i.e. ε_r, κ_x and κ_y, are determined by enforcing horizontal and rotational equilibrium on the cross-section:

$$N_e = N_i \tag{6.2a}$$

$$M_{xe} = M_{xi} \tag{6.2b}$$

$$M_{ye} = M_{yi} \tag{6.2c}$$

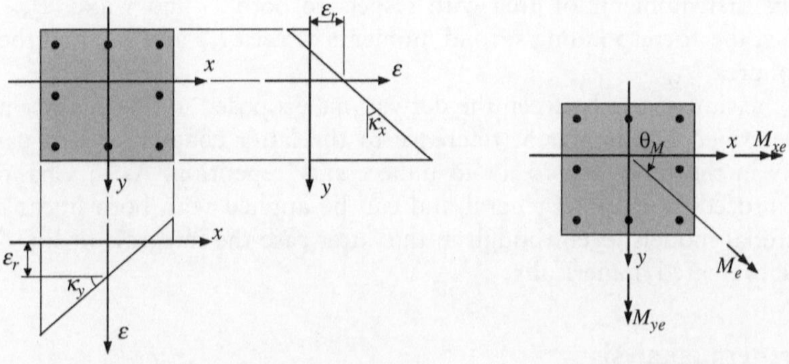

(a) Generic cross-section and strain diagram

(b) Applied moments

Figure 6.1 Typical section, strain distribution and sign convention for applied moments.

where N_i, M_{xi} and M_{yi} are the internal actions resisted by the cross-section, i.e. axial force and moments with respect to the x- and y-axes, respectively, and are defined as:

$$N_i = \int_A \sigma \, dA \tag{6.3a}$$

$$M_{xi} = \int_A y\sigma \, dA \tag{6.3b}$$

$$M_{yi} = -\int_A x\sigma \, dA \tag{6.3c}$$

The negative sign included in the expression for M_{yi} has been introduced to match the sign convention adopted for the external moment M_{ye} (see Fig. 6.1b).

It is useful to evaluate the angle of inclination θ_M of the applied moment M_e measured from the x-axis (adopting clockwise rotations to be positive) using:

$$\tan\theta_M = \frac{M_{ye}}{M_{xe}} \tag{6.4}$$

The values of the three unknowns (ε_r, κ_x and κ_y) are obtained by enforcing equilibrium (Eq. 6.2), with the strain over the section expressed by Eq. 6.1 (i.e. strain compatibility) and using the appropriate constitutive relationships in the calculation of the internal stresses and actions.

For ease of notation, the derivations presented in the following are applicable to generic reinforced and prestressed concrete cross-sections, while other types of cross-sections could be easily considered, for example, the composite sections considered in Chapter 5. The reinforced or prestressed concrete cross-section is assumed to contain m_s reinforcing bars and m_p prestressing tendons. Each steel reinforcing bar is identified by its area, elastic modulus and location with respect to both x- and y-axes, $A_{s(i)}$, $E_{s(i)}$, $x_{s(i)}$ and $y_{s(i)}$, respectively. The corresponding properties for the prestressing tendons are $A_{p(i)}$, $E_{p(i)}$, $x_{p(i)}$ and $y_{p(i)}$, respectively.

The required geometric properties of the concrete part of the cross-section are its area A_c, its first moments of area with respect to both x- and y-axes B_{xc} and B_{yc}, respectively, the corresponding second moments of area I_{xc} and I_{yc}, and the product moment of area I_{xyc}.

Due to the similarities between the derivations proposed in this chapter and those already described in Chapter 5, reference to the latter chapter will be carried out extensively in the following to avoid unnecessary repetition. As in Chapter 5, the proposed procedure is quite general and can be applied with both linear and non-linear material models, even though in the latter case the integrals of Eq. 6.3 might need to be evaluated numerically.

6.2 Short-term analysis

The short-term analysis of a reinforced or prestressed concrete cross-section subjected to axial force and biaxial bending is similar to that already described for uniaxial bending in Chapter 5, with the only addition being that moment equilibrium is enforced with respect to both the x- and y-axes. Based on linear-elastic material properties (Eq. 5.4), the axial equilibrium equation (Eq. 6.2a) can be expressed similarly to Eq. 5.6:

$$N_{e,0} = N_{i,0} = N_{c,0} + N_{s,0} + N_{p,0} \tag{6.5}$$

As in Chapter 5, the subscript '0' indicates that all variables are calculated at time τ_0 before any creep and shrinkage has taken place. In the case of axial force and biaxial bending, $N_{c,0}$, $N_{s,0}$ and $N_{p,0}$ may be expressed in terms of the strain distribution of Eq. 6.1 as:

$$N_{c,0} = \int_{A_c} \sigma_{c,0}\, dA = \int_{A_c} E_{c,0}\left(\varepsilon_{r,0} + y\kappa_{x,0} - x\kappa_{y,0}\right) dA$$

$$= A_c E_{c,0}\varepsilon_{r,0} + B_{xc}E_{c,0}\kappa_{x,0} - B_{yc}E_{c,0}\kappa_{y,0} \tag{6.6a}$$

$$N_{s,0} = \sum_{i=1}^{m_s} \left(A_{s(i)}E_{s(i)}\right)\left(\varepsilon_{r,0} + y_{s(i)}\kappa_{x,0} - x_{s(i)}\kappa_{y,0}\right)$$

$$= R_{A,s}\varepsilon_{r,0} + R_{Bx,s}\kappa_{x,0} - R_{By,s}\kappa_{y,0} \tag{6.6b}$$

$$N_{p,0} = \sum_{i=1}^{m_p} \left(A_{p(i)}E_{p(i)}\right)\left(\varepsilon_{r,0} + y_{p(i)}\kappa_{x,0} - x_{p(i)}\kappa_{y,0} + \varepsilon_{p(i),init}\right)$$

$$= R_{A,p}\varepsilon_{r,0} + R_{Bx,p}\kappa_{x,0} - R_{By,p}\kappa_{y,0} + \sum_{i=1}^{m_p}\left(A_{p(i)}E_{p(i)}\varepsilon_{p(i),init}\right) \tag{6.6c}$$

where $R_{A,s}$ and $R_{A,p}$ represent the axial rigidity of the non-prestressed steel and prestressing tendons, respectively, and $R_{Bx,s}$ and $R_{By,s}$ ($R_{Bx,p}$ and $R_{By,p}$) are the first moments of the rigidity of the steel reinforcement (prestressing tendons) with respect to the x- and y-axes, respectively, and are given by:

$$R_{A,s} = \sum_{i=1}^{m_s} \left(A_{s(i)} E_{s(i)}\right) \tag{6.7a}$$

$$R_{Bx,s} = \sum_{i=1}^{m_s} \left(y_{s(i)} A_{s(i)} E_{s(i)}\right) \tag{6.7b}$$

$$R_{By,s} = \sum_{i=1}^{m_s} \left(x_{s(i)} A_{s(i)} E_{s(i)}\right) \tag{6.7c}$$

$$R_{A,p} = \sum_{i=1}^{m_p} \left(A_{p(i)} E_{p(i)}\right) \tag{6.7d}$$

$$R_{Bx,p} = \sum_{i=1}^{m_p} \left(y_{p(i)} A_{p(i)} E_{p(i)}\right) \tag{6.7e}$$

$$R_{By,p} = \sum_{i=1}^{m_p} \left(x_{p(i)} A_{p(i)} E_{p(i)}\right) \tag{6.7f}$$

In a similar way, moment equilibrium with respect to the x-axis is enforced using Eq. 6.2b as:

$$M_{xe,0} = M_{xi,0} = M_{xc,0} + M_{xs,0} + M_{xp,0} \tag{6.8}$$

in which the internal actions are calculated using:

$$M_{xc,0} = \int_{A_c} y\sigma_{c,0}\, \mathrm{d}A = B_{xc} E_{c,0} \varepsilon_{r,0} + I_{xc} E_{c,0} \kappa_{x,0} - I_{xyc} E_{c,0} \kappa_{y,0} \tag{6.9a}$$

$$\begin{aligned} M_{xs,0} &= \sum_{i=1}^{m_s} \left(y_{s(i)} A_{s(i)} E_{s(i)}\right) \left(\varepsilon_{r,0} + y_{s(i)} \kappa_{x,0} - x_{s(i)} \kappa_{y,0}\right) \\ &= R_{Bx,s} \varepsilon_{r,0} + R_{Ix,s} \kappa_{x,0} - R_{Ixy,s} \kappa_{y,0} \end{aligned} \tag{6.9b}$$

$$\begin{aligned} M_{xp,0} &= \sum_{i=1}^{m_p} \left(y_{p(i)} A_{p(i)} E_{p(i)}\right) \left(\varepsilon_{r,0} + y_{p(i)} \kappa_{x,0} - x_{p(i)} \kappa_{y,0} + \varepsilon_{p(i),init}\right) \\ &= R_{Bx,p} \varepsilon_{r,0} + R_{Ix,p} \kappa_{x,0} - R_{Ixy,p} \kappa_{y,0} + \sum_{i=1}^{m_p} \left(y_{p(i)} A_{p(i)} E_{p(i)} \varepsilon_{p(i),init}\right) \end{aligned} \tag{6.9c}$$

and $R_{Ix,s}$ and $R_{Iy,s}$ ($R_{Ix,p}$ and $R_{Iy,p}$) are the rigidities associated with the second moments of area of the steel reinforcement (prestressing tendons) with respect to the x- and y-axes, respectively, and are given by:

$$R_{Ix,s} = \sum_{i=1}^{m_s} \left(y_{s(i)}^2 A_{s(i)} E_{s(i)} \right) \tag{6.10a}$$

$$R_{Ixy,s} = \sum_{i=1}^{m_s} \left(x_{s(i)} y_{s(i)} A_{s(i)} E_{s(i)} \right) \tag{6.10b}$$

$$R_{Ix,p} = \sum_{i=1}^{m_p} \left(y_{p(i)}^2 A_{p(i)} E_{p(i)} \right) \tag{6.10c}$$

$$R_{Ixy,p} = \sum_{i=1}^{m_p} \left(x_{p(i)} y_{p(i)} A_{p(i)} E_{p(i)} \right) \tag{6.10d}$$

Finally, considering flexural equilibrium with respect to the y-axis (Eq. 6.2c) gives:

$$M_{ye,0} = M_{yi,0} = M_{yc,0} + M_{ys,0} + M_{yp,0} \tag{6.11}$$

where:

$$M_{yc,0} = -\int_{A_c} x\sigma_{c,0}\, dA = -B_{yc}E_{c,0}\varepsilon_{r,0} - I_{xyc}E_{c,0}\kappa_{x,0} + I_{yc}E_{c,0}\kappa_{y,0} \tag{6.12a}$$

$$\begin{aligned} M_{ys,0} &= -\sum_{i=1}^{m_s} \left(x_{s(i)} A_{s(i)} E_{s(i)} \right) \left(\varepsilon_{r,0} + y_{s(i)}\kappa_{x,0} - x_{s(i)}\kappa_{y,0} \right) \\ &= -R_{By,s}\varepsilon_{r,0} - R_{Ixy,s}\kappa_{x,0} + R_{Iy,s}\kappa_{y,0} \end{aligned} \tag{6.12b}$$

$$\begin{aligned} M_{yp,0} &= -\sum_{i=1}^{m_p} \left(x_{p(i)} A_{p(i)} E_{p(i)} \right) \left(\varepsilon_{r,0} + y_{p(i)}\kappa_{x,0} - x_{p(i)}\kappa_{y,0} + \varepsilon_{p(i),init} \right) \\ &= -R_{By,p}\varepsilon_{r,0} - R_{Ixy,p}\kappa_{x,0} + R_{Iy,p}\kappa_{y,0} - \sum_{i=1}^{m_p} \left(x_{p(i)} A_{p(i)} E_{p(i)} \varepsilon_{p(i),init} \right) \end{aligned} \tag{6.12c}$$

and

$$R_{Iy,s} = \sum_{i=1}^{m_s} \left(x_{s(i)}^2 A_{s(i)} E_{s(i)} \right) \tag{6.13a}$$

$$R_{Iy,p} = \sum_{i=1}^{m_p} \left(x_{p(i)}^2 A_{p(i)} E_{p(i)} \right) \tag{6.13b}$$

Eqs 6.5, 6.8 and 6.11 can be therefore re-written as:

$$N_{e,0} = R_{A,0}\varepsilon_{r,0} + R_{Bx,0}\kappa_{x,0} - R_{By,0}\kappa_{y,0} + \sum_{i=1}^{m_p} \left(A_{p(i)} E_{p(i)} \varepsilon_{p(i),init} \right) \tag{6.14a}$$

$$M_{xe,0} = R_{Bx,0}\varepsilon_{r,0} + R_{Ix,0}\kappa_{x,0} - R_{Ixy,0}\kappa_{y,0} + \sum_{i=1}^{m_p} \left(y_{p(i)}A_{p(i)}E_{p(i)}\varepsilon_{p(i),init}\right) \tag{6.14b}$$

$$M_{ye,0} = -R_{By,0}\varepsilon_{r,0} - R_{Ixy,0}\kappa_{x,0} + R_{Iy,0}\kappa_{y,0} - \sum_{i=1}^{m_p} \left(x_{p(i)}A_{p(i)}E_{p(i)}\varepsilon_{p(i),init}\right) \tag{6.14c}$$

where the cross-sectional rigidities are given by:

$$R_{A,0} = A_c E_{c,0} + R_{A,s} + R_{A,p} \tag{6.15a}$$

$$R_{Ixy,0} = I_{xyc} E_{c,0} + R_{Ixy,s} + R_{Ixy,p} \tag{6.15b}$$

$$R_{Bx,0} = B_{xc} E_{c,0} + R_{Bx,s} + R_{Bx,p} \tag{6.15c}$$

$$R_{By,0} = B_{yc} E_{c,0} + R_{By,s} + R_{By,p} \tag{6.15d}$$

$$R_{Ix,0} = I_{xc} E_{c,0} + R_{Ix,s} + R_{Ix,p} \tag{6.15e}$$

$$R_{Iy,0} = I_{yc} E_{c,0} + R_{Iy,s} + R_{Iy,p} \tag{6.15f}$$

This governing system of equations (Eqs 6.14) can be expressed in compact form as:

$$\mathbf{r}_{e,0} = \mathbf{D}_0 \boldsymbol{\varepsilon}_0 + \mathbf{f}_{p,init} \tag{6.16}$$

and using notation similar to that adopted in Chapter 5:

$$\mathbf{r}_{e,0} = \begin{bmatrix} N_{e,0} \\ M_{xe,0} \\ M_{ye,0} \end{bmatrix} \tag{6.17a}$$

$$\mathbf{D}_0 = \begin{bmatrix} R_{A,0} & R_{Bx,0} & -R_{By,0} \\ R_{Bx,0} & R_{Ix,0} & -R_{Ixy,0} \\ -R_{By,0} & -R_{Ixy,0} & R_{Iy,0} \end{bmatrix} \tag{6.17b}$$

$$\boldsymbol{\varepsilon}_0 = \begin{bmatrix} \varepsilon_{r,0} \\ \kappa_{x,0} \\ \kappa_{y,0} \end{bmatrix} \tag{6.17c}$$

$$\mathbf{f}_{p,init} = \sum_{i=1}^{m_p} \begin{bmatrix} A_{p(i)}E_{p(i)} \\ y_{p(i)}A_{p(i)}E_{p(i)} \\ -x_{p(i)}A_{p(i)}E_{p(i)} \end{bmatrix} \varepsilon_{p(i),init} \tag{6.17d}$$

The unknown strain vector $\boldsymbol{\varepsilon}_0$ is obtained by solving Eq. 6.16 and is given by:

$$\boldsymbol{\varepsilon}_0 = \mathbf{D}_0^{-1}\left(\mathbf{r}_{e,0} - \mathbf{f}_{p,init}\right) = \mathbf{F}_0\left(\mathbf{r}_{e,0} - \mathbf{f}_{p,init}\right) \tag{6.18}$$

in which

$$\mathbf{F}_0 = \frac{1}{R_0}\begin{bmatrix} R_{Ix,0}R_{Iy,0} - R_{Ixy,0}^2 & R_{By,0}R_{Ixy,0} - R_{Bx,0}R_{Iy,0} & R_{By,0}R_{Ix,0} - R_{Bx,0}R_{Ixy,0} \\ R_{By,0}R_{Ixy,0} - R_{Bx,0}R_{Iy,0} & R_{A,0}R_{Iy,0} - R_{By,0}^2 & R_{A,0}R_{Ixy,0} - R_{Bx,0}R_{By,0} \\ R_{By,0}R_{Ix,0} - R_{Bx,0}R_{Ixy,0} & R_{A,0}R_{Ixy,0} - R_{Bx,0}R_{By,0} & R_{A,0}R_{Ix,0} - R_{Bx,0}^2 \end{bmatrix} \tag{6.19a}$$

and

$$R_0 = R_{A,0}\left(R_{Ix,0}R_{Iy,0} - R_{Ixy,0}^2\right) - R_{Bx,0}^2 R_{Iy,0} + 2R_{Bx,0}R_{By,0}R_{Ixy,0} - R_{By,0}^2 R_{Ix,0} \tag{6.19b}$$

The strain at any point (x, y) is obtained by substituting Eq. 6.18 into Eq. 6.1:

$$\varepsilon_0 = \varepsilon_{r,0} + y\kappa_{x,0} - x\kappa_{y,0} = \begin{bmatrix} 1 & y & -x \end{bmatrix} \boldsymbol{\varepsilon}_0 \tag{6.20}$$

and the stress at any point can then be determined from the calculated strain (Eq. 6.20) using the appropriate constitutive relationship (Eqs 5.4).

By setting Eq. 6.20 equal to zero, the equation of the line (in the x-y plane) defining the neutral axis may be expressed as

$$y = \frac{\kappa_{y,0}}{\kappa_{x,0}}x - \frac{\varepsilon_{r,0}}{\kappa_{x,0}} \tag{6.21}$$

Equation 6.21 is not defined when $\kappa_{x,0} = 0$, in which case the problem degenerates to one of axial force and uniaxial bending about the y-axis and can be solved using the methods of Chapter 5.

The slope of the neutral axis at time τ_0 may be obtained by differentiating Eq. 6.21 and (adopting clockwise rotations to be positive similarly to $\theta_{M,0}$) is expressed as a function of the applied moments and flexural rigidities:

$$\tan\theta_{NA,0} = \frac{\kappa_{y,0}}{\kappa_{x,0}} \tag{6.22}$$

The neutral axis (Eq. 6.21) intersects the y-axis at:

$$y(x=0) = -\frac{\varepsilon_{r,0}}{\kappa_{x,0}} \tag{6.23}$$

If the cross-section is doubly-symmetric (as is often the case in practice), the expression for $\mathbf{F}_0$ can be simplified by placing the origin of the coordinate system at the centroid of the cross-section and specifying the orthogonal axes to be parallel to the axes of symmetry (leading to $R_{Ixy,0} = R_{Bx,0} = R_{By,0} = 0$). Therefore:

$$\mathbf{F}_0 = \frac{1}{R_{A,0}R_{Ix,0}R_{Iy,0}} \begin{bmatrix} R_{Ix,0}R_{Iy,0} & 0 & 0 \\ 0 & R_{A,0}R_{Iy,0} & 0 \\ 0 & 0 & R_{A,0}R_{Ix,0} \end{bmatrix} \tag{6.24}$$

Without prestressing (usually the case for columns), using Eq. 6.24 in Eq. 6.18, the expression for the strain at a point ε_0 simplifies to:

$$\varepsilon_0 = \varepsilon_{r,0} + y\kappa_{x,0} - x\kappa_{y,0} = \frac{N_{e,0}}{R_{A,0}} + y\frac{M_{xe,0}}{R_{Ix,0}} - x\frac{M_{ye,0}}{R_{Iy,0}} \tag{6.25}$$

In this particular case (i.e. no prestressing), the equation of the line (in the x-y plane) defining the neutral axis may be expressed in terms of external loads as:

$$y = \frac{R_{Ix,0}}{R_{Iy,0}}\frac{M_{ye,0}}{M_{xe,0}}x - \frac{R_{Ix,0}}{R_{A,0}}\frac{N_{e,0}}{M_{xe,0}} \tag{6.26}$$

with corresponding slope and intersect with the y-axis calculated using:

$$\tan\theta_{NA,0} = \frac{R_{Ix,0}}{R_{Iy,0}}\frac{M_{ye,0}}{M_{xe,0}} = \frac{R_{Ix,0}}{R_{Iy,0}}\tan\theta_{M,0} \tag{6.27a}$$

and

$$y(x=0) = -\frac{R_{Ix,0}}{R_{A,0}}\frac{N_{e,0}}{M_{xe,0}} \tag{6.27b}$$

Evidently, the angle of the applied moment $M_{e,0}$ (i.e. $\theta_{M,0}$) and of the neutral axis (i.e. $\theta_{NA,0}$) coincide only when the flexural rigidities calculated with respect to both the x- and y-axes are identical, i.e. when $R_{Ix,0} = R_{Iy,0}$ (Eq. 6.27a). In all other cases, these two angles differ from each other (Fig. 6.2). Moreover, the angle of the neutral axis $\theta_{NA,0}$ may vary with time, as will be seen subsequently.

The cross-sectional rigidities could have also been calculated from the transformed section, where the areas of the reinforcing bars and bonded prestressing tendons ($A_{s(i)}$ and $A_{p(i)}$, respectively) are transformed into equivalent areas of concrete ($n_{s(i),0}A_{s(i)}$ and $n_{p(i),0}A_{p(i)}$, respectively), where $n_{s(i),0} = E_{s(i)}/E_{c,0}$ is the modular ratio of the i-th steel bar and $n_{p(i),0} = E_{p(i)}/E_{c,0}$ is the modular ratio of the i-th tendon. In this case, the cross-sectional rigidities are given by:

$$R_{A,0} = A_0E_{c,0} \tag{6.28a}$$

$$R_{Bx,0} = B_{x,0}E_{c,0} \tag{6.28b}$$

$$R_{By,0} = B_{y,0}E_{c,0} \tag{6.28c}$$

$$R_{Ix,0} = I_{x,0}E_{c,0} \tag{6.28d}$$

$$R_{Iy,0} = I_{y,0}E_{c,0} \tag{6.28e}$$

$$R_{Ixy,0} = I_{xy,0}E_{c,0} \tag{6.28f}$$

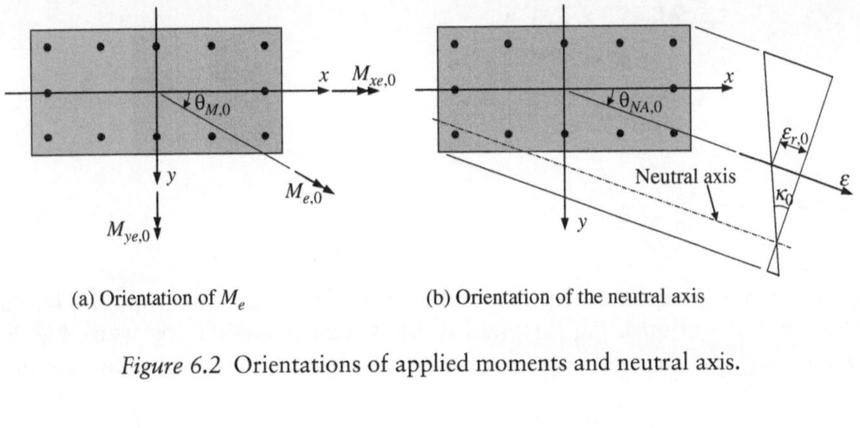

(a) Orientation of M_e

(b) Orientation of the neutral axis

Figure 6.2 Orientations of applied moments and neutral axis.

where A_0 is the area of the transformed concrete section; $B_{x,0}$ and $I_{x,0}$ are the transformed first and second moments of area calculated with respect to the x-axis, respectively; $B_{y,0}$ and $I_{y,0}$ represent the corresponding moments of area with respect to the y-axis; and $I_{xy,0}$ is the product moment of area. Using Eqs 6.28, the matrix $\mathbf{F}_0$, previously introduced in Eq. 6.19a, can be expressed as:

$$\mathbf{F}_0 = \frac{1}{S_0 E_{c,0}} \begin{bmatrix} I_{x,0}I_{y,0} - I_{xy,0}^2 & B_{y,0}I_{xy,0} - B_{x,0}I_{y,0} & B_{y,0}I_{x,0} - B_{x,0}I_{xy,0} \\ B_{y,0}I_{xy,0} - B_{x,0}I_{y,0} & A_0 I_{y,0} - B_{y,0}^2 & A_0 I_{xy,0} - B_{x,0}B_{y,0} \\ B_{y,0}I_{x,0} - B_{x,0}I_{xy,0} & A_0 I_{xy,0} - B_{x,0}B_{y,0} & A_0 I_{x,0} - B_{x,0}^2 \end{bmatrix} \tag{6.29a}$$

and

$$S_0 = A_0\left(I_{x,0}I_{y,0} - I_{xy,0}^2\right) - B_{x,0}^2 I_{y,0} + 2B_{x,0}B_{y,0}I_{xy,0} - B_{y,0}^2 I_{x,0} \tag{6.29b}$$

Considering the particular case of a doubly-symmetric section with the origin at the centroid of the section and orthogonal axes parallel to the axes of symmetry (for which $R_{Ixy,0} = R_{Bx,0} = R_{By,0} = 0$), $\mathbf{F}_0$ simplifies to:

$$\mathbf{F}_0 = \frac{1}{A_0 I_{x,0} I_{y,0} E_{c,0}} \begin{bmatrix} I_{x,0}I_{y,0} & 0 & 0 \\ 0 & A_0 I_{y,0} & 0 \\ 0 & 0 & A_0 I_{x,0} \end{bmatrix} \tag{6.30}$$

Without prestressing effects, the strain at any point (x, y) can be calculated using:

$$\varepsilon_0 = \varepsilon_{r,0} + y\kappa_{x,0} - x\kappa_{y,0} = \frac{N_{e,0}}{A_0 E_{c,0}} + y\frac{M_{xe,0}}{I_{x,0}E_{c,0}} - x\frac{M_{ye,0}}{I_{y,0}E_{c,0}} \tag{6.31}$$

By setting $\varepsilon_0 = 0$ in Eq. 6.31, the equation of the neutral axis, its slope and its intersection with the y-axis are obtained as functions of the transformed second moments of area and are given by:

$$y = \frac{I_{x,0}}{I_{y,0}}\frac{M_{ye,0}}{M_{xe,0}}x - \frac{I_{x,0}}{A_0}\frac{N_{e,0}}{M_{xe,0}} \tag{6.32a}$$

$$\tan\theta_{NA,0} = \frac{I_{x,0}}{I_{y,0}}\tan\theta_{M,0} \tag{6.32b}$$

$$y(x=0) = -\frac{I_{x,0}}{A_0}\frac{N_{e,0}}{M_{xe,0}} \tag{6.32c}$$

As discussed in Chapter 5, the calculation of the cross-sectional rigidities based on the transformed section using Eqs 6.28 is equivalent to calculation using Eqs 6.15. Either approach can be used depending on the preference of the structural designer.

Example 6.1

The doubly-symmetric reinforced concrete cross-section shown in Fig. 6.3 is subjected to biaxial bending, with $M_{xe} = 5\,\text{kNm}$ and $M_{ye} = 10\,\text{kNm}$. In this example, the axial force is zero (i.e. $N_{e,0} = 0$ kN). The stress and strain distributions are to be determined at time τ_0, assuming the origin of the coordinate system is located at the centroid of the cross-section and the x- and y-axes are parallel to the axes of symmetry (i.e. $d_{x.ref} = b/2$ and $d_{y.ref} = D/2$ measured from the left and top of the section, respectively, as shown in Fig. 6.3). Both concrete and reinforcement are assumed to be linear-elastic with $E_{c,0} =$ 32 GPa and $E_s = 200$ GPa. The modular ratio for the reinforcing steel is therefore $n_{s,0} = n_{s(j),0} = E_s/E_{c,0} = 6.25$ (with $j = 1, \ldots, 4$).

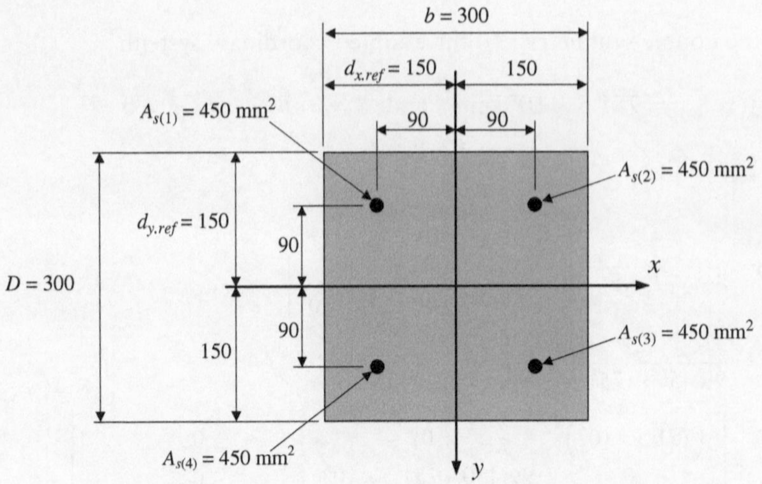

Figure 6.3 Reinforced concrete cross-section for Example 6.1 (dimensions in mm).

Based on the adopted reference system the coordinates of each of the four steel reinforcing bars are:

$$x_{s(1)} = -90 \text{ mm}$$
$$y_{s(1)} = -90 \text{ mm}$$
$$x_{s(2)} = 90 \text{ mm}$$
$$y_{s(2)} = -90 \text{ mm}$$
$$x_{s(3)} = 90 \text{ mm}$$
$$y_{s(3)} = 90 \text{ mm}$$
$$x_{s(4)} = -90 \text{ mm}$$
$$y_{s(4)} = 90 \text{ mm}$$

For this reinforced concrete section, the vectors $\mathbf{r}_{e,0}$ and $\mathbf{f}_{p,init}$ are

$$\mathbf{r}_{e,0} = \begin{bmatrix} 0\,\text{N} \\ 5 \times 10^6\,\text{Nmm} \\ 10 \times 10^6\,\text{Nmm} \end{bmatrix} \text{ and } \mathbf{f}_{p,init} = \begin{bmatrix} 0\,\text{N} \\ 0\,\text{Nmm} \\ 0\,\text{Nmm} \end{bmatrix}$$

Based on the transformed section, the relevant cross-sectional properties are calculated as:

$$A_0 = bD + (n_{s,0} - 1)[A_{s(1)} + A_{s(2)} + A_{s(3)} + A_{s(4)}]$$
$$= 300 \times 300 + (6.25 - 1) \times (4 \times 450) = 99{,}450 \text{ mm}^2;$$

$$I_{x,0} = \frac{bD^3}{12} + bD\left(\frac{D}{2} - d_{y.ref}\right)^2 + (n_{s,0} - 1)\left[A_{s(1)}y_{s(1)}^2 + A_{s(2)}y_{s(2)}^2 + A_{s(3)}y_{s(3)}^2\right.$$
$$\left. + A_{s(4)}y_{s(4)}^2\right] = \frac{300 \times 300^3}{12} + 300^2(150 - 150)^2$$
$$+ 5.25 \times [2 \times 450 \times (-90)^2 + 2 \times 450 \times 90^2] = 751.5 \times 10^6 \text{ mm}^4.$$

From the double symmetry and the adopted coordinate system:

$$I_{y,0} = I_{x,0} = 751.5 \times 10^6 \text{ mm}^4 \text{ and } B_{x,0} = B_{y,0} = I_{xy,0} = 0.$$

From Eq. 6.30:

$$\mathbf{F}_0 = \frac{1}{A_0 I_{x,0} I_{y,0} E_{c,0}} \begin{bmatrix} I_{x,0}I_{y,0} & 0 & 0 \\ 0 & A_0 I_{y,0} & 0 \\ 0 & 0 & A_0 I_{x,0} \end{bmatrix}$$
$$= \frac{1}{99{,}450 \times (751.5 \times 10^6)^2 \times 32{,}000}$$
$$\times \begin{bmatrix} (751.5 \times 10^6)^2 & 0 & 0 \\ 0 & 99{,}450 \times 751.5 \times 10^6 & 0 \\ 0 & 0 & 99{,}450 \times 751.5 \times 10^6 \end{bmatrix}$$
$$= \begin{bmatrix} 3.142 \times 10^{-10}\text{N}^{-1} & 0 & 0 \\ 0 & 4.158 \times 10^{-14}\text{N}^{-1}\text{mm}^{-2} & 0 \\ 0 & 0 & 4.158 \times 10^{-14}\text{N}^{-1}\text{mm}^{-2} \end{bmatrix}$$

The strain $\boldsymbol{\varepsilon}_0$ is then determined based on Eq. 6.18:

$$\boldsymbol{\varepsilon}_0 = \mathbf{F}_0\left(\mathbf{r}_{e,0} - \mathbf{f}_{p,init}\right) = \begin{bmatrix} 3.142 \times 10^{-10} & 0 & 0 \\ 0 & 4.158 \times 10^{-14} & 0 \\ 0 & 0 & 4.158 \times 10^{-14} \end{bmatrix}$$
$$\times \left\{\begin{bmatrix} 0 \\ 5 \times 10^6 \\ 10 \times 10^6 \end{bmatrix} - \begin{bmatrix} 0 \\ 0 \\ 0 \end{bmatrix}\right\}$$
$$= \begin{bmatrix} 0 \\ 0.208 \times 10^{-6}\text{mm}^{-1} \\ 0.416 \times 10^{-6}\text{mm}^{-1} \end{bmatrix}$$

The strain at the origin and the curvatures with respect to the x- and y-axes, respectively, are:

$$\varepsilon_{r,0} = 0$$
$$\kappa_{x,0} = 0.208 \times 10^{-6}\ \text{mm}^{-1}$$
$$\kappa_{y,0} = 0.416 \times 10^{-6}\ \text{mm}^{-1}$$

and, from Eq. 6.1, the strains at the corners of the cross-section become:

$$\varepsilon_0(x = -b/2, y = -D/2) = (0 - 150 \times 0.208 - (-150) \times 0.416) \times 10^{-6} = 31.2 \times 10^{-6}$$
$$\varepsilon_0(x = b/2, y = -D/2) = (0 - 150 \times 0.208 - 150 \times 0.416) \times 10^{-6} = -93.6 \times 10^{-6}$$
$$\varepsilon_0(x = b/2, y = D/2) = (0 + 150 \times 0.208 - 150 \times 0.416) \times 10^{-6} = -31.2 \times 10^{-6}$$
$$\varepsilon_0(x = -b/2, y = D/2) = (0 + 150 \times 0.208 - (-150) \times 0.416) \times 10^{-6} = 93.6 \times 10^{-6}$$

The stresses in the concrete at the four corners of the cross-section are obtained from Eq. 5.16a:

$$\sigma_{c,0(top\text{-}left)} = E_{c,0}\varepsilon_{0(top\text{-}left)} = 32 \times 10^3 \times 31.2 \times 10^{-6} = 1.00\ \text{MPa}$$
$$\sigma_{c,0(top\text{-}right)} = -2.99\ \text{MPa}$$
$$\sigma_{c,0(btm\text{-}right)} = -1.00\ \text{MPa}$$
$$\sigma_{c,0(btm\text{-}left)} = 2.99\ \text{MPa}$$

Similarly, the stress in each reinforcement bar is calculated using Eq. 5.16b:

$$\sigma_{s(1),0} = E_s \left[1 \;\; y_{s(1)} \;\; -x_{s(1)} \right] \boldsymbol{\varepsilon}_0$$
$$= 200 \times 10^3 \times [1 - 90 - (-90)] \times \begin{bmatrix} 0 \\ 0.208 \times 10^{-6} \\ 0.416 \times 10^{-6} \end{bmatrix}$$
$$= 3.7\ \text{MPa};$$
$$\sigma_{s(2),0} = -11.2\ \text{MPa}$$
$$\sigma_{s(3),0} = -3.7\ \text{MPa}$$
$$\sigma_{s(4),0} = 11.2\ \text{MPa}$$

The slopes of both applied moment M_e and of the neutral axis are calculated as (Eqs 6.4 and 6.32b):

$$\tan\theta_{M,0} = \frac{M_{ye,0}}{M_{xe,0}} = \frac{10 \times 10^6}{5 \times 10^6} = 2.0 \text{ from which } \theta_{M,0} = 63.43^\circ \text{ and}$$

$$\tan\theta_{NA,0} = \frac{I_{x,0}}{I_{y,0}} \tan\theta_{M,0} = \frac{751.5 \times 10^6}{751.5 \times 10^6} \times 2.0 = 2.0$$

and therefore $\theta_{NA,0} = 63.43^\circ$.

As expected, $\theta_{M,0} = \theta_{NA,0}$ because $I_{x,0} = I_{y,0}$.

The strain and stress distributions are plotted in Fig. 6.4.
The cross-sectional rigidities could have also been calculated based on Eqs 6.15:

$$R_{A,0} = 3182 \times 10^6 \text{ N}$$
$$R_{Ix,0} = R_{Iy,0} = 24.05 \times 10^{12} \text{ Nmm}^2$$
$$R_{Bx,0} = R_{By,0} = R_{Ixy,0} = 0$$

In this case, $\mathbf{F}_0$ is obtained from Eq. 6.24 and is identical to that calculated above.

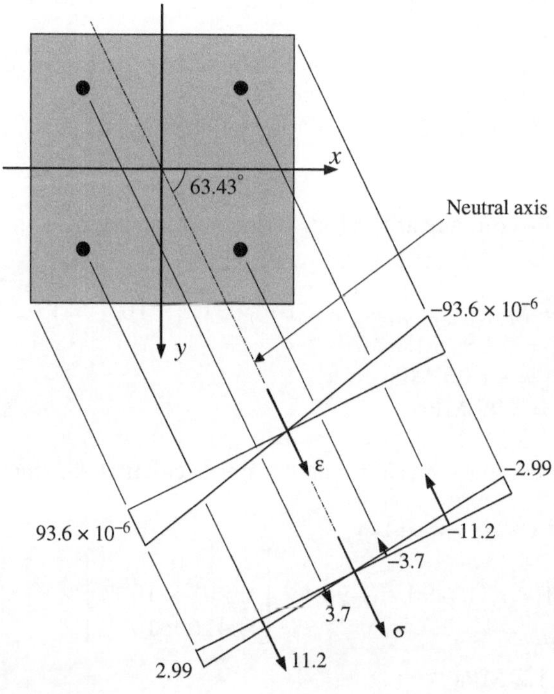

Figure 6.4 Initial strain and stress diagrams at time τ_0 for Example 6.1 (all units in mm, MPa).

Example 6.2

The strain and stress distributions on the reinforced concrete cross-section shown in Fig. 6.5 are to be determined at time τ_0 when the applied actions, $N_e = -500$ kN (compression), $M_{xe} = 25$ kNm and $M_{ye} = 25$ kNm, are first applied. The origin of the coordinate system is taken at the centroid of the section and the x- and y-axes are assumed to be parallel to the axes of symmetry (i.e. $d_{x.ref} = b/2$ and $d_{y.ref} = D/2$ measured from the left and top of the section, respectively).

The material behaviour is linear-elastic with $E_{c,0} = 32$ GPa and $E_s = 200$ GPa. The modular ratio for all reinforcement is therefore $n_{s,0} = E_s/E_{c,0} = 6.25$.

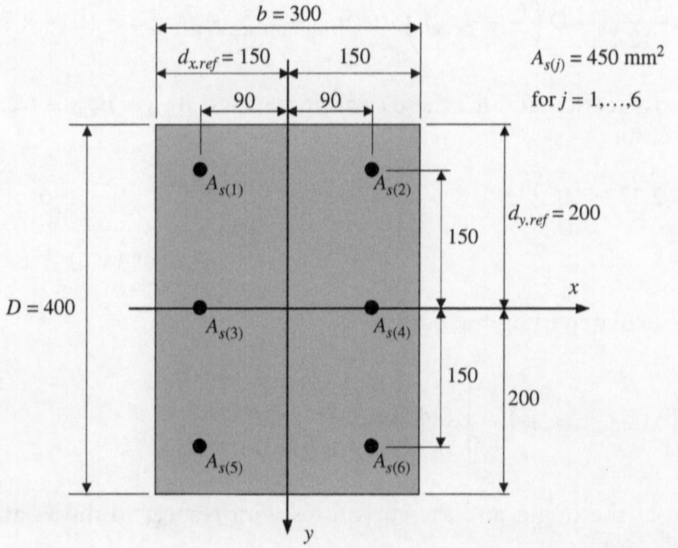

Figure 6.5 Reinforced concrete cross-section for Example 6.2 (dimensions in mm).

Based on the adopted reference system, the coordinates of the steel reinforcement are:

$x_{s(1)} = -90$ mm
$y_{s(1)} = -150$ mm
$x_{s(2)} = 90$ mm
$y_{s(2)} = -150$ mm
$x_{s(3)} = -90$ mm
$y_{s(3)} = 0$ mm
$x_{s(4)} = 90$ mm
$y_{s(4)} = 0$ mm
$x_{s(5)} = -90$ mm
$y_{s(5)} = 150$ mm
$x_{s(6)} = 90$ mm
$y_{s(6)} = 150$ mm

The vectors of external loads and prestressing are:

$$\mathbf{r}_{e,0} = \begin{bmatrix} -500 \times 10^3\,\text{N} \\ 25 \times 10^6\,\text{Nmm} \\ 25 \times 10^6\,\text{Nmm} \end{bmatrix} \quad \text{and} \quad \mathbf{f}_{p,init} = \begin{bmatrix} 0\,\text{N} \\ 0\,\text{Nmm} \\ 0\,\text{Nmm} \end{bmatrix}$$

The properties of the transformed cross-section are:

$$A_0 = \text{bD} + (n_{s,0} - 1)\sum_{i=1}^{6} A_{s(i)} = 134.1 \times 10^3\ \text{mm}^2$$

$$I_{x,0} = \frac{bD^3}{12} + bD\left(\frac{D}{2} - d_{y.ref}\right)^2 + (n_{s,0} - 1)\sum_{i=1}^{6} A_{s(i)}y_{s(i)}^2 = 1812 \times 10^6 \text{ mm}^4$$

$$I_{y,0} = \frac{Db^3}{12} + bD\left(\frac{b}{2} - d_{x.ref}\right)^2 + (n_{s,0} - 1)\sum_{i=1}^{6} A_{s(i)}x_{s(i)}^2 = 1014 \times 10^6 \text{ mm}^4$$

For this cross-section and for these coordinate axes: $B_{x,0} = B_{y,0} = I_{xy,0} = 0$. From Eq. 6.30:

$$\mathbf{F}_0 = \begin{bmatrix} 2.329 \times 10^{-10}\text{N}^{-1} & 0 & 0 \\ 0 & 1.724 \times 10^{-14}\text{N}^{-1}\text{mm}^{-2} & 0 \\ 0 & 0 & 3.080 \times 10^{-14}\text{N}^{-1}\text{mm}^{-2} \end{bmatrix}$$

and $\boldsymbol{\varepsilon}_0$ is calculated using Eq. 6.18:

$$\boldsymbol{\varepsilon}_0 = \mathbf{F}_0\left(\mathbf{r}_{e,0} - \mathbf{f}_{p,init}\right) = \begin{bmatrix} -116.4 \times 10^{-6} \\ 0.431 \times 10^{-6}\,\text{mm}^{-1} \\ 0.770 \times 10^{-6}\,\text{mm}^{-1} \end{bmatrix}$$

The strain at the origin and the curvatures with respect to the x- and y-axes, respectively, are:

$$\varepsilon_{r,0} = -116.4 \times 10^{-6}$$

$$\kappa_{x,0} = 0.431 \times 10^{-6} \text{ mm}^{-1}$$

$$\kappa_{y,0} = 0.770 \times 10^{-6} \text{ mm}^{-1}.$$

The strains at the four corners of the cross-section are (Eq. 6.1):

$$\varepsilon_{0(top\text{-}left)}(x = -b/2, y = -D/2) = -87.2 \times 10^{-6}$$
$$\varepsilon_{0(top\text{-}right)}(x = b/2, y = -D/2) = -318.1 \times 10^{-6}$$
$$\varepsilon_{0(btm\text{-}right)}(x = b/2, y = D/2) = -145.7 \times 10^{-6}$$
$$\varepsilon_{0(btm\text{-}left)}(x = -b/2, y = D/2) = 85.2 \times 10^{-6}$$

and the concrete and steel stress distributions are calculated using Eq. 5.16:

$$\sigma_{c,0(top\text{-}left)} = E_{c,0}\varepsilon_{0(top\text{-}left)} = -2.79 \text{ MPa}$$
$$\sigma_{c,0(top\text{-}right)} = -10.18 \text{ MPa}$$
$$\sigma_{c,0(btm\text{-}right)} = -4.66 \text{ MPa}$$
$$\sigma_{c,0(btm\text{-}left)} = 2.73 \text{ MPa}$$
$$\sigma_{s(1),0} = E_s\left[1\ \ y_{s(1)}\ \ -x_{s(1)}\right]\boldsymbol{\varepsilon}_0$$
$$= 200 \times 10^3 \times [1 - 150 - (-90)] \times \begin{bmatrix} -116.4 \times 10^{-6} \\ 0.431 \times 10^{-6} \\ 0.770 \times 10^{-6} \end{bmatrix}$$
$$= -22.4 \text{ MPa};$$
$$\sigma_{s(2),0} = -50.1 \text{ MPa}$$

$\sigma_{s(3),0} = -9.4$ MPa
$\sigma_{s(4),0} = -37.1$ MPa
$\sigma_{s(5),0} = 3.50$ MPa
$\sigma_{s(6),0} = -24.2$ MPa

The slopes of both applied moment $M_{e,0}$ and of the neutral axis are (Eqs 6.4 and 6.32b):

$$\tan\theta_{M,0} = \frac{M_{ye,0}}{M_{xe,0}} = \frac{25 \times 10^6}{25 \times 10^6} = 1.0 \text{ and therefore } \theta_{M,0} = 45° \text{ and}$$

$$\tan\theta_{NA,0} = \frac{I_{x,0}}{I_{y,0}} \tan\theta_M = \frac{1812 \times 10^6}{1014 \times 10^6} \times 1 = 1.79$$

and therefore $\theta_{NA,0} = 60.76°$.

The calculated stresses and strains are shown in Fig. 6.6.

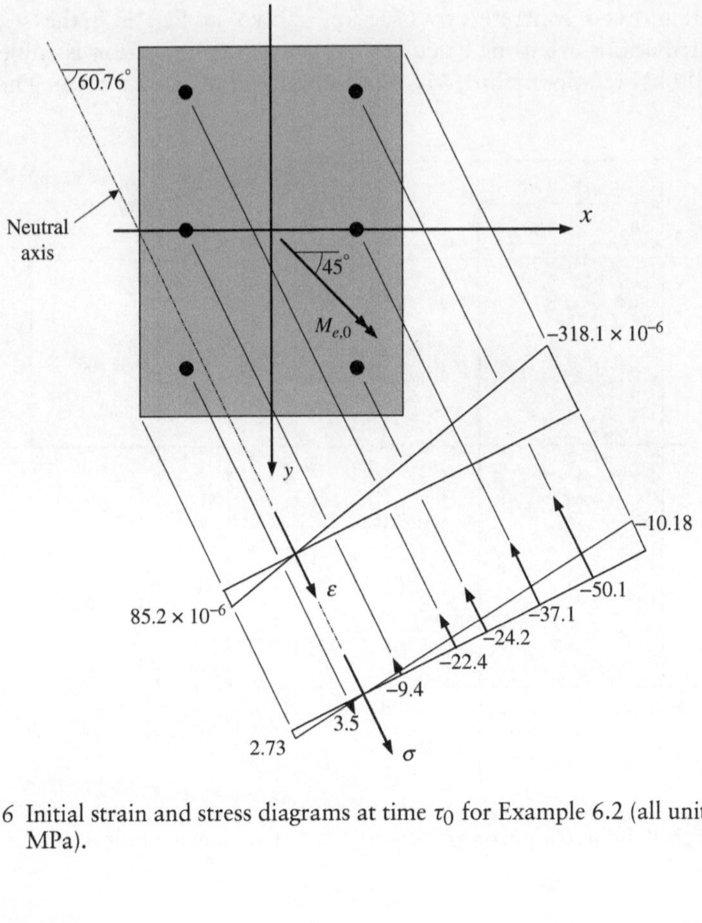

Figure 6.6 Initial strain and stress diagrams at time τ_0 for Example 6.2 (all units in mm, MPa).

In this case $\theta_{M,0} \neq \theta_{NA,0}$ (since $I_{x,0} \neq I_{y,0}$), and the neutral axis passes through the y-axis at (Eq. 6.32c):

$$y = -\frac{I_{x,0}}{A_0}\frac{N_{e,0}}{M_{xe,0}} = -\frac{1812 \times 10^6}{134.1 \times 10^3}\frac{-500 \times 10^3}{25 \times 10^6} = 270.2 \text{ mm}$$

The cross-sectional rigidities could have also been calculated based on Eqs 6.15:

$$R_{A,0} = 4293 \times 10^6 \text{ N}$$
$$R_{Ix,0} = 58 \times 10^{12} \text{ Nmm}^2$$
$$R_{Iy,0} = 32.47 \times 10^{12} \text{ Nmm}^2$$
$$R_{Bx,0} = R_{By,0} = R_{Ixy,0} = 0$$

In this case, $\mathbf{F}_0$ is obtained from Eq. 6.24 and is identical to that calculated above.

Example 6.3

For the reinforced concrete cross-section shown in Fig. 6.7, the strain and stress distributions are to be calculated at time τ_0. The section is subjected to: $N_e = -800$ kN (compression), $M_{xe} = 50$ kNm and $M_{ye} = 25$ kNm. The origin

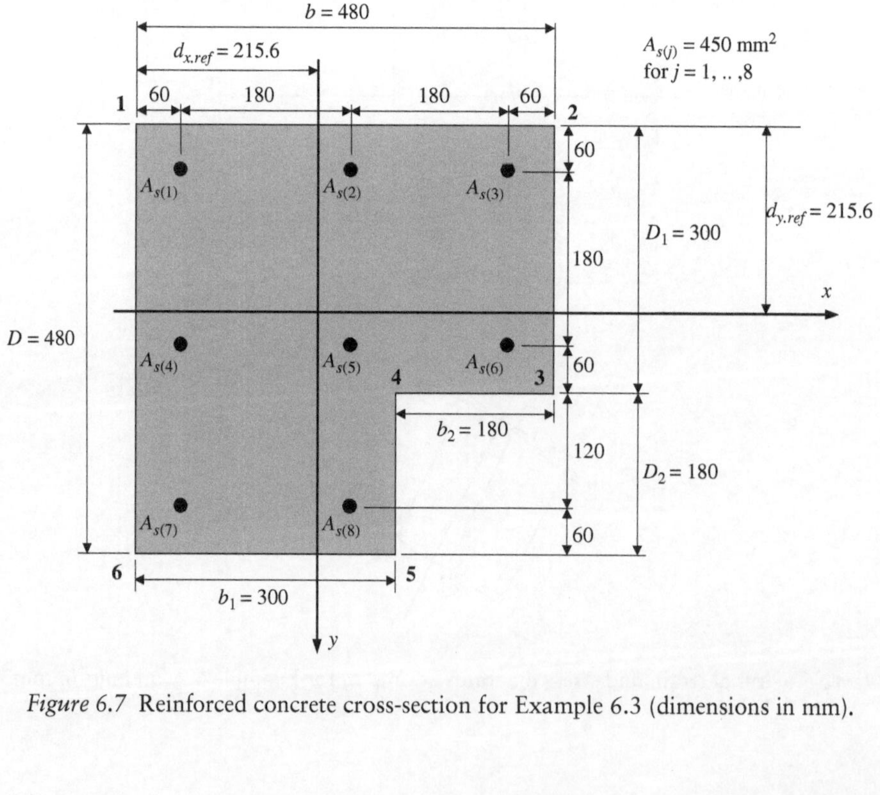

Figure 6.7 Reinforced concrete cross-section for Example 6.3 (dimensions in mm).

of the coordinate system is taken at the centroid of the section (calculated based on instantaneous material properties) located at $d_{x.ref} = d_{y.ref} = 215.6$ mm, as shown in Fig. 6.7. The material behaviour is linear-elastic with $E_{c,0} = 32$ GPa and $E_s = 200$ GPa. The modular ratio for all reinforcement is therefore $n_{s,0} = E_s/E_{c,0} = 6.25$.
Based on the adopted reference system, the coordinates of the steel reinforcement are:

$$x_{s(1)} = -155.6 \text{ mm};\ y_{s(1)} = -155.6 \text{ mm}$$
$$x_{s(2)} = 24.4 \text{ mm};\ y_{s(2)} = -155.6 \text{ mm}$$
$$x_{s(3)} = 204.4 \text{ mm};\ y_{s(3)} = -155.6 \text{ mm}$$
$$x_{s(4)} = -155.6 \text{ mm};\ y_{s(4)} = 24.4 \text{ mm}$$
$$x_{s(5)} = 24.4 \text{ mm};\ y_{s(5)} = 24.4 \text{ mm}$$
$$x_{s(6)} = 204.4 \text{ mm};\ y_{s(6)} = 24.4 \text{ mm}$$
$$x_{s(7)} = -155.6 \text{ mm};\ y_{s(7)} = 204.4 \text{ mm}$$
$$x_{s(8)} = 24.4 \text{ mm};\ y_{s(8)} = 204.4 \text{ mm}$$

and the coordinates of the corners of the concrete section (numbered 1–6 in Fig. 6.7) are:

$$x_{(1)} = -215.6 \text{ mm};\ y_{(1)} = -215.6 \text{ mm}$$
$$x_{(2)} = 264.4 \text{ mm};\ y_{(2)} = -215.6 \text{ mm}$$
$$x_{(3)} = 264.4 \text{ mm};\ y_{(3)} = 84.4 \text{ mm}$$
$$x_{s(4)} = 84.4 \text{ mm};\ y_{s(4)} = 84.4 \text{ mm}$$
$$x_{s(5)} = 84.4 \text{ mm};\ y_{s(5)} = 264.4 \text{ mm}$$
$$x_{s(6)} = -215.6 \text{ mm};\ y_{s(6)} = 264.4 \text{ mm}$$

The vectors of external loads and prestressing are:

$$\mathbf{r}_{e,0} = \begin{bmatrix} -800 \times 10^3 \text{ N} \\ 50 \times 10^6 \text{ Nmm} \\ 25 \times 10^6 \text{ Nmm} \end{bmatrix} \quad \text{and} \quad \mathbf{f}_{p,init} = \begin{bmatrix} 0 \text{N} \\ 0 \text{Nmm} \\ 0 \text{Nmm} \end{bmatrix}$$

The properties of the transformed cross-section are:

$$A_0 = bD - b_2D_2 + (n_{s,0} - 1)\sum_{i=1}^{8} A_{s(i)} = 216.9 \times 10^3 \text{ mm}^2$$

$$B_{x,0} = bD_1\left(\frac{D_1}{2} - d_{y.ref}\right) + b_1D_2\left(D_1 + \frac{D_2}{2} - d_{y.ref}\right) + (n_{s,0} - 1)\sum_{i=1}^{8} A_{s(i)}y_{s(i)} = 0 \text{ mm}^3$$

$$B_{y,0} = bD_1\left(\frac{b}{2} - d_{x.ref}\right) + b_1D_2\left(\frac{b_1}{2} - d_{x.ref}\right) + (n_{s,0} - 1)\sum_{i=1}^{8} A_{s(i)}x_{s(i)} = 0 \text{ mm}^3$$

$$I_{x,0} = \frac{bD_1^3}{12} + bD_1\left(\frac{D_1}{2} - d_{y.ref}\right)^2 + \frac{b_1D_2^3}{12} + b_1D_2\left(D_1 + \frac{D_2}{2} - d_{y.ref}\right)^2 + (n_{s,0} - 1)\sum_{i=1}^{8} A_{s(i)}y_{s(i)}^2 = 3861 \times 10^6 \text{ mm}^4$$

$$I_{y,0} = \frac{D_1 b^3}{12} + bD_1\left(\frac{b}{2} - d_{x.ref}\right)^2 + \frac{D_2 b_1^3}{12} + b_1 D_2\left(\frac{b_1}{2} - d_{x.ref}\right)^2 + (n_{s,0} - 1)\sum_{i=1}^{8} A_{s(i)} x_{s(i)}^2 = 3861 \times 10^6 \text{ mm}^4;$$

$$I_{xy,0} = bD_1\left(\frac{D_1}{2} - d_{y.ref}\right)\left(\frac{b}{2} - d_{x.ref}\right) + b_1 D_2\left(D_1 + \frac{D_2}{2} - d_{y.ref}\right) \times \left(\frac{b_1}{2} - d_{x.ref}\right) + (n_{s,0} - 1)\sum_{i=1}^{8} A_{s(i)} x_{s(i)} y_{s(i)} = -934.3 \times 10^6 \text{ mm}^4.$$

The fact that $B_{x,0} = B_{y,0} = 0$ could have also been concluded by noting that the origin of the coordinate system coincides with the centroid of the cross-section. Considering the first moments of area about both axes are nil (i.e. $B_{x,0} = B_{y,0} = 0$), Eq. 6.29a simplifies to:

$$\mathbf{F}_0 = \frac{1}{A_0\left(I_{x,0} I_{y,0} - I_{xy,0}^2\right) E_{c,0}} \begin{bmatrix} I_{x,0} I_{y,0} - I_{xy,0}^2 & 0 & 0 \\ 0 & A_0 I_{y,0} & A_0 I_{xy,0} \\ 0 & A_0 I_{xy,0} & A_0 I_{x,0} \end{bmatrix}$$

$$= \begin{bmatrix} 1.441 \times 10^{-10} & 0 & 0 \\ 0 & 8.597 \times 10^{-15} & -2.080 \times 10^{-15} \\ 0 & -2.080 \times 10^{-15} & 8.597 \times 10^{-15} \end{bmatrix}$$

and $\boldsymbol{\varepsilon}_0$ is calculated using Eq. 6.18:

$$\boldsymbol{\varepsilon}_0 = \mathbf{F}_0\left(\mathbf{r}_{e,0} - \mathbf{f}_{p,init}\right) = \begin{bmatrix} -115.3 \times 10^{-6} \\ 0.378 \times 10^{-6}\,\text{mm}^{-1} \\ 0.111 \times 10^{-6}\,\text{mm}^{-1} \end{bmatrix}$$

The strain at the origin and the curvatures with respect to the x- and y-axes, respectively, are:

$$\varepsilon_{r,0} = -115.3 \times 10^{-6}$$
$$\kappa_{x,0} = 0.378 \times 10^{-6} \text{ mm}^{-1}$$
$$\kappa_{y,0} = 0.111 \times 10^{-6} \text{ mm}^{-1}$$

The strains at the corners of the cross-section are obtained from Eq. 6.1:

$$\varepsilon_{0(corner\ 1)} = -172.8 \times 10^{-6}$$
$$\varepsilon_{0(corner\ 2)} = -226.0 \times 10^{-6}$$
$$\varepsilon_{0(corner\ 3)} = -112.6 \times 10^{-6}$$
$$\varepsilon_{0(corner\ 4)} = -92.7 \times 10^{-6}$$
$$\varepsilon_{0(corner\ 5)} = -24.6 \times 10^{-6}$$
$$\varepsilon_{0(corner\ 6)} = 8.55 \times 10^{-6}$$

and the corresponding concrete stresses are obtained using Eq. 5.16a:

$$\sigma_{c,0(corner\,1)} = E_{c,0}\varepsilon_{0(corner\,1)} = -5.53 \text{ MPa}$$
$$\sigma_{c,0(corner\,2)} = -7.23 \text{ MPa}$$
$$\sigma_{c,0(corner\,3)} = -3.60 \text{ MPa}$$
$$\sigma_{c,0(corner\,4)} = -2.97 \text{ MPa}$$
$$\sigma_{c,0(corner\,5)} = -0.79 \text{ MPa}$$
$$\sigma_{c,0(corner\,6)} = 0.27 \text{ MPa}$$

The strains in the steel reinforcing bars are:

$$\varepsilon_{s(1),0} = -156.8 \times 10^{-6}$$
$$\varepsilon_{s(2),0} = -176.8 \times 10^{-6}$$
$$\varepsilon_{s(3),0} = -196.7 \times 10^{-6}$$
$$\varepsilon_{s(4),0} = -88.8 \times 10^{-6}$$
$$\varepsilon_{s(5),0} = -108.7 \times 10^{-6}$$
$$\varepsilon_{s(6),0} = -128.7 \times 10^{-6}$$
$$\varepsilon_{s(7),0} = -20.8 \times 10^{-6}$$
$$\varepsilon_{s(8),0} = -40.7 \times 10^{-6}$$

and from Eq. 5.16b, the steel stresses are:

$$\sigma_{s(1),0} = -31.4 \text{ MPa}$$
$$\sigma_{s(2),0} = -35.4 \text{ MPa}$$
$$\sigma_{s(3),0} = -39.3 \text{ MPa}$$
$$\sigma_{s(4),0} = -17.8 \text{ MPa}$$
$$\sigma_{s(5),0} = -21.7 \text{ MPa}$$
$$\sigma_{s(6),0} = -25.7 \text{ MPa}$$
$$\sigma_{s(7),0} = -4.15 \text{ MPa}$$
$$\sigma_{s(8),0} = -8.14 \text{ MPa}$$

The slope of the applied moment $M_{e,0}$ is calculated based on Eq. 6.4:

$$\tan\theta_{M,0} = \frac{M_{ye,0}}{M_{xe,0}} = \frac{25 \times 10^6}{50 \times 10^6} = 0.5 \text{ and therefore } \theta_{M,0} = 26.56^\circ$$

while the slope of the neutral axis is obtained using Eq. 6.22:

$$\tan\theta_{NA,0} = \frac{\kappa_{y,0}}{\kappa_{x,0}} = \frac{0.111 \times 10^{-6}}{0.378 \times 10^{-6}} = 0.293 \text{ and therefore } \theta_{NA,0} = 16.31^\circ.$$

The intersection of the neutral axis with the y-axis is calculated from Eq. 6.23:

$$y(x=0) = -\frac{\varepsilon_{r,0}}{\kappa_{x,0}} = -\frac{-115.3 \times 10^{-6}}{0.378 \times 10^{-6}} = +304.8 \text{ mm}.$$

The calculated stresses and strains are shown in Fig. 6.8.

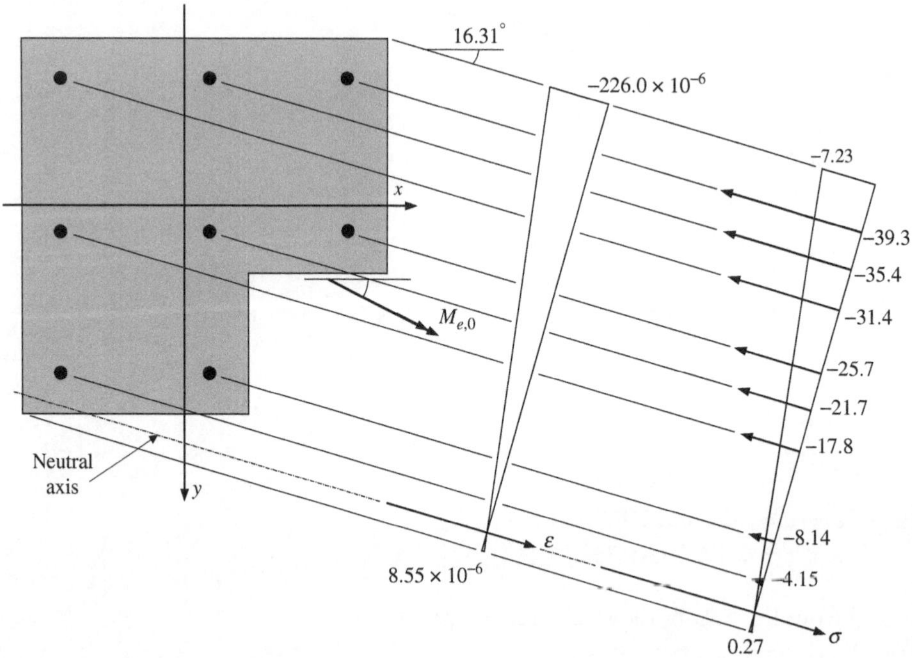

Figure 6.8 Initial strain and stress diagrams at time τ_0 for Example 6.3 (all units in mm, MPa).

6.3 Long-term analysis using the age-adjusted effective modulus method (AEMM)

The time analysis based on the AEMM to determine stresses and strains on a cross-section at time τ_k after a period of sustained loading was presented in Chapters 4 and 5 and only the steps relevant to the case of biaxial bending will be presented here. The constitutive relationships adopted for the concrete and steel are those already defined for the AEMM in Eqs 5.22.

With the instantaneous stress and strain distributions determined at time τ_0 using the analysis of the previous section, the strain diagram at time τ_k is defined by the strain at the origin and the curvatures about the orthogonal reference axes, i.e. $\varepsilon_{r,k}$, $\kappa_{x,k}$ and $\kappa_{y,k}$. These are obtained by solving the equilibrium equations defined by:

$$\mathbf{r}_{e,k} = \mathbf{r}_{i,k} \tag{6.33}$$

where

$$\mathbf{r}_{e,k} = \left[N_{e,k}\ M_{xe,k}\ M_{ye,k} \right]^T \tag{6.34a}$$

and

$$\mathbf{r}_{i,k} = \left[N_{i,k} \; M_{xi,k} \; M_{yi,k} \right]^T \tag{6.34b}$$

Based on the material constitutive relationships (Eqs 5.22), the axial force resisted by each component of the cross-section at τ_k can be determined from:

$$N_{c,k} = \int_{A_c} \sigma_{c,k} \, dA = \int_{A_c} \left[\overline{E}_{e,k} \left(\varepsilon_{r,k} + y\kappa_{x,k} - x\kappa_{y,k} - \varepsilon_{sh,k} \right) + \overline{F}_{e,0}\sigma_{c,0} \right] dA$$

$$= A_c \overline{E}_{e,k} \varepsilon_{r,k} + B_{xc} \overline{E}_{e,k} \kappa_{x,k} - B_{yc} \overline{E}_{e,k} \kappa_{y,k} - A_c \overline{E}_{e,k} \varepsilon_{sh,k} + \overline{F}_{e,0} N_{c,0} \tag{6.35a}$$

$$N_{s,k} = R_{A,s} \varepsilon_{r,k} + R_{Bx,s} \kappa_{x,k} - R_{By,s} \kappa_{y,k} \tag{6.35b}$$

$$N_{p,k} = R_{A,p} \varepsilon_{r,k} + R_{Bx,p} \kappa_{x,k} - R_{By,p} \kappa_{y,k} + \sum_{i=1}^{m_p} A_{p(i)} E_{p(i)} \left(\varepsilon_{p(i),init} - \varepsilon_{p.rel(i),k} \right) \tag{6.35c}$$

These expressions can be summed to give the axial force resisted by the whole section $N_{i,k}$:

$$N_{i,k} = R_{A,k} \varepsilon_{r,k} + R_{Bx,k} \kappa_{x,k} - R_{By,k} \kappa_{y,k} - A_c \overline{E}_{e,k} \varepsilon_{sh,k} + \overline{F}_{e,0} N_{c,0} + \sum_{i=1}^{m_p} A_{p(i)} E_{p(i)} \left(\varepsilon_{p(i),init} - \varepsilon_{p.rel(i),k} \right) \tag{6.36a}$$

and, in a similar manner, the internal moments about the *x*- and *y*-axes can be determined from:

$$M_{xi,k} = R_{Bx,k} \varepsilon_{r,k} + R_{Ix,k} \kappa_{x,k} - R_{Ixy,k} \kappa_{y,k} - B_{xc} \overline{E}_{e,k} \varepsilon_{sh,k} + \overline{F}_{e,0} M_{xc,0} + \sum_{i=1}^{m_p} y_{p(i)} A_{p(i)} E_{p(i)} \left(\varepsilon_{p(i),init} - \varepsilon_{p.rel(i),k} \right) \tag{6.36b}$$

$$M_{yi,k} = -R_{By,k} \varepsilon_{r,k} - R_{Ixy,k} \kappa_{x,k} + R_{Iy,k} \kappa_{y,k} + B_{yc} \overline{E}_{e,k} \varepsilon_{sh,k} + \overline{F}_{e,0} M_{yc,0} - \sum_{i=1}^{m_p} x_{p(i)} A_{p(i)} E_{p(i)} \left(\varepsilon_{p(i),init} - \varepsilon_{p.rel(i),k} \right) \tag{6.36c}$$

where the time-dependent cross-sectional rigidities at time τ_k are determined from expressions similar to those at time τ_0 (Eqs 6.15) except that the age-adjusted effective modulus of the concrete $\overline{E}_{e,k}$ is used instead of the elastic modulus $E_{c,0}$. That is:

$$R_{A,k} = A_c \overline{E}_{e,k} + R_{A,s} + R_{A,p} \tag{6.37a}$$

$$R_{Ixy,k} = I_{xyc} \overline{E}_{e,k} + R_{Ixy,s} + R_{Ixy,p} \tag{6.37b}$$

$$R_{Bx,k} = B_{xc} \overline{E}_{e,k} + R_{Bx,s} + R_{Bx,p} \tag{6.37c}$$

$$R_{By,k} = B_{yc}\overline{E}_{e,k} + R_{By,s} + R_{By,p} \tag{6.37d}$$

$$R_{Ix,k} = I_{xc}\overline{E}_{e,k} + R_{Ix,s} + R_{Ix,p} \tag{6.37e}$$

$$R_{Iy,k} = I_{yc}\overline{E}_{e,k} + R_{Iy,s} + R_{Iy,p} \tag{6.37f}$$

Substituting Eqs 6.36 into Eq. 6.33 produces the equilibrium equations describing the structural response at τ_k:

$$\mathbf{r}_{e,k} = \mathbf{D}_k\boldsymbol{\varepsilon}_k + \mathbf{f}_{cr,k} - \mathbf{f}_{sh,k} + \mathbf{f}_{p,init} - \mathbf{f}_{p.rel,k} \tag{6.38}$$

where

$$\boldsymbol{\varepsilon}_k = \begin{bmatrix} \varepsilon_{r,k} \\ \kappa_{x,k} \\ \kappa_{y,k} \end{bmatrix} \tag{6.39a}$$

$$\mathbf{D}_k = \begin{bmatrix} R_{A,k} & R_{Bx,k} & -R_{By,k} \\ R_{Bx,k} & R_{Ix,k} & -R_{Ixy,k} \\ -R_{By,k} & -R_{Ixy,k} & R_{Iy,k} \end{bmatrix} \tag{6.39b}$$

$$\mathbf{f}_{cr,k} = \overline{F}_{e,0}\begin{bmatrix} N_{c,0} \\ M_{xc,0} \\ M_{yc,0} \end{bmatrix} = \overline{F}_{e,0}E_{c,0}\begin{bmatrix} A_c\varepsilon_{r,0} + B_{xc}\kappa_{x,0} - B_{yc}\kappa_{y,0} \\ B_{xc}\varepsilon_{r,0} + I_{xc}\kappa_{x,0} - I_{xyc}\kappa_{y,0} \\ -B_{yc}\varepsilon_{r,0} - I_{xyc}\kappa_{x,0} + I_{yc}\kappa_{y,0} \end{bmatrix} \tag{6.39c}$$

$$\mathbf{f}_{sh,k} = \begin{bmatrix} A_c \\ B_{xc} \\ -B_{yc} \end{bmatrix}\overline{E}_{e,k}\varepsilon_{sh,k} \tag{6.39d}$$

The variables describing the strain diagram at τ_0, i.e. $\varepsilon_{r,0}$, $\kappa_{x,0}$ and $\kappa_{y,0}$, are assumed to be known from the instantaneous analysis.

The initial prestressing is included in $\mathbf{f}_{p,init}$ (already defined for the instantaneous analysis in Eq. 6.17d) and the steel relaxation is considered by means of $\mathbf{f}_{p.rel,k}$ calculated as:

$$\mathbf{f}_{p.rel,k} = \sum_{i=1}^{m_p}\begin{bmatrix} A_{p(i)}E_{p(i)} \\ y_{p(i)}A_{p(i)}E_{p(i)} \\ -x_{p(i)}A_{p(i)}E_{p(i)} \end{bmatrix}\varepsilon_{p(i),init}\varphi_p(\tau_k, \sigma_{p(i),init}) \tag{6.40}$$

Solving Eq. 6.38, the strain diagram at time τ_k is given by:

$$\boldsymbol{\varepsilon}_k = \mathbf{F}_k\left(\mathbf{r}_{e,k} - \mathbf{f}_{cr,k} + \mathbf{f}_{sh,k} - \mathbf{f}_{p,init} + \mathbf{f}_{p.rel,k}\right) \tag{6.41}$$

where

$$\mathbf{F}_k = \frac{1}{R_k}\begin{bmatrix} R_{Ix,k}R_{Iy,k} - R_{Ixy,k}^2 & R_{By,k}R_{Ixy,k} - R_{Bx,k}R_{Iy,k} & R_{By,k}R_{Ix,k} - R_{Bx,k}R_{Ixy,k} \\ R_{By,k}R_{Ixy,k} - R_{Bx,k}R_{Iy,k} & R_{A,k}R_{Iy,k} - R_{By,k}^2 & R_{A,k}R_{Ixy,k} - R_{Bx,k}R_{By,k} \\ R_{By,k}R_{Ix,k} - R_{Bx,k}R_{Ixy,k} & R_{A,k}R_{Ixy,k} - R_{Bx,k}R_{By,k} & R_{A,k}R_{Ix,k} - R_{Bx,k}^2 \end{bmatrix} \tag{6.42a}$$

and

$$R_k = R_{A,k}\left(R_{Ix,k}R_{Iy,k} - R_{Ixy,k}^2\right) - R_{Bx,k}^2 R_{Iy,k} + 2R_{Bx,k}R_{By,k}R_{Ixy,k} - R_{By,k}^2 R_{Ix,k} \tag{6.42b}$$

The stress distribution at time τ_k can then be calculated using equations similar to Eqs 5.41 as follows:

$$\sigma_{c,k} = \overline{E}_{e,k}\left(\varepsilon_k - \varepsilon_{sh,k}\right) + \overline{F}_{e,0}\sigma_{c,0} = \overline{E}_{e,k}\left\{\begin{bmatrix}1 & y & -x\end{bmatrix}\boldsymbol{\varepsilon}_k - \varepsilon_{sh,k}\right\} + \overline{F}_{e,0}\sigma_{c,0} \tag{6.43a}$$

$$\sigma_{s,k(i)} = E_{s(i)}\varepsilon_k = E_{s(i)}\begin{bmatrix}1 & y_{s(i)} - x_{s(i)}\end{bmatrix}\boldsymbol{\varepsilon}_k \tag{6.43b}$$

$$\sigma_{p(i),k} = E_{p(i)}\left(\varepsilon_k + \varepsilon_{p(i),init} - \varepsilon_{p.rel(i),k}\right) = E_{p(i)}\begin{bmatrix}1 & y_{p(i)} & -x_{p(i)}\end{bmatrix}\boldsymbol{\varepsilon}_k + E_{p(i)}\varepsilon_{p(i),init} - E_{p(i)}\varepsilon_{p.rel(i),k} \tag{6.43c}$$

where at any point (x, y) on the cross-section $\varepsilon_k = \varepsilon_{r,k} + y\kappa_{x,k} - x\kappa_{y,k} = \begin{bmatrix}1 & y & -x\end{bmatrix}\boldsymbol{\varepsilon}_k$.

Similar to Eq. 6.21, the position of the neutral axis at time τ_k is given by:

$$y = \frac{\kappa_{y,k}}{\kappa_{x,k}}x - \frac{\varepsilon_{r,k}}{\kappa_{x,k}} \tag{6.44}$$

while its slope and its intersect with the y-axis may be calculated using:

$$\tan\theta_{NA,k} = \frac{\kappa_{y,k}}{\kappa_{x,k}} \tag{6.45a}$$

$$y(x=0) = -\frac{\varepsilon_{r,k}}{\kappa_{x,k}} \tag{6.45b}$$

The expression for $\mathbf{F}_k$ can be significantly simplified for doubly symmetric cross-sections when the origin is located at the centroid of the section and the orthogonal axes are parallel to the axes of symmetry. In this case, $R_{Ixy,k} = R_{Bx,k} = R_{By,k} = 0$ and $\mathbf{F}_k$ is given by:

$$\mathbf{F}_k = \frac{1}{R_{A,k}R_{Ix,k}R_{Iy,k}}\begin{bmatrix} R_{Ix,k}R_{Iy,k} & 0 & 0 \\ 0 & R_{A,k}R_{Iy,k} & 0 \\ 0 & 0 & R_{A,k}R_{Ix,k}\end{bmatrix} \tag{6.46}$$

As discussed previously for the instantaneous analysis, the cross-sectional rigidities included in $\mathbf{F}_k$ can be calculated based on the properties of the age-adjusted transformed cross-section. For this purpose, modular ratios of $\bar{n}_{s(i),k} = E_{s(i)}/\overline{E}_{e,k}$ and $\bar{n}_{p(i),k} = E_{p(i)}/\overline{E}_{e,k}$ are to be used for the i-th steel bar and the i-th tendon, respectively.

Example 6.4

For the reinforced concrete section and coordinate system of Example 6.2 (Fig. 6.5), if the external actions (i.e. $N_e = -500$ kN, $M_{xe} = 25$ kNm, $M_{ye} = 25$ kNm) are sustained over a period of time (τ_0 to τ_k), the long-term effects caused by creep and shrinkage are to be determined. As for Example 6.2, $E_{c,0} = 32$ GPa, $E_s = 200$ GPa, and $n_{s,0} = E_s/E_{c,0} = 6.25$. The creep coefficient, the ageing coefficient and the shrinkage strain associated with the time period under consideration are $\varphi(\tau_k, \tau_0) = 2.0$; $\chi(\tau_k, \tau_0) = 0.65$ and $\varepsilon_{sh}(\tau_k) = -400 \times 10^{-6}$.

From Example 6.2 at τ_0: $\varepsilon_{r,0} = -116.4 \times 10^{-6}$; $\kappa_{x0} = 0.431 \times 10^{-6}$ mm^{-1}; $\kappa_{y0} = 0.770 \times 10^{-6}$ mm^{-1}.

From Eqs 4.35 and 4.46:

$$\bar{E}_{e,k} = \frac{E_{c,0}}{1 + \chi(\tau_k, \tau_0)\varphi(\tau_k, \tau_0)} = \frac{32{,}000}{1 + 0.65 \times 2.0} = 13{,}910 \text{ MPa}$$

and therefore $\bar{n}_{s,k} = 14.38$

$$\bar{F}_{e,0} = \frac{\varphi(\tau_k, \tau_0)[\chi(\tau_k, \tau_0) - 1]}{1 + \chi(\tau_k, \tau_0)\varphi(\tau_k, \tau_0)} = \frac{2.0 \times (0.65 - 1.0)}{1.0 + 0.65 \times 2.0} = -0.304$$

For this reinforced concrete section, the vectors $\mathbf{r}_{e,k}$, $\mathbf{f}_{p,init}$ and $\mathbf{f}_{p.rel,k}$ in Eq. 6.41 are:

$$\mathbf{r}_{e,k} = \begin{bmatrix} -500 \times 10^3 \text{ N} \\ 25 \times 10^6 \text{ Nmm} \\ 25 \times 10^6 \text{ Nmm} \end{bmatrix}$$

$$\mathbf{f}_{p,init} = \begin{bmatrix} 0 \text{ N} \\ 0 \text{ Nmm} \\ 0 \text{ Nmm} \end{bmatrix}$$

$$\mathbf{f}_{p.rel,k} = \begin{bmatrix} 0 \text{ N} \\ 0 \text{ Nmm} \\ 0 \text{ Nmm} \end{bmatrix}$$

The properties of the age-adjusted transformed cross-section based on $\bar{E}_{e,k}$ are:

$$\bar{A}_k = bD + (\bar{n}_{s,k} - 1)\sum_{i=1}^{6} A_{s(i)} = 156.1 \times 10^3 \text{ mm}^2$$

$$\bar{I}_{x,k} = \frac{bD^3}{12} + bD\left(\frac{D}{2} - d_{y.ref}\right)^2 + (\bar{n}_{s,k} - 1)\sum_{i=1}^{6} A_{s(i)} y_{s(i)}^2 = 2141 \times 10^6 \text{ mm}^4$$

$$\bar{I}_{y,k} = \frac{Db^3}{12} + bD\left(\frac{b}{2} - d_{x.ref}\right)^2 + (\bar{n}_{s,k} - 1)\sum_{i=1}^{6} A_{s(i)} x_{s(i)}^2 = 1192 \times 10^6 \text{ mm}^4$$

$$\bar{B}_{x,k} = \bar{B}_{y,k} = \bar{I}_{xy,k} = 0$$

The geometric properties of the concrete are required for the calculation of $\mathbf{f}_{cr,k}$ and $\mathbf{f}_{sh,k}$ and are obtained as follows:

$$A_c = bD - \sum_{i=1}^{6} A_{s(i)} = 117.3 \times 10^3 \text{ mm}^2$$

$$I_{xc} = \frac{bD^3}{12} + bD\left(\frac{D}{2} - d_{y.ref}\right)^2 - \sum_{i=1}^{6} A_{s(i)} y_{s(i)}^2 = 1559 \times 10^6 \text{ mm}^4$$

$$I_{yc} = \frac{Db^3}{12} + bD\left(\frac{b}{2} - d_{x.ref}\right)^2 - \sum_{i=1}^{6} A_{s(i)} x_{s(i)}^2 = 878.1 \times 10^6 \text{ mm}^4$$

$$B_{xc} = B_{yc} = I_{xyc} = 0$$

From Eqs 6.39c and d:

$$\mathbf{f}_{cr,k} = \overline{F}_{e,0} E_{c,0} \begin{bmatrix} A_c \varepsilon_{r,0} + B_{xc}\kappa_{x,0} - B_{yc}\kappa_{y,0} \\ B_{xc}\varepsilon_{r,0} + I_{xc}\kappa_{x,0} - I_{xyc}\kappa_{y,0} \\ -B_{yc}\varepsilon_{r,0} - I_{xyc}\kappa_{x,0} + I_{yc}\kappa_{y,0} \end{bmatrix} = \begin{bmatrix} 133.0 \times 10^3 \text{ N} \\ -6.55 \times 10^6 \text{ Nmm} \\ -6.58 \times 10^6 \text{ Nmm} \end{bmatrix}$$

$$\mathbf{f}_{sh,k} = \begin{bmatrix} 117.3 \times 10^3 \\ 0 \\ 0 \end{bmatrix} \times 13{,}910 \times \left(-400 \times 10^{-6}\right) = \begin{bmatrix} -652.8 \times 10^3 \text{ N} \\ 0 \text{ Nmm} \\ 0 \text{ Nmm} \end{bmatrix}$$

Based on properties of the age-adjusted transformed cross-section, $\mathbf{F}_k$ is calculated using Eq. 6.46 as:

$$\mathbf{F}_k = \begin{bmatrix} 4.604 \times 10^{-10} \text{N}^{-1} & 0 & 0 \\ 0 & 3.356 \times 10^{-14} \text{N}^{-1}\text{mm}^{-2} & 0 \\ 0 & 0 & 6.027 \times 10^{-14} \text{N}^{-1}\text{mm}^{-2} \end{bmatrix}$$

The strain $\boldsymbol{\varepsilon}_k$ at time τ_k is obtained using Eq. 6.41:

$$\boldsymbol{\varepsilon}_k = \mathbf{F}_k(\mathbf{r}_{e,k} - \mathbf{f}_{cr,k} + \mathbf{f}_{sh,k}) = \begin{bmatrix} -592.0 \times 10^{-6} \\ 1.058 \times 10^{-6} \text{ mm}^{-1} \\ 1.903 \times 10^{-6} \text{ mm}^{-1} \end{bmatrix}$$

The strain at the origin and the curvatures with respect to the *x*- and *y*-axes, respectively, are:

$$\varepsilon_{r,k} = -592.0 \times 10^{-6}$$
$$\kappa_{x,k} = 1.058 \times 10^{-6} \text{ mm}^{-1}$$
$$\kappa_{y,k} = 1.903 \times 10^{-6} \text{ mm}^{-1}$$

The strains at the corners of the cross-section are (Eq. 6.1):

$$\varepsilon_{k(top\text{-}left)}(x = -b/2, y = -D/2) = -518.2 \times 10^{-6}$$
$$\varepsilon_{k(top\text{-}right)}(x = b/2, y = -D/2) = -1089 \times 10^{-6}$$
$$\varepsilon_{k(btm\text{-}right)}(x = b/2, y = D/2) = -665.8 \times 10^{-6}$$
$$\varepsilon_{k(btm\text{-}left)}(x = -b/2, y = D/2) = -94.7 \times 10^{-6}$$

The stresses in the concrete and in the steel reinforcement are calculated using Eqs 5.16:

$$\sigma_{c,k(top\text{-}left)} = \overline{E}_{e,k}\left\{\begin{bmatrix} 1 & y_{c,(top\text{-}left)} & -x_{c,(top\text{-}left)}\end{bmatrix}\boldsymbol{\varepsilon}_k - \varepsilon_{sh,k}\right\} + \overline{F}_{e,0}\sigma_{c,0(top\text{-}left)}$$

$$= 13{,}910 \times \left\{\begin{bmatrix} 1 & -200 & -(-150)\end{bmatrix}\begin{bmatrix} -592.0\times10^{-6} \\ 1.058\times10^{-6} \\ 1.903\times10^{-6}\end{bmatrix} - (-400\times10^{-6})\right\}$$

$$+ (-0.304)(-2.79) = -0.79 \text{ MPa}$$

$$\sigma_{c,k(top\text{-}right)} = -6.49 \text{ MPa}$$
$$\sigma_{c,k(btm\text{-}right)} = -2.28 \text{ MPa}$$
$$\sigma_{c,k(btm\text{-}left)} = 3.42 \text{ MPa}$$
$$\sigma_{s(1),k} = E_s\begin{bmatrix} 1 & y_{s(1)} & -x_{s(1)}\end{bmatrix}\boldsymbol{\varepsilon}_k$$

$$= 200\times10^3 \times \begin{bmatrix} 1 & -150 & -(-90)\end{bmatrix} \times \begin{bmatrix} -592.0\times10^{-6} \\ 1.058\times10^{-6} \\ 1.903\times10^{-6}\end{bmatrix}$$

$$= -115.9 \text{ MPa};$$
$$\sigma_{s(2),k} = -184.4 \text{ MPa}$$
$$\sigma_{s(3),k} = -84.1 \text{ MPa}$$
$$\sigma_{s(4),k} = -152.7 \text{ MPa}$$
$$\sigma_{s(5),k} = -52.4 \text{ MPa}$$
$$\sigma_{s(6),k} = -120.8 \text{ MPa}$$

Stresses and strains calculated at time τ_k are shown in Fig. 6.9.

Significant transfer of axial compressive load took place from the concrete to the steel with time, with the maximum compressive stress in the concrete reducing from −10.18 MPa at τ_0 to −6.49 MPa at τ_k and the compressive stress in the most heavily stressed reinforcing bar increasing from −50.1 MPa to −184.4 MPa. It is also noted that the tensile stress in the concrete in the bottom left corner of the cross-section has increased from 2.73 MPa to 3.42 MPa, indicating that cracking may well occur during a period of sustained loading (primarily due to shrinkage).

The slopes of the applied moment $M_{e,k}$ and the neutral axis are (Eqs 6.4 and 6.45a):

$$\tan\theta_{M,k} = \tan\theta_{M,0} = 1.0 \text{ and therefore } \theta_{M,k} = 45^\circ \text{ and}$$

$$\tan\theta_{NA,k} = \frac{\kappa_{y,k}}{\kappa_{x,k}} = \frac{1.903\times10^{-6}}{1.058\times10^{-6}} = 1.80 \text{ and therefore } \theta_{NA,k} = 60.91^\circ.$$

The neutral axis passes through the y-axis at (Eq. 6.45b):

$$y = -\frac{\varepsilon_{r,k}}{\kappa_{x,k}} = -\frac{-592.0\times10^{-6}}{1.058\times10^{-6}} = 559.2 \text{ mm}$$

This indicates a translation of the neutral axis with a slight rotation from the results previously obtained at time τ_0. In fact, the slope of the neutral axis varies

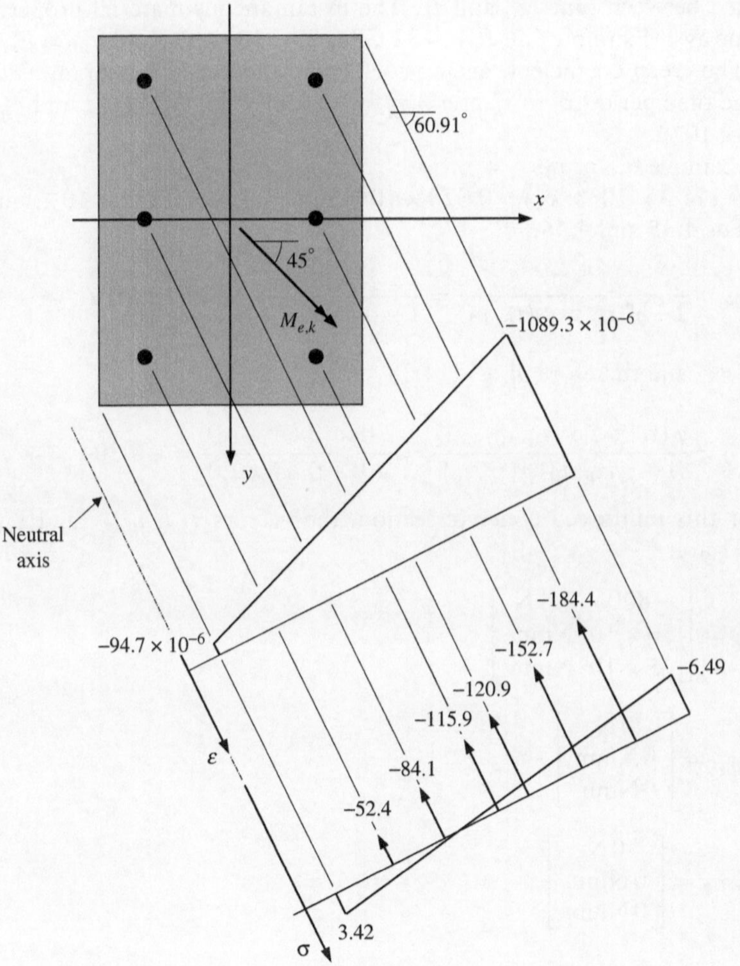

Figure 6.9 Final strain and stress diagrams at time τ_k (Example 6.4) (all units in mm, MPa).

from $\theta_{NA,0} = 60.76°$ to $\theta_{NA,k} = 60.91°$ while the intersection with the y-axis moves from $y = 270.2$ mm at τ_0 to $y = 559.2$ mm at τ_k. The translation of the neutral axis is mostly due to the effects of shrinkage. If the example was repeated with shrinkage set to zero, the intersection with the y-axis would change only slightly from $y = 270.2$ mm at τ_0 to $y = 275.3$ mm at τ_k.

Example 6.5

The long-term strains and stresses on the cross-section of Fig. 6.7 are to be determined using the AEMM assuming that the external loads applied in Example 6.3 (i.e. $N_e = -800$ kN, $M_{xe} = 50$ kNm, $M_{ye} = 25$ kNm) are to be sustained between times τ_0 and τ_k. The instantaneous material properties are the same as in Example 6.3: $E_{c,0} = 32$ GPa, $E_s = 200$ GPa, and $n_{s,0} = E_s/E_{c,0} = 6.25$. The creep coefficient, ageing coefficient and shrinkage strain associated with the time period τ_0 to τ_k are: $\varphi(\tau_k, \tau_0) = 2.0$; $\chi(\tau_k, \tau_0) = 0.65$ and $\varepsilon_{sh}(\tau_k) = -400 \times 10^{-6}$.

From Example 6.3 at τ_0:

$\varepsilon_{r,0} = -115.3 \times 10^{-6}$; $\kappa_{x0} = 0.378 \times 10^{-6}$ mm^{-1}; $\kappa_{y0} = 0.111 \times 10^{-6}$ mm^{-1}.

From Eqs 4.35 and 4.46:

$$\overline{E}_{e,k} = \frac{E_{c,0}}{1 + \chi(\tau_k, \tau_0)\varphi(\tau_k, \tau_0)} = \frac{32{,}000}{1 + 0.65 \times 2.0} = 13{,}910 \text{ MPa}$$

and therefore $\bar{n}_{s,k} = 14.38$

$$\overline{F}_{e,0} = \frac{\varphi(\tau_k, \tau_0)[\chi(\tau_k, \tau_0) - 1]}{1 + \chi(\tau_k, \tau_0)\varphi(\tau_k, \tau_0)} = \frac{2.0 \times (0.65 - 1.0)}{1.0 + 0.65 \times 2.0} = -0.304$$

and for this reinforced concrete section, the vectors $\mathbf{r}_{e,k}$, $\mathbf{f}_{p,init}$ and $\mathbf{f}_{p.rel,k}$ in Eq. 6.41 are:

$$\mathbf{r}_{e,k} = \begin{bmatrix} -800 \times 10^3 \text{ N} \\ 50 \times 10^6 \text{ Nmm} \\ 25 \times 10^6 \text{ Nmm} \end{bmatrix}$$

$$\mathbf{f}_{p,init} = \begin{bmatrix} 0 \text{ N} \\ 0 \text{ Nmm} \\ 0 \text{ Nmm} \end{bmatrix}$$

$$\mathbf{f}_{p.rel,k} = \begin{bmatrix} 0 \text{ N} \\ 0 \text{ Nmm} \\ 0 \text{ Nmm} \end{bmatrix}$$

The properties of the age-adjusted transformed cross-section based on $\overline{E}_{e,k}$ are:

$$\bar{A}_k = bD - b_2 D_2 + (\bar{n}_{s,k} - 1) \sum_{i=1}^{8} A_{s(i)} = 246.2 \times 10^3 \text{ mm}^2$$

$$\bar{B}_{x,k} = bD_1 \left(\frac{D_1}{2} - d_{y.ref} \right) + b_1 D_2 \left(D_1 + \frac{D_2}{2} - d_{y.ref} \right) + (\bar{n}_{s,k} - 1) \sum_{i=1}^{8} A_{s(i)} y_{s(i)} = 62.69 \times 10^3 \text{ mm}^3$$

$$\bar{B}_{y,k} = bD_1 \left(\frac{b}{2} - d_{x.ref} \right) + b_1 D_2 \left(\frac{b_1}{2} - d_{x.ref} \right) + (\bar{n}_{s,k} - 1) \sum_{i=1}^{8} A_{s(i)} x_{s(i)} = 62.69 \times 10^3 \text{ mm}^3$$

$$\bar{I}_{x,k} = \frac{bD_1^3}{12} + bD_1 \left(\frac{D_1}{2} - d_{y.ref} \right)^2 + \frac{b_1 D_2^3}{12} + b_1 D_2 \left(D_1 + \frac{D_2}{2} - d_{y.ref} \right)^2 + (\bar{n}_{s,k} - 1) \sum_{i=1}^{8} A_{s(i)} y_{s(i)}^2 = 4439 \times 10^6 \text{ mm}^4$$

$$\bar{I}_{y,k} = \frac{D_1 b^3}{12} + bD_1\left(\frac{b}{2} - d_{x.ref}\right)^2 + \frac{D_2 b_1^3}{12} + b_1 D_2\left(\frac{b_1}{2} - d_{x.ref}\right)^2$$
$$+ (\bar{n}_{s,k} - 1)\sum_{i=1}^{8} A_{s(i)} x_{s(i)}^2 = 4439 \times 10^6 \text{ mm}^4$$

$$\bar{I}_{xy,k} = bD_1\left(\frac{D_1}{2} - d_{y.ref}\right)\left(\frac{b}{2} - d_{x.ref}\right) + b_1 D_2\left(D_1 + \frac{D_2}{2} - d_{y.ref}\right)$$
$$\times\left(\frac{b_1}{2} - d_{x.ref}\right) + (\bar{n}_{s,k} - 1)\sum_{i=1}^{8} A_{s(i)} x_{s(i)} y_{s(i)} = -1067 \times 10^6 \text{ mm}^4$$

The geometric properties of the concrete are required to evaluate $\mathbf{f}_{cr,k}$ and $\mathbf{f}_{sh,k}$:

$$A_c = bD - b_2 D_2 - \sum_{i=1}^{8} A_{s(i)} = 194.4 \times 10^3 \text{ mm}^2$$

$$B_{xc} = bD_1\left(\frac{D_1}{2} - d_{y.ref}\right) + b_1 D_2\left(D_1 + \frac{D_2}{2} - d_{y.ref}\right) - \sum_{i=1}^{8} A_{s(i)} y_{s(i)}$$
$$= -35.64 \times 10^3 \text{ mm}^3$$

$$B_{yc} = bD_1\left(\frac{b}{2} - d_{x.ref}\right) + b_1 D_2\left(\frac{b_1}{2} - d_{x.ref}\right) - \sum_{i=1}^{8} A_{s(i)} x_{s(i)}$$
$$= -35.64 \times 10^3 \text{ mm}^3$$

$$I_{xc} = \frac{bD_1^3}{12} + bD_1\left(\frac{D_1}{2} - d_{y.ref}\right)^2 + \frac{b_1 D_2^3}{12} + b_1 D_2\left(D_1 + \frac{D_2}{2} - d_{y.ref}\right)^2$$
$$- \sum_{i=1}^{8} A_{s(i)} y_{s(i)}^2 = 3417 \times 10^6 \text{ mm}^4$$

$$I_{yc} = \frac{D_1 b^3}{12} + bD_1\left(\frac{b}{2} - d_{x.ref}\right)^2 + \frac{D_2 b_1^3}{12} + b_1 D_2\left(\frac{b_1}{2} - d_{x.ref}\right)^2$$
$$- \sum_{i=1}^{8} A_{s(i)} x_{s(i)}^2 = 3417 \times 10^6 \text{ mm}^4$$

$$I_{xyc} = bD_1\left(\frac{D_1}{2} - d_{y.ref}\right)\left(\frac{b}{2} - d_{x.ref}\right) + b_1 D_2\left(D_1 + \frac{D_2}{2} - d_{y.ref}\right)$$
$$\times\left(\frac{b_1}{2} - d_{x.ref}\right) - \sum_{i=1}^{8} A_{s(i)} x_{s(i)} y_{s(i)} = -831.9 \times 10^6 \text{ mm}^4$$

From Eqs 6.39c and d:

$$\mathbf{f}_{cr,k} = \overline{F}_{e,0} E_{c,0} \begin{bmatrix} A_c \varepsilon_{r,0} + B_{xc}\kappa_{x,0} - B_{yc}\kappa_{y,0} \\ B_{xc}\varepsilon_{r,0} + I_{xc}\kappa_{x,0} - I_{xyc}\kappa_{y,0} \\ -B_{yc}\varepsilon_{r,0} - I_{xyc}\kappa_{x,0} + I_{yc}\kappa_{y,0} \end{bmatrix} = \begin{bmatrix} 218.3 \times 10^3 \text{ N} \\ -13.52 \times 10^6 \text{ Nmm} \\ -6.705 \times 10^6 \text{ Nmm} \end{bmatrix}$$

$$f_{sh,k} = \begin{bmatrix} A_c \\ B_{xc} \\ -B_{yc} \end{bmatrix} \overline{E}_{e,k}\varepsilon_{sh,k} = \begin{bmatrix} 194.4 \times 10^3 \\ -35.64 \times 10^3 \\ 35.64 \times 10^3 \end{bmatrix} \times 13{,}910 \times \left(-400 \times 10^{-6}\right)$$

$$= \begin{bmatrix} -1082 \times 10^3\,\text{N} \\ 0.198 \times 10^6\,\text{Nmm} \\ -0.198 \times 10^6\,\text{Nmm} \end{bmatrix}$$

Based on properties of the age-adjusted transformed cross-section, $\mathbf{F}_k$ is calculated using Eq. 6.42a, except that the age-adjusted modulus $\overline{E}_{e,k}$ is used instead of $E_{c,0}$:

$$\mathbf{F}_k = \begin{bmatrix} 2.920\times10^{-10} & -5.429\times10^{-15} & 5.429\times10^{-15} \\ -5.429\times10^{-15} & 1.719\times10^{-14} & -4.133\times10^{-15} \\ 5.429\times10^{-15} & -4.133\times10^{-15} & 1.719\times10^{-14} \end{bmatrix}$$

The strain ε_k at time τ_k is obtained using Eq. 6.41:

$$\boldsymbol{\varepsilon}_k = \mathbf{F}_k(\mathbf{r}_{e,k} - \mathbf{f}_{cr,k} + \mathbf{f}_{sh,k}) = \begin{bmatrix} -613.4 \times 10^{-6} \\ 0.976 \times 10^{-6}\,\text{mm}^{-1} \\ 0.267 \times 10^{-6}\,\text{mm}^{-1} \end{bmatrix}$$

The strain at the origin and the curvatures with respect to the *x*- and *y*-axes, respectively, are:

$$\varepsilon_{r,k} = -613.4 \times 10^{-6}$$
$$\kappa_{x,k} = 0.976 \times 10^{-6}\ \text{mm}^{-1}$$
$$\kappa_{y,k} = 0.267 \times 10^{-6}\ \text{mm}^{-1}$$

The strains at the corners of the cross-section are obtained from Eq. 6.1:

$$\varepsilon_{k(corner\ 1)} = -766.4 \times 10^{-6}$$
$$\varepsilon_{k(corner\ 2)} = -894.4 \times 10^{-6}$$
$$\varepsilon_{k(corner\ 3)} = -601.6 \times 10^{-6}$$
$$\varepsilon_{k(corner\ 4)} = -553.5 \times 10^{-6}$$
$$\varepsilon_{k(corner\ 5)} = -377.8 \times 10^{-6}$$
$$\varepsilon_{k(corner\ 6)} = -297.8 \times 10^{-6}$$

and the corresponding concrete stresses are obtained using Eq. 6.43a:

$$\sigma_{c,k(corner\ 1)} = \overline{E}_{e,k}\left\{\begin{bmatrix} 1 & y_{corner\ 1} & -x_{corner\ 1} \end{bmatrix} \boldsymbol{\varepsilon}_k - \varepsilon_{sh,k}\right\} + \overline{F}_{e,0}\sigma_{c,0(corner\ 1)}$$
$$= -3.41\ \text{MPa}$$

$$\sigma_{c,k(corner\ 2)} = -4.68\ \text{MPa}$$
$$\sigma_{c,k(corner\ 3)} = -1.71\ \text{MPa}$$

$$\sigma_{c,k(corner\ 4)} = -1.23 \text{ MPa}$$
$$\sigma_{c,k(corner\ 5)} = 0.55 \text{ MPa}$$
$$\sigma_{c,k(corner\ 6)} = 1.34 \text{ MPa}$$

The strains in the steel reinforcing bars are:

$$\varepsilon_{s(1),k} = -723.8 \times 10^{-6}$$
$$\varepsilon_{s(2),k} = -771.9 \times 10^{-6}$$
$$\varepsilon_{s(3),k} = -819.9 \times 10^{-6}$$
$$\varepsilon_{s(4),k} = -548.1 \times 10^{-6}$$
$$\varepsilon_{s(5),k} = -596.1 \times 10^{-6}$$
$$\varepsilon_{s(6),k} = -644.1 \times 10^{-6}$$
$$\varepsilon_{s(7),k} = -372.4 \times 10^{-6}$$
$$\varepsilon_{s(8),k} = -420.4 \times 10^{-6}$$

and from Eq. 6.43b, the steel stresses are:

$$\sigma_{s(1),k} = -144.8 \text{ MPa}$$
$$\sigma_{s(2),k} = -154.4 \text{ MPa}$$
$$\sigma_{s(3),k} = -164.0 \text{ MPa}$$
$$\sigma_{s(4),k} = -109.6 \text{ MPa}$$
$$\sigma_{s(5),k} = -119.2 \text{ MPa}$$
$$\sigma_{s(6),k} = -128.8 \text{ MPa}$$
$$\sigma_{s(7),k} = -74.5 \text{ MPa}$$
$$\sigma_{s(8),k} = -84.1 \text{ MPa}$$

The slope of the applied moment $M_{e,k}$ is calculated based on Eq. 6.4 at time τ_k:

$$\tan\theta_{M,k} = \frac{M_{ye,k}}{M_{xe,k}} = \frac{25 \times 10^6}{50 \times 10^6} = 0.5 \text{ and therefore } \theta_{M,k} = 26.56°$$

The equation of the line of the neutral axis at time τ_k (in the x-y plane) is from Eq. 6.44:

$$y = \frac{\kappa_{y,k}}{\kappa_{x,k}}x - \frac{\varepsilon_{r,k}}{\kappa_{x,k}} = \frac{0.267 \times 10^{-6}}{0.976 \times 10^{-6}}x - \frac{-613.4 \times 10^{-6}}{0.976 \times 10^{-6}} = 0.273x + 628.3$$

and with $\tan\theta_{NA,k} = 0.273$, the inclination of the neutral axis is $\theta_{NA,k} = 15.28°$ and it passes through the y-axis at y = 628.3 mm. Without shrinkage, the intersection with the y-axis occurs at $y(x = 0) = 307.9$ mm (close to the value of 304.8 mm calculated at time τ_0).

The final stresses and strains on the cross-section are shown in Fig. 6.10.

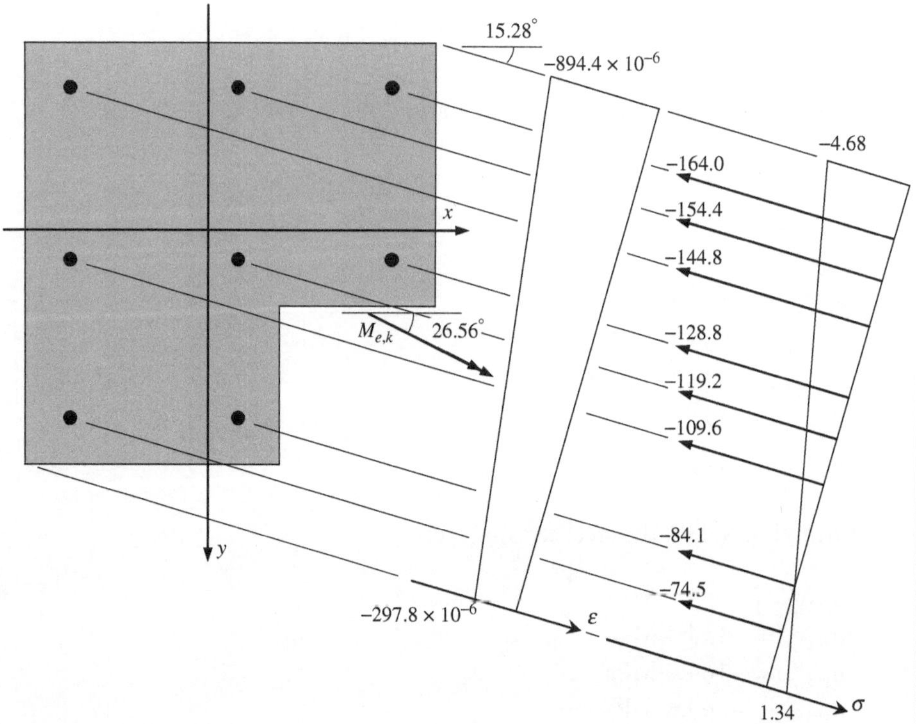

Figure 6.10 Final strain and stress diagrams at time τ_k for Example 6.5 (all units in mm, MPa).

6.4 Long-term analysis using the step-by-step method

The use of the SSM is particularly useful for the long-term analysis of reinforced concrete columns as it enables complex loading histories to be considered, such as those that occur during different construction stages. It also permits improved accuracy with finer time discretisation. In the following, only the steps peculiar to combined axial force and biaxial moments are presented as the use of the SSM has been extensively discussed in Chapters 4 and 5.

When using the SSM the time domain is discretised into a number of instants τ_j, with $j = 1, \ldots, k$ and with τ_k being the instant at which the long-term response is sought. The constitutive relationships for the concrete and the steel reinforcement and tendons are defined in Eqs 4.25 and 5.44.

The solution process is similar to that already outlined for both the instantaneous analysis and time analysis using the AEMM. At any time instant τ_j, the problem is expressed in terms of the three variables that define the strain diagram ($\varepsilon_{r,j}$, $\kappa_{x,j}$ and $\kappa_{y,j}$) and these are determined based on equilibrium considerations written as:

$$\mathbf{r}_{e,j} = \mathbf{r}_{i,j} \tag{6.47}$$

where $\mathbf{r}_{e,j}$ and $\mathbf{r}_{i,j}$ represent the external and internal actions, and similarly to Eq. 5.46, may be expressed by:

$$\mathbf{r}_{e,j} = \left[N_{e,j} \; M_{xe,j} \; M_{ye,j} \right]^T \tag{6.48a}$$

and

$$\mathbf{r}_{i,j} = \left[N_{i,j} \; M_{xi,j} \; M_{yi,j} \right]^T \tag{6.48b}$$

The internal actions $\mathbf{r}_{i,j}$ are expressed in terms of the strain measured at the origin of the reference system ($\varepsilon_{r,j}$), and the curvatures calculated with respect to the x- and y-axes ($\kappa_{x,j}$ and $\kappa_{y,j}$). Using the constitutive relationships specified in Eqs 4.25 and 5.44, the contribution to the internal axial force carried by the concrete, the steel reinforcement and the tendons at time τ_j may be written as:

$$N_{c,j} = \int_{A_c} \sigma_{c,j}\, dA = \int_{A_c} \left[E_{c,j}\left(\varepsilon_{r,j} + y\kappa_{x,j} - x\kappa_{y,j} - \varepsilon_{sh,j}\right) + \sum_{i=0}^{j-1} F_{e,j,i}\sigma_{c,i} \right] dA$$

$$= A_c E_{c,j}\varepsilon_{r,j} + B_{xc}E_{c,j}\kappa_{x,j} - B_{yc}E_{c,j}\kappa_{y,j} - A_c E_{c,j}\varepsilon_{sh,j} + \sum_{i=0}^{j-1} F_{e,j,i} N_{c,i} \tag{6.49a}$$

$$N_{s,j} = R_{A,s}\varepsilon_{r,j} + R_{Bx,s}\kappa_{x,j} - R_{By,s}\kappa_{y,j} \tag{6.49b}$$

$$N_{p,j} = R_{A,p}\varepsilon_{r,j} + R_{Bx,p}\kappa_{x,j} - R_{By,p}\kappa_{y,j} + \sum_{i=1}^{m_p} A_{p(i)}E_{p(i)}\left(\varepsilon_{p(i),init} - \varepsilon_{p.rel(i),j}\right) \tag{6.49c}$$

where the cross-sectional rigidities have been introduced previously in Eqs 6.7, 6.10 and 6.13. The expression for $N_{i,j}$ (included in $\mathbf{r}_{i,j}$) can be obtained by summing Eqs 6.49:

$$N_{i,j} = R_{A,j}\varepsilon_{r,j} + R_{Bx,j}\kappa_{x,j} - R_{By,j}\kappa_{y,j} - A_c E_{c,j}\varepsilon_{sh,j} + \sum_{i=0}^{j-1} F_{e,j,i} N_{c,i}$$
$$+ \sum_{i=1}^{m_p} A_{p(i)}E_{p(i)}\left(\varepsilon_{p(i),init} - \varepsilon_{p.rel(i),j}\right) \tag{6.50a}$$

Similarly, the expressions for the internal moments are given by:

$$M_{xi,j} = R_{Bx,j}\varepsilon_{r,j} + R_{Ix,j}\kappa_{x,j} - R_{Ixy,j}\kappa_{y,j} - B_{xc}E_{c,j}\varepsilon_{sh,j} + \sum_{i=0}^{j-1} F_{e,j,i} M_{xc,i}$$
$$+ \sum_{i=1}^{m_p} y_{p(i)}A_{p(i)}E_{p(i)}\left(\varepsilon_{p(i),init} - \varepsilon_{p.rel(i),j}\right) \tag{6.50b}$$

$$M_{yi,j} = -R_{By,j}\varepsilon_{r,j} - R_{Ixy,j}\kappa_{x,j} + R_{Iy,j}\kappa_{y,j} + B_{yc}E_{c,j}\varepsilon_{sh,j} + \sum_{i=0}^{j-1} F_{e,j,i}M_{yc,i}$$

$$- \sum_{i=1}^{m_p} x_{p(i)}A_{p(i)}E_{p(i)}\left(\varepsilon_{p(i),init} - \varepsilon_{p.rel(i),j}\right) \tag{6.50c}$$

where the cross-sectional rigidities are determined using Eqs 6.15 (proposed for the instantaneous analysis) except that the concrete elastic modulus at time τ_j, $E_{c,j}$, is used instead of the instantaneous value at time τ_0, $E_{c,0}$.

The equilibrium equations of Eqs 6.47 can be expressed in compact form at time τ_j as:

$$\mathbf{r}_{e,j} = \mathbf{D}_j\boldsymbol{\varepsilon}_j + \mathbf{f}_{cr,j} - \mathbf{f}_{sh,j} + \mathbf{f}_{p,init} - \mathbf{f}_{p.rel,j} \tag{6.51}$$

where:

$$\boldsymbol{\varepsilon}_j = \begin{bmatrix} \varepsilon_{r,j} \\ \kappa_{x,j} \\ \kappa_{y,j} \end{bmatrix} \tag{6.52a}$$

$$\mathbf{D}_j = \begin{bmatrix} R_{A,j} & R_{Bx,j} & -R_{By,j} \\ R_{Bx,j} & R_{Ix,j} & -R_{Ixy,j} \\ -R_{By,j} & -R_{Ixy,j} & R_{Iy,j} \end{bmatrix} \tag{6.52b}$$

$$\mathbf{f}_{cr,j} = \sum_{i=0}^{j-1} F_{e,j,i}\mathbf{r}_{c,i} \tag{6.52c}$$

$$\mathbf{f}_{sh,j} = \begin{bmatrix} A_c \\ B_{xc} \\ -B_{yc} \end{bmatrix} E_{c,j}\varepsilon_{sh,j} \tag{6.52d}$$

$$\mathbf{f}_{p,init} = \sum_{i=1}^{m_p} \begin{bmatrix} A_{p(i)}E_{p(i)}\varepsilon_{p(i),init} \\ y_{p(i)}A_{p(i)}E_{p(i)}\varepsilon_{p(i),init} \\ -x_{p(i)}A_{p(i)}E_{p(i)}\varepsilon_{p(i),init} \end{bmatrix} \tag{6.52e}$$

$$\mathbf{f}_{p.rel,j} = \sum_{i=1}^{m_p} \begin{bmatrix} A_{p(i)}E_{p(i)}\varepsilon_{p(i),init}\varphi_p(\tau_j, \sigma_{p(i),init}) \\ y_{p(i)}A_{p(i)}E_{p(i)}\varepsilon_{p(i),init}\varphi_p(\tau_j, \sigma_{p(i),init}) \\ -x_{p(i)}A_{p(i)}E_{p(i)}\varepsilon_{p(i),init}\varphi_p(\tau_j, \sigma_{p(i),init}) \end{bmatrix} \tag{6.52f}$$

where $\mathbf{r}_{c,i}$ included in the calculation of $\mathbf{f}_{cr,j}$ depicts the internal actions resisted by the concrete part of the cross-section at time τ_i (for which $i < j$) and is calculated based on:

$$\mathbf{r}_{c,i} = \begin{bmatrix} N_{c,i} \\ M_{xc,i} \\ M_{yc,i} \end{bmatrix} = \mathbf{D}_{c,i}\boldsymbol{\varepsilon}_i + \sum_{n=0}^{i-1} F_{e,i,n}\mathbf{r}_{c,n} - \begin{bmatrix} A_c \\ B_{xc} \\ -B_{yc} \end{bmatrix} E_{c,i}\varepsilon_{sh,i} = \mathbf{D}_{c,i}\boldsymbol{\varepsilon}_i + \mathbf{f}_{cr,i} - \mathbf{f}_{sh,i} \tag{6.53}$$

in which

$$\mathbf{D}_{c,i} = \begin{bmatrix} A_c & B_{xc} & -B_{yc} \\ B_{xc} & I_{xc} & -I_{xyc} \\ -B_{yc} & -I_{xyc} & I_{yc} \end{bmatrix} E_{c,i} \quad (6.54)$$

The strain diagram at time τ_j can then be determined based on:

$$\boldsymbol{\varepsilon}_j = \mathbf{F}_j \left(\mathbf{r}_{e,j} - \mathbf{f}_{cr,j} + \mathbf{f}_{sh,j} - \mathbf{f}_{p,init} + \mathbf{f}_{p.rel,j} \right) \quad (6.55)$$

in which $\mathbf{F}_j$ is identical to $\mathbf{F}_0$ (defined in Eqs 6.19), except that concrete elastic modulus $E_{c,j}$ replaces $E_{c,0}$.

The stress distribution at time τ_k can then be calculated substituting the expression for the strain $\varepsilon_k = \varepsilon_{r,k} + y\kappa_{x,k} - x\kappa_{y,k} = \begin{bmatrix} 1 & y & -x \end{bmatrix} \boldsymbol{\varepsilon}_k$ into Eqs 5.60.

The simplifications proposed for doubly-symmetric columns previously introduced for the instantaneous analysis and for the time analysis using the AEMM, can also be applied at time τ_j when using the SSM.

As in the previous analyses, the cross-sectional rigidities included in $\mathbf{F}_j$ could be calculated based on the transformed concrete cross-section obtained using the appropriate modular ratios $n_{s(i),j} = E_{s(i)}/E_{c,j}$ and $n_{p(i),j} = E_{p(i)}/E_{c,j}$ for the i-th steel bar and the i-th tendon, respectively.

7 Cracked sections

7.1 Introductory remarks

In the cross-sectional analyses of Chapters 4, 5 and 6, it was assumed that concrete can carry imposed stresses, both compressive and tensile. However, in reality concrete is not able to carry large tensile stress. If the tensile stress at a point reaches the tensile strength of concrete (Eq. 2.1), cracking occurs. Cracking is irreversible. On a cracked cross-section, tensile stress of any magnitude cannot be carried normal to the crack surface at any time after cracking and tensile forces can only be carried across a crack by steel reinforcement. Therefore, on a cracked cross-section, internal actions can be carried only by the steel reinforcement (and tendons) and the uncracked parts of the concrete section.

In members subjected only to axial tension, caused either by external loads or by restraint to shrinkage or temperature change, full-depth cracks occur when the tensile stress is exceeded (i.e. at each crack location, the entire cross-section is cracked). When the axial tension is caused by restraint to shrinkage, cracking causes a loss of stiffness and a consequent decrease in the internal tension. The crack width and the magnitude of the restraining force, as well as the spacing between the cracks, depend on the amount of bonded reinforcement. Between the cracks in a member subjected to axial tension, the concrete continues to carry tensile stress and hence continues to contribute to the member stiffness. This is the tension stiffening effect and was discussed in Section 3.6.2.

When cracking occurs in flexural members and the tensile strength of the concrete on the tensile surface of the member is reached, primary cracks develop at regular spacing on the tensile side of the member, as shown in Fig. 3.4a. A sudden loss of stiffness occurs at first cracking and the short-term moment–curvature relationship becomes non-linear. The primary cracks penetrate spontaneously to a height h_o (see Fig. 3.4a) which depends on the quantity of steel and the magnitude of any axial force or prestress. For reinforced concrete members in pure bending with no axial force, the height of the primary cracks h_o immediately after cracking is usually relatively high (0.6 to 0.9 times the depth of the member) and remains approximately constant under increasing bending moments until either the steel reinforcement yields or the concrete stress–strain relationship in the compressive region becomes non-linear. For prestressed members and members subjected to bending plus axial compression, h_o may be relatively small initially and gradually increases as the applied moment increases, even when material behaviour is linear-elastic.

The stress in the tensile reinforcement and the stress in the concrete at the steel level for the cracked member of Fig. 3.4a are illustrated in Figs 3.4b and c, respectively. Immediately after first cracking, the intact concrete between adjacent primary cracks carries considerable tensile force, mainly in the direction of the reinforcement, due to the bond between the steel and the concrete. The average tensile stress in the concrete is a significant percentage of the tensile strength of concrete. The steel stress is a maximum at a crack, where the steel carries the entire tensile force, and drops to a minimum between the cracks, as shown (Fig. 3.4b). The bending stiffness of the member is considerably greater than that based on a fully-cracked section, where concrete in tension is assumed to carry zero stress. This *tension stiffening* effect may be significant in the service-load performance of beams and even more so for lightly reinforced slabs.

The Euler–Bernoulli assumption that plane sections remain plane is not strictly true for a cross-section in the cracked region of a beam. However, if strains are measured over a gauge length containing several primary cracks, the average strain diagram is linear over the depth of the member.

As the load increases above the cracking moment M_{cr}, and after the primary cracks have developed, secondary cracks (or cover-controlled cracks) form around the reinforcement between the primary cracks, the average concrete tensile stress drops and the tension stiffening effect gradually reduces. A typical moment versus average curvature relationship for a reinforced concrete cross-section in pure bending is shown in Fig. 7.1a as the solid line OAB in which M_s represents the applied service moment. Also shown as the dashed line OC of slope $(EI)_{cr}$ is the moment–curvature relationship for a fully-cracked cross-section. As moment increases after first cracking, the flexural rigidity gradually reduces from that of the uncracked section $(EI)_{uncr}$ at first cracking and approaches that of the fully-cracked section, $(EI)_{cr}$, as the moment becomes large, as shown.

If an uncracked singly reinforced member such as that shown in Fig. 1.14a begins to shrink prior to loading, as is commonly the case, shrinkage warping occurs and a shrinkage-induced curvature $\kappa_{sh.uncr}$ develops on the uncracked cross-section when the applied moment is still zero (i.e. $M_s = 0$) as shown in Fig. 7.1b as point O′. The curvature $\kappa_{sh.uncr}$ and the tensile stress caused by shrinkage in the extreme fibre of the uncracked cross-section σ_{cs} were illustrated in Fig. 1.14b. Because of the initial tensile stress σ_{cs} in the concrete, the moment required to cause first cracking $M_{cr.sh0}$ will be less than M_{cr} (as indicated in Fig. 7.1) and the moment-curvature relationship is now represented by curve O′A′B′ in Fig. 7.1b. The initial curvature due to early shrinkage on a fully-cracked cross-section ($\kappa_{sh.cr}$) where the concrete is assumed to carry no tension is significantly larger that that of the uncracked member ($\kappa_{sh.uncr}$), as illustrated in Fig. 1.14. Therefore, early shrinkage before loading causes the dashed line representing the fully-cracked response to move further to the right, shown as line O″C′ in Fig. 7.1b.

For a prestressed concrete member, or a reinforced concrete member in combined bending and compression, the effect of tension stiffening is less pronounced because the loss of stiffness of the cracked section is less dramatic. As the applied moment increases, the depth of the primary cracks increases gradually (in contrast to the sudden crack propagation in a reinforced member in pure bending) and the depth of the concrete compressive zone is significantly greater than would be the case if no axial force was present. A typical short-term moment-average curvature relationship for a

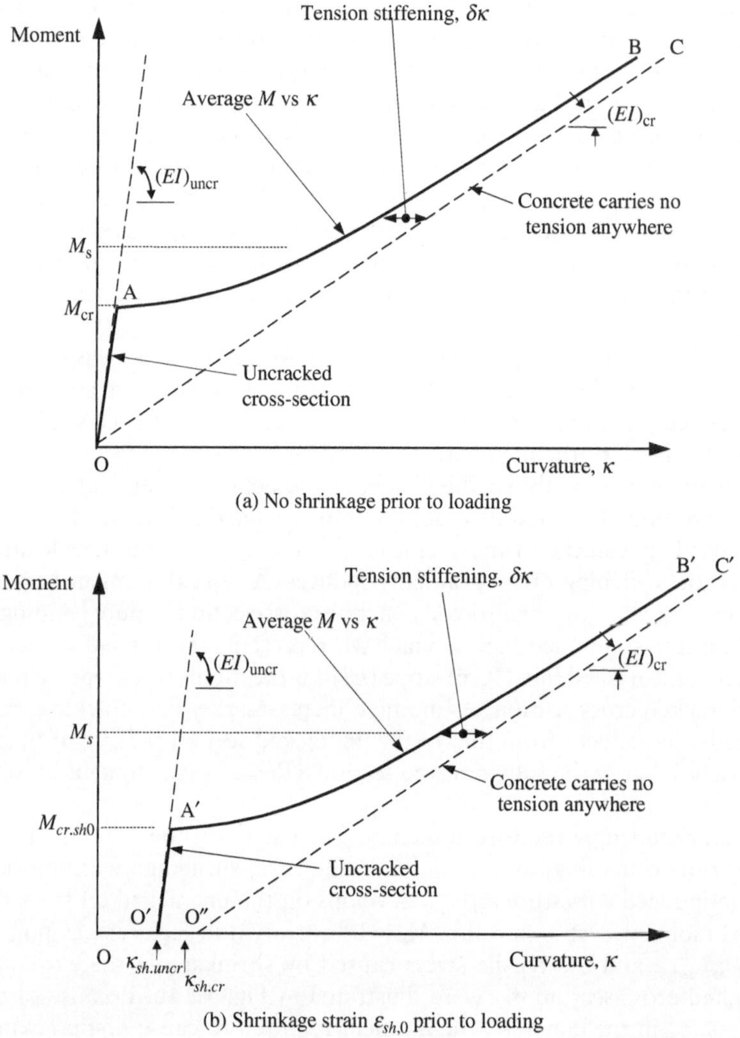

Figure 7.1 Moment-average curvature relationship for a reinforced concrete cross-section at first loading, τ_0.

prestressed concrete cross-section containing bonded tensile reinforcement is shown in Fig. 7.2, together with the moment–curvature relationship for the cross-section containing a crack. The internal moment caused by the resultant prestressing force about the centroidal axis of the uncracked section is designated *Pe*.

As can be seen by comparing the curves in Figs 7.1a and 7.2, the presence of the axial prestress significantly changes the shape of curve after cracking. The prestressing force (or axial compression) increases the post-cracking stiffness and greatly improves the in-service behaviour of a concrete member after the onset of flexural cracking.

In Section 3.6.3, modular ratio theory was presented for the determination of the stresses and deformations of a singly-reinforced rectangular cross-section in pure

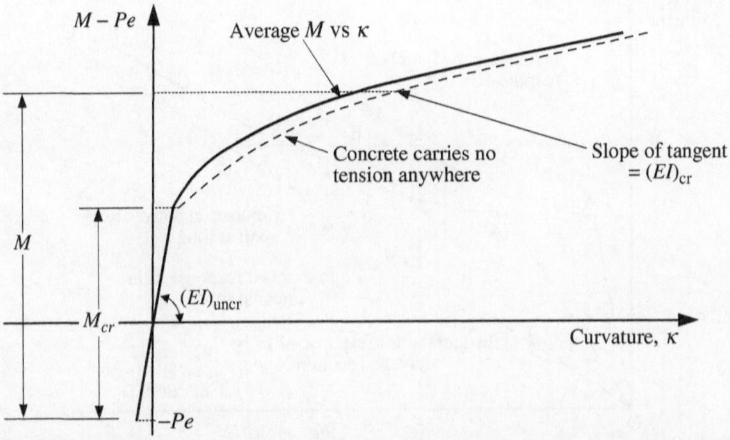

Figure 7.2 Short-term moment-average curvature relationship for a prestressed concrete cross-section.

bending at a primary crack location. The properties of the fully-cracked cross-section, including the depth to the neutral axis kd (Eq. 3.14) and the second moment of area of the cracked section about its centroidal axis I_{cr} (Eq. 3.17) were also derived. For a more general cross-section, the properties of the cracked section may be determined from a transformed section analysis in which the tensile concrete is ignored. Such an analysis is presented in Section 7.2. Cracking also influences the time-dependent behaviour of reinforced and prestressed concrete members. In Section 1.2.7, the effects of creep on a fully-cracked cross-section were compared with the corresponding effects on an uncracked cross-section (see Fig. 1.12), while the stresses and deformations caused by shrinkage on a cracked section were discussed in Section 1.3.3 and illustrated in Fig. 1.14b.

For a cross-section subjected to constant sustained moment over the time period τ_0 to t, if no shrinkage has occurred prior to loading, the instantaneous moment versus curvature response of the cross-section is shown as curve OAB in Fig. 7.3a (identical to curve OAB in Fig. 7.1a). The instantaneous fully-cracked section response (calculated ignoring the tensile concrete) is shown as dashed line OC in Fig. 7.3a. If the cross-section does not shrink with time (i.e. ε_{sh} remains at zero), creep causes an increase in curvature with time at all moment levels and the time-dependent $M-\kappa$ response shifts to curve OA′B′ in Fig. 7.3a. The creep-induced increase in curvature with time may be expressed as $\Delta\kappa_{cr}(t) = \kappa_i\varphi(t, \tau_0)/\alpha$, where κ_i is the instantaneous curvature, $\varphi(t, \tau_0)$ is the creep coefficient and α is a factor that depends on the amount of cracking and the reinforcement quantity and location. Approximate expressions for α were given in Eqs 3.32 and, for typical reinforcement ratios for beams and slabs, α is in the range 1.0–1.3 prior to cracking and in the range 4–6 when cracking is extensive. With regard to the response of a fully-cracked cross-section (calculated ignoring the tensile concrete), creep causes a softening of the response shown as dashed line OC′ in Fig. 7.3a, with the slope of the line decreasing by the factor $1/(1+\varphi(t, \tau_0)/\alpha)$, where α is in the range 4–6 depending on the reinforcement ratio.

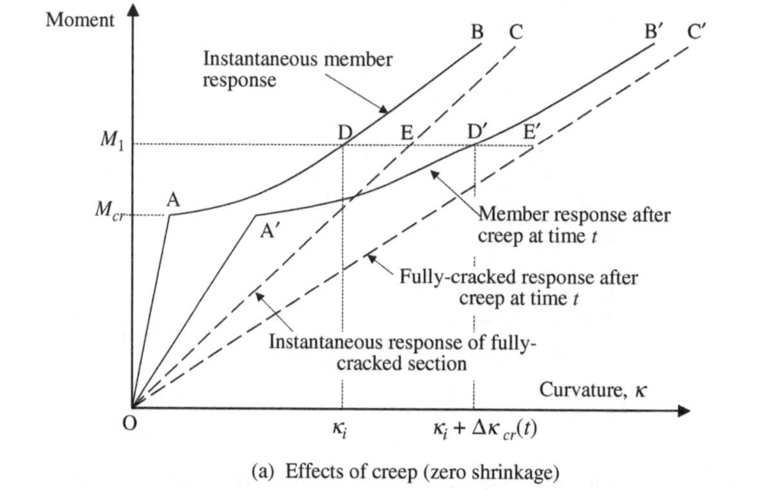

Figure 7.3 Moment-average curvature relationship for a reinforced concrete cross-section under sustained loads at time t.

When shrinkage before and after first loading is included, the curvature increases even further with time and the time-dependent response of the cross-section is shown as curve O′A′B′ in Fig. 7.3b. At $M = 0$, the curvature increases due to shrinkage of the uncracked cross-section and the point O moves horizontally to O′. Due to the restraint to shrinkage provided by the bonded reinforcement, tensile stress is induced with time and this has the effect of lowering the cracking moment from M_{cr} to $M_{cr.sh}$, as shown in Fig. 7.3b. For any cross-section subjected to a sustained moment in the range $M_{cr.sh} < M \leq M_{cr}$, cracking will occur with time and the increase in curvature will be exacerbated by the loss of stiffness caused by time-dependent cracking. In practice, critical sections of many lightly reinforced slabs are loaded in this range.

The response of the cracked section (ignoring the tensile concrete) after creep and shrinkage is shown as line O″C′ in Fig. 7.3b. The shrinkage-induced curvature of the fully-cracked cross-section when $M = 0$ is greater than that of the uncracked cross-section and the cracked section response is shifted horizontally from point O to point O″, as shown. The slope of the cracked section response in Fig. 7.3b is softened by creep and the slope of the line O″C′ in Fig. 7.3b is the same as the slope of line OC′ in Fig. 7.3a.

At a typical in-service moment M_1, the instantaneous curvature due to tension stiffening $\kappa_{ts.0}$ is DE in Fig. 7.3b and the time-dependent tension stiffening curvature after the period of sustained loadings $\kappa_{ts}(t)$ is D′E′ in Fig. 7.3b. Tension stiffening reduces under sustained loading, primarily due to time-dependent cracking, shrinkage-induced degradation of bond at the concrete-reinforcement interface and tensile creep between the cracks in the tensile concrete. It is generally believed that tension stiffening reduces rapidly after first loading and reduces to about half its instantaneous value with time (Refs 1–4).

When a fully-cracked cross-section is subjected to a period of sustained loading, creep causes a change in position of the neutral axis. In general, the depth to the neutral axis increases with time and, hence, so too does the area of concrete in compression. To account accurately for this gradual change of the properties of the cross-section, an iterative numerical solution procedure is required, in which the time period is divided into small increments and the cross-sectional properties at the end of each time increment are modified. While such a procedure is routinely implemented in finite-element models of cracked reinforced concrete, often using a layered cross-section, it is not suitable for manual solution.

The time analysis of a cracked cross-section using the age-adjusted effective modulus method is presented in Section 7.3. In this analysis, the dimensions of the fully-cracked cross-section are assumed to remain constant throughout the time analysis, i.e. the depth of the concrete in compression and the extent of cracking are assumed to remain constant with time. This assumption is in fact necessary, if the short-term and time-dependent stress and strain increments are to be calculated separately and added together to obtain final stresses and deformations, i.e. if the principle of superposition is to be applied to fully-cracked sections in the same way as it has been applied to uncracked cross-sections. The assumption also greatly simplifies the analysis and usually results in relatively little error in the estimation of time-dependent deformations.

7.2 Short-term analysis

In this section, the short-term responses of fully-cracked reinforced and partially-prestressed concrete cross-sections subjected to combined bending and axial force are considered. The time-dependent effects of creep and shrinkage are considered in Section 7.3.

As for the analyses of uncracked cross-sections in Chapters 4, 5 and 6, the analysis presented here is based on the following assumptions:

- Plane sections remain plane and, as a consequence, the strain distribution is linear over the depth of the section.
- Perfect bond exists between the steel and the concrete, i.e. steel and concrete strains are assumed to be compatible. This is usually a reasonable assumption at service loads in members containing deformed steel reinforcing bars and strands.

- Tensile stress in the concrete is ignored, and therefore the tensile concrete does not contribute to the cross-sectional properties.
- Material behaviour is linear-elastic. This includes concrete in compression, and both the non-prestressed and prestressed reinforcement.

The two cases of axial force with uniaxial bending and axial force with biaxial bending are dealt with separately in the following sections.

7.2.1 Axial force and uniaxial bending

In the short-term analysis of the section at first loading (at time τ_0), it is assumed that the axial force and bending moment about the x-axis, $N_{e,0}$ and $M_{e,0}$, respectively, produce tension of sufficient magnitude to cause cracking in the bottom fibres of the cross-section and compression at the top of the section.

Consider the cracked partially-prestressed concrete cross-section shown in Fig. 7.4. The section is symmetric about the y-axis and, for convenience, the orthogonal x-axis is selected as the reference axis. Also shown in Fig. 7.4 are the initial stress and strain distributions when the section is subjected to combined bending and axial force ($M_{e.0}$ and $N_{e.0}$) sufficient to cause cracking in the bottom fibres.

The numbers of layers of non-prestressed and prestressed reinforcement for the cross-section in Fig. 7.4 are $m_s = 3$ and $m_p = 2$, respectively. As for the analysis of uncracked cross-sections in Section 5.3, the properties of each layer of non-prestressed reinforcement are defined by its area, elastic modulus and location with respect to the arbitrarily chosen x-axis, i.e. $A_{s(i)}$, $E_{s(i)}$ and $y_{s(i)}(= d_{s(i)} - d_{ref})$, respectively. Similarly, $A_{p(i)}$, $E_{p(i)}$ and $y_{p(i)}(= d_{p(i)} - d_{ref})$ represent the area, elastic modulus and location of the prestressing steel with respect to the x-axis, respectively.

The strain at any depth y below the reference x-axis at time τ_0 is given by:

$$\varepsilon_0 = \varepsilon_{r,0} + y\kappa_0 \tag{7.1}$$

The stresses in the concrete and in the bonded reinforcement are:

$$\sigma_{c,0} = E_{c,0}\varepsilon_0 = E_{c,0}(\varepsilon_{r,0} + y\kappa_0) \text{ for } y \leq y_{n,0} \tag{7.2a}$$

$$\sigma_{c,0} = 0 \text{ for } y > y_{n,0} \tag{7.2b}$$

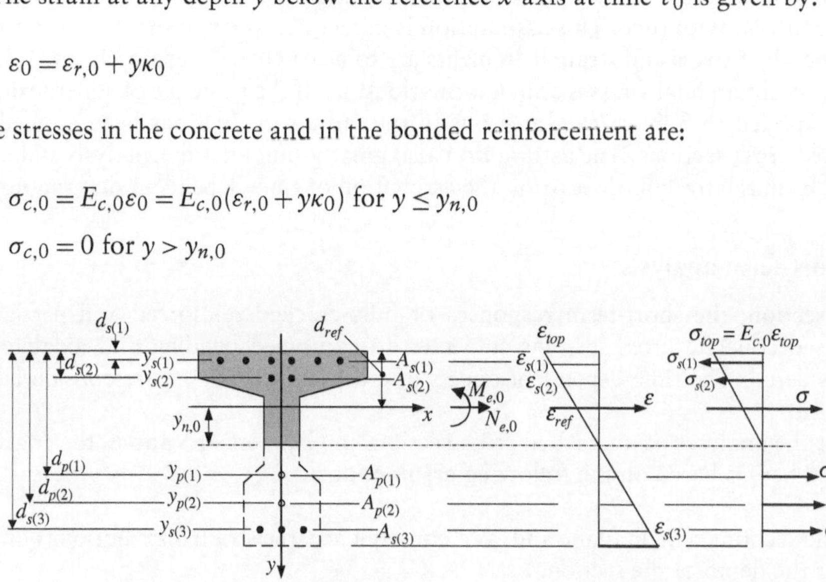

Figure 7.4 Fully-cracked reinforced concrete cross-section.

$$\sigma_{s(i),0} = E_{s(i)}\varepsilon_0 = E_{s(i)}(\varepsilon_{r,0} + y_{s(i)}\kappa_0) \tag{7.3}$$

$$\sigma_{p(i),0} = E_{p(i)}\left(\varepsilon_0 + \varepsilon_{p(i),init}\right) = E_{p(i)}\left(\varepsilon_{r,0} + y_{p(i)}\kappa_0 + \varepsilon_{p(i),init}\right) \tag{7.4}$$

where $y_{n,0}$ is the distance from the reference axis to neutral axis, as shown in Fig. 7.4a, and $\varepsilon_{p(i),init}$ is the strain in the i-th layer of prestressing steel immediately before the transfer of prestress to the concrete as expressed in Eq. 5.5 (reproduced and renumbered here for convenience):

$$\varepsilon_{p(i),init} = \frac{P_{p(i),init}}{A_{p(i)}E_{p(i)}} \tag{7.5}$$

The internal axial force $N_{i,0}$ on the cracked cross-section is the sum of the axial forces resisted by the various materials forming the cross-section and is given by:

$$N_{i,0} = N_{c,0} + N_{s,0} + N_{p,0} \tag{7.6}$$

where $N_{c,0}$, $N_{s,0}$ and $N_{p,0}$ represent the axial forces resisted by the concrete, the non-prestressed reinforcement and the prestressing steel, respectively, and are calculated using:

$$N_{c,0} = \int_{A_c} \sigma_{c,0}\,\mathrm{d}A = \int_{A_c} E_{c,0}(\varepsilon_{r,0} + y\kappa_0)\,\mathrm{d}A = A_c E_{c,0}\varepsilon_{r,0} + B_c E_{c,0}\kappa_0 \tag{7.7a}$$

$$N_{s,0} = \sum_{i=1}^{m_s}\left(A_{s(i)}E_{s(i)}\right)\left(\varepsilon_{r,0} + y_{s(i)}\kappa_0\right) = \sum_{i=1}^{m_s}\left(A_{s(i)}E_{s(i)}\right)\varepsilon_{r,0} + \sum_{i=1}^{m_s}\left(y_{s(i)}A_{s(i)}E_{s(i)}\right)\kappa_0 \tag{7.7b}$$

$$N_{p,0} = \sum_{i=1}^{m_p}\left(A_{p(i)}E_{p(i)}\right)\varepsilon_{r,0} + \sum_{i=1}^{m_p}\left(y_{p(i)}A_{p(i)}E_{p(i)}\right)\kappa_0 + \sum_{i=1}^{m_p}\left(A_{p(i)}E_{p(i)}\varepsilon_{p(i),init}\right) \tag{7.7c}$$

where A_c and B_c are the area and the first moment of area about the x-axis of the compressive concrete above the neutral axis (i.e. the properties of the intact compressive concrete). Adopting the following notation:

$$R_{A,s} = \sum_{i=1}^{m_s}\left(A_{s(i)}E_{s(i)}\right);\ R_{B,s} = \sum_{i=1}^{m_s}\left(y_{s(i)}A_{s(i)}E_{s(i)}\right);\ R_{I,s} = \sum_{i=1}^{m_s}\left(y_{s(i)}^2 A_{s(i)}E_{s(i)}\right)$$

$$R_{A,p} = \sum_{i=1}^{m_p}\left(A_{p(i)}E_{p(i)}\right);\ R_{B,p} = \sum_{i=1}^{m_p}\left(y_{p(i)}A_{p(i)}E_{p(i)}\right);\ R_{I,p} = \sum_{i=1}^{m_p}\left(y_{p(i)}^2 A_{p(i)}E_{p(i)}\right) \tag{7.8}$$

Eqs 7.7 become

$$N_{c,0} = \int_{A_c} \sigma_{c,0}\mathrm{d}A = \int_{A_c} E_{c,0}(\varepsilon_{r,0} + y\kappa_0)\mathrm{d}A \tag{7.9a}$$

$$N_{s,0} = R_{A.s}\varepsilon_{r,0} + R_{B.s}\kappa_0 \tag{7.9b}$$

$$N_{p,0} = R_{A,p}\varepsilon_{r,0} + R_{B,p}\kappa_0 + \sum_{i=1}^{m_p} \left(A_{p(i)}E_{p(i)}\varepsilon_{p(i),init}\right) \tag{7.9c}$$

Remembering that for equilibrium $N_{e,0} = N_{i,0}$ and that substituting Eqs 7.9 into Eq. 7.6 the expression for $N_{i,0}$ can be re-written in terms of the actual geometry and the elastic moduli of the materials forming the cross-section:

$$N_{i,0} = \int_{A_c} E_{c,0}(\varepsilon_{r,0} + y\kappa_0)\mathrm{d}A + (R_{A,s} + R_{A,p})\varepsilon_{r,0} + (R_{B,s} + R_{B,p})\kappa_0$$

$$+ \sum_{i=1}^{m_p} \left(A_{p(i)}E_{p(i)}\varepsilon_{p(i),init}\right) = N_{e,0} \tag{7.10}$$

Eq. 7.10 can be re-expressed as:

$$N_{e,0} - \sum_{i=1}^{m_p} \left(A_{p(i)}E_{p(i)}\varepsilon_{p(i),init}\right) = \int_{A_c} E_{c,0}(\varepsilon_{r,0} + y\kappa_0)\mathrm{d}A + (R_{A,s} + R_{A,p})\varepsilon_{r,0}$$

$$+ (R_{B,s} + R_{B,p})\kappa_0 \tag{7.11}$$

Similarly, the following expression based on moment equilibrium can be derived as:

$$M_{e,0} - \sum_{i=1}^{m_p} \left(y_{p(i)}A_{p(i)}E_{p(i)}\varepsilon_{p(i),init}\right) = \int_{A_c} E_{c,0}(\varepsilon_{r,0} + y\kappa_0)y\mathrm{d}A + (R_{B,s} + R_{B,p})\varepsilon_{r,0}$$

$$+ (R_{I,s} + R_{I,p})\kappa_0 \tag{7.12}$$

For a reinforced concrete section comprising rectangular components (e.g. rectangular flanges and webs) loaded in pure bending (i.e. $N_{e,0} = N_{i,0} = 0$) and with no prestress, Eq. 7.10 becomes a quadratic equation that can be solved to obtain the location of the neutral axis $y_{n,0}$.

If the cross-section is prestressed or the axial load $N_{e,0}$ is not equal to zero, dividing Eq. 7.12 by Eq. 7.11 gives:

$$\frac{M_{e,0} - \sum_{i=1}^{m_p} \left(y_{p(i)}A_{p(i)}E_{p(i)}\varepsilon_{p(i),inst}\right)}{N_{e,0} - \sum_{i=1}^{m_p} \left(A_{p(i)}E_{p(i)}\varepsilon_{p(i),inst}\right)}$$

$$= \frac{\int_{y=-d_{ref}}^{y=y_{n,0}} E_{c,0}(\varepsilon_{r,0} + y\kappa_0)y\mathrm{d}A + (R_{B,s} + R_{B,p})\varepsilon_{r,0} + (R_{I,s} + R_{I,p})\kappa_0}{\int_{y=-d_{ref}}^{y=y_{n,0}} E_{c,0}(\varepsilon_{r,0} + y\kappa_0)\mathrm{d}A + (R_{A,s} + R_{A,p})\varepsilon_{r,0} + (R_{B,s} + R_{B,p})\kappa_0}$$

Dividing top and bottom of the right hand side by κ_0 and recognising that at the axis of zero strain $y = y_{n,0} = -\varepsilon_{r,0}/\kappa_0$, the above expression becomes:

$$\frac{M_{e,0} - \sum_{i=1}^{m_p} (y_{p(i)} A_{p(i)} E_{p(i)} \varepsilon_{p(i),inst})}{N_{e,0} - \sum_{i=1}^{m_p} (A_{p(i)} E_{p(i)} \varepsilon_{p(i),inst})} = \frac{\int_{y=-d_{ref}}^{y=y_{n,0}} E_{c,0}(-y_{n,0} + y) y \, dA - (R_{B,s} + R_{B,p}) y_{n,0} + (R_{I,s} + R_{I,p})}{\int_{y=-d_{ref}}^{y=y_{n,0}} E_{c,0}(-y_{n,0} + y) \, dA - (R_{A,s} + R_{A,p}) y_{n,0} + (R_{B,s} + R_{B,p})} \tag{7.13}$$

and Eq. 7.13 may be solved for $y_{n,0}$ relatively quickly using a simple trial and error search.

When $y_{n,0}$ is determined, and the depth of the intact compressive concrete above the cracked tensile zone is known, the properties of the compressive concrete (A_c, B_c and I_c) with respect to the reference axis may be readily calculated. Using the same notation as in the short-term analysis of the uncracked section of Section 5.3 (Eqs 5.8–5.11), the expressions for $N_{i,0}$ and $M_{i,0}$ (Eqs 7.10 and 7.12) can be rewritten as:

$$N_{i,0} = R_{A,0} \varepsilon_{r,0} + R_{B,0} \kappa_0 + \sum_{i=1}^{m_p} (A_{p(i)} E_{p(i)} \varepsilon_{p(i),init}) \tag{7.14a}$$

$$M_{i,0} = R_{B,0} \varepsilon_{r,0} + R_{I,0} \kappa_0 + \sum_{i=1}^{m_p} (y_{p(i)} A_{p(i)} E_{p(i)} \varepsilon_{p(i),init}) \tag{7.14b}$$

and $R_{A,0}$, $R_{B,0}$ and $R_{I,0}$ are the axial rigidity and the stiffness related to the first and second moments of area of the cracked section about the reference axis, respectively, calculated at time τ_0 and are given by:

$$R_{A,0} = A_c E_{c,0} + \sum_{i=1}^{m_s} A_{s(i)} E_{s(i)} + \sum_{i=1}^{m_p} A_{p(i)} E_{p(i)} = A_c E_{c,0} + R_{A,s} + R_{A,p} \tag{7.15a}$$

$$R_{B,0} = B_c E_{c,0} + \sum_{i=1}^{m_s} y_{s(i)} A_{s(i)} E_{s(i)} + \sum_{i=1}^{m_p} y_{p(i)} A_{p(i)} E_{p(i)} = B_c E_{c,0} + R_{B,s} + R_{B,p} \tag{7.15b}$$

$$R_{I,0} = I_c E_{c,0} + \sum_{i=1}^{m_s} y_{s(i)}^2 A_{s(i)} E_{s(i)} + \sum_{i=1}^{m_p} y_{p(i)}^2 A_{p(i)} E_{p(i)} = I_c E_{c,0} + R_{I,s} + R_{I,p} \tag{7.15c}$$

The system of equilibrium equations governing the problem (Eq. 5.12) is rewritten here as:

$$\mathbf{r}_{e,0} = \mathbf{D}_0 \boldsymbol{\varepsilon}_0 + \mathbf{f}_{p,init} \tag{7.16}$$

and solving gives the vector of unknown strains:

$$\varepsilon_0 = \mathbf{D}_0^{-1}\left(\mathbf{r}_{e,0} - \mathbf{f}_{p,init}\right) = \mathbf{F}_0\left(\mathbf{r}_{e,0} - \mathbf{f}_{p,init}\right) \tag{7.17}$$

As defined previously:

$$\mathbf{r}_{e,0} = \begin{bmatrix} N_{e,0} \\ M_{e,0} \end{bmatrix} \tag{7.18a}$$

$$\mathbf{D}_0 = \begin{bmatrix} R_{A,0} & R_{B,0} \\ R_{B,0} & R_{I,0} \end{bmatrix} \tag{7.18b}$$

$$\boldsymbol{\varepsilon}_0 = \begin{bmatrix} \varepsilon_{r,0} \\ \kappa_0 \end{bmatrix} \tag{7.18c}$$

$$\mathbf{f}_{p,init} = \sum_{i=1}^{m_p} \begin{bmatrix} A_{p(i)}E_{p(i)}\varepsilon_{p(i),init} \\ y_{p(i)}A_{p(i)}E_{p(i)}\varepsilon_{p(i),init} \end{bmatrix} \tag{7.18d}$$

$$\mathbf{F}_0 = \frac{1}{R_{A,0}R_{I,0} - R_{B,0}^2} \begin{bmatrix} R_{I,0} & -R_{B,0} \\ -R_{B,0} & R_{A,0} \end{bmatrix} \tag{7.18e}$$

The stress distribution in the concrete and reinforcement can then be calculated from Eqs 7.2 to 7.4.

As an alternative approach, the solution may also be conveniently obtained using the cross-sectional properties of the transformed section. For example, for the cross-section of Fig. 7.4, the transformed cross-section in equivalent areas of concrete for the short-term analysis is shown in Fig. 7.5. The cross-sectional rigidities of the transformed section defined in Eqs 7.15 can be re-calculated as:

$$R_{A,0} = A_0 E_{c,0} \tag{7.19a}$$

$$R_{B,0} = B_0 E_{c,0} \tag{7.19b}$$

$$R_{I,0} = I_0 E_{c,0} \tag{7.19c}$$

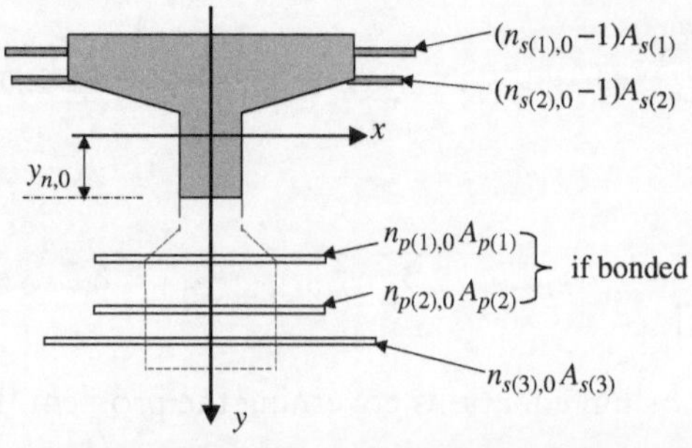

Figure 7.5 Transformed cracked section with bonded reinforcement (transformed into equivalent areas of concrete).

where A_0 is the area of the transformed cracked concrete section, and B_0 and I_0 are the first and second moments of the transformed area about the reference x-axis at first loading. Substituting Eqs 7.19 into Eq. 7.18e, the matrix $\mathbf{F}_0$ becomes:

$$\mathbf{F}_0 = \frac{1}{E_{c,0}(A_0 I_0 - B_0^2)} \begin{bmatrix} I_0 & -B_0 \\ -B_0 & A_0 \end{bmatrix} \tag{7.20}$$

Example 7.1

Assuming that the concrete can carry no tension, the position of the neutral axis and the stress and strain distributions immediately after first loading are to be calculated on the reinforced concrete cross-section shown in Fig. 7.6 for each of the following load cases:

(1) $N_{e,0} = 0$ and $M_{e,0} = 300$ kNm;

(2) $N_{e,0} = -1000$ kN and $M_{e,0} = 300$ kNm.

Both the concrete and the reinforcement are assumed to be linear-elastic with $E_{c,0} = 25$ GPa and $E_s = 200$ GPa. The modular ratio for the reinforcing steel is therefore $n_{s,0} = 8$.

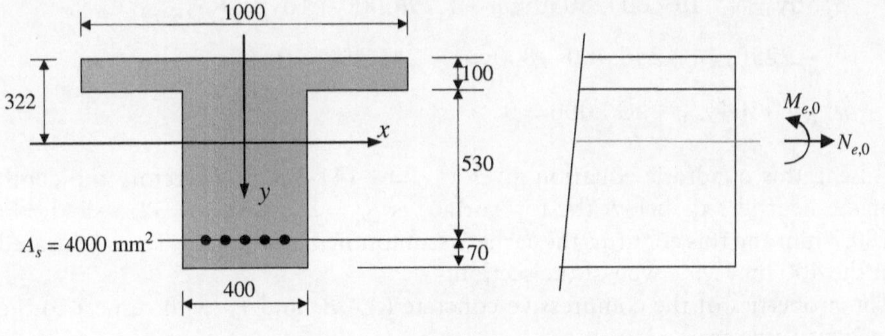

Figure 7.6 Reinforced concrete cross-section for Example 7.1

Load case 1: $N_{e,0} = 0$ and $M_{e,0} = M_{i,0} = 300$ kNm

Initially, it is assumed that the neutral axis is located in the 400 mm wide web of the cross-section.
From Eq. 7.8:

$$R_{A,s} = A_s E_s = 4000 \times 200{,}000 = 800 \times 10^6 \text{ N}$$

$$R_{B,s} = y_s A_s E_s = (630 - 322) \times 4000 \times 200{,}000 = 246.4 \times 10^9 \text{ Nmm}$$

$$R_{I,s} = y_s^2 A_s E_s = (630 - 322)^2 \times 4000 \times 200{,}000 = 75.89 \times 10^{12} \text{ Nmm}^2$$

With $N_{e,0} = N_{i,0} = 0$, Eq. 7.10 becomes:

$$N_{e,0} = \int_{A_c} E_{c,0}(\varepsilon_{r,0} + y\kappa_0)\mathrm{d}A + R_{A,s}\varepsilon_{r,0} + R_{B,s}\kappa_0 = 0$$

Dividing by κ_0 and remembering that $y_{r,0} = -\varepsilon_{r,0}/\kappa_0$ gives:

$$\int_{A_c} E_{c,0}(-y_{n,0} + y)\mathrm{d}A - R_{A,s}y_{n,0} + R_{B,s} = 0$$

which can be used to obtain a parabolic expressions in $y_{n,0}$:

$$\int_{y=-322}^{y=-222} 25{,}000(-y_{n,0} + y) \times 1000\ \mathrm{d}y + \int_{y=-222}^{y=y_{n,0}} 25{,}000(-y_{n,0} + y) \times 400\ \mathrm{d}y$$

$$- 800 \times 10^6 y_{n,0} + 246.4 \times 10^9 = 0$$

$$\left|25 \times 10^6(-y_{n,0}y + 0.5y^2)\right|_{-322}^{-222} + \left|10 \times 10^6(-y_{n,0}y + 0.5y^2)\right|_{-222}^{y_{n,0}}$$

$$- 800 \times 10^6 y_{n,0} + 246.4 \times 10^9 = 0$$

$$5550y_{n,0} + 616{,}100 - 8050y_{n,0} - 1{,}296{,}000 - 10y_{n,0}^2 + 5y_{n,0}^2$$

$$- 2220y_{n,0} - 246{,}400 - 800y_{n,0} + 246{,}400 = 0$$

$$y_{n,0}^2 + 1104y_{n,0} + 136{,}000 = 0$$

Solving this quadratic equation gives $y_{n,0} = -141.3$ mm. Therefore the depth of the neutral axis below the top surface is $d_n = d_{ref} + y_{n,0} = 322 - 141.3 = 180.7$ mm and this confirms the earlier assumption that the neutral axis is located in the 400 mm wide web of the section.

The properties of the compressive concrete (A_c, B_c and I_c) with respect to the reference axis are:

$$A_c = 1000 \times 100 + 400 \times 80.7 = 132{,}300\ \mathrm{mm}^2;$$

$$B_c = 1000 \times 100 \times (50 - 322) + 400 \times 80.7 \times (140.35 - 322)$$

$$= -33.06 \times 10^6\ \mathrm{mm}^3;$$

$$I_c = \frac{1000 \times 100^3}{12} + 1000 \times 100 \times (50 - 322)^2 + \frac{400 \times 80.7^3}{12}$$

$$+ 400 \times 80.7 \times (140.35 - 322)^2 = 8564 \times 10^6\ \mathrm{mm}^4.$$

The cross-sectional rigidities $R_{A,0}$, $R_{B,0}$ and $R_{I,0}$ are obtained from Eqs 7.15:

$$R_{A,0} = A_c E_{c,0} + R_{A,s} = 132{,}300 \times 25{,}000 + 800 \times 10^6 = 4107 \times 10^6\ \mathrm{N};$$

$$R_{B,0}=B_c E_{c,0}+R_{B,s}=-33.06\times10^6\times25{,}000+246.4\times10^9=-580.2\times10^9\text{ Nmm};$$
$$R_{I,0}=I_c E_{c,0}+R_{I,s}=8564\times10^6\times25{,}000+75.89\times10^{12}=290.0\times10^{12}\text{ Nmm}^2.$$

From Eq. 7.18e:

$$\mathbf{F}_0=\frac{1}{R_{A,0}R_{I,0}-R_{B,0}^2}\begin{bmatrix} R_{I,0} & -R_{B,0}\\ -R_{B,0} & R_{A,0}\end{bmatrix}=\begin{bmatrix} 339.4\times10^{-12} & 679.1\times10^{-15}\\ 679.1\times10^{-15} & 4.807\times10^{-15}\end{bmatrix}$$

and the strain vector is obtained from Eq. 7.17:

$$\varepsilon_0=\mathbf{F}_0\,\mathbf{r}_{e,0}=\begin{bmatrix} 339.4\times10^{-12} & 679.1\times10^{-15}\\ 679.1\times10^{-15} & 4.807\times10^{-15}\end{bmatrix}\begin{bmatrix} 0\\ 300\times10^6\end{bmatrix}$$
$$=\begin{bmatrix} 203.7\times10^{-6}\\ 1.442\times10^{-6}\text{ mm}^{-1}\end{bmatrix}$$

From Eq. 7.1, the top ($y=-322$ mm) and bottom ($y=+378$ mm) fibre strains are:

$$\varepsilon_{0(top)}=\varepsilon_{r,0}-322\times\kappa_0=(203.7-322\times1.442)\times10^{-6}=-260.6\times10^{-6}$$
$$\varepsilon_{0(btm)}=\varepsilon_{r,0}+378\times\kappa_0=(203.7+378\times1.442)\times10^{-6}=+748.8\times10^{-6}$$

The top fibre stress in the concrete and the stress in the reinforcement are (Eqs 7.2a and 7.3):

$$\sigma_{c,0(top)}=E_{c,0}\varepsilon_{0(top)}=25{,}000\times(-260.6\times10^{-6})=-6.52\text{ MPa}$$
$$\sigma_{s,0}=E_s(\varepsilon_{r,0}+y_s\kappa_0)=200\times10^3\times(203.7+308\times1.442)\times10^{-6}=+129.6\text{ MPa}$$

The results are plotted in Fig. 7.7a.

Load case 2: $N_{e,0}=-1000$ kN and $M_{e,0}=300$ kN.m

As for load case 1, $R_{A,s}=800\times10^6$ N, $R_{B,s}=246.4\times10^9$ Nmm, $R_{I,s}=75.89\times10^{12}$ Nmm^2, and it is clear that for this load case the depth of the neutral axis is greater than that for load case 1, i.e. $y_{n,0}>-141.3$ mm. In this reinforced concrete section, Eq. 7.13 reduces to:

$$\frac{M_{e,0}}{N_{e,0}}=\frac{\int_{y=-322}^{y=-222}25{,}000\times1000\times(-y_{n,0}+y)y\,dy+\int_{y=-222}^{y=-y_{n,0}}25{,}000\times400\times(-y_{n,0}+y)y\,dy-246.4\times10^9 y_{n,0}+75.89\times10^{12}}{\int_{y=-322}^{y=-222}25{,}000\times1000\times(-y_{n,0}+y)\,dy+\int_{y=-222}^{y=y_{n,0}}25{,}000\times400\times(-y_{n,0}+y)\,dy-800\times10^6 y_{n,0}+246.4\times10^9}$$

$$=\frac{\left|25\times10^6(-0.5y_{n,0}y^2+0.\dot{3}y^3)\right|_{-322}^{-222}+\left|10\times10^6(-0.5y_{n,0}y^2+0.\dot{3}y^3)\right|_{-222}^{y_{n,0}}-246.4\times10^9 y_{n,0}+75.89\times10^{12}}{\left|25\times10^6(-y_{n,0}y+0.5y^2)\right|_{-322}^{-222}+\left|10\times10^6(-y_{n,0}y+0.5y^2)\right|_{-222}^{y_{n,0}}-800\times10^6 y_{r,0}+246.4\times10^9}$$

$$=\frac{-1.6\dot{6}y_{n,0}^3+680.0\times10^3 y_{n,0}+299.4\times10^6}{-5y_{n,0}^2-5520y_{n,0}-680.0\times10^3}=-300\text{ mm}$$

Solving this cubic equation (either directly or by trial) gives $y_{n,0} = 85.5$ mm. Therefore, the depth of the neutral axis below the top surface is $d_n = d_{ref} + y_{n,0} = 407.5$ mm.

The properties of the compressive concrete (A_c, B_c and I_c) with respect to the reference axis are:

$$A_c = 1000 \times 100 + 400 \times 307.5 = 223{,}000 \ \ \text{mm}^2$$

$$B_c = 1000 \times 100 \times (50 - 322) + 400 \times 307.5 \times (253.8 - 322)$$

$$= -35.59 \times 10^6 \ \text{mm}^3$$

$$I_c = \frac{1000 \times 100^3}{12} + 1000 \times 100 \times (50 - 322)^2 + \frac{400 \times 307.5^3}{12}$$

$$+ 400 \times 307.5 \times (253.8 - 322)^2 = 9024 \times 10^6 \ \text{mm}^4$$

The cross-sectional rigidities $R_{A,0}$, $R_{B,0}$ and $R_{I,0}$ are obtained from Eqs 7.15:

$$R_{A,0} = A_c E_{c,0} + R_{A,s} = 223{,}000 \times 25{,}000 + 800 \times 10^6 = 6375 \times 10^6 \ \text{N}$$

$$R_{B,0} = B_c E_{c,0} + R_{B,s} = -35.59 \times 10^6 \times 25{,}000 + 246.4 \times 10^9$$

$$= -643.5 \times 10^9 \ \text{Nmm}$$

$$R_{I,0} = I_c E_{c,0} + R_{I,s} = 9024 \times 10^6 \times 25{,}000 + 75.89 \times 10^{12}$$

$$= 301.5 \times 10^{12} \ \text{Nmm}^2$$

From Eq. 7.18e:

$$\mathbf{F}_0 = \frac{1}{R_{A,0} R_{I,0} - R_{B,0}^2} \begin{bmatrix} R_{I,0} & -R_{B,0} \\ -R_{B,0} & R_{A,0} \end{bmatrix} = \begin{bmatrix} 199.9 \times 10^{-12} & 426.7 \times 10^{-15} \\ 426.7 \times 10^{-15} & 4.228 \times 10^{-15} \end{bmatrix}$$

and the strain vector is obtained from Eq. 7.17:

$$\boldsymbol{\varepsilon}_0 = \mathbf{F}_0 \mathbf{r}_{e,0} = \begin{bmatrix} 199.9 \times 10^{-12} & 426.7 \times 10^{-15} \\ 426.7 \times 10^{-15} & 4.228 \times 10^{-15} \end{bmatrix} \begin{bmatrix} -1000 \times 10^3 \\ 300 \times 10^6 \end{bmatrix}$$

$$= \begin{bmatrix} -71.9 \times 10^{-6} \\ 0.842 \times 10^{-6} \ \text{mm}^{-1} \end{bmatrix}$$

From Eq. 7.1, the top ($y = -322$ mm) and bottom ($y = +378$ mm) fibre strains are:

$$\varepsilon_{0(top)} = \varepsilon_{r,0} - 322 \times \kappa_0 = (-71.9 - 322 \times 0.842) \times 10^{-6} = -343.0 \times 10^{-6}$$

$$\varepsilon_{0(btm)} = \varepsilon_{r,0} + 378 \times \kappa_0 = (-71.9 + 378 \times 0.842) \times 10^{-6} = +246.4 \times 10^{-6}$$

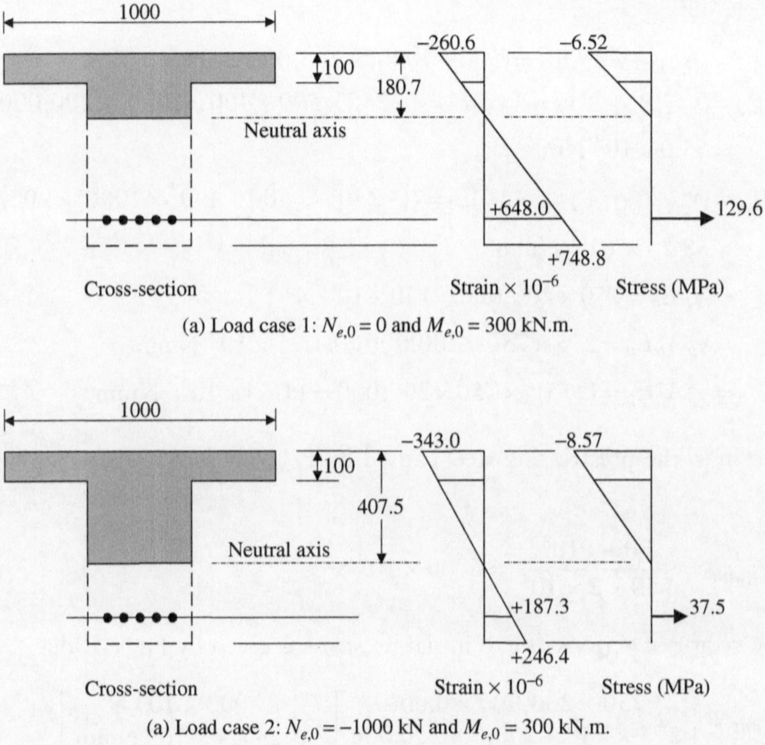

Figure 7.7 Initial stress and strain distributions in Example 7.1.

The top fibre stress in the concrete and the stress in the reinforcement are (Eqs 7.2a and 7.3):

$$\sigma_{c,0(top)} = E_{c,0}\varepsilon_{0(top)} = 25{,}000 \times (-343.0 \times 10^{-6}) = -8.57 \text{ MPa}$$

$$\sigma_{s,0} = E_s(\varepsilon_{r,0} + y_s\kappa_0) = 200 \times 10^3 \times (-71.9 + 308 \times 0.842) \times 10^{-6}$$
$$= +37.5 \text{ MPa}$$

The results are plotted in Fig. 7.7b.

Example 7.2

The depth of the concrete compression zone d_n and the short-term stress and strain distributions are to be calculated on the partially-prestressed concrete beam cross-section shown in Fig. 7.8a, when $M_{e,0} = 400$ kNm (and $N_{e,0} = 0$). The section contains two layers of non-prestressed reinforcement as shown (each with $E_s = 2 \times 10^5$ MPa) and one layer of bonded prestressing steel ($E_p = 2 \times 10^5$ MPa). The prestressing force before transfer is $P_{p,init} = 900$ kN (i.e. $\sigma_{p,init} = 1200$ MPa). The tensile strength of the concrete is $f'_{ct.f} = 3.5$ MPa and the elastic modulus is $E_{c,0} = 30{,}000$ MPa.

From Eq. 7.8:

$$R_{A,s} = (A_{s(1)} + A_{s(2)})E_s = (500 + 1000) \times 200{,}000 = 300 \times 10^6 \text{ N}$$

$$R_{B,s} = (y_{s(1)}A_{s(1)} + y_{s(2)}A_{s(2)})E_s = (-250 \times 500 + 400 \times 1000) \times 200{,}000$$
$$= 55.0 \times 10^9 \text{ Nmm}$$

$$R_{I,s} = (y_{s(1)}^2 A_{s(1)} + y_{s(2)}^2 A_{s(2)})E_s = ((-250)^2 \times 500 + 400^2 \times 1000) \times 200{,}000$$
$$= 38.25 \times 10^{12} \text{ Nmm}^2$$

$$R_{A,p} = A_p E_p = 750 \times 200{,}000 = 150 \times 10^6 \text{ N}$$

$$R_{B,p} = y_p A_p E_p = 275 \times 750 \times 200{,}000 = 41.25 \times 10^9 \text{ Nmm}$$

$$R_{I,p} = y_p^2 A_p E_p = (275)^2 \times 750 \times 200{,}000 = 11.34 \times 10^{12} \text{ Nmm}^2$$

The strain in the prestressing steel caused by $P_{p,init}$ before transfer is given by Eq. 7.5:

$$\varepsilon_{p(i),init} = \frac{900 \times 10^3}{750 \times 2 \times 10^5} = 6000 \times 10^{-6}$$

and the vector of actions due to initial prestress is given by Eq. 7.18d:

$$\mathbf{f}_{p,init} = \begin{bmatrix} 750 \times 200{,}000 \times 0.006 \\ 275 \times 750 \times 200{,}000 \times 0.006 \end{bmatrix} = \begin{bmatrix} 900 \times 10^3 \text{ N} \\ 247.5 \times 10^6 \text{ Nmm} \end{bmatrix}$$

If it is initially assumed that the section is uncracked, an analysis using the procedure outlined in Section 5.3 indicates that the tensile strength of the concrete has been exceeded in the bottom fibres of the cross-section. With the reference axis selected at d_{ref} = 300 mm below the top of the section, the depth of the neutral axis below the reference axis $y_{n,0}$ is determined from Eq. 7.13. The left hand side of Eq. 7.13 is first calculated:

$$\frac{M_{e,0} - \sum_{i=1}^{m_p} (y_{p(i)} A_{p(i)} E_{p(i)} \varepsilon_{p(i),inst})}{N_{e,0} - \sum_{i=1}^{m_p} (A_{p(i)} E_{p(i)} \varepsilon_{p(i),inst})} = \frac{400 \times 10^6 - 247.5 \times 10^6}{0 - 900 \times 10^3} = -169.\dot{4} \text{ mm}$$

and therefore:

$$-169.4 = \frac{\int_{y=-300}^{y=y_{n,0}} 30{,}000 \times 200 \times (-y_{n,0} + y) y \, dy - (55.0 + 41.25) \times 10^9 y_{n,0} + (38.25 + 11.34) \times 10^{12}}{\int_{y=-300}^{y=y_{n,0}} 30{,}000 \times 200 \times (-y_{n,0} + y) dy - (300 + 150) \times 10^6 y_{n,0} + (55.0 + 41.25) \times 10^9}$$

$$-169.4 = \frac{|6.0 \times 10^6(-0.5 y_{n,0} y^2 + 0.\dot{3} y^3)|_{-300}^{y_{n,0}} - 96.25 \times 10^9 y_{n,0} + 49.59 \times 10^{12}}{|6.0 \times 10^6(-y_{n,0} y + 0.5 y^2)|_{-300}^{y_{n,0}} - 450 \times 10^6 y_{n,0} + 96.25 \times 10^9}$$

$$-169.4 = \frac{-y_{n,0}^3 + 173{,}750 y_{n,0} + 103.6 \times 10^6}{-3y_{n,0}^2 - 2250 y_{n,0} - 173{,}750}$$

Solving gives $y_{n,0} = 208.1$ mm. Therefore the depth of the neutral axis below the top surface is $d_n = d_{ref} + y_{n,0} = 508.1$ mm.
The properties of the compressive concrete (A_c, B_c and I_c) with respect to the reference axis are:

$$A_c = 508.1 \times 200 - 500 = 101{,}100 \text{ mm}^2$$

$$B_c = 508.1 \times 200 \times (254.05 - 300) - 500 \times (50 - 300) = -4.55 \times 10^6 \text{ mm}^3$$

$$I_c = \frac{200 \times 508.1^3}{12} + 200 \times 508.1 \times (254.05 - 300)^2 - 500 \times (50 - 300)^2$$
$$= 2370 \times 10^9 \text{ mm}^4$$

The cross-sectional rigidities $R_{A,0}$, $R_{B,0}$ and $R_{I,0}$ are obtained from Eqs 7.15:

$$R_{A,0} = A_c E_{c,0} + R_{A,s} + R_{A,p} = 101{,}120 \times 30{,}000 + 300 \times 10^6$$
$$+ 150 \times 10^6 = 3484 \times 10^6 \text{ N}$$

$$R_{B,0} = B_c E_{c,0} + R_{B,s} + R_{B,p} = -4.55 \times 10^6 \times 30{,}000 + 55.0 \times 10^9$$
$$+ 41.25 \times 10^9 = -40.24 \times 10^9 \text{ Nmm}$$

$$R_{I,0} = I_c E_{c,0} + R_{I,s} + R_{I,p} = 2370 \times 10^6 \times 30{,}000 + 38.25 \times 10^{12}$$
$$+ 11.34 \times 10^{12} = 120.7 \times 10^{12} \text{ Nmm}^2$$

From Eq. 7.18e:

$$\mathbf{F}_0 = \frac{1}{R_{A,0} R_{I,0} - R_{B,0}^2} \begin{bmatrix} R_{I,0} & -R_{B,0} \\ -R_{B,0} & R_{A,0} \end{bmatrix} = \begin{bmatrix} 288.1 \times 10^{-12} & 96.06 \times 10^{-15} \\ 96.06 \times 10^{-15} & 8.317 \times 10^{-15} \end{bmatrix}$$

and the strain vector is obtained from Eq. 7.17:

$$\boldsymbol{\varepsilon}_0 = \mathbf{F}_0 \left(\mathbf{r}_{e,0} - \mathbf{f}_{p,init} \right) = \begin{bmatrix} 288.1 \times 10^{-12} & 96.06 \times 10^{-15} \\ 96.06 \times 10^{-15} & 8.317 \times 10^{-15} \end{bmatrix} \times$$
$$\times \begin{bmatrix} 0 - 900 \times 10^3 \\ 400 \times 10^6 - 247.5 \times 10^6 \end{bmatrix} = \begin{bmatrix} -244.6 \times 10^{-6} \\ 1.182 \times 10^{-6} \text{ mm}^{-1} \end{bmatrix}$$

From Eq. 7.1, the top ($y = -300$ mm) and bottom ($y = +450$ mm) fibre strains are:

$$\varepsilon_{0(top)} = \varepsilon_{r,0} - 300 \times \kappa_0 = (-244.6 - 300 \times 1.182) \times 10^{-6}$$
$$= -599.1 \times 10^{-6}$$

$$\varepsilon_{0(btm)} = \varepsilon_{r,0} + 450 \times \kappa_0 = (-244.6 + 450 \times 1.182) \times 10^{-6}$$
$$= +287.3 \times 10^{-6}$$

The distribution of strains is shown in Fig. 7.8b.

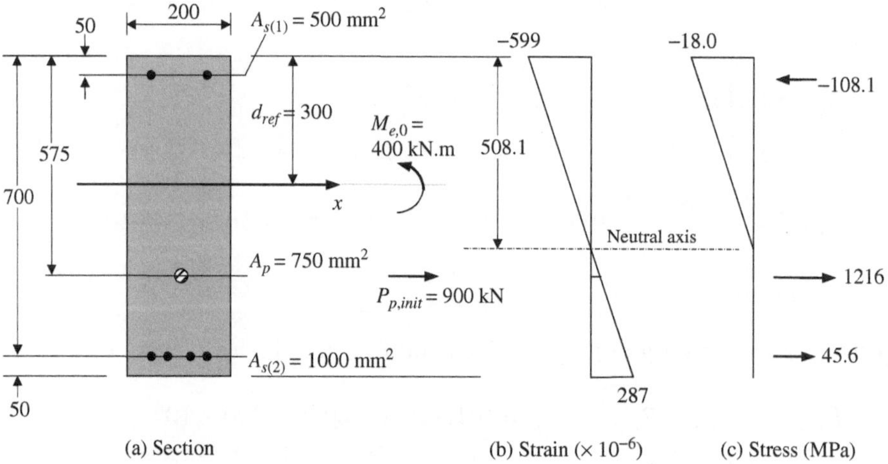

Figure 7.8 Cross-sectional details and stress and strain distributions for Example 7.2 (all dimensions in mm).

The top fibre stress in the concrete and the stress in the non-prestressed reinforcement are (Eqs 7.2a and 7.3):

$$\sigma_{c,0(top)} = E_{c,0}\varepsilon_{0(top)} = 30{,}000 \times (-599.1 \times 10^{-6}) = -18.0 \text{ MPa}$$

$$\sigma_{s(1),0} = E_s(\varepsilon_{r,0} + y_{s(1)}\kappa_0) = 200 \times 10^3 \times (-244.6 - 250 \times 1.182) \times 10^{-6}$$
$$= -108.1 \text{ MPa}$$

$$\sigma_{s(2),0} = E_s(\varepsilon_{r,0} + y_{s(2)}\kappa_0) = 200 \times 10^3 \times (-244.6 + 400 \times 1.182) \times 10^{-6}$$
$$= +45.6 \text{ MPa}$$

and the stress in the prestressing steel is given by:

$$\sigma_{p,0} = E_p\left(\varepsilon_{r,0} + y_p\kappa_0 + \varepsilon_{p,init}\right)$$
$$= 200 \times 10^3 \times (-244.6 + 275 \times 1.182 + 6000) \times 10^{-6} = 1216 \text{ MPa}$$

The stresses are plotted in Fig. 7.8c.

7.2.2 Axial force and biaxial bending

The short-term analysis of an uncracked cross-section subjected to axial force and biaxial bending was presented in Section 6.2. The procedure is readily extended to cracked sections where the extent of cracking is evaluated based on a trial and error procedure depicted in Fig. 7.9 and described below.

In the first iteration, all concrete is initially assumed to be uncracked (Fig. 7.9a) and the analysis is identical to that described in Section 6.2. If the maximum tensile

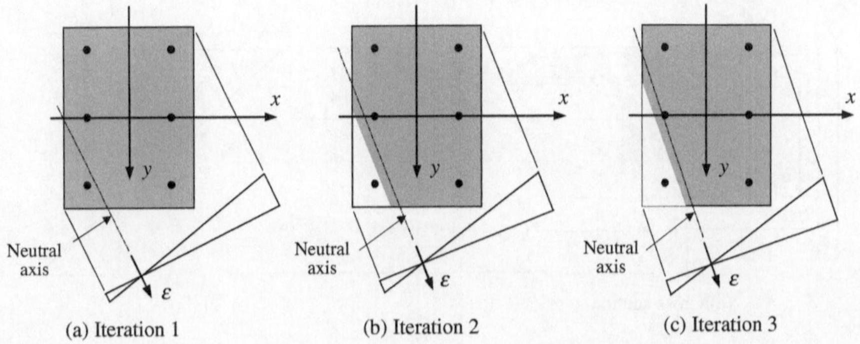

Figure 7.9 Iterative procedure for cracked sections under axial force and biaxial bending.

stress in the concrete exceeds the tensile strength of the concrete, cracking is deemed to have occurred. In this case, the area of concrete in tension identified after the first iteration is assumed to be cracked and to no longer contribute to the stiffness of the cross-section. This occurs, for example, in the bottom left hand corner of the cross-section shown in Fig. 7.9a, where the triangular part of the concrete section on the tensile side of the neutral axis is identified to have cracked. For the second iteration, the solution procedure remains unchanged except that the contribution of the concrete to the cross-sectional rigidities changes, with the cracked area of concrete identified in the first iteration (shown as the unshaded region in Fig. 7.9b) now not included in the calculation of the concrete section properties (A_c, B_{xc}, B_{yc}, I_{xc}, I_{yc}, I_{xyc}). After the second iteration, the revised neutral axis location is determined and a new area of cracked concrete is calculated. This procedure is continued in successive iterations (Fig. 7.9c) until the differences in the calculated cross-sectional rigidities between successive iterations are negligible. From a numerical viewpoint, convergence is usually assumed to have occurred when the variations of the values of the variables defining the strain diagram (i.e. ε_r, κ_x and κ_y) in consecutive iterations are negligible.

7.3 Time-dependent analysis (AEMM)

In the analyses presented here, the fully-cracked cross-sectional area (i.e. the size of the uncracked concrete compressive zone) is assumed to remain constant with time. This assumption is not correct since creep causes a gradual change of the position of the neutral axis under sustained loads. However, as discussed in Section 7.1, the assumption greatly simplifies the analysis and will usually result in relatively little error in the calculated deformations.

7.3.1 Axial force and uniaxial bending

With the depth of the concrete compression zone d_n assumed to remain constant with time, the time analysis of a fully-cracked cross-section using the AEMM is essentially the same as that outlined in Section 5.4.

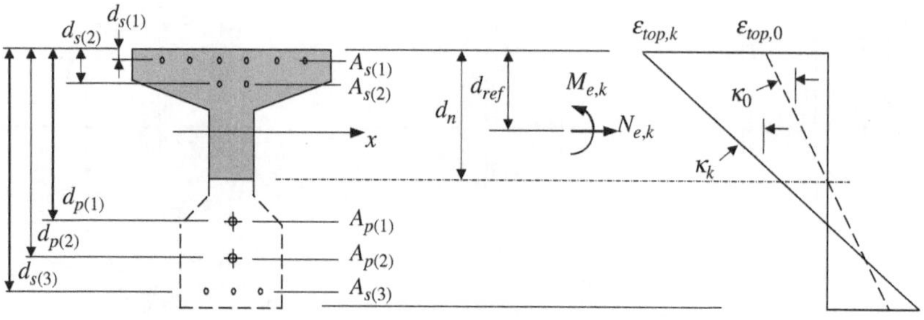

Figure 7.10 Fully-cracked cross-section – time analysis (AEMM).

The fully-cracked cross-section shown in Fig. 7.10a is subjected to a sustained external bending moment $M_{e,0}$ and axial force $N_{e,0}$. Both the short-term and time-dependent strain distributions are shown in Fig. 7.10b.

For the time analysis, the steel reinforcement and prestressing tendons (if any) are assumed to be linear-elastic (as for the short-term analysis) and the constitutive relationship for the concrete at τ_k is that given by Eq. 4.45. Therefore:

$$\sigma_{c,k} = \overline{E}_{e,k}\left(\varepsilon_k - \varepsilon_{sh,k}\right) + \overline{F}_{e,0}\sigma_{c,0} \text{ for } y \leq y_{n,0} \text{ and } \sigma_{c,k} = 0 \text{ for } y > y_{n,0} \quad (7.21a)$$

$$\sigma_{s(i),k} = E_{s(i)}\varepsilon_k \quad (7.21b)$$

$$\sigma_{p(i),k} = E_{p(i)}\left(\varepsilon_k + \varepsilon_{p(i),init} - \varepsilon_{p.rel(i),k}\right) \quad (7.21c)$$

where $\overline{E}_{e,k}$, $\overline{F}_{e,0}$ and $\varepsilon_{p.rel(i),k}$ are as defined previously (and given in Eqs 4.35, 4.46 and 5.23a).

At time τ_k, the internal axial force $N_{i,k}$ and moment $M_{i,k}$ on the cross-section are given in Eqs 5.29 and 5.31 and the axial rigidity and the stiffness related to the first and second moments of area ($R_{A,k}$, $R_{B,k}$, and $R_{I,k}$, respectively) are given by Eqs 5.30 and 5.32. The equilibrium equations are expressed in Eq. 5.34 and solving using Eq. 5.39 gives the strain vector at time τ_k. The stresses in the concrete, steel reinforcement and tendons at time τ_k are then calculated from Eqs 5.41a, b and c, respectively.

Example 7.3

The change of stress and strain with time on the cracked cross-section of Example 7.1 (Fig. 7.6) is to be calculated using AEMM. Each of the loading cases considered in Example 7.1 is to be examined here. The relevant material properties for the time period in question are: $E_{c,0} = 25$ GPa; $E_s =$ 200 GPa; $\varphi(\tau_k, \tau_0) = 3.0$; $\chi(\tau_k, \tau_0) = 0.65$; $\varepsilon_{sh}(\tau_k) = -500 \times 10^{-6}$ and the steel reinforcement is assumed to be linear elastic. To illustrate the effects of creep, the case when $\varepsilon_{sh}(\tau_k) = 0$ is also to be considered for load case 1.

Load case 1a: $N_{e,k} = 0$; $M_{e,k} = M_{i,k} = 300$ kNm and $\varepsilon_{sh}(\tau_k) = -500 \times 10^{-6}$

From load case 1 in Example 7.1, $d_n = 180.7$ mm; $\varepsilon_{r,0} = 203.7 \times 10^{-6}$; $\kappa_0 = 1.442 \times 10^{-6}$ mm^{-1}; and the shape of the cracked cross-section and the initial stress and strain distributions are illustrated in Fig. 7.7a.
From Eqs 4.35 and 4.46:

$$\overline{E}_{e,k} = \frac{E_{c,0}}{1+\chi(\tau_k,\tau_0)\varphi(\tau_k,\tau_0)} = \frac{25{,}000}{1+0.65\times 3.0} = 8475 \text{ MPa}$$

$$\overline{F}_{e,0} = \frac{\varphi(\tau_k,\tau_0)[\chi(\tau_k,\tau_0)-1]}{1+\chi(\tau_k,\tau_0)\varphi(\tau_k,\tau_0)} = \frac{3.0\times(0.65-1.0)}{1.0+0.65\times 3.0} = -0.356$$

and as calculated in Example 7.1, the properties of the concrete part of the cross-section with respect to the reference axis are $A_c = 132{,}300$ mm^2, $B_c = -33.06 \times 10^6$ mm^3 and $I_c = 8564 \times 10^6$ mm^4 and the rigidities of the steel reinforcement are $R_{A,s} = 800 \times 10^6$ N, $R_{B,s} = 246.4 \times 10^9$ Nmm and $R_{I,s} = 75.89 \times 10^{12}$ Nmm^2.
The axial force and moment resisted by the concrete part of the cross-section at time τ_0 are:

$$N_{c,0} = A_c E_{c,0}\varepsilon_{r,0} + B_c E_{c,0}\kappa_0 = -518.3 \times 10^3 \text{ N}$$

$$M_{c,0} = B_c E_{c,0}\varepsilon_{r,0} + I_c E_{c,0}\kappa_0 = +140.4 \times 10^6 \text{ Nmm}$$

The cross-sectional rigidities $R_{A,k}$, $R_{B,k}$ and $R_{I,k}$ are obtained from Eqs 5.30 and 5.32:

$$R_{A,k} = A_c\overline{E}_{e,k} + R_{A,s} = 132{,}300 \times 8475 + 800 \times 10^6 = 1921 \times 10^6 \text{ N}$$

$$R_{B,k} = B_c\overline{E}_{e,k} + R_{B,s} = -33.06 \times 10^6 \times 8475 + 246.4 \times 10^9$$
$$= -33.78 \times 10^9 \text{ Nmm}$$

$$R_{I,k} = I_c\overline{E}_{e,k} + R_{I,s} = 8564 \times 10^6 \times 8475 + 75.89 \times 10^{12}$$
$$= 148.5 \times 10^{12} \text{ Nmm}^2$$

From Eq. 5.40:

$$\mathbf{F}_k = \frac{1}{R_{A,k}R_{I,k} - R_{B,k}^2}\begin{bmatrix} R_{I,k} & -R_{B,k} \\ -R_{B,k} & R_{A,k} \end{bmatrix} = \begin{bmatrix} 522.7\times 10^{-12} & 118.9\times 10^{-15} \\ 118.9\times 10^{-15} & 6.761\times 10^{-15} \end{bmatrix}$$

and from Eqs 5.36 and 5.37:

$$\mathbf{f}_{cr,k} = \overline{F}_{e,0}\begin{bmatrix} N_{c,0} \\ M_{c,0} \end{bmatrix} = -0.356 \times \begin{bmatrix} -518.1 \times 10^3 \\ +145.4 \times 10^6 \end{bmatrix} = \begin{bmatrix} +184.5 \times 10^3 \\ -49.96 \times 10^6 \end{bmatrix}$$

$$\mathbf{f}_{sh,k} = \begin{bmatrix} A_c \\ B_c \end{bmatrix} \overline{E}_{e,k}\varepsilon_{sh,k} = \begin{bmatrix} 132{,}300 \\ -33.06 \times 10^6 \end{bmatrix} \times 8475 \times (-500 \times 10^{-6})$$

$$= \begin{bmatrix} -560.6 \times 10^3 \\ +140.1 \times 10^6 \end{bmatrix}$$

For this reinforced concrete cross-section, the strain vector is obtained from Eq. 5.39:

$$\boldsymbol{\varepsilon}_k = \mathbf{F}_k \left(\mathbf{r}_{e,k} - \mathbf{f}_{cr,k} + \mathbf{f}_{sh,k} \right)$$

$$= \begin{bmatrix} 522.7 \times 10^{-12} & 118.9 \times 10^{-15} \\ 118.9 \times 10^{-15} & 6.761 \times 10^{-15} \end{bmatrix} \begin{bmatrix} (0 - 184.5 - 560.6) \times 10^3 \\ (300 + 49.96 + 140.1) \times 10^6 \end{bmatrix}$$

$$= \begin{bmatrix} -331.0 \times 10^{-6} \\ 3.225 \times 10^{-6} \text{ mm}^{-1} \end{bmatrix}$$

From Eq. 5.20, the top ($y = -322$ mm) and bottom ($y = +378$ mm) fibre strains are:

$$\varepsilon_{k(top)} = \varepsilon_{r,k} - 322 \times \kappa_k = (-331.0 - 322 \times 3.225) \times 10^{-6} = -1370 \times 10^{-6}$$

$$\varepsilon_{k(btm)} = \varepsilon_{r,k} + 378 \times \kappa_k = (-331.0 + 378 \times 3.225) \times 10^{-6} = +888 \times 10^{-6}$$

The concrete stresses at time τ_k at the top fibre ($y = -322$ mm) and at the bottom of the compressive concrete, at $y_{n,0} = -141.3$ mm, are obtained from Eq. 7.21a:
At top of section:

$$\sigma_{c,k} = 8475 \times (-1370 + 500) \times 10^{-6} - 0.356 \times -6.52$$

$$= -5.05 \text{ MPa}$$

At $y_{n,0} = -141.3$ mm:

$$\sigma_{c,k} = 8475 \times (-331.0 - 141.3 \times 3.225 + 500) \times 10^{-6}$$

$$- 0.356 \times 0 = -2.43 \text{ MPa}$$

The stress in the reinforcement is:

$$\sigma_{s,k} = E_s(\varepsilon_{r,k} + y_s\kappa_k) = 200 \times 10^3 \times (-331.0 + 308 \times 3.225) \times 10^{-6}$$

$$= +132.5 \text{ MPa}$$

The results are plotted in Fig. 7.11a.

Load case 1b: $N_{e,k} = 0$; $M_{e,k} = 300$ kNm and $\varepsilon_{sh}(\tau_k) = 0$

The only difference from load case 1a, is that the shrinkage vector is now nil, that is, $\mathbf{f}_{sh,k} = \mathbf{0}$, and from Eq. 5.39, the strain vector is now:

$$\boldsymbol{\varepsilon}_k = \mathbf{F}_k \left(\mathbf{r}_{e,k} - \mathbf{f}_{cr,k} + \mathbf{f}_{sh,k} \right)$$

$$\varepsilon_k = \begin{bmatrix} 522.7 \times 10^{-12} & 118.9 \times 10^{-15} \\ 118.9 \times 10^{-15} & 6.761 \times 10^{-15} \end{bmatrix} \begin{bmatrix} (0 - 184.5 + 0) \times 10^3 \\ (300 + 49.96 + 0) \times 10^6 \end{bmatrix}$$

$$= \begin{bmatrix} -54.8 \times 10^{-6} \\ 2.344 \times 10^{-6}\ \text{mm}^{-1} \end{bmatrix}$$

The top and bottom fibre strains are:

$$\varepsilon_{k(top)} = \varepsilon_{r,k} - 322 \times \kappa_k = (-58.8 - 322 \times 2.344) \times 10^{-6} = -810 \times 10^{-6}$$

$$\varepsilon_{k(btm)} = \varepsilon_{r,k} + 378 \times \kappa_k = (-58.8 + 378 \times 2.344) \times 10^{-6} = +831 \times 10^{-6}$$

The concrete stresses at time τ_k at the top fibre and at the bottom of the compressive concrete, at $y_{n,0} = -141.3$ mm, are obtained from Eq. 7.21a:
At top of section:

$$\sigma_{c,k} = 8475 \times (-810) \times 10^{-6} - 0.356 \times -6.52$$
$$= -4.54 \text{ MPa}$$

At $y_{n,0} = -141.3$ mm:

$$\sigma_{c,k} = 8475 \times (-58.8 - 141.3 \times 2.344) \times 10^{-6}$$
$$- 0.356 \times 0 = -3.27 \text{ MPa}$$

The stress in the reinforcement is:

$$\sigma_{s,k} = E_s(\varepsilon_{r,k} + y_s\kappa_k) = 200 \times 10^3 \times (-58.8 + 308 \times 2.344) \times 10^{-6}$$
$$= +133.5 \text{ MPa}$$

The results are plotted in Fig. 7.11b.
The effects of shrinkage acting alone are obtained by subtracting the strains and stresses calculated for load case 1b from those for load case 1a and the results are plotted in Fig. 7.11c.

Load case 2: $N_{e,0} = -1000$ kN, $M_{e,0} = 300$ kNm and $\varepsilon_{sh}(\tau_k) = -500 \times 10^{-6}$

From load case 2 in Example 7.1, $d_n = 407.5$ mm; $\varepsilon_{r,0} = -71.9 \times 10^{-6}$; $\kappa_0 = 0.842 \times 10^{-6}$ mm^{-1} and the initial stress and strain distributions are illustrated in Fig. 7.7b.
As for load case 1: $\overline{E}_{e,k} = 8475$ MPa and $\overline{F}_{e,0} = -0.356$. As calculated in Example 7.1, the properties of the concrete part of the cross-section with respect to the reference axis are $A_c = 223{,}000$ mm^2, $B_c = -35.59 \times 10^6$ mm^3 and $I_c = 9024 \times 10^6$ mm^4 and the rigidities of the steel reinforcement are $R_{A,s} = 800 \times 10^6$ N, $R_{B,s} = 246.4 \times 10^9$ Nmm and $R_{I,s} = 75.89 \times 10^{12}$ Nmm^2.
The axial force and moment resisted by the concrete part of the cross-section at time τ_0 are:

$$N_{c,0} = A_c E_{c,0}\varepsilon_{r,0} + B_c E_{c,0}\kappa_0 = -1150 \times 10^3 \text{ N}$$

$$M_{c,0} = B_c E_{c,0}\varepsilon_{r,0} + I_c E_{c,0}\kappa_0 = +253.9 \times 10^6 \text{ Nmm}$$

The cross-sectional rigidities $R_{A,k}$, $R_{B,k}$ and $R_{I,k}$ are obtained from Eqs 5.30 and 5.32:

$$R_{A,k} = A_c\bar{E}_{e,k} + R_{A,s} = 223{,}000 \times 8475 + 800 \times 10^6 = 2690 \times 10^6 \text{ N}$$

$$R_{B,k} = B_c\bar{E}_{e,k} + R_{B,s} = -35.59 \times 10^6 \times 8475 + 246.4 \times 10^9$$
$$= -55.23 \times 10^9 \text{ Nmm}$$

$$R_{I,k} = I_c\bar{E}_{e,k} + R_{I,s} = 9024 \times 10^6 \times 8475 + 75.89 \times 10^{12}$$
$$= 152.4 \times 10^{12} \text{ Nmm}^2$$

From Eq. 5.40:

$$\mathbf{F}_k = \frac{1}{R_{A,k}R_{I,k} - R_{B,k}^2}\begin{bmatrix} R_{I,k} & -R_{B,k} \\ -R_{B,k} & R_{A,k} \end{bmatrix} = \begin{bmatrix} 374.5 \times 10^{-12} & 135.7 \times 10^{-15} \\ 135.7 \times 10^{-15} & 6.611 \times 10^{-15} \end{bmatrix}$$

and from Eqs 5.36 and 5.37:

$$\mathbf{f}_{cr,k} = \bar{F}_{e,0}\begin{bmatrix} N_{c,0} \\ M_{c,0} \end{bmatrix} = -0.356 \times \begin{bmatrix} -1150 \times 10^3 \\ +253.9 \times 10^6 \end{bmatrix} = \begin{bmatrix} +409.3 \times 10^3 \\ -90.36 \times 10^6 \end{bmatrix}$$

$$\mathbf{f}_{sh,k} = \begin{bmatrix} A_c \\ B_c \end{bmatrix}\bar{E}_{e,k}\varepsilon_{sh,k} = \begin{bmatrix} 223{,}000 \\ -35.59 \times 10^6 \end{bmatrix} \times 8475 \times -500 \times 10^{-6}$$
$$= \begin{bmatrix} -945.0 \times 10^3 \\ +150.8 \times 10^6 \end{bmatrix}$$

The strain vector is obtained from Eq. 5.39:

$$\boldsymbol{\varepsilon}_k = \mathbf{F}_k\left(\mathbf{r}_{e,k} - \mathbf{f}_{cr,k} + \mathbf{f}_{sh,k}\right)$$
$$= \begin{bmatrix} 374.5 \times 10^{-12} & 135.7 \times 10^{-15} \\ 135.7 \times 10^{-15} & 6.611 \times 10^{-15} \end{bmatrix}\begin{bmatrix} (-1000 - 409.3 - 945.0) \times 10^3 \\ (300 + 90.36 + 150.8) \times 10^6 \end{bmatrix}$$
$$= \begin{bmatrix} -808.3 \times 10^{-6} \\ 3.258 \times 10^{-6} \text{ mm}^{-1} \end{bmatrix}$$

The top ($y = -322$ mm) and bottom ($y = +378$ mm) fibre strains are therefore:

$$\varepsilon_{k(top)} = \varepsilon_{r,k} - 322 \times \kappa_k = (-808.3 - 322 \times 3.258) \times 10^{-6} = -1857 \times 10^{-6}$$

$$\varepsilon_{k(btm)} = \varepsilon_{r,k} + 378 \times \kappa_k = (-808.3 + 378 \times 3.258) \times 10^{-6} = +423 \times 10^{-6}$$

The concrete stresses at time τ_k at the top fibre ($y = -322$ mm) and at the bottom of the compressive concrete, at $y_{n,0} = +85.5$ mm, are obtained from Eq. 7.21a:
At top of section:

$$\sigma_{c,k} = 8475 \times (-1857 + 500) \times 10^{-6} - 0.356 \times -8.57$$
$$= -8.45 \text{ MPa}$$

At $y_{n,0} = +80.5$ mm:

$$\sigma_{c,k} = 8475 \times (-808.3 + 85.5 \times 3.258 + 500) \times 10^{-6} - 0.356 \times 0 = -0.39 \text{ MPa}$$

The stress in the reinforcement is:

$$\sigma_{s,k} = E_s(\varepsilon_{r,k} + y_s\kappa_k) = 200 \times 10^3 \times (-808.3 + 308 \times 3.258) \times 10^{-6} = +39.0 \text{ MPa}$$

The results are plotted in Fig. 7.11d.

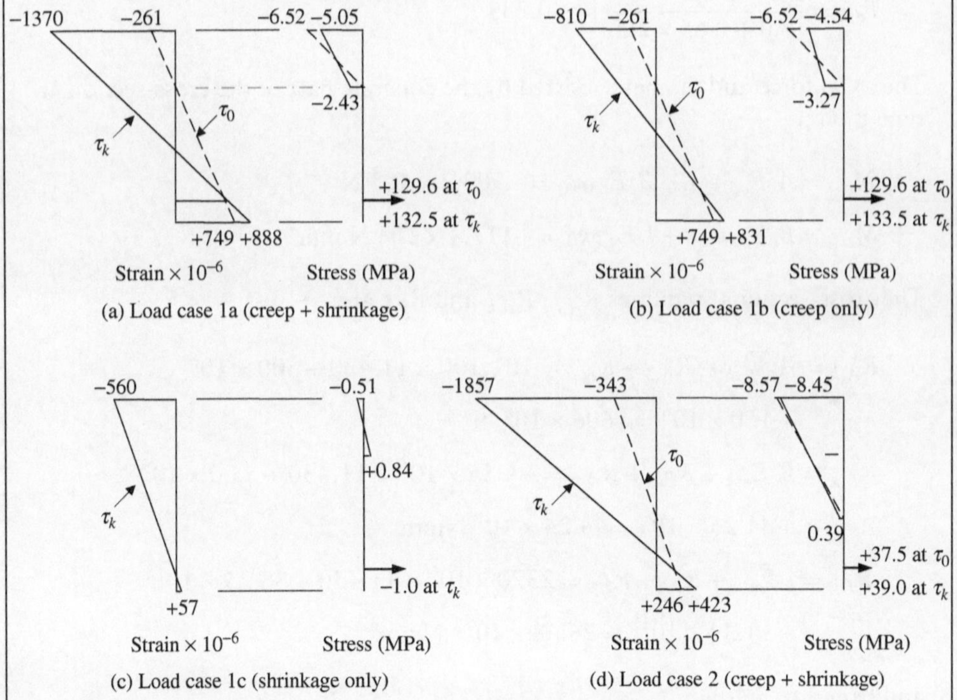

Figure 7.11 Stress and strain distributions for Example 7.3.

Example 7.4

The change of stress and strain with time on the cracked partially-prestressed cross-section of Example 7.2 (Fig. 7.8) is to be calculated using AEMM. The actions on the section are assumed to be constant throughout the time period under consideration (i.e. τ_0 to τ_k) and equal to $N_{e,k} = 0$; $M_{e,k} = 400$ kN.m. The relevant material properties: $E_{c,0} = 30$ GPa; $E_s = E_p = 200$ GPa; $\varphi(\tau_k, \tau_0) = 2.5$; $\chi(\tau_k, \tau_0) = 0.65$; $\varepsilon_{sh}(\tau_k) = -400 \times 10^{-6}$; $\varphi(\tau_k, \sigma_{p(i),init}) = 0.02$ and the steel reinforcement is assumed to be linear elastic.

From Example 7.2: $d_n = 508.1$ mm, $\varepsilon_{r,0} = -244.6 \times 10^{-6}$, $\kappa_0 = 1.182 \times 10^{-6}$ mm^{-1}, $A_c = 101{,}100$ mm^2, $B_c = -4.55 \times 10^6$ mm^3, $I_c = 2370 \times 10^6$ mm^4

and the rigidities of the steel reinforcement and tendons are $R_{A,s} = 300 \times 10^6$ N, $R_{B,s} = 55.0 \times 10^9$ Nmm, $R_{I,s} = 38.25 \times 10^{12}$ Nmm^2, $R_{A,p} = 150 \times 10^6$ N, $R_{B,p} = 41.25 \times 10^9$ Nmm and $R_{I,p} = 11.34 \times 10^{12}$ Nmm^2.
From Eqs 4.35 and 4.46:

$$\overline{E}_{e,k} = \frac{30{,}000}{1 + 0.65 \times 2.5} = 11{,}430 \text{ MPa}$$

and

$$\overline{F}_{e,0} = \frac{2.5 \times (0.65 - 1.0)}{1.0 + 0.65 \times 2.5} = -0.333$$

The axial force and moment resisted by the concrete part of the cross-section at time τ_0 are:

$$N_{c,0} = A_c E_{c,0} \varepsilon_{r,0} + B_c E_{c,0} \kappa_0 = -903.2 \times 10^3 \text{ N}$$

$$M_{c,0} = B_c E_{c,0} \varepsilon_{r,0} + I_c E_{c,0} \kappa_0 = +117.4 \times 10^6 \text{ N.mm}$$

The cross-sectional rigidities $R_{A,k}$, $R_{B,k}$ and $R_{I,k}$ are:

$$R_{A,k} = A_c \overline{E}_{e,k} + R_{A,s} + R_{A,p} = 101{,}100 \times 11{,}430 + 300 \times 10^6 + 150 \times 10^6 = 1606 \times 10^6 \text{ N}$$

$$R_{B,k} = B_c \overline{E}_{e,k} + R_{B,s} + R_{B,p} = -4.55 \times 10^6 \times 11{,}430 + 55.0 \times 10^9 + 41.25 \times 10^9 = 44.24 \times 10^9 \text{ Nmm}$$

$$R_{I,k} = I_c \overline{E}_{e,k} + R_{I,s} + R_{I,p} = 2370 \times 10^6 \times 11{,}430 + 38.25 \times 10^{12} + 11.34 \times 10^{12} = 76.68 \times 10^{12} \text{ Nmm}^2$$

and from Eq. 5.40:

$$\mathbf{F}_k = \frac{1}{R_{A,k} R_{I,k} - R_{B,k}^2} \begin{bmatrix} R_{I,k} & -R_{B,k} \\ -R_{B,k} & R_{A,k} \end{bmatrix} = \begin{bmatrix} 632.7 \times 10^{-12} & -365.1 \times 10^{-15} \\ -365.1 \times 10^{-15} & 13.25 \times 10^{-15} \end{bmatrix}$$

From Eqs 5.36 to 5.38:

$$\mathbf{f}_{cr,k} = \overline{F}_{e,0} \begin{bmatrix} N_{c,0} \\ M_{c,0} \end{bmatrix} = -0.33\dot{3} \times \begin{bmatrix} -903.2 \times 10^3 \\ +117.4 \times 10^6 \end{bmatrix} = \begin{bmatrix} +301.1 \times 10^3 \\ -39.13 \times 10^6 \end{bmatrix}$$

$$\mathbf{f}_{sh,k} = \begin{bmatrix} A_c \\ B_c \end{bmatrix} \overline{E}_{e,k} \varepsilon_{sh,k} = \begin{bmatrix} 101{,}100 \\ -4.55 \times 10^6 \end{bmatrix} \times 11{,}430 \times -400 \times 10^{-6}$$

$$= \begin{bmatrix} -462.2 \times 10^3 \\ +20.80 \times 10^6 \end{bmatrix}$$

$$\mathbf{f}_{p,init} = \sum_{i=1}^{m_p} \begin{bmatrix} A_{p(i)}E_{p(i)} \\ y_{p(i)}A_{p(i)}E_{p(i)} \end{bmatrix} \varepsilon_{p(i),init} = \begin{bmatrix} 900 \times 10^3 \\ 247.5 \times 10^6 \end{bmatrix} \text{(from Example 7.2)}$$

$$\mathbf{f}_{p.rel,k} = \mathbf{f}_{p.init}\varphi_p(\tau_k, \sigma_{p(i),init}) = \begin{bmatrix} 18 \times 10^3 \\ 4.95 \times 10^6 \end{bmatrix}$$

The strain vector is obtained from Eq. 5.39:

$$\boldsymbol{\varepsilon}_k = \mathbf{F}_k \left(\mathbf{r}_{e,k} - \mathbf{f}_{cr,k} + \mathbf{f}_{sh,k} - \mathbf{f}_{p,init} + \mathbf{f}_{p.rel,k} \right)$$

$$\boldsymbol{\varepsilon}_k = \begin{bmatrix} 632.7 \times 10^{-12} & -365.1 \times 10^{-15} \\ -365.1 \times 10^{-15} & 13.25 \times 10^{-15} \end{bmatrix} \times$$

$$\begin{bmatrix} (0 - 301.1 - 462.2 - 900 + 18) \times 10^3 \\ (400 + 39.13 + 20.80 - 247.5 + 4.95) \times 10^6 \end{bmatrix} = \begin{bmatrix} -1120 \times 10^{-6} \\ 3.481 \times 10^{-6} \text{ mm}^{-1} \end{bmatrix}$$

The top ($y = -300$ mm) and bottom ($y = +450$ mm) fibre strains are:

$$\varepsilon_{k(top)} = \varepsilon_{r,k} - 300 \times \kappa_k = (-1120 - 300 \times 3.481) \times 10^{-6}$$
$$= -2165 \times 10^{-6}$$

$$\varepsilon_{k(btm)} = \varepsilon_{r,k} + 450 \times \kappa_k = (-1120 + 450 \times 3.481) \times 10^{-6}$$
$$= +446.2 \times 10^{-6}$$

The concrete stresses at time τ_k at the top fibre ($y = -300$ mm) and at the bottom of the compressive concrete, at $y_{n,0} = 208.1$ mm, are obtained from Eq. 7.21a: At top of section:

$$\sigma_{c,k} = 11{,}430 \times (-2165 + 400) \times 10^{-6} - 0.333 \times -18.0$$
$$= -14.2 \text{ MPa}$$

At $y_{n,0} = +208.1$ mm:

$$\sigma_{c,k} = 11{,}430 \times (-1120 + 208.1 \times 3.481 + 400) \times 10^{-6}$$
$$- 0.33\dot{3} \times 0 = +0.05 \text{ MPa}$$

The stress in the non-prestressed reinforcement are (Eq. 7.21b):

$$\sigma_{s(1),0} = E_s(\varepsilon_{r,0} + y_{s(1)}\kappa_0) = 200 \times 10^3 \times (-1120 - 250 \times 3.481) \times 10^{-6}$$
$$= -398 \text{ MPa}$$

$$\sigma_{s(2),0} = E_s(\varepsilon_{r,0} + y_{s(2)}\kappa_0) = 200 \times 10^3 \times (-1120 + 400 \times 3.481) \times 10^{-6}$$
$$= +54.4 \text{ MPa}$$

and the stress in the prestressing steel is given by (Eq. 7.21c):

$$\sigma_{p,0} = E_p \left(\varepsilon_{r,0} + y_p\kappa_0 + \varepsilon_{p.init} + \varepsilon_{p.rel,k} \right) = 200 \times 10^3$$
$$\times (-1120 + 275 \times 3.481 + 6000 - 120) \times 10^{-6} = +1143 \text{ MPa}$$

The results are plotted in Fig. 7.12.

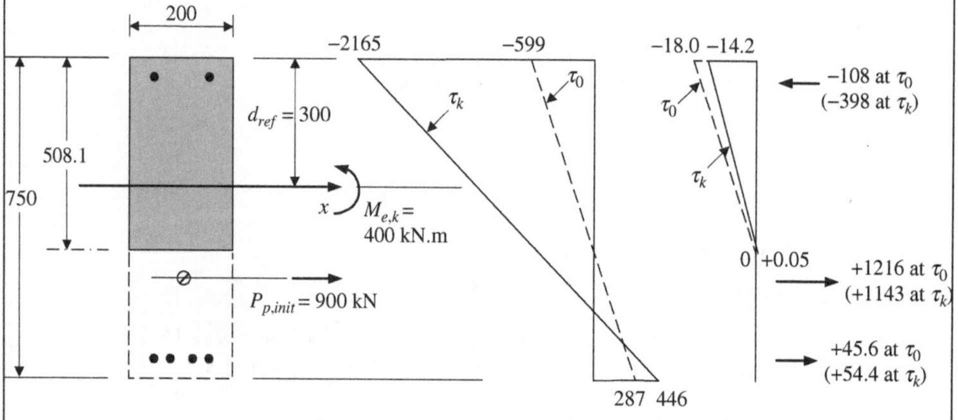

Figure 7.12 Initial and time-dependent strain and stress distributions for Example 7.4 (all dimensions in mm).

7.3.2 Axial force and biaxial bending

The procedure required for the time analysis of a cracked section in biaxial bending using the AEMM is identical to that for an uncracked section as presented in Section 6.3 and, for this reason, is not repeated here. The only difference is that the concrete cross-sectional properties are determined based on the shape of the cracked section identified in the instantaneous analysis. As already discussed, when modelling the time-dependent behaviour of a cracked section using the AEMM, it is convenient to assume that the area of uncracked concrete does not vary with time. For most purposes, this assumption is quite reasonable and does not result in significant error. More refined computer-based methods of analysis can be used to trace the time-varying position of the neutral axis and the gradual change in shape of the cracked area of the cross-section. Such methods are discussed in the following section.

7.4 Short- and long-term analysis using the step-by-step method

In some situations, time analysis of a cracked section using the AEMM may not be appropriate. For example, where the section is subjected to a complex load history, it may be necessary to use a more refined analysis. A refined method for the time analysis of a cracked section is described below for the cases of axial force combined with uniaxial and biaxial bending. To avoid unnecessary repetitions reference is made to the derivations presented in previous chapters.

7.4.1 Axial force and uniaxial bending

7.4.1.1 Instantaneous analysis

To trace the change of position of the neutral axis and the change in shape of the cracked portion of the cross-section with time, it is convenient to subdivide the concrete

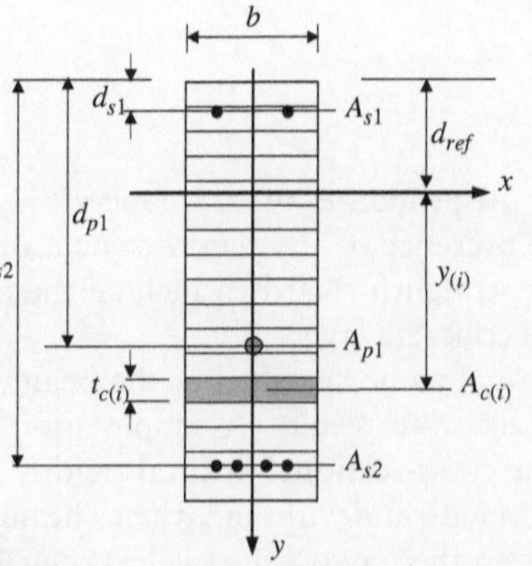

Figure 7.13 Layered cross-section subjected to axial force and uniaxial bending.

into layers as shown in Fig. 7.13. With this approach, the internal actions, i.e. axial force and moment, are calculated by summing the contributions of each of the layers.

Assuming that the cross-section is subdivided into m_c concrete layers, and initially assuming that all concrete layers are uncracked, the contribution of the concrete component to the internal axial force $N_{i,0}$ previously introduced in Eq. 7.6 can be obtained as follows:

$$N_{c,0} = \sum_{i=1}^{m_c} A_{c(i)}\sigma_{c,0} = \sum_{i=1}^{m_c} A_{c(i)}E_{c(i),0}(\varepsilon_{r,0} + y_{(i)}\kappa_0)$$
$$= \sum_{i=1}^{m_c} \left(A_{c(i)}E_{c(i),0}\right)\varepsilon_{r,0} + \sum_{i=1}^{m_c} \left(B_{c(i)}E_{c(i),0}\right)\kappa_0 \tag{7.22}$$

where $A_{c(i)}$ and $B_{c(i)}$ are the area and the first moment of area of the i-th concrete layer about the x-axis; $y_{(i)}$ is the distance between the centroid of the i-th concrete layer and the reference axis (Fig. 7.13). This approach assumes that a constant stress is resisted by each layer calculated based on the strain measured at its centroid. Although not used here, a linearly varying stress across the thickness of each layer ($t_{(i)}$) can easily be implemented.

Similarly, the contribution of the concrete to the internal moment $M_{i,0}$ is given by:

$$M_{c,0} = \sum_{i=1}^{m_c} y_{(i)}A_{c(i)}\sigma_{c,0} = \sum_{i=1}^{m_c} A_{c(i)}E_{c(i),0}(y_{(i)}\varepsilon_{r,0} + y_{(i)}^2\kappa_0)$$
$$= \sum_{i=1}^{m_c} \left(B_{c(i)}E_{c(i),0}\right)\varepsilon_{r,0} + \sum_{i=1}^{m_c} \left(I_{c(i)}E_{c(i),0}\right)\kappa_0 \tag{7.23}$$

where $I_{c(i)}$ is the second moment of area of the i-th concrete layer about the x-axis.

The concrete rigidities for the cross-section are:

$$R_{A,c} = \sum_{i=1}^{m_c} \left(A_{c(i)}E_{c(i),0}\right); R_{B,c} = \sum_{i=1}^{m_c} \left(y_{c(i)}A_{c(i)}E_{c(i),0}\right); R_{I,c} = \sum_{i=1}^{m_c} \left(y_{c(i)}^2 A_{c(i)}E_{c(i),0}\right)$$

(7.24a,b,c)

and the axial force and moment carried by the concrete are re-written as:

$$N_{c,0} = R_{A.c}\varepsilon_{r,0} + R_{B.c}\kappa_0 \tag{7.25a}$$

$$M_{c,0} = R_{B.c}\varepsilon_{r,0} + R_{I.c}\kappa_0 \tag{7.25b}$$

In the determination of the properties of each concrete layer ($A_{c(i)}$, $B_{c(i)}$ and $I_{c(i)}$), the concrete displaced by the presence of any reinforcement or tendons in a layer is often ignored, but it is usually straightforward to include the actual concrete area rather than the gross area of the concrete layer.

The number of layers (m_c) to be specified at the beginning of the analysis should be sufficient to produce accurate results. A simple approach to determine m_c is to compare the values for the cross-sectional flexural rigidity $R_{I,0}$ calculated considering the uncracked cross-section with and without layers. The number of layers is acceptable when the difference between these two values is less than 1 per cent.

The governing equilibrium equations for a typical cross-section are therefore:

$$N_{i,0} = (R_{A,c} + R_{A,s} + R_{A,p})\varepsilon_{r,0} + (R_{B,c} + R_{B,s} + R_{B,p})\kappa_0 + \sum_{i=1}^{m_p} \left(A_{p(i)}E_{p(i)}\varepsilon_{p(i),init}\right) = N_{e,0} \tag{7.26a}$$

$$M_{i,0} = (R_{B,c} + R_{B,s} + R_{B,p})\varepsilon_{r,0} + (R_{I,c} + R_{I,s} + R_{I,p})\kappa_0 + \sum_{i=1}^{m_p} \left(y_{p(i)}A_{p(i)}E_{p(i)}\varepsilon_{p(i),init}\right) = M_{e,0} \tag{7.26b}$$

where the concrete rigidities are given by Eq. 7.24 and the rigidities of the reinforcement and the tendons are defined in Eqs 7.8. This system of two equations can be used to solve for the two unknowns $\varepsilon_{r,0}$ and κ_0.

To determine the extent of cracking, the solution process is iterative. For the first iteration all concrete layers are assumed to be uncracked, that is, each concrete layer contributes to the cross-sectional stiffness. The cross-section is analysed and the concrete stress at the centroid of each layer is determined. For the second iteration, concrete layers subjected to tension in the first iteration in excess of the tensile strength of concrete are assumed to have cracked and are subsequently neglected, i.e. the value $E_{c(i),0}$ in a cracked layer is set to zero. At the end of each subsequent iteration the stresses are checked in the uncracked layers to identify whether any additional layers have cracked. The short-term analysis is terminated when no additional cracking is detected. More refined material models could be implemented, including the effects of tension-stiffening, tension-softening or other material nonlinearities, and more efficient solving strategies could be used.

7.4.1.2 *Time analysis*

The step-by-step method (SSM) of analysis is adopted. The time discretisation associated with the method was discussed in Section 4.3.1 and the analysis of uncracked cross-sections was presented in Section 5.5, with the constitutive relationships for the concrete, reinforcement and tendons given in Eqs 4.25 and 5.44.

The steps required for the long-term analysis of a cracked section are similar to those required for an uncracked section. The only difference in this case is that the concrete in a cracked layer is assumed to carry no tension.

For the layered section of Fig. 7.13, the axial force and moment carried by the concrete to be included in the expression for the internal actions (Eqs 5.50 and 5.51) and the equilibrium equations (Eq. 5.54) can be calculated from:

$$N_{c,j} = \sum_{i=1}^{m_c} A_{c(i)}\sigma_{c(i),j} = \sum_{i=1}^{m_c}\left(A_{c(i)}E_{c(i),j}\left(\varepsilon_{(i),j} - \varepsilon_{sh,j}\right) + \sum_{n=0}^{j-1} F_{e,j,n}\sigma_{c(i),n}A_{c(i)}\right)$$
$$= \sum_{i=1}^{m_c}\left(A_{c(i)}E_{c(i),j}\varepsilon_{r,j} + B_{c(i)}E_{c(i),j}\kappa_j - A_{c(i)}E_{c(i),j}\varepsilon_{sh,j} + \sum_{n=0}^{j-1} F_{e,j,n}\sigma_{c(i),n}A_{c(i)}\right) \tag{7.27a}$$

$$M_{c,j} = \sum_{i=1}^{m_c} y_{(i)}A_{c(i)}\sigma_{c(i),j} = \sum_{i=1}^{m_c}\left(y_{(i)}A_{c(i)}E_{c(i),j}\left(\varepsilon_{(i)j} - \varepsilon_{sh,j}\right) + \sum_{i=0}^{j-1} F_{e,j,n}\sigma_{c(i),n}y_{(i)}A_{c(i)}\right)$$
$$= \sum_{i=1}^{m_c}\left(B_{c(i)}E_{c(i),j}\varepsilon_{r,j} + I_{c(i)}E_{c(i),j}\kappa_j - B_{c(i)}E_{c(i),j}\varepsilon_{sh,j} + \sum_{n=0}^{j-1} F_{e,j,n}\sigma_{c(i),n}B_{c(i)}\right) \tag{7.27b}$$

where the stress history of each layer is recorded separately and the contribution of all cracked layers is zero.

As for the instantaneous analysis, an iterative procedure is adopted at each time step. During any iteration, if cracking is detected in a previously uncracked concrete layer (i.e. if the stress at the centroid of the layer exceeds the tensile strength of concrete), the contribution to the cross-sectional rigidities of the concrete in that layer is set to zero and the previous stress history for that layer is subsequently ignored. If the stress in a previously cracked layer becomes compressive, the crack is deemed to have closed and the concrete in the layer begins to contribute to the cross-sectional stiffness and, for the purposes of subsequent creep calculations, the stress history for the layer is stored from this point in time. The iterative procedure continues until the changes in the calculated strain diagram (i.e. the changes in $\varepsilon_{r,j}$ and κ_j) from one iteration to the next are sufficiently small.

7.4.2 Axial force and biaxial bending

The geometric discretisation required for the analysis of a cross-section subjected to combined axial force and biaxial bending is shown in Fig. 7.14. The cross-section is subdivided both horizontally and vertically into smaller rectangular areas (similar to the layers introduced in the previous section for the case of uniaxial bending). In each rectangular concrete area, the stress is assumed to be constant and is calculated from the strain measured at its centroid. A linearly varying stress distribution could also be considered based on strain values calculated at each corner of the subdivision.

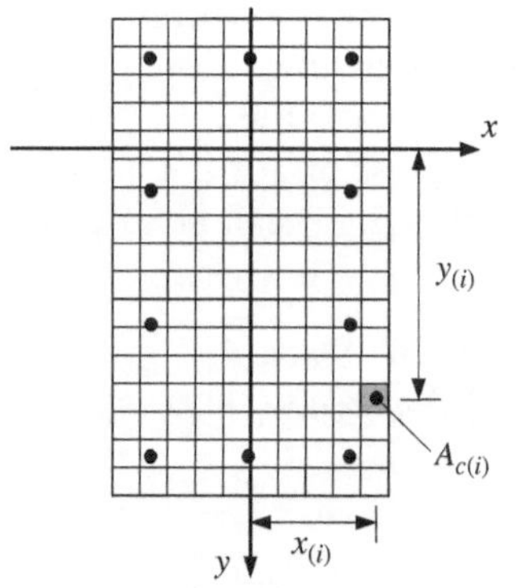

Figure 7.14 Geometric discretisation of cross-section - axial force and biaxial bending.

As suggested previously, the fineness of the subdivision to be used for the analysis can be determined by comparing the values of the uncracked flexural rigidities $R_{Ix,k}$, $R_{Iy,k}$ and $R_{Ixy,k}$ calculated with and without the subdivisions. When the differences are less than 1 per cent, the degree of discretisation is acceptable.

The solution process for cracked sections in biaxial bending is identical to that described in Section 7.4.1. In the first iteration of the instantaneous analysis, all concrete is assumed to be uncracked. The cracked areas of concrete are identified and ignored in the next and subsequent iterations. This iterative procedure is applied at each time step, where the effects of creep from the previously stored stress history in each subdivision and the effects of shrinkage are determined. Further changes in the extent of cracking are identified at the end of each iteration, as are locations where the cracks may have closed in previously cracked regions. In this way, the time-dependent changes in deformation and stresses are calculated, without the simplifying assumptions inherent in the AEMM, where a single time step is considered and the extent of cracking on the cross-section is assumed not to change with time.

7.5 References

1. Bischoff, P.H. (2001). Effects of shrinkage on tension stiffening and cracking in reinforced concrete. *Canadian Journal of Civil Engineering*, 28(3), 363–374.
2. Gilbert, R.I. and Wu, H.Q. (2009). Time-dependent stiffness of cracked reinforced concrete elements. *fib London 09, Concrete: 21st Century Superhero*, June, London.
3. Scott, R.H. and Beeby, A.W. (2005). Long-term tension stiffening effects in concrete. *ACI Structural Journal*, 102(1): 31–39.
4. Bamforth, P.B. (2007). *Early-age thermal crack control in concrete.* CIRIA C660, London.

8 Members and structures

8.1 Introductory remarks

The cross-sectional analyses developed in Chapters 4–7 form the basis for the prediction of the in-service deformation of concrete members and structures. Techniques for making such predictions are presented in this chapter.

From the point of view of serviceability, it is important that the deflection and shortening of structural members remain acceptably small under normal in-service conditions. It is also important that crack widths are such that they do not detract from the appearance or durability of the structure. In Chapter 3, some guidance was given for the design of concrete structures for serviceability. The load combinations and design criteria for the serviceability limit states that are specified in some of the more well-known codes of practice were discussed. In this chapter, procedures for the prediction of structural behaviour at service loads for a variety of structural forms are discussed and illustrated. In most cases, reference is made to the relevant cross-sectional analyses described and illustrated in Chapters 4–7.

Throughout this chapter the age-adjusted effective modulus method (AEMM) has been adopted for the analysis of the individual cross-sections, however, the member analyses could be carried out just as easily using any of the other methods of analysis introduced in Chapter 4 [such as the effective modulus method (EEM), step-by-step method (SSM) or rate of creep method (RCM)]. In Section 8.5, the effects of time-dependent displacements on the behaviour of slender reinforced concrete columns at the strength limit state are also considered.

Often the accurate prediction of structural behaviour at service loads is difficult, particularly for many-fold indeterminate structures, but useful approximations are available to simplify many of the calculations. In the analyses presented here, approximations and simplifying assumptions are introduced from time to time to reduce the computational effort. In design, a thorough understanding of the nature and implications of these assumptions is essential if a realistic and practical interpretation of the final results is to be made. As with all aspects of structural analysis and design, sound engineering judgment is an essential ingredient.

8.2 Deflection of statically determinate beams

8.2.1 Deflection and axial shortening of uncracked beams

In Section 3.6.1, the calculation of the deflection and axial shortening of a member by integration of the deformations over the length of the member was discussed. If the curvature and axial strain at the centroid of the section are known at each end of a span (i.e. at cross-sections A and B in Fig. 3.1) and at mid-span (cross-section C in Fig. 3.1) and if the distributions of axial strain and curvature along the member are parabolic, the axial deformation and mid-span deflection of the member are given by Eqs 3.7a and b, given here and renumbered for convenience:

$$e_{AB} = \frac{\ell}{6}(\varepsilon_{aA} + 4\varepsilon_{aC} + \varepsilon_{aB}) \tag{8.1a}$$

$$v_C = \frac{\ell^2}{96}(\kappa_A + 10\kappa_C + \kappa_B) \tag{8.1b}$$

Example 8.1

The axial shortening and mid-span deflection of a simply-supported, uncracked, post-tensioned concrete beam spanning 12.0 m are to be calculated, immediately after first loading and after a period of sustained loading at time τ_k. An elevation of the beam is shown in Fig. 8.1, together with details of the cross-section at mid-span. The non-prestressed reinforcement is uniform along the span.

The beam is post-tensioned by a single cable of parabolic profile, with the depth of the tendon below the top surface of the beam (d_p) equal to 450 mm at each support and 780 mm at mid-span. The duct diameter is 70 mm. The initial tensile force in the tendon prior to transfer (at time τ_0) is assumed to be constant along the length of the member and equal to $P_{p,init}$ = 1300 kN. That is, $\sigma_{p,init} = P_{p,init}/A_p = 1300$ MPa at every cross-section.

The prestressing duct is grouted soon after the transfer of prestress and immediately after the application of the sustained load, so that the prestressing tendon is unbonded for the short-term analysis but fully bonded to the concrete throughout the period of the time analysis.

Two load cases are considered. In load case 1, the beam is subjected only to its self-weight (i.e. $w_s = 6.1$ kN/m) plus prestress from first loading to time τ_k. In load case 2, the beam carries a constant sustained uniformly distributed load of $w_s = 40$ kN/m plus prestress from first loading to time τ_k. The relevant material properties for the time period under consideration are:

$E_c(\tau_0) = 30{,}000$ MPa
$E_s = E_p = 2 \times 10^5$ MPa
$\varphi(\tau_k, \tau_0) = 1.8$
$\chi(\tau_k, \tau_o) = 0.65$
$\varepsilon_{sh}(\tau_k - \tau_o) = -400 \times 10^{-6}$
and $\varphi_p(\tau_k, \sigma_{p,init}) = 0.02$

Taking the centroidal axis of the gross cross-section as the reference axis (i.e. $d_{ref} = 450$ mm), the initial and final strains at the reference axis and the initial and final curvatures are first calculated at mid-span and at each support using the

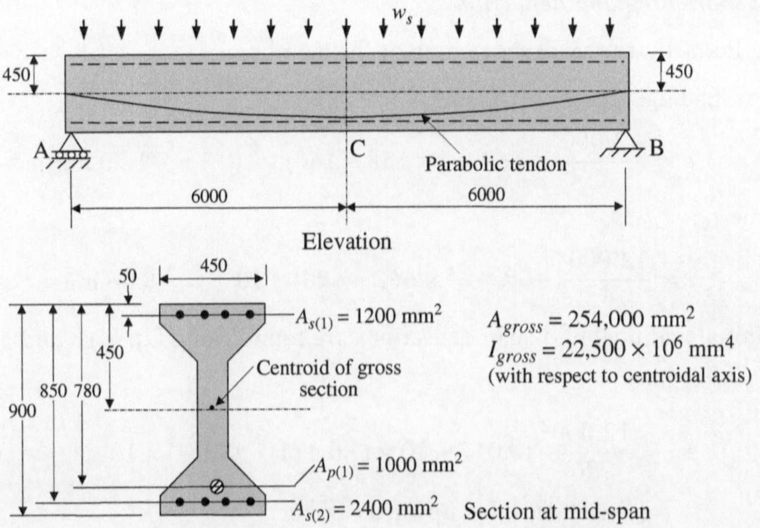

Figure 8.1 Details of beam in Example 8.1 (all dimensions in mm).

cross-section analyses outlined in Sections 5.3 and 5.4. The axial deformation and deflection are then calculated using Eqs 8.1. For the short-term analyses, the duct is empty and the prestressing steel does not contribute to the properties of the transformed section. For the long-term analysis the duct is grouted and the prestressing steel forms part of the transformed cross-section.

Load case 1

Cross-section at mid-span

The applied moment at mid-span caused by the sustained load $w_s = 6.1$ kN/m is:

$$M_s = \frac{6.1 \times 12^2}{8} = 109.8 \text{ kNm}$$

and from a cross-sectional analysis of Section 5.3, the strain at the reference axis and the curvature immediately after first loading are $\varepsilon_{r,0} = -158 \times 10^{-6}$ and $\kappa_0 = -0.411 \times 10^{-6}$ mm^{-1}. The strain at the reference axis and the curvature at mid-span at time τ_k are calculated using the analysis of Section 5.4 and are equal to $\varepsilon_{r,k} = -662 \times 10^{-6}$ and $\kappa_k = -0.641 \times 10^{-6}$ mm^{-1}. In the time analysis, the grout was initially assumed to be unloaded.

Cross-section at the supports

Similar calculations for the section at each support, where $M_s = 0$ and $d_p =$ 450 mm, give $\varepsilon_{r,0} = -160 \times 10^{-6}$; $\kappa_0 = 0.017 \times 10^{-6}$ mm^{-1}; $\varepsilon_{r,k} = -690 \times 10^{-6}$; and $\kappa_k = 0.163 \times 10^{-6}$ mm^{-1}.

Axial shortening and deflection

Using Eq. 8.1a, the axial shortening of the member is:

At first loading:

$$e_i = \frac{12{,}000}{6}(-160 - 4 \times 158 - 160) \times 10^{-6} = -1.90 \text{ mm}$$

At time τ_k:

$$e = \frac{12{,}000}{6}(-690 - 4 \times 662 - 690) \times 10^{-6} = -8.06 \text{ mm}$$

The initial and final mid-span deflections are found using Eq. 8.1b and are:

At first loading:

$$v_{ci} = \frac{12{,}000^2}{96}[0.017 + 10 \times (-0.411) + 0.017] \times 10^{-6}$$
$$= -6.11 \text{ mm (i.e. upward camber)}$$

At time $\tau_k : v_c = \dfrac{12{,}000^2}{96}[0.163 + 10 \times (-0.641) + 0.163] \times 10^{-6} = -9.13$ mm

Load case 2

The applied moment at mid-span caused by the sustained load $w_s = 40.0$ kN/m is $M_s = 40 \times 12^2/8 = 720$ kNm and the strain at the reference axis and the curvature immediately after first loading and after the period of sustained loading are $\varepsilon_{r,0} = -162 \times 10^{-6}$; $\kappa_0 = +0.392 \times 10^{-6}$ mm^{-1}; $\varepsilon_{r,k} = -727 \times 10^{-6}$; and $\kappa_k = +1.116 \times 10^{-6}$ mm^{-1}.
The strain and curvature at each support are the same as those calculated for load case 1, since at each support the applied moment M_s in both cases is zero. That is $\varepsilon_{r,0} = -160 \times 10^{-6}$; $\kappa_0 = 0.017 \times 10^{-6}$ mm^{-1}; $\varepsilon_{r,k} = -690 \times 10^{-6}$; and $\kappa_k = 0.163 \times 10^{-6}$ mm^{-1}.

Axial shortening and deflection

The axial shortening and deflection of the member are obtained from Eqs 8.1a and b:

At first loading:

$$e_i = \frac{12{,}000}{6}(-160 - 4 \times 162 - 160) \times 10^{-6} = -1.94 \text{ mm}$$

$$v_{ci} = \frac{12{,}000^2}{96}(0.017 + 10 \times 0.392 + 0.017) \times 10^{-6} = 5.93 \text{ mm}$$

At time τ_k:

$$e = \frac{12{,}000}{6}(-690 - 4 \times 727 - 690) \times 10^{-6} = -8.58 \text{ mm}$$

$$v_c = \frac{12{,}000^2}{96}(0.163 + 10 \times 1.116 + 0.163) \times 10^{-6} = 17.23 \text{ mm}$$

The stress and strain distributions on the cross-section at mid-span are shown in Fig. 8.2 for both load cases.

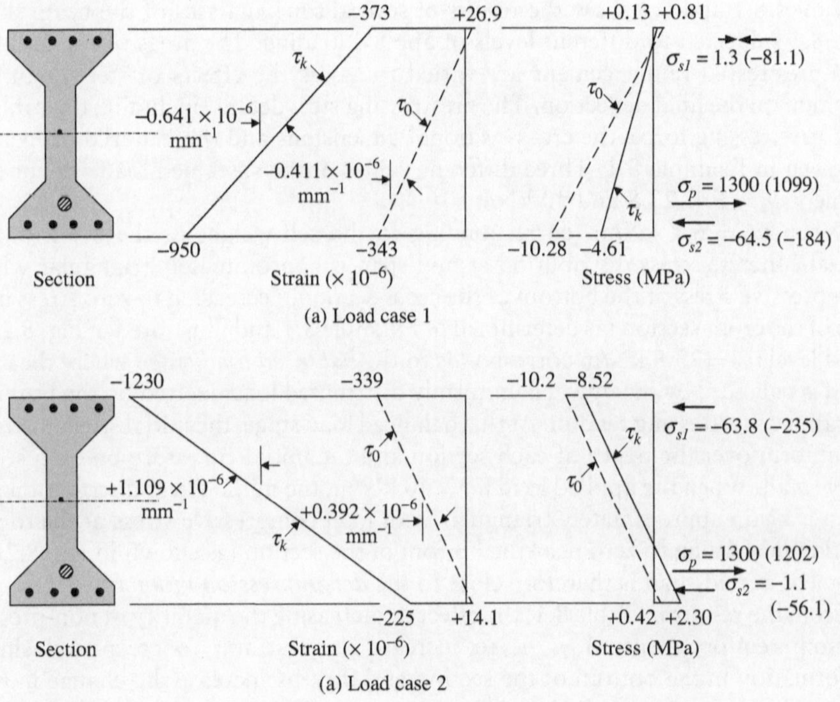

Figure 8.2 Stress and strain on cross-section at mid-span for Example 8.1.

8.2.2 Control of deflection using non-prestressed reinforcement

The change in curvature on a cross-section depends on the quantity and location of the bonded reinforcement. Consider the beam analysed in Example 8.1 and shown in Fig. 8.1. By increasing the area of top steel, $A_{s(1)}$, on the cross-section at mid-span, the rate of change of positive curvature with time would decrease. Increased amounts of top steel are particularly effective in reducing the time-dependent increase in positive curvature when the applied load is sufficiently high to induce significant compressive stresses in the top fibres, such as for load case 2 in Example 8.1. An increase in top steel would therefore reduce the final downward deflection of the beam in Example 8.1 (load case 2). However, when the initial curvature is negative, as for load case 1, an increased amount of top steel would cause an increase in the negative curvature with time and an increase in the upward deflection.

By increasing the area of bottom steel, $A_{s(2)}$, on the cross-section at mid-span, the rate of change of positive curvature with time would increase. Increased amounts of bottom steel are particularly effective in reducing the time-dependent increase in negative curvature when the applied load on a prestressed member is small and the compressive stresses in the bottom fibres are relatively high, such as for load case 1 in Example 8.1. An increase in bottom steel would therefore reduce the final upward deflection of the beam for load case 1. However, when the initial curvature is positive, as for load case 2, an increased amount of bottom steel would cause an increase in the positive curvature with time and an increase in the downward deflection.

Tables 8.1 and 8.2 show the results of several time analyses of the beam shown in Fig. 8.1 subjected to different levels of applied loading. The areas of top and bottom non-prestressed reinforcement are varied to assess the effects of steel quantity and position on the final deflection. The prestressing steel details, including the cable drape and prestressing force, the cross-sectional dimensions and the material properties are as given in Example 8.1. Three different values of the sustained load are considered, namely $w_s = 6.1$, 23.8 and 40 kN/m.

When $w_s = 6.1$ kN/m (corresponding to the self-weight of the beam only), the initial concrete stress distribution at mid-span is approximately triangular with high compressive stress at the bottom of the cross-section, decreasing to zero stress near the top of the cross-section (as determined in Example 8.1 and illustrated in Fig. 8.2a). The load level $w_s = 23.8$ kN/m corresponds to the *balanced load* stage where the external load is balanced by the upward uniformly distributed load exerted on the beam by the parabolic prestressing tendon. At this balanced load stage, the initial stress distribution is uniform over the depth of each section and the initial curvature on each section is very small. When the applied load $w_s = 40$ kN/m, the initial stress distribution at mid-span is again approximately triangular with high compressive stress at the top of the section decreasing to zero near the bottom of the section (as shown in Fig. 8.2b). The moment at mid-span is therefore close to the *decompression moment.*

From the results in Table 8.1, the effect of increasing the quantity of non-prestressed bottom reinforcement, $A_{s(2)}$, is to increase the restraint to creep and shrinkage deformation in the bottom of the section and thereby increase the change in positive or sagging curvature with time. The increase is most pronounced when the initial concrete compressive stress at the level of the steel is high, i.e. when the sustained applied moment is low and the section is initially subjected to a negative or hogging curvature. Increasing the quantity of bottom reinforcement therefore increases the change in downward deflection with time.

When the beam is subjected only to its self-weight ($w_s = 6.1$ kN/m), increasing $A_{s(2)}$ from zero to 4800 mm^2($p = A_{s(2)}/bd = 1.25\%$) changes the time-dependent deflection from −8.55 mm (upwards) to +3.77 mm (downwards). Evidently, the inclusion of additional bottom reinforcement is an excellent way of reducing or eliminating the time-dependent upward deflection or camber that often causes serviceability problems in prestressed concrete construction when the sustained load is relatively light. When the beam is initially subjected to the balanced load ($w_s = 23.8$ kN/m), increasing $A_{s(2)}$ from zero to 4800 mm^2 changes the time-dependent deflection from +2.90 mm (downwards) to +11.36 mm (downwards). When the sustained load is sufficient to decompress the bottom fibres at mid-span ($w_s = 40.0$ kN/m), increasing $A_{s(2)}$ from zero to 4800 mm^2 changes the time-dependent deflection from +13.38 mm (downward) to +18.32 mm (downwards). The inclusion of tensile non-prestressed reinforcement in the bottom of a prestressed beam may increase the downward deflection with time and may in fact cause undesirable deflection in some situations.

From the results in Table 8.2, the inclusion of non-prestressed reinforcement in the compressive zone, $A_{s(1)}$, increases the restraint to creep and shrinkage deformation in the top of the section and thereby increases the change in negative or hogging curvature with time (or reduces the change in positive curvature with time). The effect

Table 8.1 Effect of varying the area of non-prestressed tensile steel $A_{s(2)} (A_{s(1)} = 0)$

		Initial deformation of section at mid-span			*Change in deformation and steel stress with time on section at mid-span*				
w_s (*kN/m*)	$A_{s(2)}$ (*mm*^{2})	$\varepsilon_{r,0}$ $(\times 10^{-6})$	κ_0 $(\times 10^{-6}\ mm^{-1})$	*Initial deflection at mid-span (mm)*	$\Delta\varepsilon_{r,k}$ $(\times 10^{-6})$	$\Delta\kappa_k$ $(\times 10^{-6}\ mm^{-1})$	$\Delta\sigma_{s(2),k}$ (*MPa*)	$\Delta\sigma_{p,k}$ (*MPa*)	*Change in mid-span deflection with time (mm)*
	0	−175	−0.492	−7.38	−632	−0.570	–	−258	−8.55
6.1	2400	−158	−0.412	−6.06	−534	−0.115	−116	−199	−0.68
	4800	−144	−0.350	−5.06	−478	0.143	−84.3	−164	3.77
	0	−173	−0.011	0.16	−655	0.193	–	−180	2.90
23.8	2400	−165	0.027	0.52	−592	0.482	−79.9	−144	8.28
	4800	−158	0.057	1.04	−557	0.649	−59.4	−122	11.36
	0	−171	0.429	6.44	−676	0.892	–	−108	13.38
40.0	2400	−171	0.429	6.55	−646	1.029	−46.9	−93.2	16.48
	4800	−171	0.429	6.63	−628	1.113	−36.6	−84.1	18.32

Table 8.2 Effect of varying the area of non-prestressed compressive steel $A_{s(1)}$ ($A_{s(2)} = 2400\ mm^2$)

		Initial deformation of section at mid-span			*Change in deformation and steel stress with time on section at mid-span*					
w_s (kN/m)	$A_{s(1)}$ (mm^2)	$\varepsilon_{r,0}$ $(\times 10^{-6})$	κ_0 $(\times 10^{-6}\ mm^{-1})$	*Initial deflection at mid-span (mm)*	$\Delta\varepsilon_{r,k}$ $(\times 10^{-6})$	$\Delta\kappa_k$ $(\times 10^{-6}\ mm^{-1})$	$\Delta\sigma_{s(1),k}$ (MPa)	$\Delta\sigma_{s(2),k}$ (MPa)	$\Delta\sigma_{p,k}$ (MPa)	*Change in mid-span deflection with time (mm)*
	0	−158	−0.412	−6.06	−534	−0.115	–	−116	−199	−0.68
6.1	2400	−158	−0.410	−6.15	−481	−0.315	−71.1	−121	−202	−4.73
	4800	−158	−0.409	−6.22	−450	−0.433	−55.4	−125	−203	−7.08
	0	−165	0.027	0.52	−592	0.482	–	−79.9	−144	8.27
23.8	2400	−156	−0.008	−0.12	−494	0.114	−108	−89.9	−149	1.70
	4800	−150	−0.036	−0.62	−439	−0.094	−80.3	−95.3	−152	−1.46
	0	−171	0.429	6.55	−646	1.029	–	−46.9	−93.2	16.5
40.0	2400	−155	0.360	5.39	−506	0.506	−142	−60.8	−101	7.59
	4800	−142	0.306	4.50	−428	0.217	−103	−68.3	−106	2.67

is most pronounced when the initial concrete compressive stress at the level of the top steel is high, i.e. when the sustained applied moment is high and the section is initially subjected to a positive or sagging curvature. Increasing the quantity of top reinforcement therefore decreases the change in downward deflection with time.

When the beam is subjected only to self-weight (w_s = 6.1 kN/m), increasing $A_{s(1)}$ from zero to 4800 mm^2($p' = A_{s(1)}/bd$ = 1.25%) changes the time-dependent deflection from −0.68 mm (upwards) to −7.08 mm (upwards). The inclusion of top reinforcement therefore increases the upward camber in a prestressed concrete member with time. When the beam is initially subjected to the *balanced load* (w_s = 23.8 kN/m), increasing $A_{s(1)}$ from zero to 4800 mm^2 changes the time-dependent deflection from +8.27 mm (downward) to −1.46 mm (upwards). When the sustained load is sufficient to decompress the bottom fibres at mid-span (w_s = 40.0 kN/m), increasing $A_{s(1)}$ from zero to 4800 mm^2 reduces the time-dependent deflection from +16.5 mm (downward) to +2.67 mm (downwards). The inclusion of non-prestressed compressive reinforcement in the top of a reinforced or prestressed beam substantially reduces the long-term downward deflection.

The significant reduction with time of the resultant compressive force carried by the concrete should also be noted. In Tables 8.1 and 8.2, the time-dependent changes in the bonded steel stress on the cross-section at mid-span are reported. This gradual build up of compression in the steel at each level of bonded reinforcement is balanced by an equal and opposite tensile force acting on the concrete at that level of magnitude $A_{s(i)}\Delta\sigma_{s(i)}$ or $A_p\Delta\sigma_p$. In Table 8.2, for example, when w_s = 23.8 kN/m and when $A_{s(1)} = A_{s(2)}$ = 2400 mm^2, the concrete on the section at mid-span is subjected to three gradually increasing tensile forces (i.e. $A_{s(1)}\Delta\sigma_{s(1)}$ = 259 kN, $A_{s(2)}\Delta\sigma_{s(2)}$ = 216 kN and $A_p\Delta\sigma_p$ = 149 kN). Initially the concrete was subjected to a net compressive prestressing force of 1300 kN, but this reduces by 624 kN with time. About 48 per cent of the initial compression in the concrete is shed into the bonded reinforcement with time. The loss of prestress in the tendon is only 149 kN (11.5 per cent).

Clearly, a reliable picture of the time-dependent behaviour of a partially prestressed concrete beam cannot be obtained unless the restraint provided to creep and shrinkage by the non-prestressed reinforcement is adequately accounted for. It is also evident that the presence of non-prestressed reinforcement significantly reduces the cracking moment with time and may in fact relieve the concrete of much of its initial prestress. This increases the possibility of time-dependent cracking and the effect should be included in design considerations. Although, the restraint provided by the bonded reinforcement, generally lowers the cracking moment, a judicious placement of non-prestressed reinforcement is an effective means of controlling deflection and can even be used to eliminate the time-dependent change in deflection in situations where such changes in deflection are undesirable.

8.2.3 *Deflection and axial shortening of cracked beams*

In Chapter 3, code-oriented methods for the prediction of deflection in reinforced and prestressed concrete members were presented. Where more reliable predictions of deformation are needed, more refined methods of analysis may be required such as those for uncracked cross-sections, outlined in Chapters 5 and 6, and for cracked cross-sections, discussed in Chapter 7. These methods of analysis may be used to predict the deflection of cracked members.

Clearly, for a cracked member, deflection will be underestimated if the analysis assumes every cross-section is uncracked. On the other hand, deflection will be overestimated, sometimes grossly overestimated, if every cross-section is assumed to be fully cracked. Tension stiffening, or the contribution of the tensile concrete to the member stiffness, ensures that the actual deflection of a cracked member lies somewhere between these two extremes, i.e. between the deflection in the *uncracked condition* and in the *fully-cracked condition.* According to Eurocode 2 (Ref. 1), to account for tension stiffening, the average curvature at a particular cracked cross-section may be taken from Eq. 3.23 (reproduced here as Eq. 8.2):

$$\kappa_{avge} = \zeta \kappa_{cr} + (1 - \zeta)\kappa_{uncr} \tag{8.2}$$

where ζ is the distribution coefficient presented earlier as Eq. 3.24b.

The most rigorous method for determining deflection using the cross-sectional analyses presented in Chapters 4–7 is to calculate the cracked and uncracked curvatures at frequent cross-sections along the member and then to calculate the average curvature at each section using Eq. 8.2. For this purpose the distribution coefficient ζ may be taken as:

$$\zeta = 1 - \left(\frac{M_{cr.t}}{M_s^*}\right)^2 \tag{8.3}$$

where $M_{cr.t}$ is the cracking moment at the time under consideration and M_s^* is the maximum in-service moment that has been imposed on the member at, or before, the time instant at which deflection is being determined. With the curvature diagram thus determined, the deflection can be obtained by numerical integration. If the curvatures at the critical cross-sections are calculated using Eq. 8.2, the mid-span deflection at the time under consideration may be conveniently obtained using Eq. 8.1b.

Example 8.2

The initial and long-term deflections at mid-span of a 10.4 m span simply-supported reinforced concrete T-beam are to be calculated. The beam carries a constant sustained uniformly distributed service load of 22.2 kN/m (which includes self-weight) first applied at age τ_0. Details of the cross-section and the relevant material properties are shown in Fig. 8.3. The reinforcement is uniform throughout and shrinkage is assumed to commence at τ_0. The cross-section is the same as that shown in Fig. 7.6 and analysed in Examples 7.1 and 7.3. The flexural tensile strength of the concrete is taken to be $f'_{ct.f} = 0.6\sqrt{f'_c} = 3.0$ MPa, so wherever the extreme fibre stress exceeds $f'_{ct.f}$, cracking is deemed to have occurred.

The bending moment diagram for the simply-supported span is illustrated in Fig 8.4a. From a short-term analysis of the uncracked cross-section, the stress in the bottom concrete fibres reaches $f'_{ct.f} = 3.0$ MPa when the applied moment is 149.3 kNm, i.e. at points D and D' in Fig. 8.4a, 1.514 m from each support. It is evident that, within 1.514 m from each support, the member is initially uncracked, whilst further into the span cracking occurs.

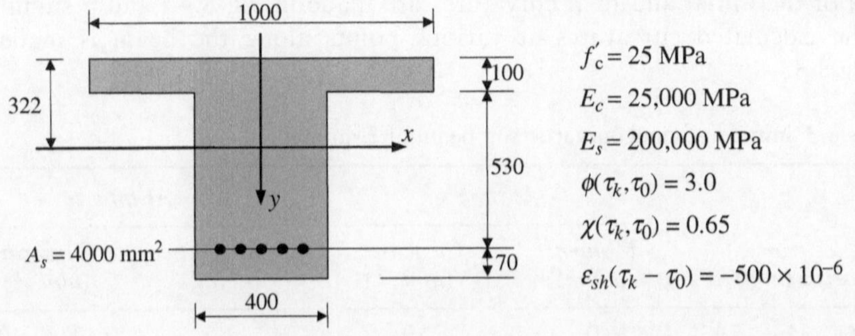

Figure 8.3 Cross-sectional details and material properties for Example 8.2 (dimensions in mm).

Using the cross-sectional analyses of Sections 5.3 (short-term) and 5.4 (time-dependent) for the uncracked portions of the span and Sections 7.2 (short-term) and 7.3 (time-dependent) for each fully-cracked cross-section, the initial and final curvatures at any cross-section along the member may be calculated. It is noted that shrinkage causes an increase in the bottom fibre tensile stress in the concrete on the uncracked cross-sections, effectively reducing the cracking moment from 149.3 kNm to 56.4 kNm with time. The extent of cracking therefore increases with time until eventually only 0.514 m at each end of the span remains uncracked. In Fig. 8.4b, the portion of the curvature diagram at time τ_k between points E and D on the span is approximated as a straight line.

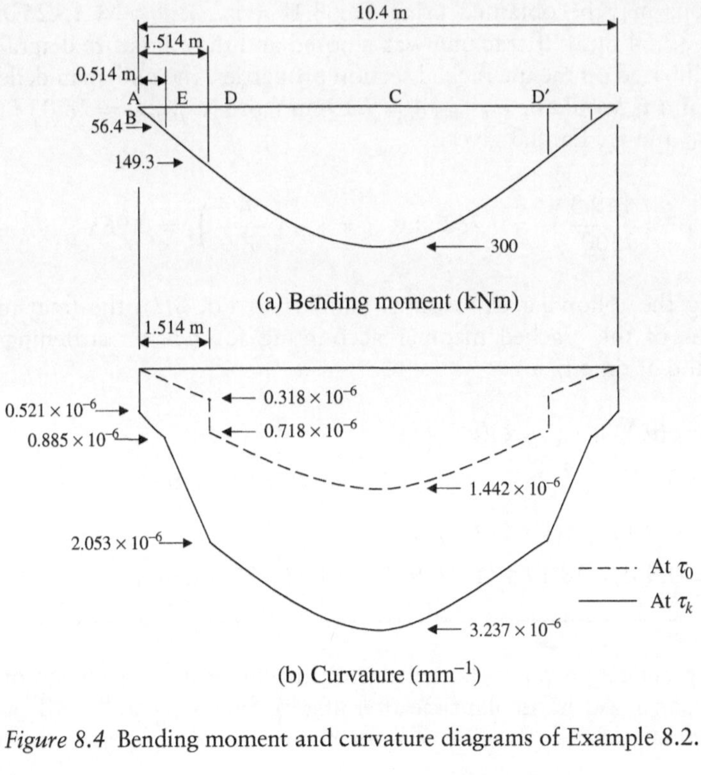

Figure 8.4 Bending moment and curvature diagrams of Example 8.2.

Plots of the initial and final curvatures are made in Fig 8.4b and a summary of the calculated curvatures at various points along the beam is made in Table 8.3.

Table 8.3 Initial and final curvatures in beam of Example 8.2

	At time τ_0		*At time* τ_k	
Distance from support, z (mm)	*Moment (kNm)*	*Curvature* (mm^{-1})	*Moment (kNm)*	*Curvature* (mm^{-1})
0	0	0	0	0.521×10^{-6}
0.514	56.4	0.120×10^{-6}	56.4	0.885×10^{-6}
1.514 –	149.3	0.318×10^{-6}	149.3	2.053×10^{-6}
1.514 +	149.3	0.718×10^{-6}	149.3	2.053×10^{-6}
3.075	250.0	1.202×10^{-6}	250.0	2.841×10^{-6}
5.2	300.0	1.442×10^{-6}	300.0	3.237×10^{-6}

The initial mid-span deflection of the cracked member at first loading τ_0 and the final deflection at time τ_k calculated by numerically integrating these curvature diagrams are:

$$(v_c)_{cr.0} = 15.98 \text{ mm and } (v_c)_{cr.k} = 38.14 \text{ mm.}$$

A close approximation of these more accurate (but considerably more tedious) calculations may be obtained using Eq. 8.1b (viz., $(v_c)_{cr.0} = 16.25$ mm and $(v_c)_{cr.k} = 37.64$ mm). If cracking was ignored and the curvature diagrams were calculated based on the uncracked section properties, the mid-span deflection at times τ_0 and τ_k would be $(v_c)_{uncr.0.} = 7.19$ mm and $(v_c)_{uncr.k} = 28.03$ mm.
At time τ_0 and τ_k, Eq. 8.3 gives:

$$\zeta_0 = 1 - \left(\frac{149.3}{300}\right)^2 = 0.752 \text{ and } \zeta_k = 1 - \left(\frac{56.4}{300}\right)^2 = 0.965$$

and using the following expressions (similar to Eq. 8.2), the final mid-span deflections of the cracked member accounting for tension stiffening at first loading and at time τ_k are:

$$v_{c.0} = \zeta(v_c)_{cr.0} + (1-\zeta)(v_c)_{uncr.0}$$
$$= 0.752 \times 15.98 + (1 - 0.752) \times 7.19 = 13.8 \text{ mm}$$
$$v_{c.k} = \zeta(v_c)_{cr.k} + (1-\zeta)(v_c)_{uncr.k}$$
$$= 0.965 \times 38.14 + (1 - 0.965) \times 28.03 = 37.8 \text{ mm}$$

For most practical purposes, knowledge of the curvature at each end of a member and at mid-span at any particular time after first loading is generally sufficient to make a reliable estimate of the deflection of a cracked reinforced concrete member.

8.3 Statically indeterminate beams and slabs

8.3.1 Discussion

In each cross-sectional analyses presented in Chapters 4–7, the internal actions were known and the changes in strain and curvature on a particular cross-section were assumed to occur without restraint from either the supports or adjacent parts of the structure. While this assumption is perfectly reasonable for statically determinate members, it may be unreasonable for continuous members.

Consider a plain concrete continuous beam subjected to a uniformly distributed load that is small enough to avoid cracking. If the concrete is assumed to be homogeneous, with the same creep characteristics throughout, the effect of creep is simply to reduce the effective modulus and to increase the strains and displacements with time. There is no change in the reactions at the supports and consequently no change in the internal actions on any cross-section. However, if the creep characteristics are not uniform throughout, creep will induce reactions at the supports and cause a change in the distribution of internal actions.

In practical concrete structures, the creep characteristics are rarely uniform. Different parts of a structure may have different ages and therefore different creep coefficients. If part of a structure has cracked, its response to creep is different from that of the uncracked parts of the structure. In addition, the quantity of reinforcement may vary from region to region and therefore the amount of internal restraint to creep will vary accordingly. Therefore, in statically indeterminate concrete structures, creep may lead to a considerable redistribution of internal actions with time.

Shrinkage also may cause significant changes in the distribution of internal actions with time in indeterminate members. If a member is not free to shorten, shrinkage produces internal tension. Shrinkage-induced curvature on each cross-section due to the restraint provided by asymmetrically placed steel reinforcement can introduce significant reactions at the supports and hence significant changes in the internal actions on each cross-section.

Consider the uncracked, propped cantilevers shown in Fig. 8.5. The member in Fig. 8.5a(i) contains a preponderance of reinforcement in the bottom fibres and shrinkage causes a positive curvature in the singly reinforced region. If the weight of the beam and any applied loads are ignored and if the support at B is removed, shrinkage would cause the member to gradually deflect upwards, as indicated. With the support at B, in fact, resisting upward movement, a downward reaction $\Delta R_{B.k}$ gradually develops with time, as shown. The initial reaction at B caused by the load on the beam $R_{B.0}$ is therefore reduced by $\Delta R_{B.k}$ and this leads to an increase in the negative moment at support A and a decrease in the positive moment at mid-span. Shrinkage therefore causes the time-dependent redistribution of moments shown in Fig. 8.5b(i). Note that the shrinkage-induced reactions at the built-in support at A (namely, a vertical reaction $\Delta R_{A.k}$, equal and opposite to $\Delta R_{B.k}$, and a moment $\Delta M_{A.k} = \Delta R_{B.k}L$) are not shown in Fig. 8.5.

For the member shown in Fig. 8.5a(ii), with a preponderance of steel in the top, shrinkage causes an upward reaction $\Delta R_{B.k}$ to develop with time at support B. As a consequence, the positive span moments increase with time and the negative moment at support A decreases, as shown in Fig. 8.5b(ii).

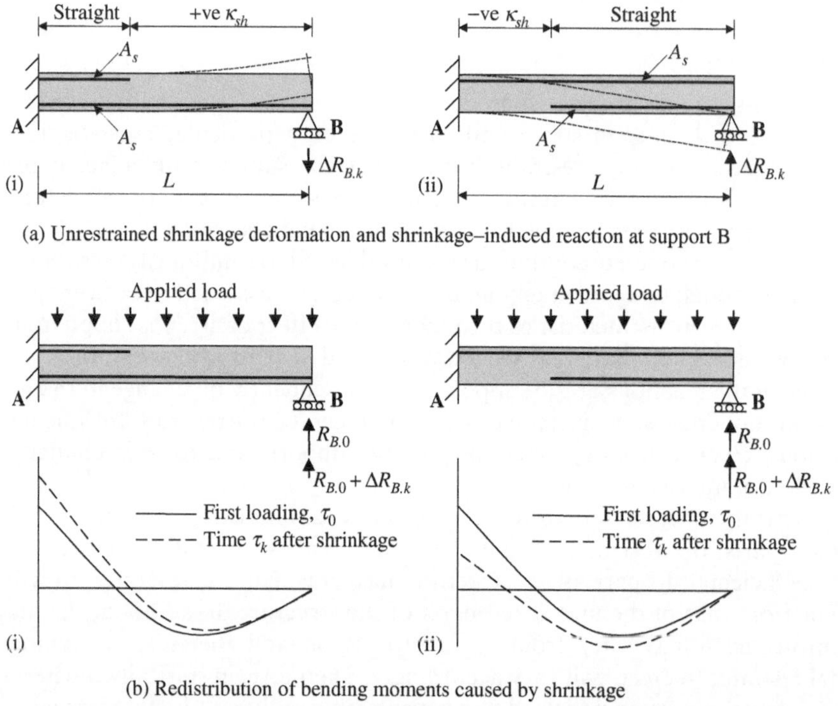

Figure 8.5 Effects of shrinkage in a one-fold indeterminate beam.

Of course, the shrinkage-induced actions occur gradually and are therefore relieved by creep. If shrinkage were not accompanied by creep, the time-dependent redistribution of internal actions would be significantly greater than it actually is.

8.3.2 Determination of redundants

A detailed discussion of the various techniques commonly employed for the analysis of statically indeterminate structures is not attempted here. There are numerous excellent texts on structural analysis, including Refs 2 and 3. In this section, a brief revision of the force method of structural analysis is made and the way in which the method can be used for the time-analysis of indeterminate structures is described and illustrated by example. In the following, the structure is initially assumed to be homogeneous.

In the force method, a n-fold indeterminate structure is converted into a statically determinate primary structure by the selection of n internal actions (or external reactions) as redundants and making the corresponding releases. Let the vector of redundant forces be $\mathbf{R}_R$.

The vector of external loads $\mathbf{P}$ applied to the primary structure produces reactions $\mathbf{R}_P$ at the supports and the external loads (or the load-independent environmental effects such as temperature change or shrinkage) cause displacements of the primary structure at each release $\mathbf{u}_P$. These displacements are incompatible with the support conditions of the indeterminate structure and must be eliminated by the unknown redundant forces.

If the primary structure is subjected to a unit virtual force at the position of the i-th redundant, i.e. $\tilde{R}_{Ri} = 1$, the resulting displacements at each of the n releases are called *flexibility coefficients* and are denoted f_{i1}, f_{i2}, ... f_{in}. If flexibility coefficients are determined for unit values of each redundant, an $n \times n$ *flexibility matrix* **F** may be established, where the term f_{ij} is the displacement of the primary structure at release j due to a unit value of the ith redundant.

The reactions at the supports of the primary structure caused by unit values of the redundants are also calculated and may be assembled into an m by n matrix $\mathbf{R}_{PR}$ where m is the number of reactions in the primary structure.

The redundant forces are obtained by solving the following set of simultaneous equations which is generated when compatibility is enforced at each release:

$$\mathbf{F}\,\mathbf{R}_R = -\mathbf{u}_P \tag{8.4}$$

where, if necessary, support movements could also be included on the right hand side of the equation. The reactions of the statically indeterminate structure are obtained by adding the reactions caused by the redundants ($\mathbf{R}_{PR}\,\mathbf{R}_R$) to the reactions calculated earlier for the released structure $\mathbf{R}_P$. That is:

$$\mathbf{R} = \mathbf{R}_P + \mathbf{R}_{PR}\,\mathbf{R}_R \tag{8.5}$$

For a structure with uniform creep characteristics throughout that is subjected to constant sustained loads, the deformation at time τ_k will be $(1+\varphi)$ times the deformation at first loading τ_0, where φ $(= \varphi(\tau_k, \tau_0))$ is the creep coefficient at time, τ_k, due to a stress first applied at τ_0. If the presence of the reinforcement is ignored and if cracking is also ignored, the displacements **u** at time τ_k are $(1+\varphi)$ times their value at τ_0 and each term of the flexibility matrix **F** is also increased by the factor $(1+\varphi)$. In addition, the reactions **R** of the statically indeterminate structure do not change and therefore there is no change in the internal actions with time.

As has been previously pointed out, this not the case, however, in most concrete structures where the reinforcement layout varies throughout the structure and the extent of cracking also varies. The propped cantilevers shown in Fig. 8.5a are a typical case. The reinforcement is not uniform throughout and, in all probability, parts of the beams are cracked and parts are not. The creep characteristics are therefore far from uniform. In such structures, some creep-induced change in the reactions (and in the internal actions) with time can be expected. In Section 8.3.1, the change in reactions of the members shown in Fig. 8.5a caused by shrinkage warping was also discussed.

To calculate the time-dependent change in reactions and internal actions during any time interval $(\tau_k - \tau_0)$, it is first necessary to calculate the change in the displacements of the primary structure at each release, $\Delta\mathbf{u}_P$. The displacements caused by both creep and shrinkage are included here, plus any other displacement that may occur during the time interval, perhaps due to support settlement, additional external loads, losses of prestress, and so on.

The AEMM may be used to include time-dependent deformations and an age-adjusted flexibility matrix $\overline{\mathbf{F}}$ can also be determined. Changes in the redundants with time are developed gradually between τ_0 and τ_k and the creep plus elastic displacements caused by these gradually applied redundants forces may be calculated using the age-adjusted modulus for concrete $\overline{E}_e$ (defined in Eq. 4.35) instead of E_c. The age-adjusted flexibility matrix $\overline{\mathbf{F}}$ is therefore formed in exactly the same way as **F**, except that the displacement (elastic + creep) at each release due to the gradually applied unit force at the position of the *i*th redundant is calculated using $\overline{E}_e$.

The time-dependent change in the redundants is obtained by solving the compatibility equations:

$$\overline{\mathbf{F}}\Delta\mathbf{R}_R = -\Delta\mathbf{u}_P \tag{8.6}$$

and the change in the reactions of the statically indeterminate structure are:

$$\Delta\mathbf{R} = \Delta\mathbf{R}_P + \mathbf{R}_{PR}\Delta\mathbf{R}_R \tag{8.7}$$

where $\Delta\mathbf{R}_P$ is the change in reactions of the primary structure due to any changes of the external loads during the time interval under consideration.

For the analysis of a statically indeterminate reinforced or prestressed concrete beam or frame, the procedures outlined in Chapters 4–7 may be used to calculate the deformation (strain and curvature, $\varepsilon_{r.k}$ and κ_k) on any cross-section of the determinate primary frame at any time. When deformations have been determined at several sections, the time-dependent changes in displacement $\Delta\mathbf{u}$ may be calculated by numerical integration or by virtual work. In this way, the effect of the reinforcement and the influence of cracking, as well as the effects of creep and shrinkage, can be included directly in the time analysis of an indeterminate member.

Example 8.3

The final shrinkage-induced internal actions at time τ_k in the unloaded and uncracked propped cantilever shown in Fig 8.6 are to be determined. The structure is, of course, one-fold indeterminate, that is, $n = 1$.

$E_c = 25{,}000$ MPa; $\varphi = 2.5$; $\chi = 0.65$;
$\varepsilon^*_{sh} = -600 \times 10^{-6}$; and $\overline{E}_e = \frac{E_c}{1+\chi\varphi} = 9{,}524$ MPa.

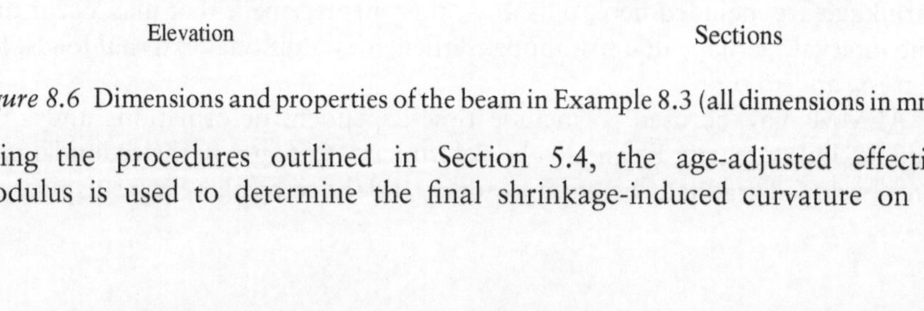

Figure 8.6 Dimensions and properties of the beam in Example 8.3 (all dimensions in mm).

Using the procedures outlined in Section 5.4, the age-adjusted effective modulus is used to determine the final shrinkage-induced curvature on an

otherwise unloaded and unrestrained cross-section in Region 2: $(\kappa_{sh})_2 = 0.663 \times 10^{-6}$ mm^{-1}. The shrinkage-induced curvature in Region 1 is zero, because each cross-section is symmetrically reinforced (and uncracked).
The vertical reaction at support B is selected as the redundant, i.e. $R_R = V_B$. The primary structure is therefore a cantilever fixed at support A. The change in the vertical displacement of the primary structure at support B (Δu_B), due to the shrinkage-induced curvature distribution shown in Fig. 8.7a, is calculated using the second moment area method as:

$$\Delta u_B = \frac{0.663 \times 10^{-6} \times 7000^2}{2} = 16.24 \text{ mm}$$

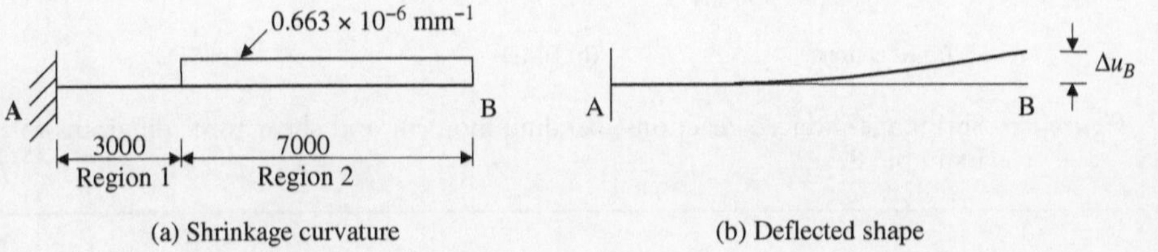

Figure 8.7 Shrinkage-induced curvature and deflected shape of primary structure in Example 8.3.

The vertical displacement at B due to a gradually applied unit force at B is $\bar{f}_{11}$ and is obtained from the curvature diagram shown in Fig. 8.8c, calculated by analysing the uncracked section in the various regions of the cantilever:

$$\bar{f}_{11} = 0.1301 \times 10^{-6} \times \frac{10{,}000^2}{3} + (0.1299 - 0.0911) \times 10^{-6} \times \frac{7000^2}{3}$$
$$= 4.97 \text{ mm/kN}$$

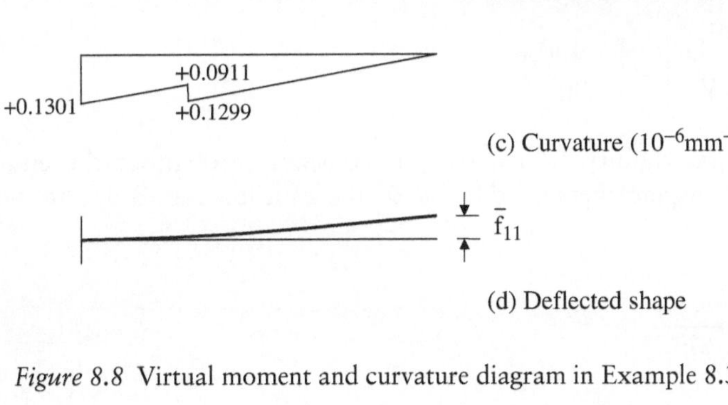

Figure 8.8 Virtual moment and curvature diagram in Example 8.3.

The redundant V_B is calculated from Eq. 8.6 as:

$$\bar{f}_{11} V_B = -\Delta u_B \quad \text{from which} \quad V_B = \frac{-16.24}{4.97} = -3.27 \text{ kN (i.e. } \downarrow)$$

The shrinkage-induced internal actions and external reactions are shown in Fig. 8.9.

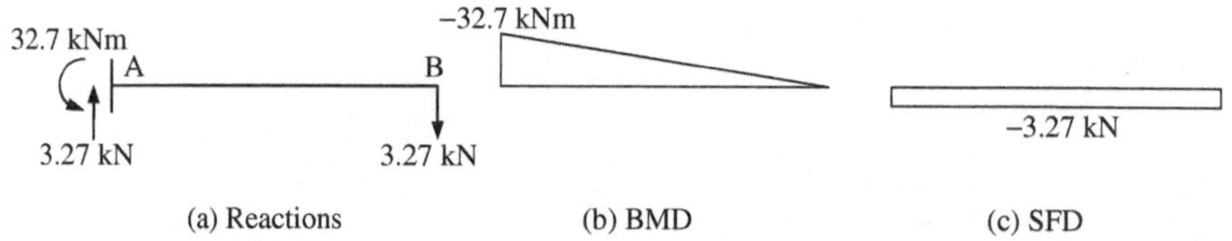

Figure 8.9 Shrinkage-induced reactions, bending moment and shear force diagrams for Example 8.3.

Example 8.4

The reactions of the one-fold indeterminate propped cantilever of Fig. 8.6 are to be determined at times τ_0 and τ_k. A uniformly distributed load $w_s = 8$ kN/m first applied at τ_0 is assumed to remain constant with time. The material properties are as for Example 8.3 and the flexural tensile strength of concrete is taken to be $f'_{ct.f} = 3.0$ MPa.

1. Calculation of the reactions at time τ_0 immediately after the application of w_s

(a) Ignoring the effects of cracking

As in the previous example, the vertical reaction at support B is selected as the redundant, i.e. $R_R = V_B$. The 8 kN/m uniformly distributed load produces the following reactions at support A of the primary beam: $M_A = -400$ kNm and $V_A = 80$ kN, therefore:

$$\mathbf{R}_P = \begin{bmatrix} M_A \\ V_A \end{bmatrix} = \begin{bmatrix} -400 \\ 80 \end{bmatrix}$$

If the flexural rigidity EI ($= M/\kappa$) is constant throughout the entire span, the vertical displacement at the end of the cantilever at B due to a uniform load w_s is:

$$u_B = -\frac{w_s L^4}{8EI}$$

and the vertical displacement at B due to a unit vertical (upward) force at B (f_{11}) and the corresponding reactions of the primary structure are:

$$f_{11} = \frac{L^3}{3EI}$$

and

$$\mathbf{R}_{PR} = \begin{bmatrix} M_A = L \\ V_A = -1 \end{bmatrix}$$

From Eq. 8.4:

$$V_B \times \frac{L^3}{3EI} = \frac{wL^4}{8EI}$$

and therefore

$$V_B = \frac{3wL}{8} = 30 \text{ kN}(\uparrow)$$

The other reactions of the statically indeterminate frame are obtained from Eq. 8.5:

$$\mathbf{R} = \begin{bmatrix} M_A \\ V_A \\ V_B \end{bmatrix} = \begin{bmatrix} -400 \\ +80 \\ 0 \end{bmatrix} + \begin{bmatrix} 10 \\ -1 \\ 1 \end{bmatrix} \times 30 = \begin{bmatrix} -100 \text{ kN.m} \\ +50 \text{ kN} \\ +30 \text{ kN} \end{bmatrix}$$

If the effects of cracking are ignored, the bending moment diagram of the indeterminate member immediately after first loading is shown in Fig. 8.10.

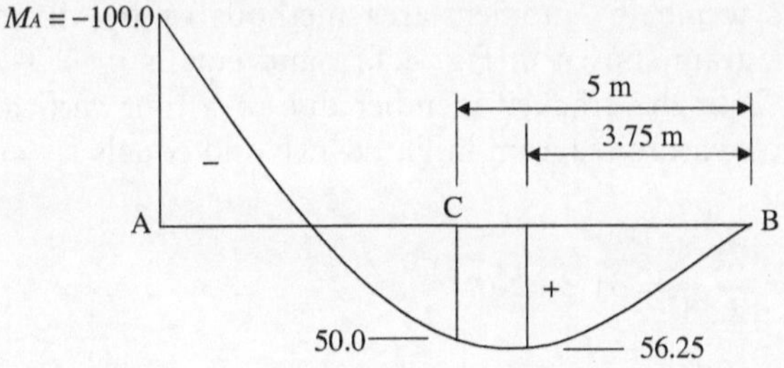

Figure 8.10 Initial bending moment diagram (ignoring cracking) in kNm for Example 8.4.

If the effects of cracking are also ignored in the calculation of deflection (which, of course may grossly underestimate the actual deflection), the initial curvature on the doubly-reinforced section at support A and on the singly-reinforced section at the mid-span C are calculated using the analysis outlined in Section 5.3 as $\kappa_{A,0} = -0.569 \times 10^{-6}$ mm^{-1} and $\kappa_{C,0} = +0.325 \times 10^{-6}$ mm^{-1}, respectively. The short-term deflection at mid-span v_C may be determined using Eq. 8.1b:

$$v_C = \frac{10{,}000^2}{96}[-0.569 + 10 \times 0.325 + 0] \times 10^{-6} = 2.79 \text{ mm}$$

(b) Including the effects of cracking

From a short-term cross-sectional analysis of the singly reinforced section in the positive moment region (Region 2), the cracking moment is 65.3 kNm. This is greater than the maximum positive moment in Fig. 8.10 and, therefore, the mid-span region is not cracked initially. For the doubly reinforced section at support A (in Region 1), the cracking moment is 70.3 kNm. Therefore, the beam is cracked in the negative moment region near support A.

The previous analysis must now be modified to include the effects of cracking at A. Tension stiffening is not considered here, but could be easily taken into account using the averaging procedure discussed in Section 8.2.3.

If the bending moment diagram in Fig. 8.10 is taken as an initial approximation, the extent of cracking is small, with the cracked region extending just 0.625 m from support A. The curvatures at sections in the various regions of the beam are calculated from short-term cross-sectional analyses (Section 5.3 for the uncracked cross-sections and Section 7.2 for the cracked cross-sections) and the instantaneous flexural rigidities ($EI = M/\kappa$) are:

In Region 1: Cracked $(EI)_{cr.1} = 70.67 \times 10^{12}$ Nmm^2
Uncracked $(EI)_{uncr.1} = 175.7 \times 10^{12}$ Nmm^2
In Region 2: Uncracked $(EI)_{uncr.2} = 154.0 \times 10^{12}$ Nmm^2

As previously calculated:

$$\mathbf{R}_P = \begin{bmatrix} M_A \\ V_A \end{bmatrix} = \begin{bmatrix} -400 \\ +80 \end{bmatrix}$$

and the vertical displacement at the end of the primary beam (at B) due to $w_s = 8$ kN/m is found, by moment-area methods or by virtual work, from the curvature diagram shown in Fig. 8.11a and equals $u_B = -78.5$ mm. The displacement at B of the cracked member due to a unit vertical force at B is found from the curvature diagram in Fig. 8.11b and equals $f_{11} = 2.51$ mm/kN. From Eq. 8.4:

$$V_B = \frac{-u_B}{f_{11}} = \frac{78.5}{2.51} = 31.3 \text{ kN}(\uparrow)$$

and the reactions of the statically indeterminate beam are obtained from Eq. 8.5:

$$\{R\} = \begin{bmatrix} M_A \\ V_A \\ V_B \end{bmatrix} = \begin{bmatrix} -400 \\ 80 \\ 0 \end{bmatrix} + \begin{bmatrix} 10 \\ -1 \\ 1 \end{bmatrix} \times 31.3 = \begin{bmatrix} -87 \text{ kN.m} \\ 48.7 \text{ kN} \\ 31.3 \text{ kN} \end{bmatrix}$$

The bending moment and curvature diagrams after cracking are illustrated in Fig. 8.12. The short-term deflection at mid-span may be obtained by virtual work from the curvature diagram in Fig. 8.12 and is $v_c = 3.17$ mm.

For this beam, cracking at, and near, the support at A causes a reduction in the negative moment at A, an increase in the positive span moments and a 14 per cent increase in mid-span deflection. After initial cracking, moment is shed to the stiffer uncracked parts of the structure. If the load w was increased,

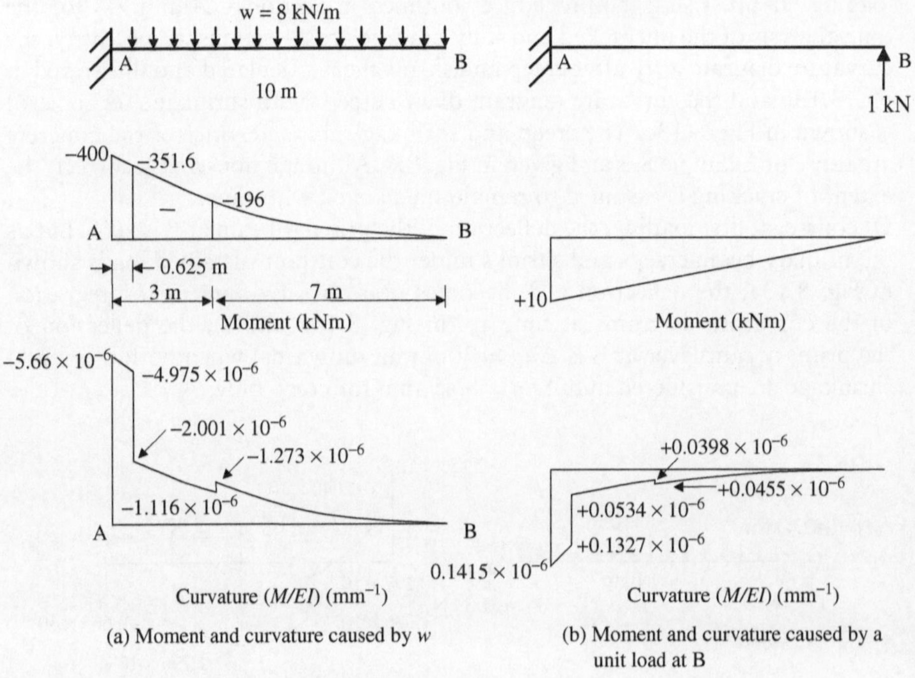

Figure 8.11 Moment and curvature diagrams for cracked primary beam of Example 8.4.

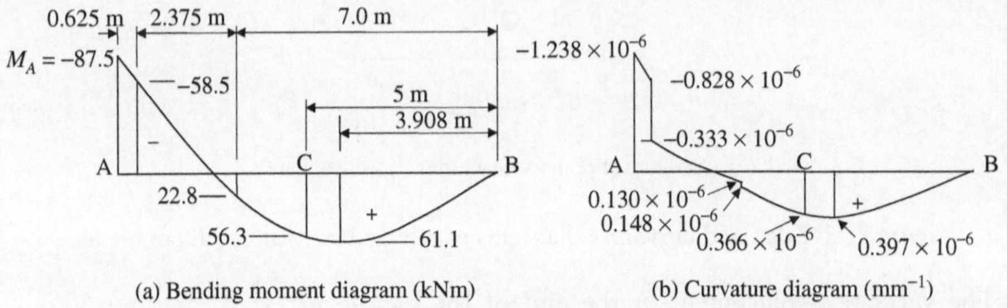

Figure 8.12 Moment and curvature diagrams after cracking in Example 8.4.

cracking would also occur in the positive moment region. With the member more uniformly cracked and the stiffness more uniform in the peak moment regions, the bending moments would redistribute back towards the distribution shown in Fig. 8.10, but the deflection would increase significantly due to the reduction in the flexural rigidity for Region 2 from $(EI)_{uncr.2}$ to $(EI)_{cr.2}$.

2. Calculation of the reactions at time τ_k after creep and shrinkage

Let the change in reaction at support B be the unknown redundant at time τ_k, i.e. $\mathbf{R}_R = \Delta V_{B.k}$. The primary structure (cantilevered beam) is shown in Fig. 8.13a, together with the external loads and the known reactions immediately after first loading (at τ_0). Using the procedures outlined in Sections 5.4 and 7.3 for the time analysis of the uncracked and fully-cracked cross-sections, respectively, the curvature diagram at τ_k after creep and shrinkage is calculated and illustrated in Fig. 8.13b and the curvature diagram due to creep (with shrinkage set to zero) is shown in Fig. 8.13c. The creep and shrinkage characteristics of the concrete are those of Example 8.3 and given in Fig. 8.6. Although not strictly correct, the extent of cracking is assumed to remain unchanged with time.

Of course, at first loading, the deflection of the primary beam at B is zero. But as the primary beam creeps and shrinks under the constant sustained loads shown in Fig. 8.13a, the deflection at B becomes non-zero. By numerical integration of the curvature diagrams at time τ_k (in Fig. 8.13b and c), the deflection of the primary cantilever at B is $\Delta u_B = 20.5$ mm (upwards) when both creep and shrinkage are considered and $\Delta u_B = 8.49$ mm for creep only.

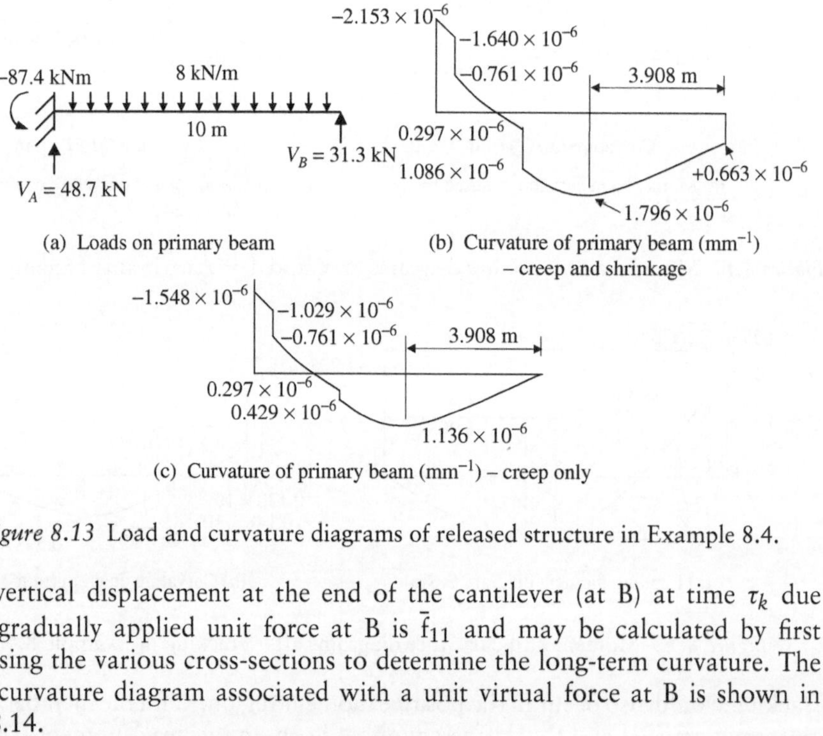

Figure 8.13 Load and curvature diagrams of released structure in Example 8.4.

The vertical displacement at the end of the cantilever (at B) at time τ_k due to a gradually applied unit force at B is $\bar{f}_{11}$ and may be calculated by first analysing the various cross-sections to determine the long-term curvature. The final curvature diagram associated with a unit virtual force at B is shown in Fig. 8.14.

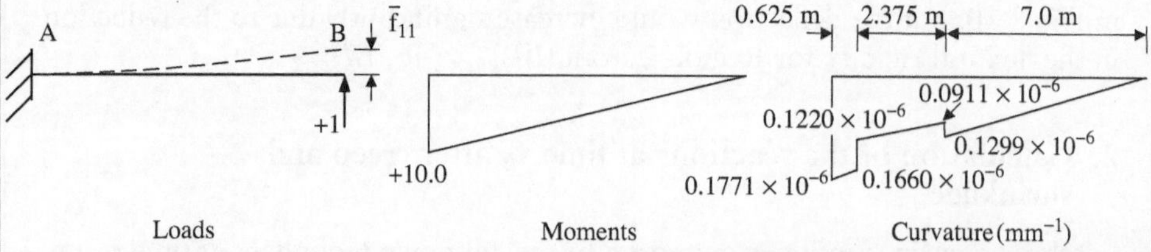

Figure 8.14 Moment and curvature at time τ_k due to gradually applied unit virtual load at B in Example 8.4.

The deflection at B due to a unit virtual force at B ($\bar{f}_{11}$) is readily obtained from the curvature diagram using moment-area methods or numerical integration:

$$\bar{f}_{11} = 0.1299 \times \frac{7.0^2}{3} + 0.0911 \times 2.375 \times 8.188 + (0.1220 - 0.0911) \times \frac{2.375}{2} \times 8.583 + 0.1660 \times 0.625 \times 9.688 + (0.1771 - 0.1660) \times \frac{0.625}{2} \times 9.792 = 5.25 \text{ mm/kN}$$

From Eq. 8.6, the change in the reaction at B with time is:

$$\Delta V_B = -\frac{\Delta u_B}{\bar{f}_{11}} = -\frac{20.5}{5.25} = -3.90 \text{ kN(i.e.} \downarrow) - \text{ for both creep and shrinkage}$$

$$\Delta V_B = -\frac{8.49}{5.25} = -1.62 \text{ kN(i.e.} \downarrow) - \text{ for creep only}$$

In Fig. 8.15, the reactions at time τ_k and the corresponding internal actions are compared with the values immediately after first loading. Note that although the

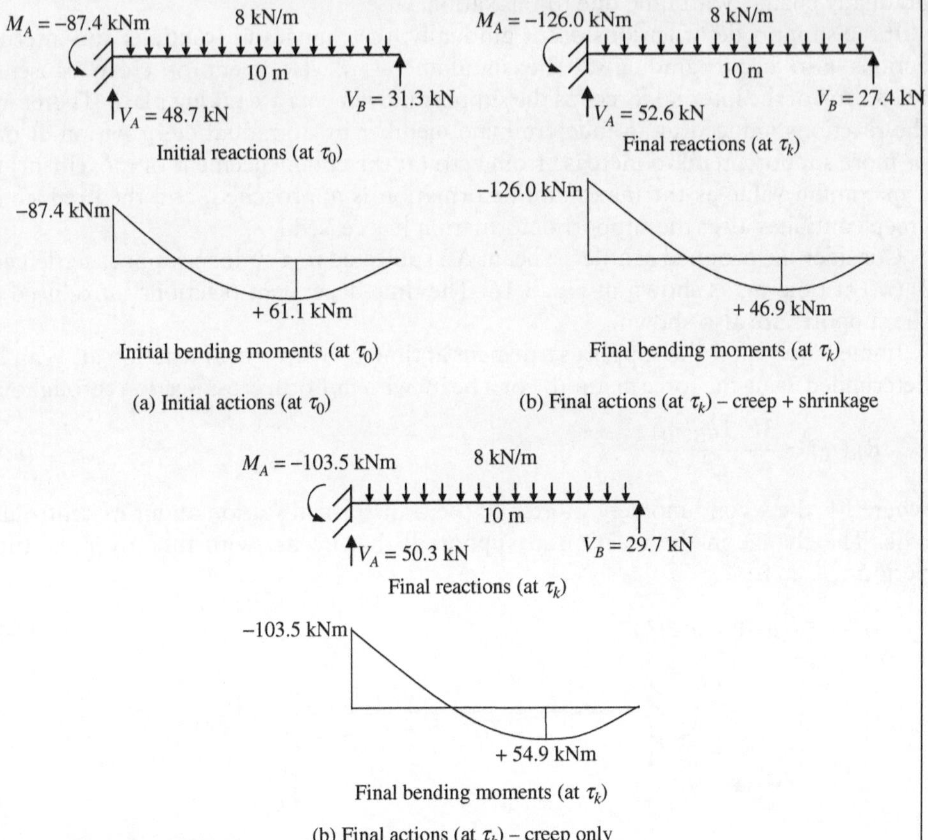

Figure 8.15 Initial and final reactions and moments in Example 8.4.

initial cracking at A caused a reduction in the peak negative moment at the fixed support, the effect of creep and shrinkage is to substantially increase the moment at support A and reduce the positive moment at mid-span. In this example, creep causes the magnitude of the moment at A to increase by about 18 per cent (from 87.4 to 103.5 kNm) and shrinkage causes a further increase of 26 per cent (from 103.5 to 126.0 kNm). In the positive moment region, shrinkage-induced tension in the concrete may cause time-dependent cracking and this will reduce the stiffness of the positive moment region resulting in a further increase in the negative support moment and a decrease in the positive span moments. The increase in negative moments will increase the extent of cracking near the support and this, in turn, will cause a slight redistribution of moments from the negative to the positive moment region.

8.3.3 Effects of deformation or settlement at the supports

If a statically indeterminate member is subjected to a sudden movement of the supports, changes in the magnitudes of the reactions and internal actions occur. If the imposed deformation at the supports is kept constant, the reactions and internal actions gradually change with time due to relaxation.

If the support deformations occur gradually, the changes in reactions and internal actions also occur gradually. The simultaneous development of creep causes a relaxation of the internal forces as the support movements are taking place. Therefore, the reactions induced in an indeterminate member by a gradual deformation at one or more supports tend to increase from zero (at the commencement of movement) to a maximum value as the maximum deformation is approached, and then reduce as creep continues after the support deformation has ceased.

Consider the propped cantilever beam AB subjected to a sudden support settlement $\delta_B(\tau_0)$ at time τ_0, as shown in Fig. 8.16. The time-dependent reactions introduced at the supports are also shown.

Immediately after the support settlement at time τ_0, the vertical reaction at B can be determined using the force method. For a beam with uniform cross-section throughout:

$$R_B(\tau_0) = \frac{3E_c I \delta_B(\tau_0)}{L^3} \tag{8.8}$$

where I is the second moment of area of the transformed section about its centroidal axis. The change in the reaction at support B that occurs with time over the time period $\tau_k - \tau_0$ is:

$$\Delta R_B = R_B(\tau_k) - R_B(\tau_0) \tag{8.9}$$

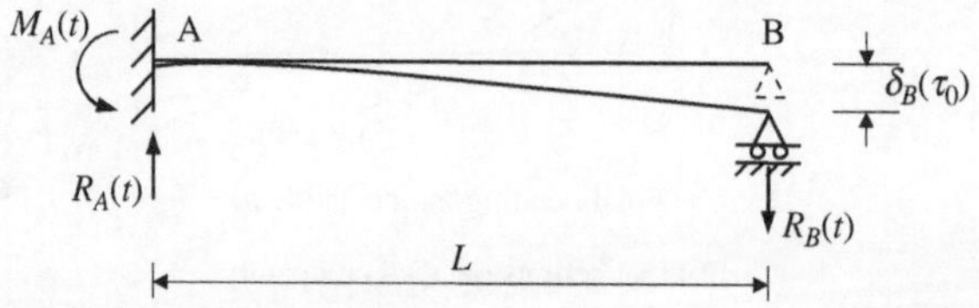

Figure 8.16 Time-dependent reactions caused by a support settlement.

and may be determined using the procedure outlined in Section 8.3.2 (and Eqs 8.6 and 8.7).

If the support settlement $\delta_B(t)$ is gradually applied, the reaction $R_B(t)$ increases gradually with time. If the settlement occurs at the same rate as the creep of the concrete, the reaction at B in a member of uniform cross-section may be approximated by:

$$R_B(t) = \frac{3\overline{E}_e \overline{I}\,\delta_B(t)}{L^3} \tag{8.10}$$

where $\overline{I}$ is the second moment of area of the age-adjusted transformed cross-section about its centroidal axis. After the support settlement reaches its maximum value at some time τ_1, the subsequent change in the reaction with time (at $t > \tau_1$), $\Delta R_B = R_B(t) - R_B(\tau_1)$, may be calculated using Eq. 8.6.

Example 8.5

The initial and final internal actions in the unloaded and uncracked member shown in Fig. 8.17 are to be determined, if the support at B suffers a vertical displacement of 20 mm, as shown.

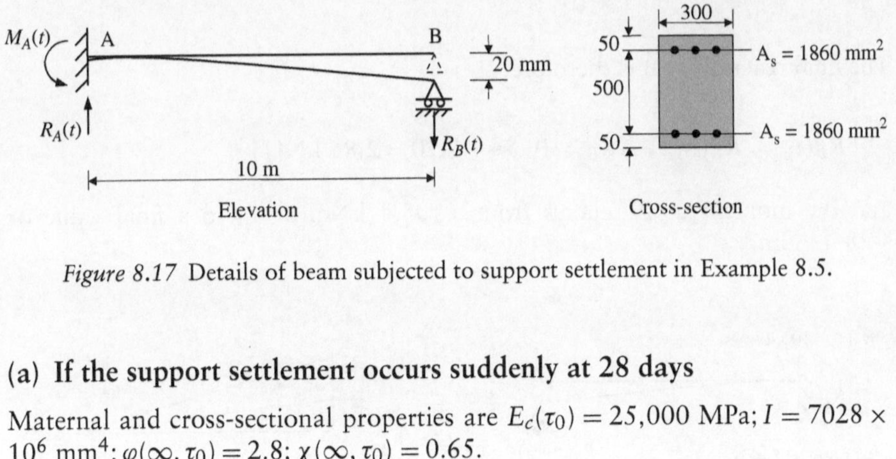

Figure 8.17 Details of beam subjected to support settlement in Example 8.5.

(a) If the support settlement occurs suddenly at 28 days

Maternal and cross-sectional properties are $E_c(\tau_0) = 25{,}000$ MPa; $I = 7028 \times 10^6$ mm^4; $\varphi(\infty, \tau_0) = 2.8$; $\chi(\infty, \tau_0) = 0.65$.
The reaction at B at age τ_0, immediately after the support displacement, is found using Eq. 8.8:

$$R_B(\tau_0) = \frac{3 \times 25{,}000 \times 7028 \times 10^6 \times 20}{10{,}000^3} = 10.54 \text{ kN} \downarrow$$

and the internal bending moment varies from zero at B to -105.4 kNm at A. The initial curvature varies from zero at B to $\kappa_{A.0} = M_A/E_c I = -0.60 \times 10^{-6}$ mm^{-1} at A as shown in Fig. 8.18b. It is assumed here that the member remains uncracked throughout. The final curvature at end A, caused by the sustained moment of −105.4 kNm, is calculated using the cross-sectional analysis of Section 5.4 and equals $\kappa_{A.k} = -1.434 \times 10^{-6}$ mm^{-1}, as shown in Fig 8.18b. The creep-induced change of curvature at end A is therefore $\kappa_{A.k} - \kappa_{A.0} = -0.834 \times 10^{-6}$ mm^{-1}. If the displaced beam is released at B, the creep induced curvature causes a gradual displacement at end B given by:

$$\Delta u_B = \frac{0.834 \times 10^{-6} \times 10{,}000^2}{3} = 27.8 \text{ mm} \downarrow$$

For this member, the age-adjusted effective modulus is $\bar{E}_e = 8865$ MPa and the second moment of area of the age-adjusted transformed cross-section about its centroidal axis is $\bar{I} = 10{,}410 \times 10^6$ mm^4.
The vertical displacement caused by a gradually applied unit load (1 kN) at B is:

$$\bar{f}_{11} = \frac{L^3}{3\bar{E}_e\bar{I}} = \frac{10{,}000^3}{3 \times 8865 \times 10{,}410 \times 10^6} = 3.61 \times 10^{-3} \text{ mm/N}$$

and the change in the reaction at B with time is:

$$\Delta R_B = \frac{\Delta u_B}{\bar{f}_{11}} = \frac{27.8}{3.61 \times 10^{-3}} = 7701 \text{ N} = 7.70 \text{ kN} \uparrow$$

The final reaction at B is therefore:

$$R_B(\tau_k) = R_B(\tau_0) + \Delta R_B = 10.54 - 7.70 = 2.84 \text{ kN}(\downarrow)$$

and the moment at A relaxes from −105.4 kNm at τ_0 to a final value of −28.4 kNm.

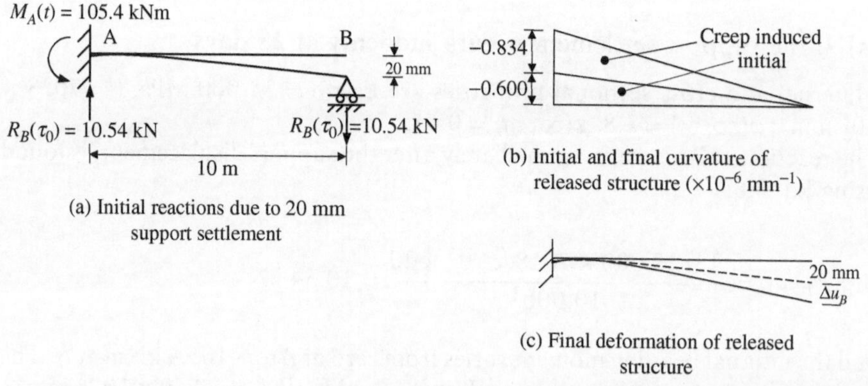

Figure 8.18 Actions and deformations of released primary beam in Example 8.5.

(b) If support settlement of 20 mm occurs gradually between $\tau_0 = 28$ days and $\tau_1 = 180$ days

It is here assumed that the rate of settlement is similar to the rate of creep and the following material properties are assumed:

For concrete loaded at 28 days: $E_c(28) = 25{,}000$ MPa; $\varphi(180, 28) = 1.5$; $\varphi(\infty, 28) = 2.8$; $\chi(180, 28) = \chi(\infty, 28) = 0.65$
For concrete loaded at 180 days: $E_c(180) = 30{,}000$ MPa; $\varphi(\infty, 180) = 1.0$; $\chi(\infty, 180) = 0.65$.

The age-adjusted effective modulus at age 180 days due to a stress first applied at 28 days is $\bar{E}_e = 12{,}660$ MPa and the second moment of area of the corresponding age-adjusted transformed cross-section is $\bar{I} = 8841 \times 10^6$ mm^4. The reaction at B at $\tau_1 = 180$ days, when δ_B reaches 20 mm, is calculated using Eq. 8.10:

$$R_B(180) = \frac{3 \times 12{,}660 \times 8841 \times 10^6 \times 20}{10{,}000^3} = 6720 \text{ N} = 6.72 \text{ kN}(\downarrow)$$

and the corresponding moment at the fixed support A is -67.2 kNm. If the displacement at B remains constant at 20 mm after $\tau_1 = 180$ days, the reactions and internal actions will gradually decrease, in the same manner as in case (a).
The change in reaction at B that occurs after τ_1, ΔR_B, is selected as the redundant. The creep-induced curvature $\Delta\kappa_A$ that develops at A in the primary beam due to a sustained moment of -67.2 kNm (after $t = 180$ days) is calculated using the cross-sectional analysis presented in Section 5.4. With $\varphi(\infty, 180) = 1.0$ and $\chi(\infty, 180) = 0.65$, $\Delta\kappa_A$ is equal to -0.223×10^{-6} mm^{-1}. The resulting time-dependent change in deflection of the primary beam at B is:

$$\Delta u_B = \frac{-0.223 \times 10^{-6} \times 10{,}000^2}{3} = -7.43 \text{ mm (i.e. } \downarrow)$$

The appropriate properties of the age-adjusted transformed section are $\bar{E}_e = 18{,}180$ MPa and $\bar{I} = 7725 \times 10^6$ mm^4. Due to a gradually applied unit load at B:

$$\bar{f}_{11} = \frac{10{,}000^3}{3 \times 18{,}180 \times 7725 \times 10^6} = 2.37 \times 10^{-3} \text{ mm/N}$$

The change in the reaction at B, which occurs after $\tau_1 = 180$ days, is therefore:

$$\Delta R_B = -\frac{-7.43}{2.37 \times 10^{-3}} = 3140 \text{ N} = 3.14 \text{ kN}(\uparrow)$$

and the final reaction at B is therefore $R_B(\infty) = 6.72 - 3.14 = 3.58$ kN ($\downarrow$). The corresponding bending moment at the fixed end of the member $M_A(\infty) = -35.8$ kNm.

Plots of M_A versus time for both the sudden support settlement and the gradual support settlement are shown in Fig. 8.19.

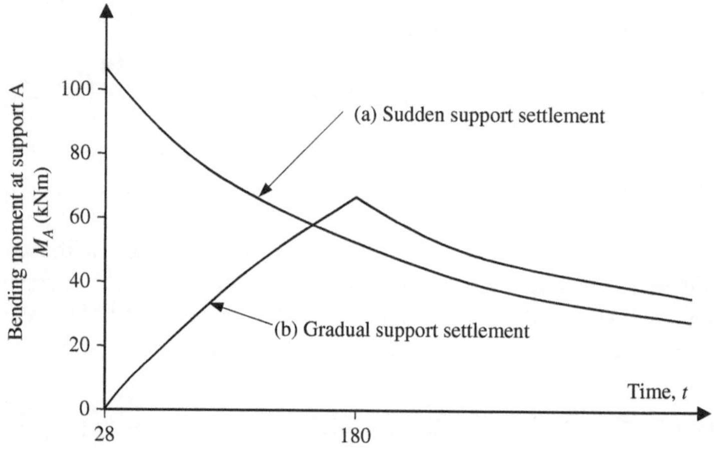

Figure 8.19 Time variation of moment due to sudden and gradual support settlement in Example 8.5.

8.3.4 Further effects of creep in prestressed construction

Although creep in continuous members causes a dramatic reduction of the internal actions caused by imposed deformation (as shown in curve (a) of Fig. 8.19), the effects of creep on the redistribution of internal actions in continuous members subjected to imposed loads is less pronounced, but still significant (as demonstrated in Example 8.4). In general, internal actions are redistributed from the regions with a higher creep rate to the regions with the lower creep rate. In Example 8.4, with the rate of change of curvature due to creep in the singly-reinforced mid-span Region 2 greater than the rate of change of curvature in the doubly-reinforced Region 1, creep causes a decrease in the positive span moments with time and an increase in the negative moments at the continuous support (see Fig. 8.15).

The internal actions caused by prestress are also affected by creep of concrete. They are obviously affected by the losses of prestress caused by creep and these losses can vary from near zero, when the member is cracked at the level of the tendon, to as much as 20 per cent or more, when the member is heavily prestressed at an early age and the concrete compressive stress at the level of the bonded tendon is high. The hyperstatic reactions induced by prestress in indeterminate structures are in fact imposed loads applied to the structure at the supports and the resulting secondary moments are affected by creep.

If the structural system changes after the application of some of the prestress, creep may cause a change in the hyperstatic reactions. As an example, consider the two-span beam shown in Fig. 8.20. The beam is fabricated by erecting two simply-supported prismatic precast concrete girders (of constant cross-section) over the two spans of length L. Each girder contains a single layer of straight pretensioned strands at a

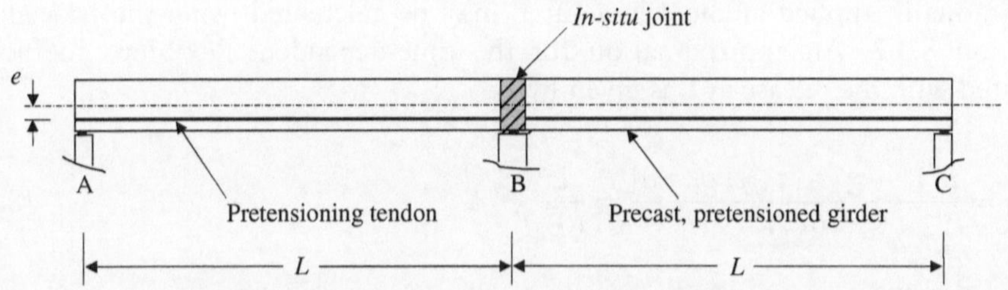

Figure 8.20 Providing continuity at a continuous support.

constant eccentricity *e* below the centroidal axis of the cross-section. The structure is then rendered continuous by casting an *in-situ* joint over the interior support. Any time-dependent defomations in the members will be restrained by the imposed fixity at the interior support and hyperstatic reactions will develop with time, together with the associated secondary moments and shears.

Before the *in-situ* joint in Fig. 8.20 is cast, the two-precast girders are simply supported, with zero deflection but some non-zero slope at the interior support at B. Immediately after the joint is fabricated and continuity is established, the internal primary moment imposed by the prestress on each cross-section is Pe about the centroidal axis, but the initial secondary moment at B (and elsewhere) is zero. With time, creep will cause a gradual change in the curvature on each cross-section.

If the support at B was released so that a vertical displacement at B was possible, the member ABC would gradually deflect upwards at B due to the creep-induced hogging curvature associated with the primary moment Pe on each cross-section. If it is assumed that the creep characteristics are uniform and the prestressing force is constant throughout, the time-dependent curvature caused by creep on each cross-section can be determined using the procedure presented in Section 5.4. For lightly reinforced cross-sections, the creep-induced curvature $(\kappa_k)_{cr}$ may be taken as a product of the instantaneous curvature and the creep coefficient (Eq. 8.11a) and the upward displacement at B is given by Eq. 8.11b:

$$(\kappa_k)_{cr} = \frac{Pe}{E_c I}\,\varphi(\tau_k, \tau_0) \tag{8.11a}$$

$$\Delta u_B(\tau_k) = \frac{(\kappa_k)_{cr} L^2}{2} \tag{8.11b}$$

where I is the second moment of area of the transformed section about its centroidal axis.

In Eq. 8.11a, it is assumed that the restraint offered to creep by the bonded reinforcement is insignificant. If the amount of bonded reinforcement is significant, a more accurate approximation may be made by dividing the creep coefficient in Eqs 8.11 by the parameter α defined in Section 3.6.6 and, for uncracked cross-sections, specified in Eq. 3.32b.

The deflection at B caused by a unit value of the vertical redundant reaction force gradually applied at the release at B may be calculated using the procedures of Section 8.3.2. An approximation for the time-dependent flexibility coefficient associated with the release at B is given by:

$$\bar{f}_{11} = \frac{L^3[1+\chi(\tau_k,\tau_0)\varphi(\tau_k,\tau_0)]}{6E_cI} = \frac{L^3}{6\bar{E}_eI} \tag{8.12}$$

The redundant force at B that gradually develops with time $R_B(\tau_k)$ is obtained from Eq. 8.6:

$$\begin{aligned} R_B(\tau_k) &= -\frac{\Delta u_B(\tau_k)}{\bar{f}_{11}} \approx -\frac{3Pe}{L}\frac{\varphi(\tau_k,\tau_0)}{1+\chi(\tau_k,\tau_0)\varphi(\tau_k,\tau_0)} \\ &= R_B\frac{\varphi(\tau_k,\tau_0)}{1+\chi(\tau_k,\tau_0)\varphi(\tau_k,\tau_0)} \end{aligned} \tag{8.13}$$

where R_B is the hyperstatic reaction that would have developed at B if the structure was initially continuous and later prestressed with a straight tendon of constant eccentricity e.

In general, if R is any hyperstatic reaction or the restrained internal action that would occur at a point due to prestress in a continuous member and $R(t)$ is the corresponding creep-induced value if the member is made continuous after the application of prestress, then:

$$R_B(t) \approx \frac{\varphi(\tau_k,\tau_0)}{1+\chi(\tau_k,\tau_0)\varphi(\tau_k,\tau_0)}R \tag{8.14}$$

If the creep characteristics of the concrete are uniform throughout the structure, then Eq. 8.14 may be applied to systems with any number of redundants.

Providing continuity at the interior supports of a series of simple precast beams, not only restrains the time-dependent deformation caused by prestress, but also restrains the deformation caused by external loads. For all the external loads applied after the continuity is established, the effects can be calculated by moment distribution or an equivalent method of structural analysis. Under the loads applied to the simply-supported beams prior to casting the joints (such as the self-weight of the precast beams), the moments at each interior support are initially zero. However, after the joint has been cast, the creep-induced deformations resulting from these *self-weight* moments are restrained and moments develop with time at the supports. For the two-span beam shown in Fig. 8.20, the initial moment at the support B due to self-weight is zero. However, if the beam had initially been continuous over the interior support, the moment at B would have been $M_B = w_{sw}L^2/8$. The moment that develops with time due to creep at support B due to creep and self-weight is:

$$M_B(t) \approx \frac{\varphi(\tau_k,\tau_0)}{1+\chi(\tau_k,\tau_0)\varphi(\tau_k,\tau_0)}M_B$$

Example 8.6

Consider two simply-supported pretensioned concrete planks, erected over two adjacent spans as shown in Fig. 8.20. An *in-situ* reinforced concrete joint is cast at the interior support, to provide continuity. Each plank is 1000 mm wide by 150 mm thick and pretensioned with straight strands at a constant depth of 110 mm below the top fibre, with A_p = 400 mm and $P_{p,init}$ = 500 kN. A typical cross-section is shown in Fig. 8.21. The span of each plank is $L = 6$ m. If it is assumed that the continuity is provided immediately after the transfer of prestress, the reactions that develop with time due to creep are to be determined. For convenience, and to better illustrate the effects of creep in this situation, shrinkage is not considered and the self-weight of the planks is also ignored. The material properties are: $E_c = 30{,}000$ MPa; $f_{ct} = 3.0$ MPa; $\varphi(\tau_k, \tau_0) = 2.0$; $\chi(\tau_k, \tau_0) = 0.65$; $\varepsilon_{sh} = 0$; and $\bar{E}_e = 13{,}040$ MPa.

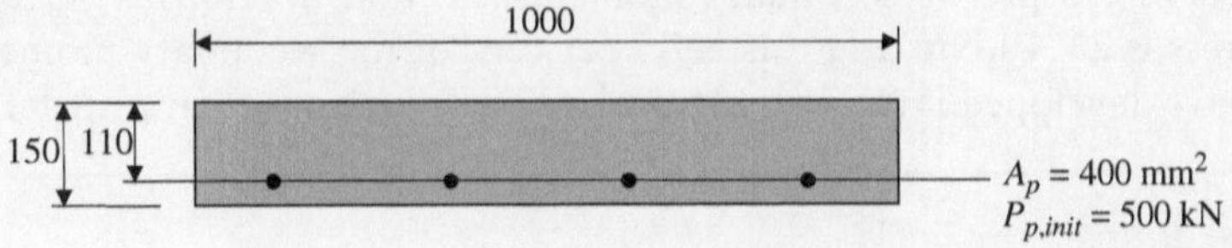

Figure 8.21 Cross-section of planks in Example 8.6 (all dimensions in mm).

Immediately after transfer, and before any creep has taken place, the curvature on each cross-section caused by the eccentric prestress may be calculated using the procedure outlined in Section 5.3 giving $\kappa_0 = -2.02 \times 10^{-6}$ mm^{-1}. The top and bottom fibre concrete stresses are $\sigma_{top,0} = 1.30$ MPa and $\sigma_{top,0} = -7.81$ MPa, so that cracking has not occurred.
In the absence of any restraint at the supports, the curvature on each cross-section would change with time due to creep from κ_0 to $\kappa_k = -5.81 \times 10^{-6}$ mm^{-1} (calculated using the procedure outlined in Section 5.4). If the support at B in Fig. 8.20 was removed, the deflection at B ($\Delta u_B(\tau_k)$) that would occur with time due to the change in curvature caused by creep ($(\kappa_k)_{cr} = \kappa_k - \kappa_0 = -3.79 \times 10^{-6}$ mm^{-1}) can be calculated from Eq. 8.1b for the 12 m long planks and is also given by Eq. 8.11b:

$$\Delta u_B(\tau_k) = \frac{-3.79 \times 10^{-6} \times 6000^2}{2} = -68.2 \text{ mm (i.e. upward)}$$

The deflection at B caused by a unit value of the vertical redundant reaction force gradually applied at the release at B is calculated using the procedures of Section 8.3.2 and for this example is given by Eq. 8.12. With the second moment of area of the age-adjusted transformed cross-section $\bar{I} = 288.3 \times 10^6$ mm^4, Eq. 8.12 gives:

$$\bar{f}_{11} = \frac{L^3}{6\bar{E}_e\bar{I}} = \frac{6000^3}{6 \times 13{,}040 \times 288.3 \times 10^6} = 9.58 \times 10^{-3}$$

and the redundant force at B that gradually develops with time, $R_B(\tau_k)$ is therefore:

$$R_B(\tau_k) = -\frac{\Delta u_B(\tau_k)}{\bar{f}_{11}} = -\frac{-68.2}{0.00958} = 7124 \text{ N}$$

The reaction that would have developed at B due to prestress immediately after transfer if the member had initially been continuous is R_B = 8621 kN and the approximation of Eq. 8.14 gives:

$$R_B(\tau_B) \approx \frac{2.0}{1+0.65\times 2.0} \times 8621 = 7497 \text{ N}$$

In this example, Eq. 8.14 gives a value for $R_B(\tau_k)$ that is within 6 per cent of the value determined above and provides a quick and reasonable estimate of the effects of creep. The secondary moment at B that develops with time due to prestress is 21.4 kNm and this is 83 per cent of the secondary moment that would have developed if the two planks had been continuous at transfer.

8.4 Two-way slab systems

8.4.1 Discussion

The design of suspended concrete slab systems is complicated by the difficulties involved in estimating the service load behaviour. It is relatively easy to find a load path and reinforce a slab to satisfy the requirements of adequate strength. Slabs are usually very ductile, highly indeterminate members that are capable of considerable moment redistribution as the ultimate load is approached. The distribution of moments in a statically indeterminate slab at failure is very much dependent on the reinforcement pattern and, therefore, on the load path assumed in design. However, in the design for serviceability, the distribution of moments and the prediction of deflection at service loads are far more uncertain. The distribution of moments varies with time and depends on the extent of cracking, the post-cracking stiffness of the various parts of the slab, the level of creep and shrinkage and the restraint provided at the supports.

As outlined in the preceding pages, the initial and time-dependent deformations and the extent of cracking in concrete structures depend primarily on the non-linear and inelastic properties of concrete and, as such, are difficult to predict with confidence. The problem is exacerbated in the case of slabs which are typically thin in relation to their spans and are therefore more deflection sensitive. It is stiffness rather than strength that controls the design of most reinforced and prestressed concrete slab systems.

The first step in the design of a two-way slab panel is the initial selection of the slab thickness. A reasonable first estimate is desirable since, in many cases, the slab self-weight is a large portion of the total service load. The initial estimate is often based on personal experience or on recommended maximum span-to-depth ratios or minimum thickness requirements, such as those presented in Section 3.5. While providing a useful starting point in design, such a selection does not necessarily ensure serviceability. Deflections at all critical stages in the history of the slab must be calculated and the magnitudes limited to acceptable design values. Failure to adequately predict deflections has frequently resulted in serviceability problems (Ref. 4).

The procedures discussed so far in this chapter have been concerned with the prediction of the in-service behaviour of line members, such as beams, one-way slabs and frames. For the calculation of two-way slab deflections, several additional problems need to be overcome. The three-dimensional nature of the slab, the less well defined influence of cracking and tension stiffening and the development of biaxial creep and shrinkage strains must all be modelled adequately.

In addition, the final deflection of a slab depends very much on the extent of initial cracking which, in turn, depends on the construction procedure (the shoring and reshoring sequence), the amount of early shrinkage, the temperature gradients in the first few weeks after casting, the degree of curing and so on. Many of these parameters are, to a large extent, outside the control of the designer. In field measurements of the deflection of many identical slab panels (Refs 5 and 6), a large variability was reported. Deflections of identical panels after one year differed by over 100 per cent in some cases. These differences can be attributed to the different conditions that existed in the first few weeks after casting of each slab, including differences in both the applied load and the environmental conditions. In more recent laboratory tests on large-scale flat slabs, a change in deflection at least as large as that caused by the full design live load was observed after exposing the top surface of a dry slab to rainwater (Ref. 7).

Nevertheless, various approximate methods are available that may be used to predict ball-park estimates of the deflection of two-way slabs, making use of the cross-sectional analyses presented in Chapters 4–7. By varying the amount of early cracking, for example by varying the tensile strength of concrete, upper and lower bounds on deflection can be obtained. Some approximate procedures for deflection calculation are reviewed in the following sections.

8.4.2 Slab deflection models

In view of the complexities and uncertainties involved in estimating the service load behaviour of two-way slab systems, great accuracy in the calculation of deflection is neither possible nor warranted. Nevertheless, numerous approximate techniques are available for the prediction of moments and deformations in concrete slab systems. These range from relatively simple procedures to sophisticated research models.

The small deflection theory of elastic plates can be used to predict slab deflections. Deflection coefficients for elastic slabs with ideal boundary conditions and subjected to full panel loading have been presented by Timoshenko and Woinowsky-Krieger (Ref. 8). Many more simple models involve the analysis of orthogonal slab strips. The deflection of the slab panel is approximated by the sum of the deflection components of one or more slab strips (see Refs 9–15).

The finite-element method is perhaps the most powerful and potentially the most accurate tool for the analysis of concrete slabs. The basic method is well established and has been described in many text books. Since the early 1970s, many investigators have developed non-linear finite-element models to study the short-term service load behaviour of reinforced concrete slabs. A number of researchers extended their models to handle the time-dependent effects of creep and shrinkage (e.g. Refs 16–18). Today numerous commercial finite element software packages are available to undertake non-linear analysis of slabs at service loads. The treatment of time effects in these computer packages ranges from rational and reliable to crude and unreliable. Users of any commercial program should be aware of the assumptions made in the modelling of material non-linearity, particularly relating to cracking, creep and shrinkage, and make a rational assessment of the reliability of the output.

However, finite element modelling of two-way slab systems, even with the most reliable of software, is time consuming and generally unsuitable for routine use in structural design. However, it is a useful research tool to examine the effects of various parameters on slab behaviour and to generate the parametric data necessary for the development of more simple, design-oriented procedures for the estimation of slab deflections. Such parametric studies have been reported elsewhere (Ref. 17) and have led to the development of simple, design-oriented methods for the deflection control of slabs (Ref. 15). The finite-element model described in Ref. 17 has been used for the calibration of slab stiffness factors in the following section.

8.4.3 Two-way edge-supported slabs

This section deals with the prediction of the deflection of two-way rectangular slab panels that are continuously supported on all four edges by beams or walls and that carry a uniformly distributed load. The slab panel may be either continuous or discontinuous on each edge. A simple and useful estimate of deflection may be made using the so-called *crossing beam analogy*. This involves the consideration of a pair of orthogonal slab strips of unit width spanning through the centre of the panel as shown in Fig. 8.22.

Each slab strip carries only part of the transverse load w on the slab. By equating the mid-span deflection of each slab strip and assuming that the entire transverse load is carried by bending in the two orthogonal directions, the fraction of load carried in the short-span direction γ^* is easily determined. If each strip is initially assumed to be uncracked:

$$\gamma^* = \frac{L_y^4}{\lambda L_x^4 + L_y^4} \tag{8.15}$$

where λ depends on the support conditions of the panel and is given by:

$\lambda = 1.0$ for 4 edges continuous discontinuous
$\lambda = 1.0$ for 2 adjacent edges discontinuous
$\lambda = 2.0$ for 1 long edge discontinuous
$\lambda = 0.5$ for 1 short edge discontinuous
$\lambda = 2.5$ for 2 long + 1 short edge discontinuous
$\lambda = 0.4$ for 2 short + 1 long edge discontinuous
$\lambda = 5.0$ for 2 long edges discontinuous
$\lambda = 0.2$ for 2 short edges discontinuous.

The deflection of the slab may be estimated conservatively by calculating the initial and time-dependent deflection of the shorter span slab strip subjected to a uniformly distributed load, $\gamma^* w$. However, this ignores the torsional stiffness of the slab and violates compatibility at all but the slab centre. As a consequence, the predicted deflection may be significantly greater than the actual deflection. Using the non-linear

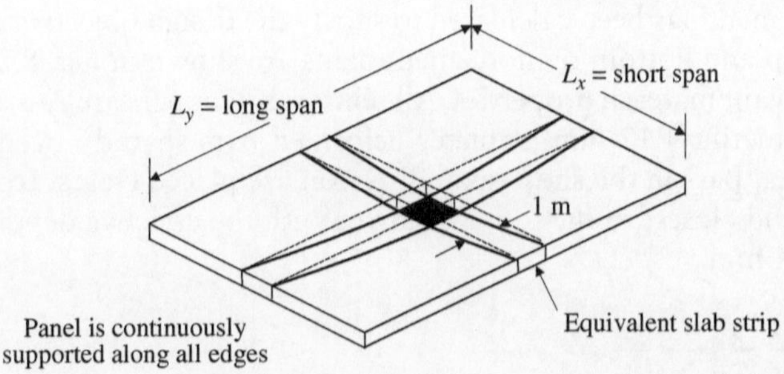

Figure 8.22 Orthogonal slab strips in an edge-supported slab.

finite-element model described in Ref. 17, a slab stiffness factor has been calibrated which may be used to adjust the stiffness of the shorter span beam, so that the deflection of the slab panel is predicted more accurately. In Table 8.4, this slab stiffness factor is combined with the value of γ^* (as given by Eq. 8.15). The deflection of the panel may be estimated by analysing a unit wide one-way slab strip in the short span direction with the same boundary conditions as the relevant slab edges and carrying uniformly distributed load of $\gamma^* w$. For each possible boundary condition, γ^* depends on the aspect ratio L_y/L_x and may be interpolated from Table 8.4.

If the slab edges are supported by beams, the average deflection of the supporting beams along the long edges must be calculated and added to the slab strip deflection to obtain the total deflection at the centre of the panel.

Table 8.4 Values of γ^* for two-way edge-supported slabs

	Value of γ^*				
	L_y/L_x				
	1.0	1.25	1.5	2.0	5.0
4 edges continuous	0.48	0.70	0.84	0.94	1.00
1 short edge discontinuous	0.60	0.78	0.91	0.97	1.00
1 long edge discontinuous	0.31	0.50	0.69	0.87	1.00
2 short edges discontinuous	0.74	0.87	0.95	0.99	1.00
2 long edges discontinuous	0.15	0.28	0.41	0.65	0.97
2 adjacent edges discontinuous	0.45	0.66	0.79	0.93	1.00
2 short + 1 long edge discontinuous	0.55	0.71	0.81	0.94	1.00
2 long + 1 short edge discontinuous	0.22	0.36	0.50	0.72	1.00
4 edges discontinuous	0.36	0.50	0.60	0.80	1.00

Example 8.7

The maximum final deflection at the mid-span of a 6 by 8 m exterior panel of a beam and slab floor system is to be calculated. The slab panel is 150 mm thick and forms part of the floor of a retail store. The panel edges are continuously supported by stiff beams. The design moments in the slab have been determined from moment coefficients specified in the local building code and the reinforcement has been calculated to satisfy the design objective of adequate strength. Top and bottom reinforcement plans are shown in Fig. 8.23, together with the relevant material properties. All reinforcement bars are Australian Class N (normal ductility) 12 mm diameter deformed bars spaced at centres shown in millimetres. Bars in the short-span direction are placed closest to the bottom of the slab and closest to the top, as shown, with the effective depth $d = 150 - 20 - 6 = 124$ mm.

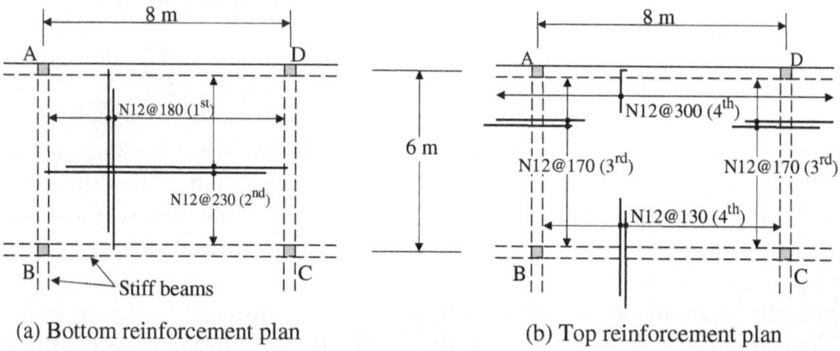

$f'_c = 25$ MPa ; $E_c = 25{,}000$ MPa ; $\phi^* = 2.8$; $\chi^* = 0.8$;
$\varepsilon^*_{sh} = -0.0006$; $f_y = 500$ MPa ; $E_s = 200{,}000$ MPa ; $f_t = 2.0$ MPa ;
Minimum concrete cover = 20 mm

Figure 8.23 Details of 150 mm thick slab panel of Example 8.7.

The slab carries a superimposed dead load of $g_{sup} = 1.5$ kPa, in addition to its own self-weight g_{sw}, and the specified live load q is 5.0 kPa. For a retail store, the service load factors are taken as (refer to Section 3.3) $\psi_1 = 0.6$ (for short-term loads) and $\psi_2 = 0.3$ (for long-term loads).
If reinforced concrete is taken to weigh 24 kN/m^3, then the sustained load is:

$$w_{sus} = g_{sw} + g_{sup} + \psi_2 q = 24 \times 0.15 + 1.5 + 0.3 \times 5.0 = 6.6 \text{ kPa}$$

and the variable load is:

$$w_{var} = (\psi_1 - \psi_2)q = (0.6 - 0.3) \times 5.0 = 1.5 \text{ kPa}$$

The slab panel is discontinuous on one long edge only, with an aspect ratio of $L_y/L_x = 1.33$, and from Table 8.4 $\gamma^* = 0.56$.
A 1 m wide slab strip spanning in the short-span direction through the slab centre is to be analysed. The slab strip is considered to be continuous at one end and simply-supported at the other, as shown in Fig. 8.24a. The analysis of such a member was presented in Example 8.4. Assume the bending moment produced by the maximum short-term load on the strip (i.e. $\gamma^*(w_{sus} + w_{var}) = 4.54$ kN/m) is as shown in Fig. 8.24b.

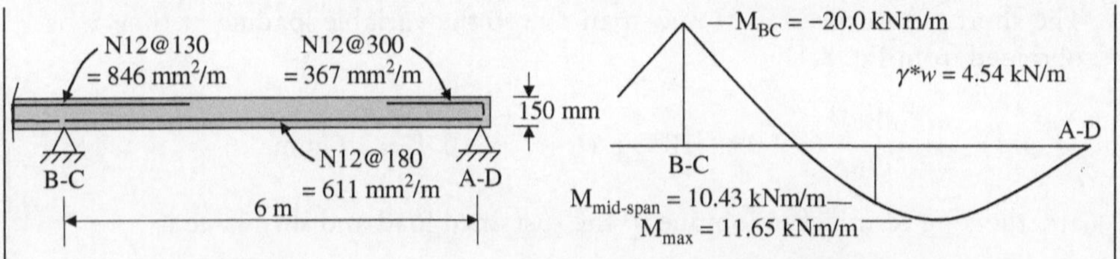

(a) Elevation (b) Bending moment diagram

Figure 8.24 Elevation of slab strip and moment diagram in Example 8.7.

From a short-term cross-sectional analysis using the approach outlined in Section 5.3, the cracking moments for this 150 mm thick slab at mid-span at first loading (with zero shrinkage) and at the interior support are +9.89 and −10.09 kNm/m. It is evident that both the positive and negative moment regions of the slab will crack under the full service load.

The initial and time-dependent curvatures at each support and at mid-span caused by the sustained load $\gamma^* w_{sus} = 3.70$ kN/m and the short-term curvatures caused by the variable load $\gamma^* w_{var} = 0.84$ kN/m are calculated using the cross-sectional analyses of Sections 5.3 and 5.4 for the uncracked sections and Sections 7.2 and 7.3 for the cracked sections and are summarised in Table 8.5. From the uncracked analyses, the moments required to produce an extreme fibre tensile stress in the concrete of $f_{ct} = 2.5$ MPa at first loading τ_0 and at τ_k are readily determined and given in Table 8.5.

Tension stiffening is accounted for using the distribution factor ζ given in Eq. 8.3 and the average curvature κ_{avge} at each critical section due to each loading is calculated using Eq. 8.2 and is also given in Table 8.5.

Table 8.5 Uncracked, cracked and average curvature at critical sections in Example 8.7.

	At continuous support		*At mid-span*		*At discontinuous support*	
	At first loading, τ_0	*Final at τ_k*	*At first loading, τ_0*	*Final at τ_k*	*At first loading, τ_0*	*Final at τ_k*
Cracking moment, M_{cr} (kNm/m)	−10.09	−4.64	9.89	5.99	–	–
Max. moment, M_s^* (kNm/m)	−20.0	−20.0	11.65	11.65	0	0
Distribution factor, ζ (Eq. 8.3)	0.746	0.946	0.279	0.736	0	0
Due to $\gamma^* w_{var}$:						
κ_{uncr} (mm^{-1})	-0.50×10^{-6}	-0.50×10^{-6}	0.27×10^{-6}	0.27×10^{-6}	0	0
κ_{cr} (mm^{-1})	-2.18×10^{-6}	-2.18×10^{-6}	1.48×10^{-6}	1.48×10^{-6}	0	0
κ_{avge} (mm^{-1})	-1.75×10^{-6}	-2.09×10^{-6}	0.61×10^{-6}	1.16×10^{-6}	0	0
Due to $\gamma^* w_{sus}$ + shrinkage:						
κ_{uncr} (mm^{-1})	-2.21×10^{-6}	-9.20×10^{-6}	1.17×10^{-6}	6.52×10^{-6}	0	0.77×10^{-6}
κ_{cr} (mm^{-1})	-9.60×10^{-6}	-20.0×10^{-6}	5.33×10^{-6}	14.8×10^{-6}	0	–
κ_{avge} (mm^{-1})	-7.42×10^{-6}	-19.4×10^{-6}	2.33×10^{-6}	$12.6\times^{-6}$	0	0.77×10^{-6}

The short-term deflection at mid-span due to the variable loading at time τ_k is obtained from Eq. 8.1b:

$$(v_c)_{var} = \frac{6000^2}{96}(-2.09 + 10 \times 1.16 + 0) \times 10^{-6} = 3.6 \text{ mm}$$

and the long-term deflection due to the sustained load and shrinkage is

$$(v_c)_{sus} = \frac{6000^2}{96}(-19.4 + 10 \times 12.6 - 0.77) \times 10^{-6} = 39.6 \text{ mm}$$

The maximum total panel deflection is therefore

$$(v_c)_{\max} = (v_c)_{\text{var}} + (v_c)_{sus} = 43.2 \text{ mm.}$$

With a maximum deflection of span/130, this slab is unserviceable for most applications and a thicker slab would almost certainly be required for this floor system.

8.4.4 Flat slabs

In this section, a simple method is presented for predicting the deflection at the mid-panel of a uniformly loaded two-way column-supported flat slab. The basic procedure is known as the wide beam method (or equivalent frame method) and was formalised by Nilson and Walters (Ref. 19). It is assumed here that the design bending moments in each direction are known and that the quantity and layout of the reinforcement in each region of the slab are also known: that is, the slab has been designed for strength and an estimate of deflection is now required.

The basis of the method is illustrated in Fig. 8.25. For the deflection calculation, the deformation of a slab panel in one direction at a time is considered. The contributions in each direction are then added to obtain the total deflection.

In Fig. 8.25a, the slab is considered to act as a wide, shallow beam of width equal to the panel dimension L_y and span equal to L_x and carrying the entire load in the x-direction. This wide beam is assumed to rest on unyielding supports. Because of variations in moment and flexural rigidity across the width of the slab, all unit strips in the x-direction will not deform identically. Moments, and hence curvatures, in the regions near the column lines (the column strip) are greater than in the middle strips. The deflection on the column line is therefore greater than that at the panel centre. This is particularly so, when, as is usually the case, the column strips are cracked and the middle strips are uncracked.

The slab is next considered to act as a wide shallow beam carrying the entire load in the y-direction as shown in Fig 8.25b. Once again, the effect of variation of moment and flexural rigidity across the panel is shown.

The mid-panel deflection is taken to be the sum of the average mid-span deflection of the column strips in one direction (usually taken as the long-span direction) and the mid-span deflection of the middle strip in the other direction (Fig. 8.25c). That is:

$$v_{\max} = \left(\frac{v_{cx1} + v_{cx2}}{2}\right) + v_{my} \tag{8.16}$$

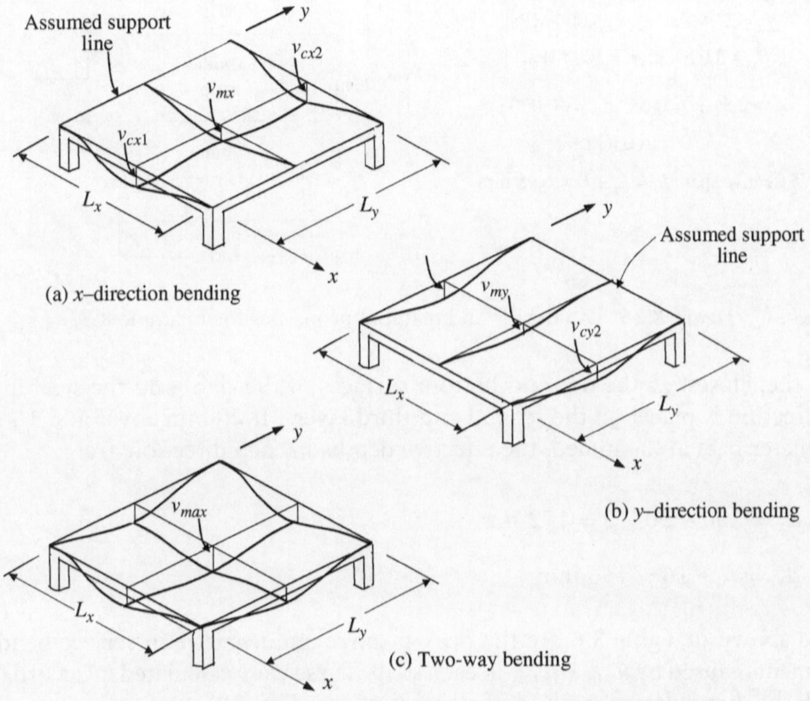

Figure 8.25 Deflection components of a flat slab (Ref. 19).

The method is fully compatible with the equivalent frame method of moment analysis (Ref. 20). The definition of column and middle strips, the longitudinal moment distribution, lateral moment distribution coefficients and other details are the same as for the moment analysis, so that in design most of the information required for the estimation of deflection has been previously calculated.

With the moment diagrams for the column and middle strips known, it is a simple matter to determine the curvatures at the supports and at the mid-span of each strip, using the cross-sectional analyses of Chapters 5 and 7. The initial and time-dependent strip deflections can be calculated using Eq. 8.1b. For flat slabs, the span in Eq. 8.1b may be taken to be the centre to centre distance between columns or the clear span (face to face of columns) plus the slab depth, whichever is smaller.

Example 8.8

The maximum total deflection at the mid-point of a square interior panel of the flat plate shown in Fig. 8.26 is to be calculated. The slab is 200 mm thick and supports a sustained service load of $w_{sus} = 6.3$ kPa (which includes self-weight) and a variable live load of $w_{var} = 3.0$ kPa.

The column and middle strips in each direction are 3575 mm wide. The tensile reinforcement quantities in the x-direction column strip and y-direction middle strip are given in Table 8.6. The steel in the x-direction is placed first and

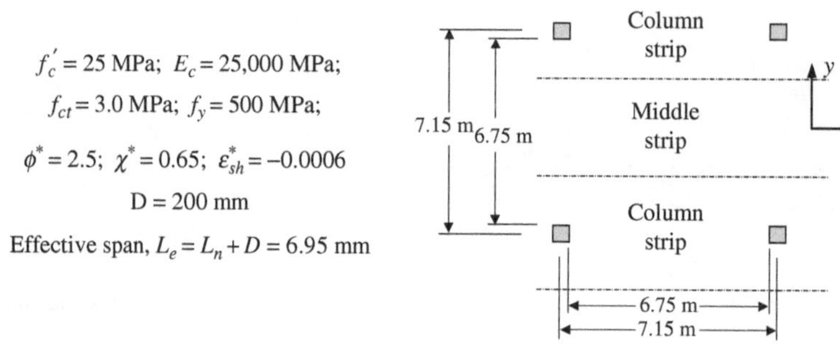

Figure 8.26 Slab details and material properties for Example 8.8.

last (i.e. closest to the top and bottom surfaces of the slab) and the steel in the y-direction is placed in the second and third layers. If 20 mm cover and 16 mm diameter bars are assumed, the effective depths in each direction are:

$$d_x = 200 - 20 - 8 = 172 \text{ mm}$$

$$d_y = d_x - 16 = 156 \text{ mm}$$

Also shown in Table 8.6 are the peak positive and negative in-service bending moments caused by $w_{sus} + w_{var}$ in each strip. These were calculated in accordance with the direct design method of slab design (see Ref. 20).

Table 8.6 Steel quantities and in-service moments for Example 8.8

	Column strip (x-direction)		*Middle strip (y-direction)*	
	Negative moment region	*Positive moment region*	*Negative moment region*	*Positive moment region*
A_{st} (mm^2)	5420	2280	1950	1650
$M_{sus+var}$ (kNm)	−196	85	−66	56

Column strip deflection (x-direction)

At the support

The maximum in-service moment in the column strip at the support (due to $w_{sus} + w_{var}$) is −196 kNm and, from the cross-section analysis of Section 5.3, the cracking moment at first loading is −80.0 kNm and after all shrinkage is −32.8 kNm. The critical cross-section at the support will obviously be cracked. The final time-dependent curvature at the support caused by the moment due to the sustained load ($M_{sus} = -132.8$ kNm) plus shrinkage and the short-term curvature caused by the moment due to the variable load ($M_{var} = -63.2$ kNm) are calculated using the cross-sectional analyses of Sections 5.3 and 5.4 for the uncracked section and Sections 7.2 and 7.3 for the cracked section and are summarised below. For the inclusion of tension stiffening in the final deflection calculations, the distribution factor ζ at time τ_k is obtained from Eq. 8.3:

$$\zeta = 1 - \left(\frac{-32.8}{-196}\right)^2 = 0.972$$

and the average curvatures at the supports of the column strip are obtained from Eq. 8.2.

Instantaneous curvature due to M_{var}:

$$\kappa_{uncr} = -0.98 \times 10^{-6}\ \text{mm}^{-1}; \kappa_{cr} = -3.19 \times 10^{-6}\ \text{mm}^{-1};\ \text{and therefore}$$

$$\kappa_{avge} = -3.13 \times 10^{-6}\ \text{mm}^{-1}$$

Final long-term curvature due to M_{sus}+ shrinkage:

$$\kappa_{uncr} = -7.85 \times 10^{-6}\ \text{mm}^{-1}; \kappa_{cr} = -14.23 \times 10^{-6}\ \text{mm}^{-1};\ \text{and therefore}$$

$$\kappa_{avge} = -14.05 \times 10^{-6}\ \text{mm}^{-1}.$$

At mid-span

The maximum in-service moment in the column strip at mid-span (due to w_{sus} + w_{var}) is 85 kNm and, from Section 5.3, the cracking moment at first loading is 75.1 kNm and after all shrinkage is 56.0 kNm. The critical cross-section at mid-span will crack. The final time-dependent curvature at mid-span caused by the moment due to the sustained load (M_{sus} = 57.6 kNm) plus shrinkage and the short-term curvature caused by the moment due to the variable load (M_{var} = 27.4 kNm) are next determined using the cross-sectional analyses of Sections 5.3 and 5.4 for the uncracked section and Sections 7.2 and 7.3 for the cracked section. The distribution factor ζ at time τ_k is obtained from Eq. 8.3:

$$\zeta = 1 - \left(\frac{56.0}{85}\right)^2 = 0.566$$

and the average curvatures at mid-span of the column strip are obtained from Eq. 8.2:

Instantaneous curvature due to M_{var}:

$$\kappa_{uncr} = 0.44 \times 10^{-6}\ \text{mm}^{-1}; \kappa_{cr} = 2.79 \times 10^{-6}\ \text{mm}^{-1};\ \text{and therefore}$$

$$\kappa_{avge} = 1.77 \times 10^{-6}\ \text{mm}^{-1}$$

Final long-term curvature due to M_{sus}+ shrinkage:

$$\kappa_{uncr} = 3.81 \times 10^{-6}\ \text{mm}^{-1}; \kappa_{cr} = 11.71 \times 10^{-6}\ \text{mm}^{-1};\ \text{and therefore}$$

$$\kappa_{avge} = 8.28 \times 10^{-6}\ \text{mm}^{-1}.$$

Deflection

The short-term deflection at mid-span of the column strip due to the variable loading is:

$$(v_{cx})_{var} = \frac{6950^2}{96}(-3.13 + 10 \times 1.77 + -3.13) \times 10^{-6} = 5.8 \text{ mm}$$

and the long-term deflection due to the sustained load and shrinkage is:

$$(v_{cx})_{sus} = \frac{6950^2}{96}(-14.05 + 10 \times 8.28 - 14.05) \times 10^{-6} = 27.5 \text{ mm}$$

The maximum total panel deflection at mid-span of the columns strip is therefore:

$$(v_{cx})_{max} = (v_{cx})_{var} + (v_{cx})_{sus} = 33.3 \text{ mm}$$

Middle strip deflection (*y*-direction)

The cracking moment for the middle strip at the support after all shrinkage has taken place is −58.4 kNm and at mid-span is 60.4 kNm, so the strip may suffer minor cracking at the support line (where from Table 8.6 the maximum in-service moment $M_{sus+var} = -66$ kNm) but will be uncracked at mid-span (where $M_{sus+var} = 56$ kNm).

At the support

At time τ_k:

$$\zeta = 1 - \left(\frac{-58.4}{-66}\right)^2 = 0.217$$

Due to M_{var}:
$\kappa_{uncr} = -0.35 \times 10^{-6}$ mm^{-1}; $\kappa_{cr} = -3.05 \times 10^{-6}$ mm^{-1};
and therefore $\kappa_{avge} = -0.94 \times 10^{-6}$ mm^{-1}

Final long-term curvature due to M_{sus}+ shrinkage:
$\kappa_{uncr} = -3.01 \times 10^{-6}$ mm^{-1}; $\kappa_{cr} = -12.77 \times 10^{-6}$ mm^{-1};
and therefore $\kappa_{avge} = -5.13 \times 10^{-6}$ mm^{-1}.

At mid-span

Due to M_{var}: $\kappa_{avge} = \kappa_{uncr} = 0.30 \times 10^{-6}$ mm^{-1}
Due to M_{sus}+ shrinkage: $\kappa_{avge} = \kappa_{uncr} = 2.58 \times 10^{-6}$ mm^{-1}.

Deflection

The short-term deflection at mid-span of the middle strip due to the variable loading is:

$$(v_{my})_{var} = \frac{6950^2}{96}(-0.94 + 10 \times 0.30 - 0.94) \times 10^{-6} = 0.6 \text{ mm}$$

and the long-term deflection due to the sustained load and shrinkage is:

$$(v_{my})_{sus} = \frac{6950^2}{96}(-5.13 + 10 \times 2.58 - 5.13) \times 10^{-6} = 7.8 \text{ mm}$$

The maximum total panel deflection at mid-span of the middle strip in the *y*-direction is:

$$(v_{my})_{max} = (v_{my})_{var} + (v_{my})_{sus} = 8.4 \text{ mm}$$

Panel deflection

Finally, the maximum deflection at the mid-point of the panel is obtained from Eq. 8.16:

$$v_{max} = \left(\frac{33.3 + 33.3}{2}\right) + 8.4 = 41.7 \text{ mm}$$

8.5 Slender reinforced concrete columns

8.5.1 Discussion

In most structural members, creep and shrinkage of concrete cause increases of deformation and redistribution of stresses, but do not affect strength. In some situations, however, the deformations caused by creep and shrinkage can lead to an increase in the loads on the structure and, therefore, a reduction in strength. A slender column under sustained, eccentric compression is such an example. Other examples include shallow concrete arches and domes.

Consider the slender pin-ended column shown in Fig. 8.27. The column is subjected to a compressive force P applied at an initial eccentricity e_0. When P is first applied at time τ_0, the column shortens and deflects laterally by an amount δ_0. The bending moment at each end of the column is Pe_0, but at the column mid-length, the moment is $P(e_0 + \delta_0)$. Therefore, the lateral deflection of the member causes an increase in the internal actions.

For long columns, the *secondary moment* $P\delta_0$ may be many times greater than the initial *primary moment* Pe_0 and the load-carrying capacity is much less than that of a short column with the same cross-section. For very long columns, an instability failure may occur under relatively small compressive loads, i.e. buckling may occur before the strength of the cross-section at mid-length is reached.

For reinforced concrete columns under sustained loads, the member suffers additional lateral deflection due to creep. This time-dependent deformation leads to additional bending in the member, which in turn causes the column to deflect still further. During a period of sustained loading, an additional deflection $\Delta\delta(t)$ will develop at the mid-length of the column. The gradual increase in the secondary moment with time $P(\delta_0 + \Delta\delta(t))$ reduces the factor of safety and, for long columns, creep buckling may occur.

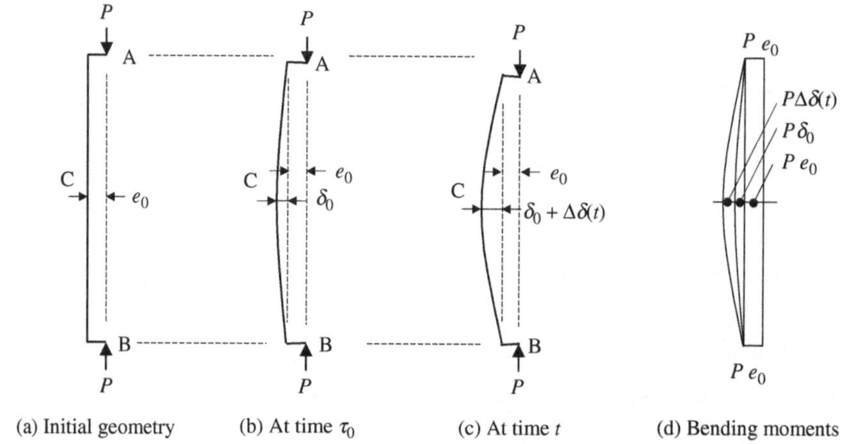

Figure 8.27 Displacements and moments in a slender pin-ended column.

In order to design slender reinforced concrete columns for the strength limit state, an accurate assessment of the creep deformation is necessary to ensure an adequate factor of safety. In the following section, an iterative computer-based procedure for the analysis of slender reinforced concrete columns is presented.

8.5.2 An iterative method of analysis

To account for the geometric non-linearity associated with the secondary actions in a slender column ($P\delta_0 + P\Delta\delta(t)$), as well as the material non-linearity associated with cracking, creep and shrinkage, an iterative solution procedure that uses the AEMM to include the time effects is described here. The analyses of cross-sections in the uncracked parts of the column are undertaken using the method described in Sections 5.3 and 6.2 (for short-term analysis) and Sections 5.4 and 6.3 (for long-term analysis), while for cross-sections in the cracked parts of the column, the analyses are undertaken in accordance with Section 7.2 (for short-term behaviour) and Section 7.3 (for time-dependent behaviour).

Consider again the pin-ended column shown in Fig. 8.27. The iterative solution procedure involves the following steps:

(i) The member is subjected initially to the axial load P and the primary moment Pe_0 on every cross-section. The initial strain distributions and curvatures are calculated at the cross-sections at each end of the member (i.e. at A and B) and at the mid-length (C).
(ii) The lateral deflection of the member at the mid-length of the column, δ_0, is determined from the initial curvatures calculated in step (i). If a parabolic variation of curvature along the member is assumed, δ_0 may be calculated using Eq. 8.1b.
(iii) The additional increment of moment caused by δ_0 at the mid-length of the columns ($P\delta_0$) is added to the primary moment and the cross-section at C is re-analysed. The curvature at C is now larger than that calculated in step (i), and so too is the

lateral deflection δ_1 at the mid-length of the column (re-calculated using Eq. 8.1b). The revised secondary moment ($P\delta_1$) is added to the primary moment at C and the cross-section is again reanalysed. The procedure continues until the lateral displacement at C converges to δ_i after the ith iteration.

The short-term response of the slender column is thus determined. The lateral deflection at C immediately after loading (δ_i) is known, as is the stress and strain distributions on the cross-sections at A, B and C.

If the member is very slender, convergence may not occur in step (iii) and a buckling instability may result. The secondary moment $P\delta$ calculated during the iterative cycle, may cause cracking on a previously uncracked cross-section and therefore result in considerable loss of stiffness and an increase in the lateral displacement and secondary moment. The number of iterations required for convergence therefore depends on the slenderness of the member, the initial eccentricity and the tensile strength of the concrete. In step (iii), convergence may be deemed to occur when the additional lateral displacement calculated at mid-length during the ith iteration ($\delta_i - \delta_{i-1}$) is less than 0.001 of the lateral displacement δ_0 caused by the primary moments and calculated in step (ii).

With the behaviour immediately after loading at τ_0 determined in step (iii), the time analysis commences:

(iv) The time-dependent changes in strain and curvature at A, B and C are calculated using the previously specified time analyses. Initially, the internal actions at C (P and $P(e_0 + \delta_i)$) are assumed to remain constant with time. The change in the lateral displacement at C ($\Delta\delta_0$) caused by the change in curvature with time at each cross-section is calculated using Eq. 8.1b. This increase in displacement with time causes an additional increment of secondary moment $P\Delta\delta_0$ to be applied gradually to the cross-section at C. This time-dependent secondary moment, in turn, produces additional lateral displacements and additional increments of secondary moment. An iterative procedure is again followed until the time-dependent lateral displacement converges to $\Delta\delta(t)$ (or until instability occurs).

To model behaviour using the AEMM, the time-dependent increments of secondary moment, which in fact occur gradually, are applied to the age-adjusted transformed cross-section at mid-span. For initially cracked cross-sections, the depth to the neutral axis is assumed to remain constant with time. As was discussed in Section 7.3, this assumption, although not correct, is a reasonable compromise between accuracy and economy of solution. A more refined analysis could be implemented to trace the development of time-dependent cracking using the SSM based on the iterative procedure outlined in Section 7.4.

Example 8.9

A pin-ended column with the cross-section shown in Fig. 8.28 is subjected to a constant sustained axial compressive load $P = 500$ kN acting at a constant eccentricity of $e_0 = 65$ mm. The behaviour of the column is to be determined,

both at first loading τ_0 and at time infinity, for each of the following three column lengths: (i) $L = 6000$ mm; (ii) $L = 7500$ mm; and $L = 9000$ mm.
Material properties are: $E_c = 25{,}000$ MPa; $E_s = 200{,}000$ MPa; $f'_c = 25$ MPa; $f_t = 2.5$ MPa; $\phi^*(\tau_0) = 2.5$; $f_{ct} = 2.5$ MPa; $\phi^*(\tau_0) = 2.5$; $\chi^*(\tau_0) = 0.65$; and $\varepsilon^*_{sh} = 0$.

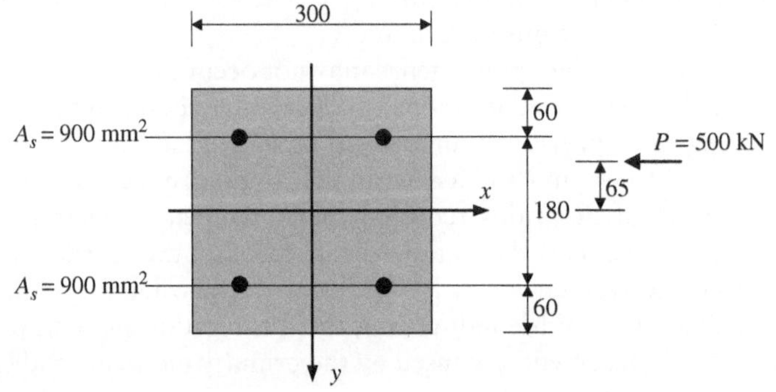

Figure 8.28 Column section for Example 8.9 (all dimensions in mm).

Case (i): $L = 6000$ mm

The initial stresses and strains on each cross-section due to the axial force $P = 500$ kN and the moment $Pe_0 = 32.5$ kNm are calculated using the cross-sectional analysis described in Section 5.3. In this case, the primary moment and axial force at each end and at mid-length are the same. On each cross-section, the compressive strain at the reference x-axis and curvature are $\varepsilon_{r,0} = -195 \times 10^{-6}$ and $\kappa_0 = 1.673 \times 10^{-6}$ mm^{-1} and the extreme fibre concrete stresses are $\sigma_{top} = -11.15$ MPa and $\sigma_{bot} = +1.40$ MPa. Since the tensile strength of the concrete is not exceeded, the member is uncracked.
The lateral displacement at the column mid-length caused by the primary moment is obtained from Eq. 8.1b:

$$\delta_o = \frac{6000^2}{96}(1.673 + 10 \times 1.673 + 1.673) = 7.53 \text{ mm}$$

and the increment of secondary moment at the column mid-length $P\delta_0 = 3.76$ kNm. This secondary moment is added to the primary moment and the cross-section at mid-length is re-analysed. The revised curvature at mid-length is $(\kappa_i)_1 = 1.841 \times 10^{-6}$ mm^{-1} and the revised displacement at mid-length is calculated using Eq. 8.1b is $\delta_1 = 8.16$ mm. The corresponding secondary moment is $P\delta_1 = 4.08$ kNm. The revised secondary moment is added to the primary moment and the section at C is again re-analysed. The procedure continues for two more iterations until the lateral deflection converges to $\delta_i = 8.33$ mm.
The stress and strain distributions at mid-length are $\varepsilon_{r,0} = -195 \times 10^{-6}$; $\kappa_0 = 1.887 \times 10^{-6}$ mm^{-1}; $\sigma_{top} = -11.95$ MPa; and $\sigma_{bot} = +2.20$ MPa. The section at mid-length therefore remains uncracked.

For the time analysis, the cross-sections at each end and at the mid-length of the column are analysed using the procedure outlined in Section 5.4. Initially, the internal actions are assumed to remain constant. The change in the lateral displacement at mid-length caused by the change in curvature on each cross-section with time is $\Delta\delta_0 = 14.23$ mm. The secondary moment $P\Delta\delta_0$ is applied to the age-adjusted transformed cross-section at mid-length, thus producing an additional increment of displacement and secondary moment. Convergence occurs after five iterations. The time-dependent change in displacement at the column mid-length is $\Delta\delta_i = 17.73$ mm. The column therefore suffers a total mid-length lateral deflection of $\delta_i + \Delta\delta_i = 26.06$ mm.

The initial and final strain distributions and concrete stresses on the cross-sections at each end and at the mid-length of the column are illustrated in Fig. 8.29. It is noted that the final bottom fibre tensile stress exceeds the tensile stress of the concrete indicating that cracking in the mid-length region of the column is likely to occur with time.

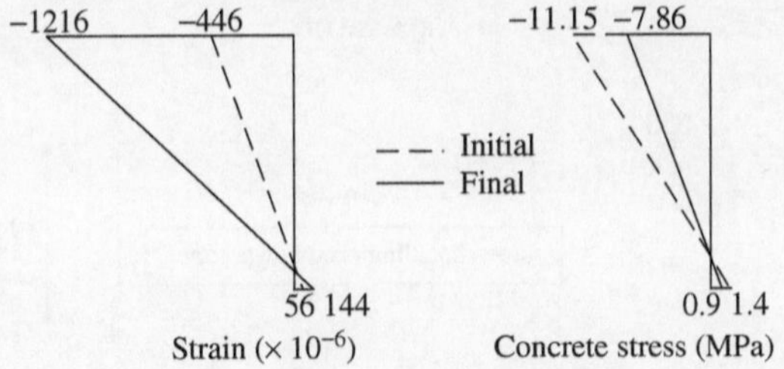

(a) At each end of the column

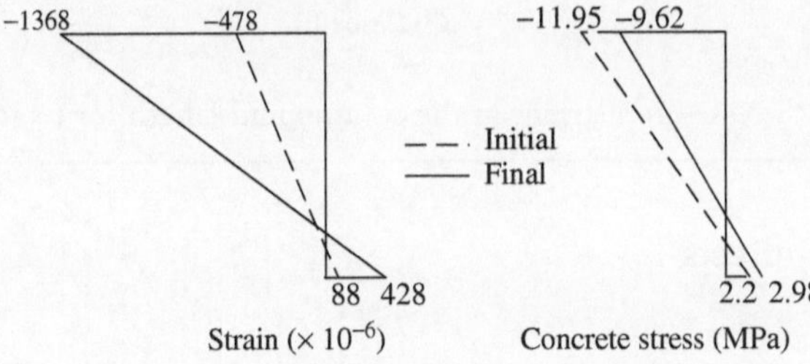

(b) At the mid-length of the column

Figure 8.29 Stresses and strains for case (i) of Example 8.9.

Case (ii): $L = 7500$ mm

The stress and strain distributions at each end of the column are the same as those calculated in case (i). However, the increased column length results in larger lateral displacements and increased secondary moments. The secondary moments in this case are large enough to cause cracking at the mid-length of the column at initial loading. The initial and final lateral displacements at the column mid-length in case (ii) are $\delta_i = 16.43$ mm and $\delta_i + \Delta\delta = 69.71$ mm and the corresponding strains and stresses on the cross-section at mid-length are shown in Fig. 8.30a.

Case (iii): $L = 9000$ mm

In this case, the initial column deflection at mid-length is $\delta_i = 30.03$ mm and the initial states of stress and strain on the cross-section at the mid-point of the column are shown in Fig. 8.30b. As time progresses, additional secondary moments caused by creep become so large that convergence does not occur and the member buckles. For such a slender member, creep deformations result in a reduction of strength with time.

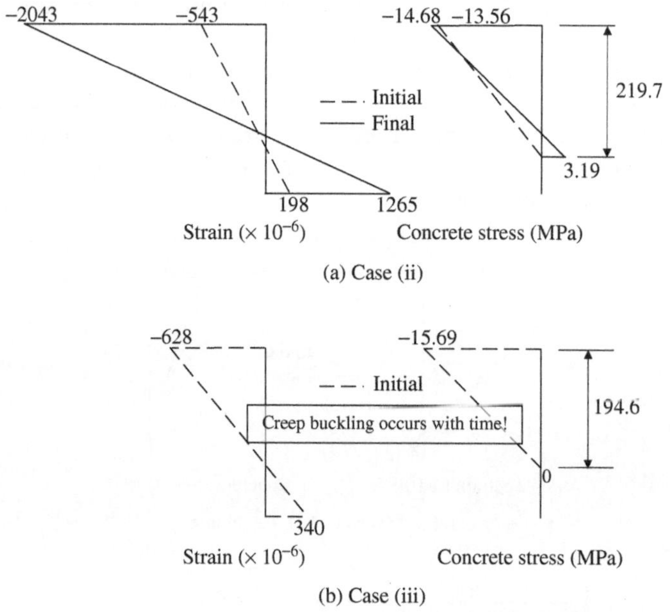

Figure 8.30 Stresses and strains at the column mid-length for Example 8.9.

8.6 Temperature effects

8.6.1 Introduction

A uniform or linearly varying temperature distribution through the depth of a statically determinate homogeneous member causes the member to deform, but no stress is produced on any cross-section. In practice, however, temperature rarely varies linearly through the depth of a member and members are usually unable to accommodate the deformation caused by temperature change without some restraining forces developing at the supports. Because plane sections tend to remain plane, a non-linear temperature gradient produces internal, self-equilibrating *eigenstresses*, as illustrated in Fig. 8.31. The free expansion (or contraction) of each concrete fibre is restrained by the adjacent fibres resulting in the internal stress distribution shown. A similar distribution of self-equilibrating internal stress is produced by non-linear shrinkage and was discussed in Section 1.3.3 and illustrated in Fig. 1.13.

Usually design engineers are concerned with temperature changes applied over a short time period. If the temperature gradient is sustained for a period of time, the

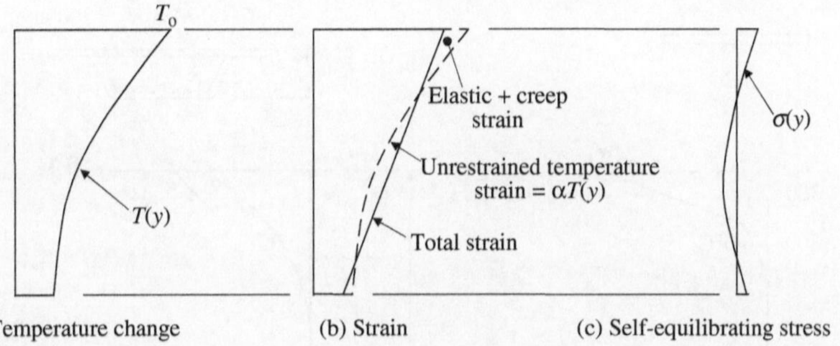

Figure 8.31 Effects of non-linear temperature distribution on cross-sectional response.

internal stresses induced by temperature are relieved, to some extent, by creep. If necessary, a time analysis using the AEMM can be used to determine the variation of cross-sectional behaviour with time.

In a statically indeterminate member, the displacement at the member's ends caused by the axial deformation and curvature induced by temperature may not be free to occur. Reactions develop at the supports and internal actions are induced at each cross-section of the member. The stresses resulting from these secondary actions can be large and must be calculated and added to the stresses illustrated in Fig. 8.31, if the total stresses on a cross-section of a statically indeterminate member are required. The determination of these continuity stresses is discussed in Section 8.6.4.

8.6.2 Temperature distributions

Temperature variations caused by solar radiation in bridge decks, for example, can produce stress levels that are of similar magnitude to those produced by the gravity loads, even if the structure is statically determinate. In the design of such structures, an analysis to determine the effects of temperature often plays an important part.

Design codes provide guidance on temperature gradients that should be considered for particular structural applications. For example, typical temperature profiles to be used in the design of composite concrete-concrete bridges are shown in Fig. 8.32 and are based on the Australian specifications (Refs 21 and 22).

8.6.3 Temperature analysis of cross-sections

In this section, the short-term analysis of a cross-section subjected to an arbitrary non-linear temperature variation at time τ_0 is presented. The overall method of analysis is similar to that presented in Sections 5.3–5.7, except that a thermal-induced strain is included in the constitutive relationship of each constituent material. For a reinforced or prestressed concrete cross-section with m_c concrete components, m_s layers of non-prestressed reinforcement and m_p layers of prestressing tendons, the constitutive relationships are:

$$\sigma_{c(i),0} = E_{c(i),0}\left(\varepsilon_0 - \varepsilon_{c(i),T}\right) \tag{8.17a}$$

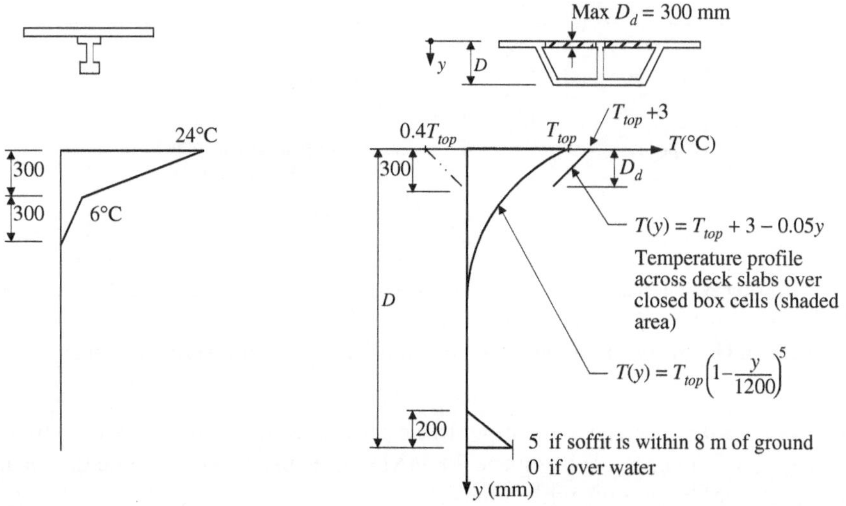

(a) Concrete deck on girder (Ref. 21) (b) Box girder bridge deck (Ref. 22)

Figure 8.32 Temperature gradients for concrete box girders (Refs 21–22).

$$\sigma_{s(i),0} = E_{s(i)}\left(\varepsilon_0 - \varepsilon_{s(i),T}\right) \tag{8.17b}$$

$$\sigma_{p(i),0} = E_{p(i)}\left(\varepsilon_0 + \varepsilon_{p(i),init} - \varepsilon_{p(i),T}\right) \tag{8.17c}$$

where $\varepsilon_{c(i),T}$, $\varepsilon_{s(i),T}$ and $\varepsilon_{p(i),T}$ are the thermal-induced strains in the concrete, non-prestressed reinforcement and prestressing tendons, respectively, and are determined from:

$$\varepsilon_{c(i),T} = \alpha_{c(i),T}T(y) \tag{8.18a}$$

$$\varepsilon_{s(i),T} = \alpha_{s(i),T}T(y) \tag{8.18b}$$

$$\varepsilon_{p(i),T} = \alpha_{p(i),T}T(y) \tag{8.18c}$$

in which $T(y)$ is the temperature distribution on the cross-section as a function of distance y from the reference axis, and $\alpha_{c(i),T}$, $\alpha_{s(i),T}$ and $\alpha_{p(i),T}$ are the coefficients of thermal expansion for the concrete, non-prestressed reinforcement and prestressing tendons, respectively.

The contributions to the internal axial force on the cross-section of the constituent materials at time τ_0 after the temperature change are:

$$N_{c,0} = \sum_{i=1}^{m_c}\int_{A_{c(i)}} \sigma_{c(i),0}\,\mathrm{d}A = \sum_{i=1}^{m_c} A_{c(i)}E_{c(i),0}\varepsilon_{r,0} + \sum_{i=1}^{m_c} B_{c(i)}E_{c(i),0}\kappa_0$$

$$-\sum_{i=1}^{m_c}\int_{A_{c(i)}} E_{c(i),0}\varepsilon_{c(i),T}\,\mathrm{d}A \tag{8.19a}$$

$$N_{s,0}=\sum_{i=1}^{m_s}\left(A_{s(i)}E_{s(i)}\right)\varepsilon_{r,0}+\sum_{i=1}^{m_s}\left(y_{s(i)}A_{s(i)}E_{s(i)}\right)\kappa_0-\sum_{i=1}^{m_s}\left(A_{s(i)}E_{s(i)}\varepsilon_{s(i),T}\right) \tag{8.19b}$$

$$N_{p,0}=\sum_{i=1}^{m_p}\left(A_{p(i)}E_{p(i)}\right)\varepsilon_{r,0}+\sum_{i=1}^{m_p}\left(y_{p(i)}A_{p(i)}E_{p(i)}\right)\kappa_0+\sum_{i=1}^{m_p}\left(A_{p(j)}E_{p(i)}\varepsilon_{p(i),init}\right)$$
$$-\sum_{i=1}^{m_p}\left(A_{p(i)}E_{p(i)}\varepsilon_{p(i),T}\right) \tag{8.19c}$$

Combining these contributions, the internal axial force and bending moment resisted by the whole cross-section are:

$$N_{i,0}=N_{c,0}+N_{s,0}+N_{p,0}=R_{A,0}\varepsilon_{r,0}+R_{B,0}\kappa_0+\sum_{j=1}^{m_p}\left(A_{p(j)}E_{p(j)}\varepsilon_{p(j),init}\right)-N_{T,0} \tag{8.20a}$$

$$M_{i,0}=M_{c,0}+M_{s,0}+M_{p,0}=R_{B,0}\varepsilon_{r,0}+R_{I,0}\kappa_0+\sum_{i=1}^{m_p}\left(y_{p(i)}A_{p(i)}E_{p(i)}\varepsilon_{p(i),init}\right)-M_{T,0} \tag{8.20b}$$

where the cross-sectional rigidities at time τ_0, i.e. $R_{A,0}$, $R_{B,0}$ and $R_{I,0}$, have already been defined in Eqs 5.72, and $N_{T,0}$ and $M_{T,0}$ are the equivalent actions required to produce the same deformation as the temperature change if each material was unrestrained and are given by:

$$N_{T,0}=\sum_{i=1}^{m_c}\int_{A_{c(i)}}E_{c(i),0}\varepsilon_{c(i),T}\,\mathrm{d}A+\sum_{i=1}^{m_s}\left(A_{s(i)}E_{s(i)}\varepsilon_{s(i),T}\right)+\sum_{i=1}^{m_p}\left(A_{p(i)}E_{p(i)}\varepsilon_{p(i),T}\right) \tag{8.21a}$$

$$M_{T,0}=\sum_{i=1}^{m_c}\int_{A_{c(i)}}yE_{c(i),0}\varepsilon_{c(i),T}\,\mathrm{d}A+\sum_{i=1}^{m_s}\left(y_{s(i)}A_{s(i)}E_{s(i)}\varepsilon_{s(i),T}\right)$$
$$+\sum_{i=1}^{m_p}\left(y_{p(i)}A_{p(i)}E_{p(i)}\varepsilon_{p(i),T}\right) \tag{8.21b}$$

Substituting $N_{i,0}$ and $M_{i,0}$ (Eqs 8.20) into Eqs 5.2, the equilibrium equations may be expressed as:

$$\mathbf{r}_{e,0}=\mathbf{D}_0\boldsymbol{\varepsilon}_0+\mathbf{f}_{p,init}-\mathbf{f}_T \tag{8.22}$$

and solving Eq. 8.22 gives the unknown strain variables $\boldsymbol{\varepsilon}_0$:

$$\boldsymbol{\varepsilon}_0=\mathbf{D}_0^{-1}\left(\mathbf{r}_{e,0}-\mathbf{f}_{p,init}+\mathbf{f}_T\right)=\mathbf{F}_0\left(\mathbf{r}_{e,0}-\mathbf{f}_{p,init}+\mathbf{f}_T\right) \tag{8.23}$$

where $\mathbf{\varepsilon}_0$, $\mathbf{r}_{e,0}$, $\mathbf{D}_0$, $\mathbf{F}_0$ and $\mathbf{f}_{p,init}$ are all defined in Eqs 5.71 and $\mathbf{f}_T$ is given by:

$$\mathbf{f}_T = \begin{bmatrix} N_{T,0} \\ M_{T,0} \end{bmatrix} \tag{8.24}$$

Evaluation of the integral expressing the thermal response of the concrete can be carried out numerically or in closed form when the temperature variation is specified with a known function.

Example 8.10

The stress and strain distributions on the section shown in Fig. 8.33a caused by the temperature gradient shown in Fig. 8.33b are to be calculated. The cross-section is of a statically determinate beam and the rise in temperature is applied over a short time period. Because temperature rises are associated with expansion, they are here taken to be positive.
The coefficient of thermal expansion for both concrete and steel is $\alpha_c = \alpha_s = 10 \times 10^{-6}/^{\circ}C$ and the prestressing steel is bonded to the surrounding concrete. Take $E_c = 30{,}000$ MPa and $E_s = E_p = 200{,}000$ MPa.

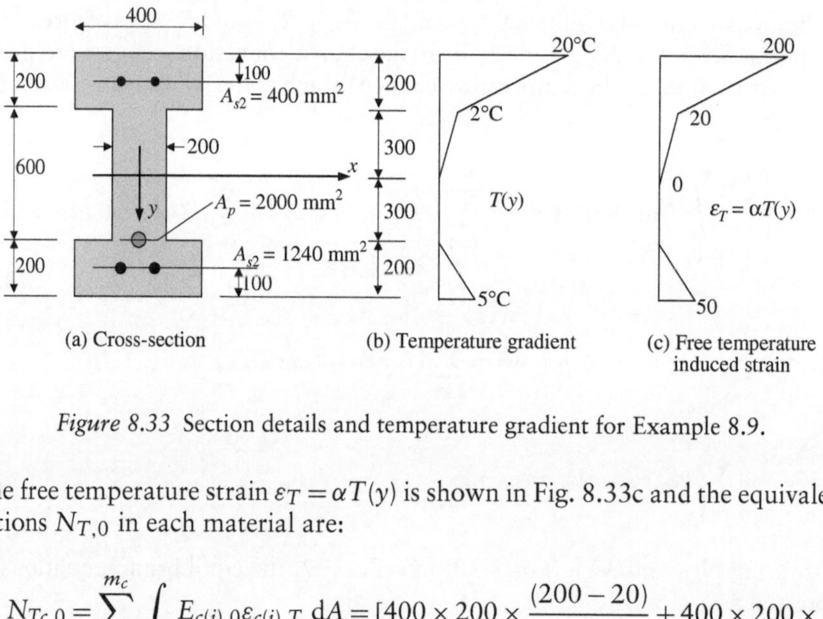

Figure 8.33 Section details and temperature gradient for Example 8.9.

The free temperature strain $\varepsilon_T = \alpha T(y)$ is shown in Fig. 8.33c and the equivalent actions $N_{T,0}$ in each material are:

$$N_{Tc,0} = \sum_{i=1}^{m_c} \int_{A_{c(i)}} E_{c(i),0}\varepsilon_{c(i),T}\, dA = [400 \times 200 \times \frac{(200-20)}{2} + 400 \times 200 \times 20$$

$$+ 200 \times 300 \times \frac{20}{2} + 400 \times 200 \times \frac{50}{2} - 400 \times 110$$

$$- 1240 \times 25] \times 10^{-6} \times 30{,}000 = 339.8 \text{ kN}$$

$$N_{Ts,0} = \sum_{i=1}^{m_s} (A_{s(i)} E_{s(i)} \varepsilon_{s(i),T}) = [400 \times 110 + 1240 \times 25] \times 10^{-6} \times 200{,}000$$

$$= 15.0 \text{ kN}$$

$$N_{Tp,0} = \sum_{i=1}^{m_p} (A_{p(i)} E_{p(i)} \varepsilon_{p(i),T}) = [2000 \times 200{,}000 \times 0] = 0$$

and the moments of these forces about the centroidal axis of the section are:

$$M_{Tc,0} = \sum_{i=1}^{m_c} \int_{A_{c(i)}} y E_{c(i),0} \varepsilon_{c(i),T} \, dA = [400 \times 200 \times \frac{(200-20)}{2} \times (-433.\dot{3})$$

$$+ 400 \times 200 \times 20 \times (-400) + 200 \times 300 \times \frac{20}{2} \times (-200)$$

$$+ 400 \times 200 \times \frac{50}{2} \times 433.3 - 400 \times 110 \times (-400)$$

$$- 1240 \times 25 \times 400] \times 10^{-6} \times 30{,}000 = -90.25 \text{ kNm}$$

$$M_{Ts,0} = \sum_{i=1}^{m_s} (y_{s(i)} A_{s(i)} E_{s(i)} \varepsilon_{s(i),T}) = [400 \times 110 \times (-400)$$

$$+ 1240 \times 25 \times 400] \times 10^{-6} \times 200{,}000 = -1.04 \text{ kNm}$$

$$N_{Tp,0} = \sum_{i=1}^{m_p} (y_{p(i)} A_{p(i)} E_{p(i)} \varepsilon_{p(i),T}) = [300 \times 2000 \times 200{,}000 \times 0] = 0$$

From Eqs 8.21:

$$N_{T,0} = N_{Tc,0} + N_{Ts,0} + N_{Tp,0} = 354.8 \text{ kN}$$

$$M_{T,0} = M_{Tc,0} + M_{Ts,0} + M_{Tp,0} = -90.25 - 1.04 = -91.29 \text{ kNm}$$

The rigidities of this cross-section are obtained from Eqs 5.72 as $R_{A,0} = 9.02 \times 10^9$ N, $R_{B,0} = 15.91 \times 10^{10}$ Nmm, $R_{I,0} = 9.67 \times 10^4$ Nmm^2, and with the $\mathbf{F}_0$ matrix for this cross-section calculated using Eq. 5.71b, the strain matrix for the otherwise unloaded cross-section is determined using Eq. 8.23:

$$\boldsymbol{\varepsilon}_0 = \mathbf{F}_0 \mathbf{f}_T = \begin{bmatrix} 1.112 \times 10^{-10} & -1.829 \times 10^{-14} \\ -1.829 \times 10^{-14} & 1.037 \times 10^{-15} \end{bmatrix} \begin{bmatrix} 354.8 \times 10^3 \\ -91.29 \times 10^6 \end{bmatrix}$$

$$= \begin{bmatrix} 41.1 \times 10^{-6} \\ -0.1012 \times 10^{-6} \text{ mm}^{-1} \end{bmatrix}$$

The top ($y = -500$ mm) and bottom ($y = +500$ mm) fibre strains are:

$$\varepsilon_{0,top} = 41.1 \times 10^{-6} + (-500) \times (-0.1012 \times 10^{-6}) = 91.7 \times 10^{-6}$$

$$\varepsilon_{0,bot} = 41.1 \times 10^{-6} + 500 \times (-0.1012 \times 10^{-6}) = -9.5 \times 10^{-6}$$

The temperature-induced changes of stress in the concrete are determined from Eq. 8.17a as follows:

At $y = -500$ mm:
$$\Delta\sigma_{c,0} = 30{,}000[41.1 + (-500) \times (-0.1012) - 200] \times 10^{-6} = -3.25 \text{ MPa}$$

At $y = -300$ mm:
$$\Delta\sigma_{c,0} = 30{,}000[41.1 + (-300) \times (-0.1012) - 20] \times 10^{-6} = 1.54 \text{ MPa}$$

At $y = 0$ mm:
$$\Delta\sigma_{c,0} = 30{,}000[41.1 + (0) \times (-0.1012) - 0] \times 10^{-6} = 1.23 \text{ MPa}$$

At $y = +300$ mm:
$$\Delta\sigma_{c,0} = 30{,}000[41.4 + (300) \times (-0.1012) - 0] \times 10^{-6} = 0.32 \text{ MPa}$$

At $y = +500$ mm:
$$\Delta\sigma_{c,0} = 30{,}000[41.4 + (500) \times (-0.1012) - 50] \times 10^{-6} = -1.79 \text{ MPa}$$

From Eqs 8.17b and c:

$$\Delta\sigma_{s(1)} = 200{,}000[41.4 + (-400) \times (-0.1012) - 110] \times 10^{-6} = -5.68 \text{ MPa}$$

$$\Delta\sigma_{s(2)} = 200{,}000[41.4 + (400) \times (-0.1012) - 25] \times 10^{-6} = -4.88 \text{ MPa}$$

$$\Delta\sigma_{p} = 200{,}000[41.4 + (300) \times (-0.1012) - 0] \times 10^{-6} = -2.15 \text{ MPa}$$

The temperature-induced strains and eigenstresses on the concrete section are shown in Fig. 8.34.

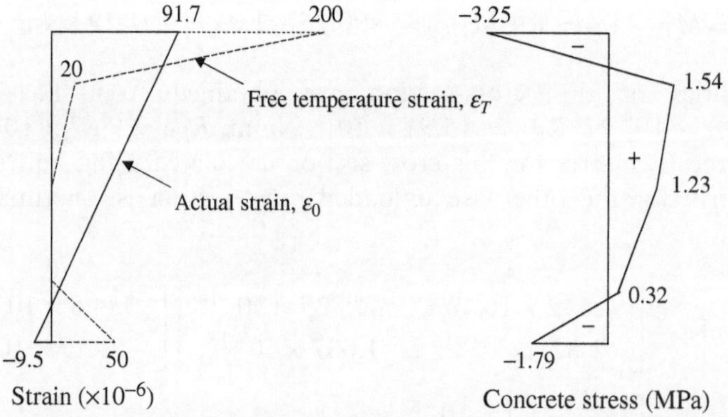

Figure 8.34 Temperature-induced stress and strain in Example 8.10.

8.6.4 *Temperature effects in members and structures*

The temperature-induced strains on individual cross-sections cause axial deformation and rotation of members and structures. In statically indeterminate members, the displacement caused by temperature changes may not be free to occur, and restraining forces may develop at the supports. The stresses and deformations caused by these restraining forces must be calculated and added to the temperature-induced eigenstresses in order to obtain the total effect of temperature. The redundant forces caused by restraint to temperature can be determined using exactly the same procedure as outlined in Section 8.3.2.

Example 8.11

A beam with cross-section shown in Fig. 8.33a is subjected to the temperature gradient of Fig. 8.33b applied over its entire length in an eight-hour period. If the beam was statically determinate, the temperature-induced stresses and strains on the cross-section were calculated in Example 8.10 and illustrated in Fig. 8.34. From Example 8.10, the temperature-induced strain at the reference axis (at the mid-depth of the cross-section) and the curvature on each cross-section of the beam are: $\varepsilon_{r,0} = 41.1 \times 10^{-6}$ and $\kappa_0 = -0.1012 \times 10^{-6}$ mm^{-1}.

(a) The beam is uncracked throughout and simply supported over a span of 14 m

The displacement of the beam (idealised in Fig. 8.35a) is to be determined. The axial deformation of the member is:

$$e_{AB} = \varepsilon_{r,0}\, L = 41.1 \times 10^{-6} \times 14{,}000 = 0.58 \text{ mm (elongation)}$$

and the deflection at mid-span is calculated using Eq. 8.1b:

$$v_C = \frac{14{,}000^2}{96}\left[12 \times (-0.1012) \times 10^{-6}\right] = -2.48 \text{ mm}$$

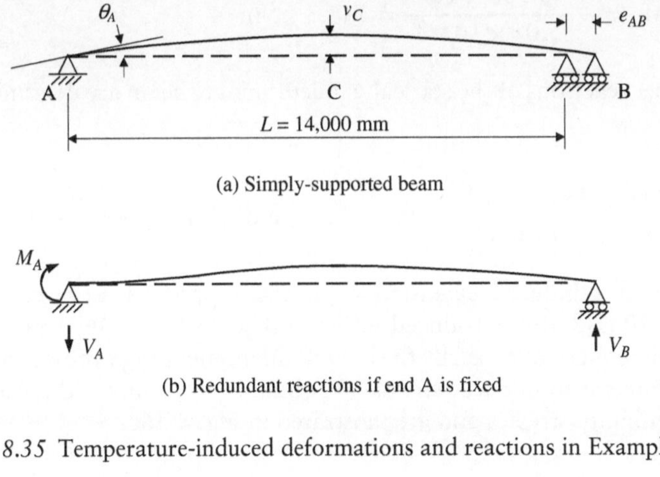

Figure 8.35 Temperature-induced deformations and reactions in Example 8.11.

From Eq. 3.7b, the slope at each end of the beam is:

$$\theta_A = \theta_B = \frac{14{,}000}{6}\left[3 \times (-0.1012 \times 10^{-6})\right] = -0.708 \times 10^{-3} \text{ rad}$$

(b) The support at A is fixed, so that θ_A is prevented

If the rotation at A cannot occur, the temperature gradient will induce the reactions shown in Fig. 8.35b at the supports. For this analysis, the moment at A is taken to be the redundant, $R_R = M_A$, and the deformation of the primary structure at the release is:

$$u_A = \theta_A = -0.708 \times 10^{-3}$$

Due to a unit moment applied at A, the rotation at A is $\bar{f}_{11}$ and the reactions at the supports of the primary beam are $\mathbf{R}_{PR}$, where:

$$\bar{f}_{11} = \frac{L}{3E_c I} \quad \text{and} \quad \mathbf{R}_{PR} = \begin{Bmatrix} V_A = -1/L \\ V_B = +1/L \end{Bmatrix}$$

For this beam, E_c = 30,000 MPa and the second moment of are of the transformed cross-section about its centroidal axis is equal to I = 31,040 $\times 10^6$ mm^4. Therefore:

$$\bar{f}_{11} = \frac{14{,}000}{3 \times 30{,}000 \times 31{,}040 \times 10^{-6}} = 5.01 \times 10^{-12}$$

From Eq. 8.5, the redundant moment at A is:

$$R_R = M_A = -\frac{-0.708 \times 10^{-3}}{5.01 \times 10^{-12}} = 141.3 \text{ kNm}$$

and the other reactions of the statically indeterminate beam are determined from statics:

$$V_B = \frac{M_A}{L} = \frac{141.3 \times 10^6}{14{,}000} = 10.09 \text{ kN} \quad \text{and} \quad V_A = -V_B = -10.09 \text{ kN}$$

The temperature-induced eigenstresses at the support A were calculated in Example 8.10 and are reproduced in Fig. 8.36a. In Fig. 8.36b, the continuity stresses in the concrete caused by the redundant moment M_A are also shown. The resultant temperature-induced stresses are obtained by summing the eigenstresses and the continuity stresses and are illustrated in Fig. 8.36c.

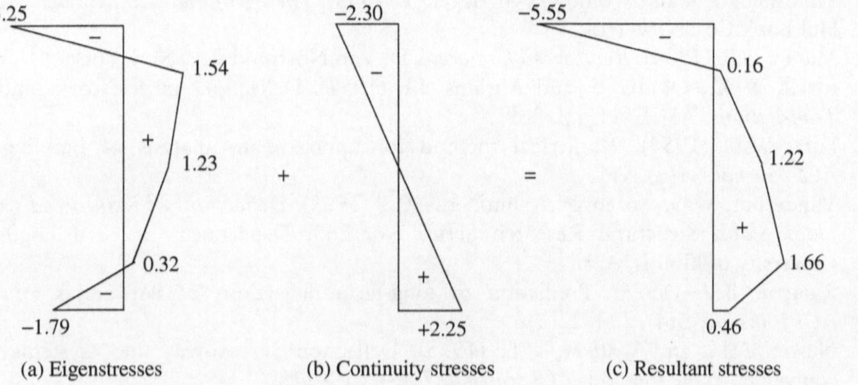

Figure 8.36 Temperature-induced concrete stresses at fixed support at A in Example 8.11.

8.7 Concluding remarks

The techniques described and illustrated in this chapter may be extended readily to include a wide range of additional structural applications. The procedures are not daunting and require little more than an elementary background in mechanics and structural analysis.

An analysis for the effects of creep and shrinkage on the behaviour of concrete structures allows the structural design engineer to better predict and control in-service performance. But perhaps of greater importance, it provides a clear picture of the interaction between concrete and reinforcement at service loads and a better understanding of why concrete structures behave as they do.

8.8 References

1. British Standards Institution (2004). Eurocode 2: Design of Concrete Structures – part 1-1: General rules and rules for buildings BS EN 1992-1-1: 2004. *European Committee for Standardization*, Brussels.
2. Ghali, A. and Neville, A.M. (1978). *Structural Analysis – A Unified Classical and Matrix Approach*. Chapman and Hall, London.
3. Hall, A.S. and Kabaila, A.P. (1977). *Basic Concepts of Structural Analysis*. Pitman Publishing, London.
4. Mayer, H. and Rüsch, H. (1967). *Building Damage Caused by Deflection of Reinforced Concrete Building Components. Technical Translation 1412*, National Research Council Ottawa, Canada (Deutscher Auschuss für Stahlbeton, Heft 193, Berlin, West Germany).
5. Sbarounis, J.A. (1984). Multi-storey flat plat buildings – measured and computed one-year deflections. *Concrete International*, 6(8), 31–35.
6. Jokinen, E.P. and Scanlon, A. (1985). Field measured two-way slab deflections. *CSCE Annual Conference*, Saskatoon, Canada, May.
7. Gilbert, R.I. and Guo, X.H. (2005). Time-dependent deflection and deformation of reinforced concrete flat slabs – an experimental study. *ACI Structural Journal*, 102(3), 363–373.

8. Timoshenko, S. and Woinowsky-Krieger, S. (1959). *Theory of Plates and Shells*. McGraw-Hill Book Co., New York.
9. Marsh, C.F. (1904). *Reinforced Concrete*. D. Van Nostrand Co., New York.
10. Ewell, W.W., Okuba, S. and Abrams, J.I. (1952). Deflections in gridworks and slabs. *Transactions, ASCE*, 117, p. 869.
11. Furr, W.L. (1959). Numerical method for approximate analysis of building slabs. *ACI Journal*, 31(6), 511.
12. Vanderbilt, M.D., Sozen, M.A. and Siess, C.P (1963). *Deflections of Reinforced Concrete Floor Slabs*. Structural Research Series No. 263, Department of Civil Engineering, University of Illinois, April.
13. Rangan, B.V. (1976). Prediction of long-term deflections of flat plates and slabs. *ACI Journal*, 73(4), 223–226.
14. Nilson. A.H. and Walters, D.B. (1975). Deflection of two-way floor systems by the equivalent frame method. *ACI Journal*, 72(5), 210–218.
15. Gilbert, R.I. (1985). Deflection control of slabs using allowable span to depth ratios. *ACI Journal*, 82(1), 67–72.
16. Scanlon, A. and Murray, D.W. (1974). Time-dependent reinforced concrete slab deflections. *Journal of the Structural Division, ASCE*, 100, 1911–1924.
17. Gilbert, R.I. (1979). Time-Dependent Behaviour of Structural Concrete Slabs. *PhD Thesis*, University of New South Wales, pp 361.
18. Chong, K.T., Foster, S.J. and Gilbert, R.I. (2008). Time-dependent modelling of RC structures using the cracked membrane model and solidification theory. *Computers and Structures*, 86(11–12), 1305–1317.
19. Nilson, A.H. and Walters, D.B. (1975). Deflection of two-way floor systems by the equivalent frame method. *ACI Journal*, 72(5), 210–218.
20. American Concrete Institute (2008). *Building Code Requirements for Reinforced Concrete ACI 318M-08 and Commentary*. ACI Committee 318, Michigan, USA.
21. National Association of State Road Authorities (1992). *NAASRA – Bridge design specification*. Sydney.
22. Standards Australia (2004). *Bridge Design Part 2: Design Loads AS5100.2-2004*. Sydney.

9 Stiffness method and finite-element modelling

9.1 Introduction

The analysis of members and frames is commonly carried out in design offices using commercial software based on either the stiffness method or the finite-element approach. This chapter outlines how these methods can be used for the long-term analyses of concrete structures, accounting for the effects of creep and shrinkage. The solution procedures required for the methods are well established (Refs 1–3) and only a brief overview of them is provided here. Throughout the chapter frequent reference is made to Appendix A, where the detailed derivations required to establish the proposed approaches are outlined. Worked examples are also presented to better illustrate the methods.

9.2 Overview of the stiffness method

The stiffness method of analysis, also referred to as the displacement method, is a matrix method for the analysis of complex structural systems using computers. A beam or frame is discretised into a number of line elements, referred to as stiffness elements, connected at their ends by nodes. A typical two-dimensional six degree of freedom (6 DOF) frame element of length L is illustrated in Fig. 9.1, where the three possible displacement components at each end node and the three corresponding nodal actions are shown. The overall structural response is evaluated by combining the contribution of each single element in resisting the applied loads or deformations. This approach relies on the fact that, for each element, it is possible to establish a stiffness relationship between nodal actions and nodal displacements, i.e.:

$$\mathbf{p} = \mathbf{k}_e \mathbf{u} \tag{9.1}$$

where $\mathbf{p}$ and $\mathbf{u}$ are the nodal actions and displacements of an element, and $\mathbf{k}_e$ is the element stiffness matrix. For any particular element, this relationship is expressed in its particular coordinate system (usually referred to as the local or member coordinate system). For example, for the 6 DOF frame element of Fig. 9.1, the stiffness matrix is derived in Appendix A as Eq. A.51 and is reproduced here as:

$$\mathbf{k}_e = \begin{bmatrix} \frac{R_A}{L} & 0 & -\frac{R_B}{L} & -\frac{R_A}{L} & 0 & \frac{R_B}{L} \\ 0 & \frac{12}{\alpha_1 L^3} & \frac{6}{\alpha_1 L^2} & 0 & -\frac{12}{\alpha_1 L^3} & \frac{6}{\alpha_1 L^2} \\ -\frac{R_B}{L} & \frac{6}{\alpha_1 L^2} & \frac{4}{\alpha_1 L} + \frac{R_B^2}{R_A L} & \frac{R_B}{L} & -\frac{6}{\alpha_1 L^2} & \frac{2}{\alpha_1 L} - \frac{R_B^2}{R_A L} \\ -\frac{R_A}{L} & 0 & \frac{R_B}{L} & \frac{R_A}{L} & 0 & -\frac{R_B}{L} \\ 0 & -\frac{12}{\alpha_1 L^3} & -\frac{6}{\alpha_1 L^2} & 0 & \frac{12}{\alpha_1 L^3} & -\frac{6}{\alpha_1 L^2} \\ \frac{R_B}{L} & \frac{6}{\alpha_1 L^2} & \frac{2}{\alpha_1 L} - \frac{R_B^2}{R_A L} & -\frac{R_B}{L} & -\frac{6}{\alpha_1 L^2} & \frac{4}{\alpha_1 L} + \frac{R_B^2}{R_A L} \end{bmatrix} \quad (9.2a)$$

and the nodal actions and displacements shown in Fig. 9.1 are:

$$\mathbf{p} = \begin{bmatrix} N_L \\ S_L \\ M_L \\ N_R \\ S_R \\ M_R \end{bmatrix} \quad (9.2b)$$

$$\mathbf{u} = \begin{bmatrix} u_L \\ v_L \\ \theta_L \\ u_R \\ v_R \\ \theta_R \end{bmatrix} \quad (9.2c)$$

The rigidities R_A, R_B and R_I have been introduced in Chapters 5 and 7 for different types of cross-sections, and the coefficients α_0 and α_1 are calculated from the rigidities using Eqs A.50a and b. The subscripts 'L' and 'R' on the nodal loads and displacements denote the left (at $z = 0$) and right (at $z = L$) member ends, respectively, when looking at the element with the z-axis horizontal and pointing towards the right.

In the proposed formulation, the coordinate system is illustrated in Fig. 9.2 for a typical reinforced concrete member. To remain consistent with the coordinate systems adopted in previous chapters, rotations are assumed to be positive when clockwise and transverse displacements are positive when in the direction of the y-axis.

In a real structure, members may have different orientations (with different local coordinate systems). It is therefore convenient to introduce a global coordinate system to be used for the whole structure. The load and displacement vectors for a particular

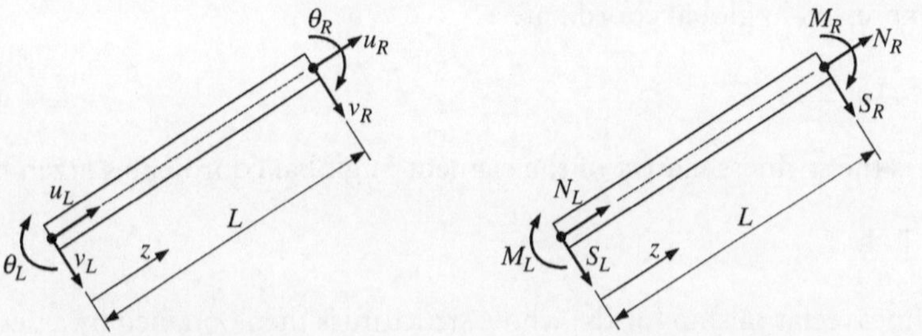

Figure 9.1 Freedoms of the 6 DOF frame element.

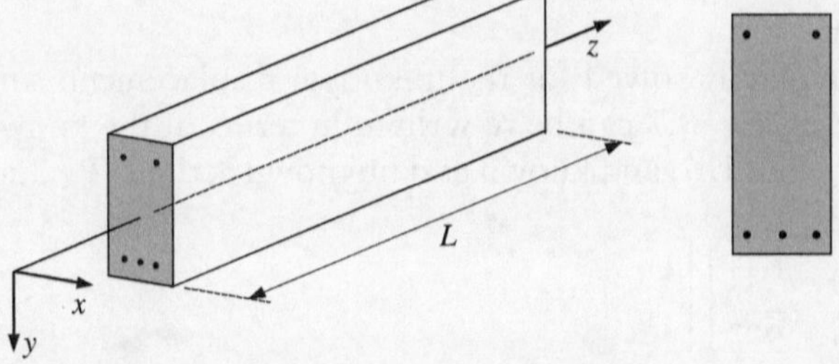

Figure 9.2 Coordinate system adopted for the member analysis.

element (defined in local coordinates in Eqs 9.2b and c, respectively) can be expressed in global coordinates by carrying out the following transformations:

$$\mathbf{u} = \mathbf{T}\mathbf{U}_e \tag{9.3a}$$

$$\mathbf{P}_e = \mathbf{T}^T\mathbf{p} \tag{9.3b}$$

where $\mathbf{P}_e$ represents the load vector in global coordinates, $\mathbf{U}_e$ is the vector of nodal displacements in global coordinates, and the transformation matrix $\mathbf{T}$ is given by:

$$\mathrm{T} = \begin{bmatrix} c & s & 0 & 0 & 0 & 0 \\ -s & c & 0 & 0 & 0 & 0 \\ 0 & 0 & 1 & 0 & 0 & 0 \\ 0 & 0 & 0 & c & s & 0 \\ 0 & 0 & 0 & -s & c & 0 \\ 0 & 0 & 0 & 0 & 0 & 1 \end{bmatrix} \tag{9.4}$$

where c and s are, respectively, the cosine and the sine of the angle between the global and local coordinate systems (measured clockwise from the global system).

Substituting Eqs 9.3 into Eq. 9.1 gives the stiffness relationship of a particular element expressed in global coordinates:

$$\mathbf{P}_e = \mathbf{K}_e \mathbf{U}_e \tag{9.5}$$

where $\mathbf{K}_e$ is the stiffness matrix of the element in global coordinates given by:

$$\mathbf{K}_e = \mathbf{T}^T \mathbf{k}_e \mathbf{T} \tag{9.6}$$

The stiffness relationship for the whole structure is then obtained by assembling the contribution of each element and can be expressed as:

$$\mathbf{P} = \mathbf{K}\mathbf{U} \tag{9.7}$$

where $\mathbf{K}$ is the structural stiffness matrix, while $\mathbf{P}$ and $\mathbf{U}$ are, respectively, the vectors of nodal actions and displacements for the whole structure expressed in the global coordinate system.

Equation 9.7 is readily solved for the unknown displacements and reactions. For illustrative purposes, Eq. 9.7 can be re-written in terms of the known and unknown displacements ($\mathbf{U}_K$ and $\mathbf{U}_U$) and known and unknown actions ($\mathbf{P}_K$ and $\mathbf{P}_U$) as follows:

$$\begin{bmatrix} \mathbf{P}_K \\ \mathbf{P}_U \end{bmatrix} = \begin{bmatrix} \mathbf{K}_{11} & \mathbf{K}_{12} \\ \mathbf{K}_{21} & \mathbf{K}_{22} \end{bmatrix} \begin{bmatrix} \mathbf{U}_U \\ \mathbf{U}_K \end{bmatrix} \tag{9.8}$$

The unknown displacements and actions are determined as follows:

$$\mathbf{U}_U = \mathbf{K}_{11}^{-1} (\mathbf{P}_K - \mathbf{K}_{12}\mathbf{U}_K) \tag{9.9a}$$

$$\mathbf{P}_U = \mathbf{K}_{21}\mathbf{U}_U + \mathbf{K}_{22}\mathbf{U}_K \tag{9.9b}$$

9.3 Member loads

Distributed member loads can be considered in the analysis by introducing particular sets of nodal actions that produce equivalent effects to the original distributed loads and, for this reason, are usually referred to as equivalent nodal loads. These correspond to the opposite of the reactions that would occur at the end of a loaded stiffness element if its ends were assumed to be fixed. Different sets of equivalent nodal loads have been derived in Appendix A. For example, in the case of uniform longitudinal and transverse distributed loads (referred to as n and p, respectively, in Fig. 9.3), the end reactions of a fixed ended member $\mathbf{p}_{F.m}$ are given in Eq. A.52 (reproduced here for ease of reference):

$$\mathbf{p}_{F.m} = \begin{bmatrix} 0 \\ -L/2 \\ -L^2/12 \\ 0 \\ -L/2 \\ L^2/12 \end{bmatrix} p + \begin{bmatrix} -L/2 \\ \alpha_3 \\ L\alpha_3/2 \\ -L/2 \\ \alpha_3 \\ L\alpha_3/2 \end{bmatrix} n \tag{9.10}$$

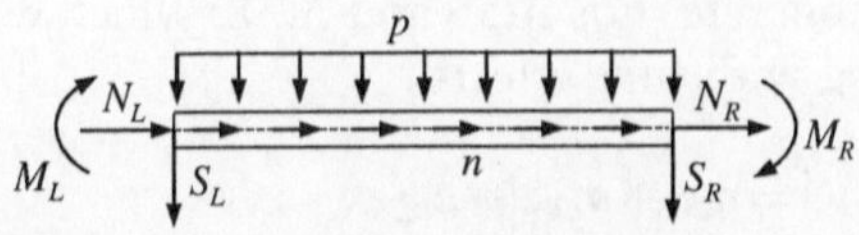

Figure 9.3 Free body diagram of a frame element highlighting uniformly distributed vertical and horizontal member loads n and p.

The subscript 'm' indicates that the reactions relate to member loads, while the subscript 'F' indicates that these actions have been determined assuming both ends of the element are fully fixed. The term α_3 is the ratio of the cross-sectional rigidities R_B and R_A and is defined in Eq. A.50d.

The equivalent nodal loads for inclusion in the stiffness analysis are the opposite of these reactions (i.e. $-\mathbf{p}_{F.m}$). The unknown displacements and reactions of the structure are obtained from Eqs 9.9 after the nodal loads from all elements are assembled.

The actions on each element (depicted in Fig. 9.1b) can be calculated by adding the forces and moments caused by the calculated displacements ($= \mathbf{ku}$) to the fixed-end reactions specified in Eq. 9.10 as:

$$\mathbf{p} = \mathbf{ku} + \mathbf{p}_{F.m} \tag{9.11}$$

Based on the Euler–Bernoulli beam assumptions, the displacements at any point along the length of an element are given by:

$$v = \alpha_1 p \frac{z^4}{24} + \overline{C}_1 \frac{z^3}{6} + \overline{C}_2 \frac{z^2}{2} + \overline{C}_3 z + \overline{C}_4 \tag{9.12a}$$

$$u = \alpha_2 p \frac{z^3}{6} + (\alpha_3 \overline{C}_1 + \alpha_4 n) \frac{z^2}{2} + \overline{C}_5 z + \overline{C}_6 \tag{9.12b}$$

where these equations are derived in Appendix A as Eqs A.53. The terms α_1, α_2, α_3 and α_4 are calculated from the cross-sectional rigidities and are given in Eqs A.50b–e.

The constants of integration $\overline{C}_1$ to $\overline{C}_6$ are obtained by specifying the appropriate boundary conditions and are expressed in terms of the nodal (element) displacements and applied member loads in Eqs A.54.

The expressions describing the variation of the axial strain (measured at the level of the x-axis) and curvature along the member length can be obtained by differentiating Eqs 9.12:

$$u' = \alpha_2 p \frac{z^2}{2} + (\alpha_3 \overline{C}_1 + \alpha_4 n) z + \overline{C}_5 \tag{9.13a}$$

$$v'' = \alpha_1 p \frac{z^2}{2} + \overline{C}_1 z + \overline{C}_2 \tag{9.13b}$$

For example, these expressions can be used to evaluate the internal axial force and moment resisted by the concrete (Eqs A.56 and A.58) which will become useful in the following sections dealing with time effects:

$$N_c = A_c E_c u' - B_c E_c v'' = a_{c0} + a_{c1} z + a_{c2} z^2 \tag{9.14a}$$

$$M_c = B_c E_c u' - I_c E_c v'' = b_{c0} + b_{c1} z + b_{c2} z^2 \tag{9.14b}$$

The concrete cross-sectional properties have been defined in Chapter 5, while the use of the parabolic polynomial to describe the variation of both N_c and M_c is justified in Appendix A for the derivation of Eqs A.56 and A.58. The coefficients of the polynomials ($a_{c0}, a_{c1}, a_{c2}, b_{c0}, b_{c1}, b_{c2}$) are given in Appendix A (Eqs A.57 and A.59).

In the following example, a simple beam subdivided into two elements is considered for illustrative purposes only. Clearly, the method is more applicable for more complex structures.

Example 9.1

The instantaneous mid-span deflection of the simply-supported beam shown in Fig. 9.4 is to be determined using the stiffness method. The beam is subjected to the uniformly distributed transverse load and the axial point load shown. The set of external actions reproduces the loading condition of Example 5.1 at mid-span of the member. The analysis is carried out using two stiffness elements of equal length numbered 1 and 2, respectively, as shown in Fig. 9.4. The geometry and material properties of the cross-section are as specified in Example 5.1 (including the use of the x-axis located at 200 mm below the top fibre of the cross-section). The member actions for both elements are to be calculated, as are all the coefficients necessary to define the expressions for the strain and for the internal actions resisted by the concrete along each element.

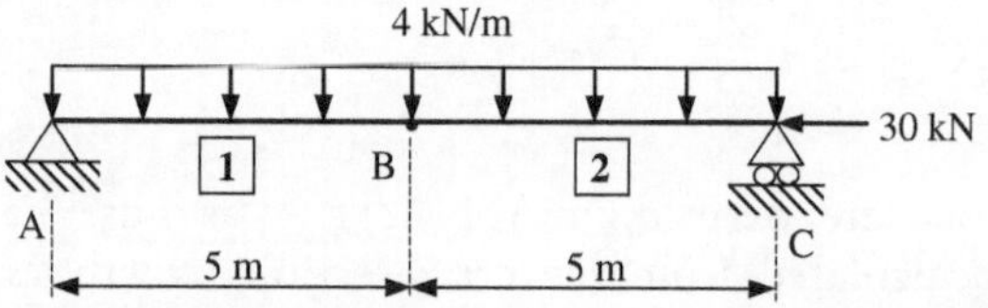

Figure 9.4 Loading, support conditions and discretisation for Example 9.1.

All units utilised in the solutions are in mm and N. For clarity, units are not displayed next to the terms in large matrices in the following solution.

An additional subscript '(1)' or '(2)' is used to specify whether a variable is calculated for elements 1 or 2, respectively. Where variables are identical for both elements, the subscript might be omitted for ease of notation. This is usually the case for the cross-sectional rigidities and material properties that are identical for both elements.

The freedoms for the problem are numbered in Fig. 9.5, where the lower numbers are used for the unrestrained freedoms. This numbering strategy is particularly useful for hand calculation as it leads to the partitioning introduced in Eqs 9.8. Also for this particular example, with the numbering proposed in Fig. 9.5, global and local coordinates coincide and there is no need to include transformations between the two coordinate systems in the solution, such as those described in Eqs 9.3 and 9.6.

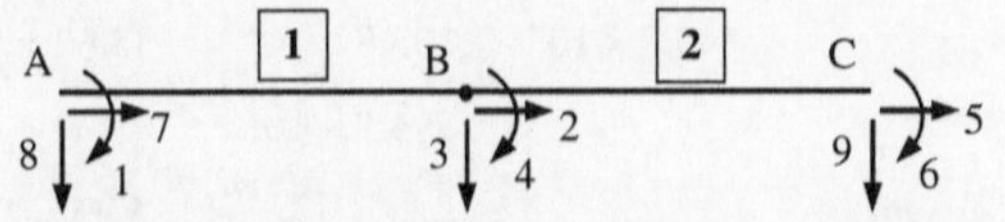

Figure 9.5 Freedom numbering for Example 9.1.

Considering that freedoms 7, 8 and 9 are restrained by the pinned and roller supports, their displacements are known, while the displacements along the remaining freedoms are unknown. Therefore:

$$\mathbf{U}_K = \begin{bmatrix} U_7 & U_8 & U_9 \end{bmatrix}^T = \begin{bmatrix} 0 & 0 & 0 \end{bmatrix}^T \text{ and } \mathbf{U}_U = \begin{bmatrix} U_1 & U_2 & U_3 & U_4 & U_5 & U_6 \end{bmatrix}^T$$

The unknown reactions are collected in P_U:

$$\mathbf{P}_U = \begin{bmatrix} P_7 & P_8 & P_9 \end{bmatrix}^T$$

From Example 5.1, the instantaneous cross-section rigidities are:

$$R_A = 4923 \times 10^6 \text{ N}$$
$$R_B = 543.9 \times 10^9 \text{ Nmm}$$
$$R_I = 221.0 \times 10^{12} \text{ Nmm}^2$$

and from Eqs A.50:

$$\alpha_0 = R_1 R_A - R_B^2 = 792.3 \times 10^{21} \text{ N}^2\text{mm}^2$$

$$\alpha_1 = \frac{R_A}{\alpha_0} = 6.214 \times 10^{-15} \text{ N}^{-1}\text{mm}^{-2}$$

$$\alpha_2 = \frac{R_B}{\alpha_0} = 686.5 \times 10^{-15} \text{ N}^{-1}\text{mm}^{-1}$$

$$\alpha_3 = \frac{R_B}{R_A} = 110.5 \text{ mm}$$

$$\alpha_4 = -\frac{1}{R_A} = -203.1 \times 10^{-12} \text{ N}^{-1}$$

The stiffness matrices for elements 1 and 2 are identical (as both possess the same geometric and cross-sectional properties) and are given by Eq. 9.2a:

$$
\mathbf{k} = \begin{bmatrix}
984.7\times10^{3} & 0 & -108.8\times10^{6} & -984.7\times10^{3} & 0 & 108.8\times10^{6} \\
 & 15.45\times10^{3} & 38.62\times10^{6} & 0 & -15.45\times10^{3} & 38.62\times10^{6} \\
 & & 140.7\times10^{9} & 108.8\times10^{6} & -38.62\times10^{6} & 52.35\times10^{9} \\
 & & & 984.7\times10^{3} & 0 & -108.8\times10^{6} \\
 & sym & & & 15.45\times10^{3} & -38.62\times10^{6} \\
 & & & & & 140.7\times10^{9}
\end{bmatrix}
$$

The fixed end forces on each element due to the transverse load $p = 4$ N/mm are obtained from Eq. 9.10:

$$
\mathbf{p}_{F.m(1)} = \mathbf{p}_{F.m(2)} = \begin{bmatrix} 0 \\ -\dfrac{L}{2} \\ -\dfrac{L^2}{12} \\ 0 \\ -\dfrac{L}{2} \\ \dfrac{L^2}{12} \end{bmatrix} p = \begin{bmatrix} 0\text{ N} \\ -10\times10^{3}\text{ N} \\ -8.3\times10^{6}\text{ Nmm} \\ 0\text{ N} \\ -10\times10^{3}\text{ N} \\ 8.3\times10^{6}\text{ Nmm} \end{bmatrix}
$$

Assembling the contribution of the two elements into the stiffness matrix of the structure enables the determination of the various sub-matrices of **K** given in Eq. 9.8:

$$
\mathbf{K}_{11} = \begin{bmatrix}
140.7\times10^{9} & 108.8\times10^{6} & -38.62\times10^{6} & 52.35\times10^{9} & 0 & 0 \\
 & 1.969\times10^{6} & 0 & -217.6\times10^{6} & -984.7\times10^{3} & 108.8\times10^{6} \\
 & & 30.89\times10^{3} & 0 & 0 & 38.62\times10^{6} \\
 & & & 281.5\times10^{9} & 108.8\times10^{6} & 52.35\times10^{9} \\
 & sym & & & 984.7\times10^{3} & -108.8\times10^{6} \\
 & & & & & 140.7\times10^{9}
\end{bmatrix}
$$

$$
\mathbf{K}_{12} = \begin{bmatrix}
-108.8\times10^{6} & 38.62\times10^{6} & 0 \\
-984.7\times10^{3} & 0 & 0 \\
0 & -15.45\times10^{3} & -15.45\times10^{3} \\
108.8\times10^{6} & 38.62\times10^{6} & -38.62\times10^{6} \\
0 & 0 & 0 \\
0 & 0 & -38.62\times10^{6}
\end{bmatrix}
$$

$$\mathbf{K}_{22} = \begin{bmatrix} 984.7 \times 10^3 & 0 & 0 \\ & 15.45 \times 10^3 & 0 \\ sym & & 15.45 \times 10^3 \end{bmatrix}$$

$$\mathbf{K}_{21} = \mathbf{K}_{12}^T$$

Assembling the load vector for the whole structure (including the axial load of −30 kN applied at freedom 5) produces:

$$\mathbf{P} = \begin{bmatrix} 8.3 \times 10^6 \text{ Nmm} \\ 0 \text{ N} \\ 20 \times 10^3 \text{ N} \\ 0 \text{ Nmm} \\ -30 \times 10^3 \text{ N} \\ -8.3 \times 10^6 \text{ Nmm} \\ \mathbf{P}_7 \\ \mathbf{P}_8 + 10 \times 10^3 \text{ N} \\ \mathbf{P}_9 + 10 \times 10^3 \text{ N} \end{bmatrix}$$

Unknown displacements $\mathbf{U}_U$ and reactions $\mathbf{P}_U$ are then calculated using Eqs 9.9:

$$\mathbf{U}_U = \mathbf{K}_{11}^{-1}(\mathbf{P}_K - \mathbf{K}_{12}\mathbf{U}_K) = \begin{bmatrix} 1.138 \times 10^{-3} \\ -156.2 \times 10^{-3} \text{ mm} \\ 3.494 \text{ mm} \\ 0 \\ -312.5 \times 10^{-3} \text{ mm} \\ -1.138 \times 10^{-3} \end{bmatrix}$$

$$\mathbf{P}_U = \begin{bmatrix} P_7 \\ P_8 \\ P_9 \end{bmatrix} = \begin{bmatrix} 30 \times 10^6 \text{ N} \\ -20 \times 10^6 \text{ N} \\ -20 \times 10^6 \text{ N} \end{bmatrix}$$

The mid-span deflection is the displacement related to freedom 3) and is equal to 3.49 mm.

Post-processing

The solution is post-processed below to determine the variables defining the strain diagram and the internal actions along each element.

Element 1

The nodal displacements undergone by the end nodes of element 1 are collected in vector $\mathbf{u}_{(1)}$:

$$\mathbf{u}_{(1)} = \begin{bmatrix} u_{L(1)} \\ v_{L(1)} \\ v'_{L(1)} \\ u_{R(1)} \\ v_{R(1)} \\ v'_{R(1)} \end{bmatrix} = \begin{bmatrix} 0 \text{ mm} \\ 0 \text{ mm} \\ 1.138 \times 10^{-3} \\ -156.2 \times 10^{-3} \text{ mm} \\ 3.494 \text{ mm} \\ 0 \end{bmatrix}$$

The member actions are obtained from Eq. 9.11:

$$\mathbf{p}_{(1)} = \mathbf{k}\mathbf{u}_{(1)} + \mathbf{p}_{F.m(1)}$$

$$= \begin{bmatrix} 30 \times 10^3 \text{N} & -20 \times 10^3 \text{N} & 0 \text{Nmm} & 30 \times 10^3 \text{N} & 0 \text{N} & -50 \times 10^6 \text{Nmm} \end{bmatrix}^T$$

The relevant constants of integration are then obtained from Eqs A.54:

$$\overline{C}_{1(1)} = -124.3 \times 10^{-12} \text{ mm}^{-2}$$

$$\overline{C}_{2(1)} = -20.59 \times 10^{-9} \text{ mm}^{-1}$$

$$\overline{C}_{3(1)} = 1.138 \times 10^{-3}$$

$$\overline{C}_{4(1)} = 0 \text{ mm}$$

$$\overline{C}_{5(1)} = -8.368 \times 10^{-6}$$

$$\overline{C}_{6(1)} = 0 \text{ mm}$$

These can be substituted in Eqs 9.13 to determine the strain distribution along member 1. For example, the axial strain measured at the level of the reference axis $\varepsilon_{r(1)}$ $(= u'_{(1)})$ and the curvature $\kappa_{(1)}$ $(= -v''_{(1)})$ at mid-span of the beam can be obtained by substituting $z = 5000$ mm in Eqs 9.13:

$$\varepsilon_{r(1)} = u'_{(1)}(z = 5000) = -42.7 \times 10^{-6} \text{ and } \kappa_{(1)} = -v''_{(1)}(z = 5000)$$

$$= 0.331 \times 10^{-6} \text{ mm}^{-1}$$

which, as expected, corresponds to the same solution obtained in Example 5.1.

The coefficients defining the expressions for the concrete internal axial force and moment are determined from Eqs A.57 and A.59:

$$a_{c0(1)} = -28.16 \times 10^3 \text{ N}$$

$$a_{c1(1)} = -6.701 \text{ Nmm}^{-1}$$

$$a_{c2(1)} = 670.2 \times 10^{-6} \text{ Nmm}^{-2}$$

$$b_{c0(1)} = -66.96 \times 10^3 \text{ Nmm}$$

$$b_{c1(1)} = 15.65 \times 10^3 \text{ N}$$

$$b_{c2(1)} = -1.564 \text{ Nmm}^{-1}$$

Element 2

Similar to element 1, the terms describing the response of element 2 are calculated as:

$$\mathbf{u}_{(2)} = \begin{bmatrix} u_{L(2)} & v_{L(2)} & v'_{L(2)} & u_{R(2)} & v_{R(2)} & v'_{R(2)} \end{bmatrix}$$

$$= \begin{bmatrix} 156.2 \times 10^{-3} & 3.494 & 0 & -312.5 \times 10^{-3} & 0 & -1.138 \times 10^{-3} \end{bmatrix}$$

$$\mathbf{p}_{(2)} = \mathbf{ku}_{(2)} + \mathbf{p}_{F.m(2)}$$

$$= \begin{bmatrix} 30 \times 10^3 \text{N} & 0\text{N} & 50 \times 10^6 \text{Nmm} & -30 \times 10^3 \text{N} & -20 \times 10^3 \text{N} & 0\text{Nmm} \end{bmatrix}^T$$

$$\overline{C}_{1(2)} = 0 \text{ mm}^{-2}$$

$$\overline{C}_{2(2)} = -331.3 \times 10^{-9} \text{ mm}^{-1}$$

$$\overline{C}_{3(2)} = 0$$

$$\overline{C}_{4(2)} = 3.494 \text{ mm}$$

$$\overline{C}_{5(2)} = -42.69 \times 10^{-6}$$

$$\overline{C}_{6(2)} = -156.2 \times 10^{-3} \text{ mm}$$

The mid-span values of the strain variables (already calculated with element 1) could also have been determined with element 2 using the expressions of Eqs 9.13 at $z = 0$ mm.

9.4 Time analysis using AEMM

When using the AEMM, deformations and stresses are calculated at two instants, immediately after loading at τ_0 and after the effects of creep and shrinkage have taken place at τ_k. The instantaneous analysis at τ_0 was presented in the previous section. The time analysis at time τ_k is outlined in the following.

Creep and shrinkage effects are included in the stiffness analysis using equivalent nodal actions. As already discussed for the member loads, the set of equivalent nodal actions to be used in the stiffness method correspond to the opposite of the reactions obtained when the supports of the element under consideration are assumed to be fixed at both ends. The reactions produced by creep and shrinkage effects and by member loads are derived in Appendix A (Eqs A.66) and are reproduced below.

$$\mathbf{p}_{F.cr,k} = \begin{bmatrix} -\overline{F}_{e,0}\left(\dfrac{a_{c2,0}L^2}{3} + \dfrac{a_{c1,0}L}{2} + a_{c0,0}\right) \\ \overline{F}_{e,0}\left(\alpha_{3,k}a_{c1,0} + L\alpha_{3,k}a_{c2,0} - b_{c1,0} - b_{c2,0}L\right) \\ -\overline{F}_{e,0}\left(-\dfrac{a_{c1,0}+a_{c2,0}L}{2}\alpha_{3,k}L - b_{c0,0} + b_{c2,0}\dfrac{L^2}{6}\right) \\ \overline{F}_{e,0}\left(\dfrac{a_{c2,0}L^2}{3} + \dfrac{a_{c1,0}L}{2} + a_{c0,0}\right) \\ -\overline{F}_{e,0}\left(\alpha_{3,k}a_{c1,0} + L\alpha_{3,k}a_{c2,0} - b_{c1,0} - b_{c2,0}L\right) \\ -\overline{F}_{e,0}\left(-\dfrac{a_{c1,0}+a_{c2,0}L}{2}\alpha_{3,k}L + b_{c0,0} + b_{c1,0}L + b_{c2,0}\dfrac{5L^2}{6}\right) \end{bmatrix} \tag{9.15a}$$

$$\mathbf{p}_{F.sh.k} = \begin{bmatrix} \overline{E}_{e,k}A_c\varepsilon_{sh,k} \\ 0 \\ -\overline{E}_{e,k}B_c\varepsilon_{sh,k} \\ -\overline{E}_{e,k}A_c\varepsilon_{sh,k} \\ 0 \\ \overline{E}_{e,k}B_c\varepsilon_{sh,k} \end{bmatrix} \tag{9.15b}$$

$$\mathbf{p}_{F.m,k} = \begin{bmatrix} 0 \\ -L/2 \\ -L^2/12 \\ 0 \\ -L/2 \\ L^2/12 \end{bmatrix} p + \begin{bmatrix} -L/2 \\ \alpha_{3,k} \\ L\alpha_{3,k}/2 \\ -L/2 \\ -\alpha_{3,k} \\ L\alpha_{3,k}/2 \end{bmatrix} n \tag{9.15c}$$

where the constants $\alpha_{0,k}$, $\alpha_{1,k}$, $\alpha_{2,k}$, $\alpha_{3,k}$ and $\alpha_{4,k}$ are obtained from the cross-sectional rigidities at time τ_k using Eqs A.50a–e, respectively. The subscripts '*m*', '*cr*' and '*sh*' used in $\mathbf{p}_{F.m,k}$, $\mathbf{p}_{F.cr,k}$ and $\mathbf{p}_{F.sh,k}$ indicate that the fixed end reactions relate to either member loads, creep effects or shrinkage effects, respectively. The calculation at time τ_k of the cross-sectional rigidities $R_{A,k}$, $R_{B,k}$ and $R_{I,k}$ required for input into Eqs A.50 has already been outlined in Chapter 5 for different types of cross-sections. The time-dependent behaviour of concrete is represented by $\overline{E}_{e,k}$ and $\overline{F}_{e,0}$ (given in Eqs 4.35 and 4.46), while $\varepsilon_{sh,k}$ denotes the shrinkage strain. The coefficients $a_{cj,0}$ and $b_{cj,0}$ (with $j = 0$ to 2) that are necessary to define the expressions of the concrete internal actions are specified in Eqs A.57 and A.59.

The time analysis can then be carried out by assembling the opposite of the vectors $\mathbf{p}_{F.m,k}$, $\mathbf{p}_{F.cr,k}$ and $\mathbf{p}_{F.sh,k}$ for each element into the loading vector of the structure, $\mathbf{P}_k$, while using the cross-sectional properties calculated at τ_k in the determination of the stiffness coefficients of $\mathbf{K}_k$. When the unknown displacements and reactions are evaluated based on Eqs 9.9, the end actions of each element (i.e. member actions) can be obtained by combining the results of the stiffness analysis with those related to member loads, creep and shrinkage as follows:

$$\mathbf{p}_k = \mathbf{k}_k \mathbf{u}_k + \mathbf{p}_{F.m,k} + \mathbf{p}_{F.cr,k} + \mathbf{p}_{F.sh,k} \tag{9.16}$$

The axial displacement u_k and deflection v_k can be calculated along the member axis using Eqs A.63 (repeated here for convenience):

$$u_k = (\alpha_{2,k} p + \beta_{2,k}) \frac{z^3}{6} + (\alpha_{3,k} \overline{C}_{1,k} + \alpha_{4,k} n + \beta_{3,k}) \frac{z^2}{2} + \overline{C}_{5,k} z + \overline{C}_{6,k} \tag{9.17a}$$

$$v_k = (\alpha_{1,k} p + \beta_{1,k}) \frac{z^4}{24} + \overline{C}_{1,k} \frac{z^3}{6} + \overline{C}_{2,k} \frac{z^2}{2} + \overline{C}_{3,k} z + \overline{C}_{4,k} \tag{9.17b}$$

where the constants $\alpha_{j,k}$ (with $j = 0$ to 4) and $\beta_{j,k}$ (with $j = 1$ to 3) are given in Eqs A.50 and A.64, respectively and the constants of integration $\overline{C}_{1,k}$ to $\overline{C}_{6,k}$ are given in Eqs A.67.

The strain diagram at any point along an element can be obtained by differentiating Eqs 9.17 to give:

$$u'_k = (\alpha_{2,k} p + \beta_{2,k}) \frac{z^2}{2} + (\alpha_{3,k} \overline{C}_{1,k} + \alpha_{4,k} n + \beta_{3,k}) z + \overline{C}_{5,k} \tag{9.18a}$$

$$v_k = (\alpha_{1,k} p + \beta_{1,k}) \frac{z^2}{2} + \overline{C}_{1,k} z + \overline{C}_{2,k} \tag{9.18b}$$

from which the stresses may be evaluated following the procedures illustrated in Chapter 5.

Example 9.2

The deflection at time τ_k and the variables defining the strain diagram ($\varepsilon_{r,k}$ and κ_k) at mid-span of the simply-supported beam analysed in Example 9.1 are to be determined, as well as the actions resisted by the two elements. The applied loads are assumed to remain constant with time. The cross-sectional and material properties of the beam are those of the reinforced concrete section considered in Example 5.3. The analysis is to be carried out using the stiffness method assuming the same two elements and freedom numbering as adopted in Example 9.1.
All units are in mm and N, unless noted otherwise. From Example 5.3:

$$\overline{E}_{e,k} = 9524 \text{ MPa}$$

$$\overline{F}_{e,0} = -0.333$$

$$\varphi(\tau_k, \tau_0) = 2.5$$

$$\chi(\tau_k, \tau_0) = 0.65$$

$$\varepsilon_{sh}(\tau_k) = -600 \times 10^{-6}$$

$$A_c = 177.6 \times 10^3 \text{ mm}^2$$

$$B_c = 17.46 \times 10^6 \text{ mm}^3$$

$$I_c = 6966 \times 10^6 \text{ mm}^4$$

$$R_{A,k} = 2175 \times 10^6 \text{ N}$$

$$R_{B,k} = 273.7 \times 10^9 \text{ Nmm}$$

$$R_{I,k} = 113.2 \times 10^{12} \text{ Nmm}^2$$

and from Example 5.1:

$$R_{A,0} = 4923 \times 10^6 \text{ N}$$

$$R_{B,0} = 543.9 \times 10^9 \text{ Nmm}$$

$$R_{I,0} = 221.0 \times 10^{12} \text{ Nmm}^2$$

As for Example 9.1:

$$\mathbf{U}_{U,k} = \begin{bmatrix} U_{1,k} & U_{2,k} & U_{3,k} & U_{4,k} & U_{5,k} & U_{6,k} \end{bmatrix}^T$$

$$\mathbf{U}_{K,k} = \begin{bmatrix} U_{7,k} & U_{8,k} & U_{9,k} \end{bmatrix}^T = \begin{bmatrix} 0 & 0 & 0 \end{bmatrix}^T$$

$$\mathbf{P}_{U,k} = \begin{bmatrix} P_{7,k} & P_{8,k} & P_{9,k} \end{bmatrix}^T$$

From Eqs A.50:

$$\alpha_{0,k} = 171.4 \times 10^{21} \text{ N}^2\text{mm}^2$$

$$\alpha_{1,k} = 12.69 \times 10^{-15} \text{ N}^{-1}\text{mm}^{-2}$$

$$\alpha_{2,k} = 1597 \times 10^{-15} \text{ N}^{-1}\text{mm}^{-1}$$

$$\alpha_{3,k} = 125.8 \text{ mm}$$

$$\alpha_{4,k} = -459.7 \times 10^{-12} \text{ N}^{-1}$$

The stiffness matrix for both elements 1 and 2 are identical and calculated at time τ_k as (Eq. 9.2a):

$$\mathbf{k}_e = \begin{bmatrix} 435.0 \times 10^3 & 0 & -54.74 \times 10^6 & -435.0 \times 10^3 & 0 & -54.74 \times 10^6 \\ & 7.563 \times 10^3 & 18.91 \times 10^6 & 0 & -7.563 \times 10^3 & 18.91 \times 10^6 \\ & & 69.92 \times 10^9 & 54.74 \times 10^6 & -18.91 \times 10^6 & 24.62 \times 10^9 \\ & & & 435.0 \times 10^3 & 0 & -54.74 \times 10^6 \\ & sym & & & 7.563 \times 10^3 & -18.91 \times 10^6 \\ & & & & & 69.92 \times 10^9 \end{bmatrix}$$

The set of reactions required to account for the member load $p = 4$ N/mm and for shrinkage are (Eqs 9.15b and c):

$$\mathbf{p}_{F.m,k(1)} = \mathbf{p}_{F.m,k(2)} = \begin{bmatrix} 0 \\ -L/2 \\ -L^2/12 \\ 0 \\ -L/2 \\ L^2/12 \end{bmatrix} p = \begin{bmatrix} 0\text{ N} \\ -10 \times 10^3\text{ N} \\ -8.3 \times 10^6\text{ Nmm} \\ 0\text{ N} \\ -10 \times 10^3\text{ N} \\ 8.3 \times 10^6\text{ Nmm} \end{bmatrix}$$

$$\mathbf{p}_{F.sh,k(1)} = \mathbf{p}_{F.sh,k(2)} = \begin{bmatrix} a_{sh,k} \\ 0 \\ -b_{sh,k} \\ -a_{sh,k} \\ 0 \\ b_{sh,k} \end{bmatrix} = \begin{bmatrix} \bar{E}_{e,k} A_c \varepsilon_{sh,k} \\ 0 \\ -\bar{E}_{e,k} B_c \varepsilon_{sh,k} \\ -\bar{E}_{e,k} A_c \varepsilon_{sh,k} \\ 0 \\ \bar{E}_{e,k} B_c \varepsilon_{sh,k} \end{bmatrix} = \begin{bmatrix} -1014 \times 10^6\text{ N} \\ 0\text{ N} \\ 99.78 \times 10^6\text{ Nmm} \\ 1014 \times 10^6\text{ N} \\ 0\text{ N} \\ -99.78 \times 10^6\text{ Nmm} \end{bmatrix}$$

The vectors $\mathbf{p}_{F.cr,k(1)}$ and $\mathbf{p}_{F.cr,k(2)}$ need to be evaluated for the two elements separately.
From Example 9.1:

$$a_{c0(1),0} = -28.16 \times 10^3\text{ N}$$

$$a_{c1(1),0} = -6.701\text{ Nmm}^{-1}$$

$$a_{c2(1),0} = 670.2 \times 10^{-6}\text{ Nmm}^{-2}$$

$$b_{c0(1),0} = -66.96 \times 10^3\text{ Nmm}$$

$$b_{c1(1),0} = 15.65 \times 10^3\text{ N}$$

$$b_{c2(1),0} = -1.564\text{ Nmm}^{-1}$$

$$a_{c0(2),0} = -44.91 \times 10^3\text{ N}$$

$$a_{c1(2),0} = 2.788 \times 10^{-12}\text{ Nmm}^{-1}$$

$$a_{c2(2),0} = 670.2 \times 10^{-6}\text{ Nmm}^{-2}$$

$$b_{c0(2),0} = 39.05 \times 10^6\text{ Nmm}$$

$$b_{c1(2),0} = -6.509 \times 10^{-12}\text{ N}$$

$$b_{c2(2),0} = -1.564\text{ Nmm}^{-1}$$

From Eq. 9.15a:

$$\mathbf{p}_{F.cr(1),k} = \begin{bmatrix} -13.11\times 10^3\ \text{N} \\ 2.748\times 10^3\ \text{N} \\ -1.799\times 10^6\ \text{Nmm} \\ 13.11\times 10^3\ \text{N} \\ -2.748\times 10^3\ \text{N} \\ 15.54\times 10^6\ \text{Nmm} \end{bmatrix} \quad \text{and } \mathbf{p}_{F.cr(2),k} = \begin{bmatrix} -13.11\times 10^3\ \text{N} \\ -2.748\times 10^3\ \text{N} \\ -15.54\times 10^6\ \text{Nmm} \\ 13.11\times 10^3\ \text{N} \\ 2.748\times 10^3\ \text{N} \\ 1.799\times 10^6\ \text{Nmm} \end{bmatrix}$$

After assembly, the sub-matrices of $\mathbf{K}_k$ and the loading vector $\mathbf{P}_k$ of the whole structure at τ_k are:

$$\mathbf{K}_{11,k} = \begin{bmatrix} 69.92\times 10^9 & 54.74\times 10^6 & -18.91\times 10^6 & 24.63\times 10^9 & 0 & 0 \\ & 870.1\times 10^3 & 0 & -109.5\times 10^6 & -435.0\times 10^3 & 54.74\times 10^6 \\ & & 15.13\times 10^3 & 0 & 0 & 18.91\times 10^6 \\ & & & 139.8\times 10^9 & 54.74\times 10^6 & 24.63\times 10^9 \\ & sym & & & 435.0\times 10^3 & -54.74\times 10^6 \\ & & & & & 69.92\times 10^9 \end{bmatrix}$$

$$\mathbf{K}_{12,k} = \begin{bmatrix} -54.74\times 10^6 & 18.91\times 10^6 & 0 \\ -435.0\times 10^3 & 0 & 0 \\ 0 & -7.56\times 10^3 & -7.56\times 10^3 \\ 54.74\times 10^6 & 18.91\times 10^6 & -18.91\times 10^6 \\ 0 & 0 & 0 \\ 0 & 0 & -18.91\times 10^6 \end{bmatrix}$$

$$\mathbf{K}_{22,k} = \begin{bmatrix} 435.0\times 10^3 & 0 & 0 \\ & 7.563\times 10^3 & 0 \\ sym & & 7.563\times 10^3 \end{bmatrix}$$

$$\mathbf{K}_{21,k} = \mathbf{K}_{12,k}^T$$

$$\mathbf{P}_k = [89.66\times 10^6 \quad 0 \quad 25.50\times 10^3 \quad 0 \quad -1057\times 10^3 \quad 89.66\times 10^6 \quad (P_7 + 1027\times 10^3) \quad (P_8 + 7.252\times 10^3) \quad (P_7 + 7.252\times 10^3)]^T$$

From Eqs 9.9, unknown displacements $\mathbf{U}_{U,k}$ and reactions $\mathbf{P}_{U,k}$ are obtained:

$$\mathbf{U}_{U,k} = \mathbf{K}_{11,k}^{-1}(\mathbf{P}_{K,k} - \mathbf{K}_{12,k}\mathbf{U}_{K,k}) = \begin{bmatrix} 4.781\times 10^{-3} \\ -3.033\ \text{mm} \\ 13.65\ \text{mm} \\ 0 \\ -6.066\ \text{mm} \\ -4.780\times 10^{-3} \end{bmatrix} \quad \text{and}$$

$$\mathbf{p}_{U,k} = \begin{bmatrix} P_{7,k} \\ P_{8,k} \\ P_{9,k} \end{bmatrix} = \begin{bmatrix} 30 \times 10^6 \text{ N} \\ -20 \times 10^6 \text{ N} \\ -20 \times 10^6 \text{ N} \end{bmatrix}$$

The mid-span deflection at τ_k (associated with freedom 3) is 13.65 mm which has increased from its initial value of 3.49 mm.

Post-processing

The solution is post-processed below to determine the variables defining the strain diagram and the internal actions along each element.

Element 1

The nodal displacements and the member actions associated with the end nodes of element 1 are:

$$\mathbf{u}_{(1),k} = \begin{bmatrix} u_{L(1),k} \\ v_{L(1),k} \\ v'_{L(1),k} \\ u_{R(1),k} \\ v_{R(1),k} \\ v'_{R(1),k} \end{bmatrix} = \begin{bmatrix} 0 \text{ mm} \\ 0 \text{ mm} \\ 4.781 \times 10^{-3} \\ -3.033 \text{ mm} \\ 13.65 \text{ mm} \\ 0 \end{bmatrix}$$

$$\mathbf{p}_{(1),k} = \mathbf{k}_k \mathbf{u}_{(1),k} + \mathbf{p}_{F.m(1),k} + \mathbf{p}_{F.cr(1),k} + \mathbf{p}_{F.sh(1),k}$$
$$= \begin{bmatrix} 30 \times 10^3 & -20 \times 10^3 & 0 & -30 \times 10^3 & 0 & -50 \times 10^6 \end{bmatrix}^T$$

As expected, the member actions are identical to the instantaneous values as the structure is statically determinate.
From Eqs A.67:

$$\overline{C}_{1(1),k} = -323.6 \times 10^{-12} \text{ mm}^{-2}$$
$$\overline{C}_{2(1),k} = -416.7 \times 10^{-9} \text{ mm}^{-1}$$
$$\overline{C}_{3(1),k} = 4.780 \times 10^{-3}$$
$$\overline{C}_{4(1),k} = 0 \text{ mm}$$
$$\overline{C}_{5(1),k} = -539.6 \times 10^{-6}$$
$$\overline{C}_{6(1),k} = 0 \text{ mm}$$

The axial strain measured at the level of the reference axis and the curvature at mid-span are determined by substituting the calculated constants of integration into Eqs 9.18 with $z = 5000$ mm: i.e. $\varepsilon_{r(1),k} = u'_{(1),k} = -641.4 \times 10^{-6}$ and $\kappa_{(1),k} = -v''_{(1),k} = 1.225 \times 10^{-6}$ mm^{-1}. As expected, these results are identical to those calculated in Example 5.3.

Element 2

The nodal displacements and the member actions associated with the end nodes of element 1 are:

$$\mathbf{u}_{(2),k} = \begin{bmatrix} u_{L(2),k} \\ v_{L(2),k} \\ v'_{L(2),k} \\ u_{R(2),k} \\ v_{R(2),k} \\ v'_{R(2),k} \end{bmatrix} = \begin{bmatrix} -3.033 \text{ mm} \\ 13.65 \text{ mm} \\ 0 \\ -6.066 \text{ mm} \\ 0 \\ -4.780 \times 10^{-3} \end{bmatrix}$$

$$\mathbf{p}_{(2),k} = \mathbf{k}_k \mathbf{u}_{(1),k} + \mathbf{p}_{F.m(1),k} + \mathbf{p}_{F.cr(1),k} + \mathbf{p}_{F.sh(1),k}$$
$$= \begin{bmatrix} 30 \times 10^3 & 0 & 50 \times 10^6 & -30 \times 10^3 & -20 \times 10^3 & 0 \end{bmatrix}^T$$

9.5 Time analysis using SSM

When using the SSM, the time domain is discretised into a number of time instants τ_i (with $i = 1, \ldots, k$) as shown in Fig. 4.3. As for the AEMM discussed in the previous section, the time-dependent behaviour of the concrete is included in the analysis using equivalent nodal actions that correspond to the opposite of the reactions obtained when the ends of the element are fixed and the member is loaded and subject to creep and shrinkage. These reactions are determined at a particular instant in time, here referred to as τ_k, and are expressed as the vectors $\mathbf{p}_{F.m,k}$, $\mathbf{p}_{F.cr,k}$ and $\mathbf{p}_{F.sh,k}$ in Eqs A.91.

Most of the notation used in the expressions for $\mathbf{p}_{F.m,k}$, $\mathbf{p}_{F.cr,k}$ and $\mathbf{p}_{F.sh,k}$ has already been defined in the previous section. The procedure for the SSM and the various constants and coefficients are derived in the Section A.5.3 and presented in Eqs A.71–A.91.

The loading vector $\mathbf{P}_k$ is assembled, including the contributions of $\mathbf{p}_{F.m,k}$, $\mathbf{p}_{F.cr,k}$ and $\mathbf{p}_{F.sh,k}$ for every element, and the stiffness coefficients for $\mathbf{K}_k$ are calculated based on the cross-sectional rigidities determined at time τ_k.

The solution at time τ_k is obtained using Eqs 9.9. Post-processing to obtain stresses and strains on particular cross-sections is as described previously. The element end actions are evaluated from Eq. 9.16 and the axial displacement and deflection at any distance z along the element are expressed by Eqs A.77. The variables describing the strain diagram at time τ_k ($\varepsilon_{r,k} = u'_k$ and $\kappa_k = -v''_k$) are obtained from Eqs A.80.

Example 9.3

The simply-supported beam analysed in Example 9.2 is to be re-analysed using the stiffness method and using the SSM to handle the time effects. The time domain is discretised into three time instants: $\tau_0 = 28$ days, $\tau_1 = 100$ days and $\tau_2 = 30{,}000$ days. The deflection and the axial strain at the level of the

reference axis and curvature at mid-span are to be calculated at each of these time instants. In addition, the member actions at each time step are determined for both elements. As in Examples 9.1 and 9.2, two elements are considered with freedom numbering as shown in Fig. 9.5.
The cross-section of the beam at mid-span was previously analysed using the SSM in Example 5.5, where:

$$E_{c,0} = 25 \text{ GPa}$$
$$E_{c,1} = 28 \text{ GPa}$$
$$E_{c,2} = 30 \text{ GPa}$$
$$F_{e,1,0} = -1.8$$
$$F_{e,2,0} = -0.986$$
$$F_{e,2,1} = -2.214$$
$$\varepsilon_{sh}(\tau_0) = \varepsilon_{sh,0} = 0$$
$$\varepsilon_{sh}(\tau_1) = \varepsilon_{sh,1} = -300 \times 10^{-6}$$
$$\varepsilon_{sh}(\tau_2) = \varepsilon_{sh,2} = -600 \times 10^{-6}$$
$$A_c = 177.6 \times 10^3 \text{ mm}^2$$
$$B_c = 17.46 \times 10^6 \text{ mm}^3$$
$$I_c = 6966 \times 10^6 \text{ mm}^4$$

Similarly to Example 9.1, the sets of known and unknown displacements and unknown actions for the three time steps τ_j (with $j = 1, 2, 3$) are:

$$\mathbf{U}_{U,j} = \begin{bmatrix} U_{1,j} & U_{2,j} & U_{3,j} & U_{4,j} & U_{5,j} & U_{6,j} \end{bmatrix}^T$$

$$\mathbf{U}_{K,j} = \begin{bmatrix} U_{7,j} & U_{8,j} & U_{9,j} \end{bmatrix}^T = \begin{bmatrix} 0 & 0 & 0 \end{bmatrix}^T$$

$$\mathbf{P}_{U,j} = \begin{bmatrix} P_{7,j} & P_{8,j} & P_{9,j} \end{bmatrix}^T$$

From Examples 5.1 and 5.5:

$$R_{A,0} = 4923 \times 10^6 \text{ N}$$
$$R_{B,0} = 543.9 \times 10^9 \text{ Nmm}$$
$$R_{I,0} = 221.0 \times 10^{12} \text{ Nmm}^2$$
$$R_{A,1} = 5456 \times 10^6 \text{ N}$$
$$R_{B,1} = 596.4 \times 10^9 \text{ Nmm}$$
$$R_{I,1} = 241.9 \times 10^{12} \text{ Nmm}^2$$
$$R_{A,2} = 5811 \times 10^6 \text{ N}$$
$$R_{B,2} = 631.3 \times 10^9 \text{ Nmm}$$
$$R_{I,2} = 255.8 \times 10^{12} \text{ Nmm}^2$$

From Eqs A.50:

$$\alpha_{0,1} = 792.3 \times 10^{21}\ \text{N}^2\text{mm}^2$$
$$\alpha_{1,1} = 6.214 \times 10^{-15}\ \text{N}^{-1}\text{mm}^{-2}$$
$$\alpha_{2,1} = 686.6 \times 10^{-15}\ \text{N}^{-1}\text{mm}^{-1}$$
$$\alpha_{3,1} = 110.5\ \text{mm}$$
$$\alpha_{4,1} = -203.1 \times 10^{-12}\ \text{N}^{-1}$$
$$\alpha_{0,2} = 964.4 \times 10^{21}\ \text{N}^2\text{mm}^2$$
$$\alpha_{1,2} = 5.658 \times 10^{-15}\ \text{N}^{-1}\text{mm}^{-2}$$
$$\alpha_{2,2} = 618.4 \times 10^{-15}\ \text{N}^{-1}\text{mm}^{-1}$$
$$\alpha_{3,2} = 109.3\ \text{mm}$$
$$\alpha_{4,2} = -193.3 \times 10^{-12}\ \text{N}^{-1}$$

Short-term analysis ($\tau_0 = 28$ days)

The solution at time τ_0 is identical to that presented in Example 9.1.

Time analysis – SSM ($\tau_1 = 100$ days)

The stiffness matrix for elements 1 and 2 is determined using Eq. 9.2a with the rigidities calculated at time τ_1:

$$\mathbf{k}_1 = \begin{bmatrix} 1091\times10^3 & 0 & -119.3\times10^6 & -1091\times10^3 & 0 & 119.3\times10^6 \\ & 16.97\times10^3 & 42.42\times10^6 & 0 & -16.97\times10^3 & 42.42\times10^6 \\ & & 154.4\times10^9 & 119.3\times10^6 & -42.42\times10^6 & 57.66\times10^9 \\ & & & 1091\times10^3 & 0 & -119.3\times10^6 \\ & sym & & & 16.97\times10^3 & -42.42\times10^6 \\ & & & & & 154.4\times10^9 \end{bmatrix}$$

Since, the external loads are constant in time, the sets of reactions required to account for the member load $p = 4$ N/mm (i.e. $\mathbf{p}_{F,m(1),1}$ and $\mathbf{p}_{F,m(2),1}$) are identical to those determined in Example 9.1 at τ_0. Shrinkage is accounted for using Eq. A.91b with

$$\mathbf{p}_{F.sh(1),1} = \mathbf{p}_{F.sh(2),1} = \begin{bmatrix} E_{c,1}A_c\varepsilon_{sh,1} \\ 0 \\ -E_{c,1}B_c\varepsilon_{sh,1} \\ -E_{c,1}A_c\varepsilon_{sh,1} \\ 0 \\ E_{c,1}B_c\varepsilon_{sh,1} \end{bmatrix} = \begin{bmatrix} -1491\times10^6 \\ 0 \\ 146.7\times10^6 \\ 1491\times10^6 \\ 0 \\ -146.7\times10^6 \end{bmatrix}$$

From Example 9.1, for element 1:

$$a_{c0(1),0} = -28.16 \times 10^3 \text{ Nmm}^2$$

$$a_{c1(1),0} = -6.701 \text{ Nmm}^{-1}$$

$$a_{c2(1),0} = 670.2 \times 10^{-6} \text{ Nmm}^{-2}$$

$$b_{c0(1),0} = -66.96 \times 10^3 \text{ Nmm}$$

$$b_{c1(1),0} = 15.65 \times 10^3 \text{ N}$$

$$b_{c2(1),0} = -1.564 \text{ Nmm}^{-1}$$

and for element 2:

$$a_{c0(2),0} = -44.91 \times 10^3 \text{ Nmm}^2$$

$$a_{c1(2),0} = 2.788 \times 10^{-12} \text{ Nmm}^{-1}$$

$$a_{c2(2),0} = 670.2 \times 10^{-6} \text{ Nmm}^{-2}$$

$$b_{c0(2),0} = 39.05 \times 10^6 \text{ Nmm}$$

$$b_{c1(2),0} = -6.509 \times 10^{-12} \text{ N}$$

$$b_{c2(2),0} = -1.564 \text{ Nmm}^{-1}$$

From Eq. A.91, the creep effects are included in $\mathbf{p}_{F.cr(1),1}$ and $\mathbf{p}_{F.cr(2),1}$:

$$\mathbf{p}_{F.cr(1),1} = \begin{bmatrix} -70.8 \times 10^3 \\ 14.74 \times 10^3 \\ -9.968 \times 10^6 \\ 70.8 \times 10^3 \\ -14.74 \times 10^3 \\ 83.67 \times 10^6 \end{bmatrix} \text{ and } \mathbf{p}_{F.cr(2),1} = \begin{bmatrix} -70.8 \times 10^3 \\ -14.74 \times 10^3 \\ -83.67 \times 10^6 \\ 70.8 \times 10^3 \\ 14.74 \times 10^3 \\ 9.968 \times 10^6 \end{bmatrix}$$

After assembly, the sub-matrices of $\mathbf{K}_1$ and the load vector $\mathbf{P}_1$ of the whole structure at τ_1 are:

$$\mathbf{K}_{11,1} = \begin{bmatrix} 154.4 \times 10^9 & 119.3 \times 10^6 & -42.42 \times 10^6 & 57.66 \times 10^9 & 0 & 0 \\ & 2182 \times 10^3 & 0 & -238.5 \times 10^6 & -1091 \times 10^3 & 119.3 \times 10^6 \\ & & 119.3 \times 10^3 & 0 & 0 & 42.42 \times 10^6 \\ & & & 308.9 \times 10^9 & 119.3 \times 10^6 & 57.66 \times 10^9 \\ & sym & & & 1091 \times 10^3 & -119.3 \times 10^6 \\ & & & & & 154.4 \times 10^9 \end{bmatrix}$$

$$\mathbf{K}_{12,1} = \begin{bmatrix} -119.3\times10^6 & 42.42\times10^6 & 0 \\ -1091\times10^3 & 0 & 0 \\ 0 & -16.97\times10^3 & -16.97\times10^3 \\ 119.3\times10^6 & 42.42\times10^6 & -42.42\times10^6 \\ 0 & 0 & 0 \\ 0 & 0 & -42.42\times10^6 \end{bmatrix}$$

$$\mathbf{K}_{22,1} = \begin{bmatrix} 1091\times10^3 & 0 & 0 \\ & 16.97\times10^3 & 0 \\ sym & & 16.97\times10^3 \end{bmatrix}$$

$$\mathbf{K}_{21,1} = \mathbf{K}_{12,1}^T$$

$$\mathbf{P}_1 = [-128.4\times10^6 \quad 0 \quad 49.48\times10^3 \quad 0 \quad -1592\times10^3 \quad 128.4\times10^6 \quad (P_7+1562\times10^3) \quad (P_8-4.742\times10^3) \quad (P_7-4.742\times10^3)]^T$$

Solving for the unknown displacements and reactions using Eqs 9.9 gives:

$$\mathbf{U}_{U,1} = \mathbf{K}_{11,1}^{-1}(\mathbf{P}_{K,1} - \mathbf{K}_{12,1}\mathbf{U}_{K,1}) = \begin{bmatrix} 3.042\times10^{-3} \\ -1.792 \\ 9.063 \\ 0 \\ -3.583 \\ -3.042\times10^{-3} \end{bmatrix}$$

$$\mathbf{P}_{U,1} = \begin{bmatrix} P_{7,1} \\ P_{8,1} \\ P_{9,1} \end{bmatrix} = \begin{bmatrix} 30\times10^6 \\ -20\times10^6 \\ -20\times10^6 \end{bmatrix}$$

Post-processing

The solution is post-processed below to determine the variables defining the strain diagram and the internal actions along each element.

Element 1

The nodal displacements and the member actions associated with the end nodes of element 1 are:

$$\mathbf{u}_{(1),1} = [0 \quad 0 \quad 3.042\times10^{-3} \quad -1.792 \quad 9.063 \quad 0]^T$$

$$\mathbf{p}_{(1),1} = \mathbf{k}_1\mathbf{u}_{(1),1} + \mathbf{p}_{F.m(1),1} + \mathbf{p}_{F.cr(1),1} + \mathbf{p}_{F.sh(1),1}$$
$$= [30\times10^3 \quad -20\times10^3 \quad 0 \quad -30\times10^3 \quad 0 \quad -50\times10^6]^T$$

Using Eqs A.90:

$$\beta_{1(1),1} = 33.36 \times 10^{-15} \text{ mm}$$

$$\beta_{2(1),1} = 4.088 \times 10^{-12} \text{ mm}^2$$

$$\beta_{3(1),1} = -2.211 \times 10^{-9} \text{ mm}^{-1}$$

From Eqs A.79:

$$\overline{C}_{1(1),1} = -280 \times 10^{-12} \text{ mm}^{-2}$$

$$\overline{C}_{2(1),1} = -141.7 \times 10^{-9} \text{ mm}^{-1}$$

$$\overline{C}_{3(1),1} = 3.042 \times 10^{-3}$$

$$\overline{C}_{4(1),1} = 0 \text{ mm}$$

$$\overline{C}_{5(1),1} = -303.7 \times 10^{-6}$$

$$\overline{C}_{6(1),1} = 0 \text{ mm}$$

The axial strain and curvature are calculated at mid-span by substituting $z = 5000$ mm into Eqs A.68 giving $\varepsilon_{r(1),1} = u'_{(1),1} = -385.7 \times 10^{-6}$ and $\kappa_{(1),1} = -v''_{(1),1} = 0.841 \times 10^{-6}$ mm^{-1}, which are the same as the results of Example 5.5 at τ_1.

From Eqs A.82 and A.84:

$$a_{c0(1),1} = 101.7 \times 10^3 \text{ Nmm}^2$$

$$a_{c1(1),1} = -14.19 \text{ Nmm}^{-1}$$

$$a_{c2(1),1} = 1419 \times 10^{-6} \text{ Nmm}^{-2}$$

$$b_{c0(1),1} = 25.97 \times 10^6 \text{ Nmm}$$

$$b_{c1(1),1} = 10.39 \times 10^3 \text{ N}$$

$$b_{c2(1),1} = -1.039 \text{ Nmm}^{-1}$$

Element 2

$$\mathbf{u}_{(2),1} = \begin{bmatrix} -1.792 & 9.063 & 0 & -3.583 & 0 & -3.042 \times 10^{-3} \end{bmatrix}^T$$

$$\mathbf{p}_{(2),1} = \begin{bmatrix} 30 \times 10^3 & 0 & 50 \times 10^6 & -30 \times 10^3 & -20 \times 10^3 & 0 \end{bmatrix}^T$$

Time analysis – SSM ($\tau_2 = 30{,}000$ days)

Following the procedure outlined for the time analysis at τ_1, the stiffness matrix for elements 1 and 2 is based on the rigidities at τ_2:

$$\mathbf{k}_2 = \begin{bmatrix} 1162\times10^3 & 0 & -126.3\times10^6 & -1162\times10^3 & 0 & 126.3\times10^6 \\ & 17.97\times10^3 & 44.95\times10^6 & 0 & -17.98\times10^3 & 44.95\times10^6 \\ & & 163.5\times10^9 & 126.3\times10^6 & -44.95\times10^6 & 61.19\times10^9 \\ & & & 1162\times10^3 & 0 & -126.3\times10^6 \\ & sym & & & 19.97\times10^3 & -44.95\times10^6 \\ & & & & & 163.5\times10^9 \end{bmatrix}$$

The equivalent nodal loads to account for shrinkage and creep in each element are:

$$\mathbf{p}_{F.sh(1),2} = \mathbf{p}_{F.sh(2),2} = \begin{bmatrix} E_{c,2}A_c\varepsilon_{sh,2} \\ 0 \\ -E_{c,2}B_c\varepsilon_{sh,2} \\ -E_{c,2}A_c\varepsilon_{sh,2} \\ 0 \\ E_{c,2}B_c\varepsilon_{sh,2} \end{bmatrix} = \begin{bmatrix} -3196\times10^6 \\ 0 \\ 314.3\times10^6 \\ 3196\times10^6 \\ 0 \\ -314.3\times10^6 \end{bmatrix}$$

$$\mathbf{p}_{F.cr(1),2} = \begin{bmatrix} 134.2\times10^3 \\ 21.29\times10^3 \\ -68.29\times10^6 \\ -134.2\times10^3 \\ -21.29\times10^3 \\ 174.7\times10^6 \end{bmatrix}$$

$$\mathbf{p}_{F.cr(2),2} = \begin{bmatrix} 134.2\times10^3 \\ -21.29\times10^3 \\ -174.7\times10^6 \\ -134.2\times10^3 \\ 21.29\times10^3 \\ 68.3\times10^6 \end{bmatrix}$$

Assembling of the stiffness matrix $\mathbf{K}_2$ and load vector $\mathbf{P}_2$ for the whole structure:

$$\mathbf{K}_{11,2} = \begin{bmatrix} 162.5\times10^9 & 126.3\times10^6 & -44.95\times10^6 & 61.19\times10^9 & 0 & 0 \\ & 2324\times10^3 & 0 & -252.5\times10^6 & -1162\times10^3 & 126.3\times10^6 \\ & & 35.96\times10^3 & 0 & 0 & 44.95\times10^6 \\ & & & 327.1\times10^9 & 126.3\times10^6 & 61.19\times10^9 \\ & sym & & & 1162\times10^3 & -126.3\times10^6 \\ & & & & & 163.5\times10^9 \end{bmatrix}$$

$$\mathbf{K}_{12,2} = \begin{bmatrix} -126.3\times10^6 & 44.95\times10^6 & 0 \\ -1162\times10^3 & 0 & 0 \\ 0 & -17.98\times10^3 & -17.98\times10^3 \\ 126.3\times10^6 & 44.95\times10^6 & -44.95\times10^6 \\ 0 & 0 & 0 \\ 0 & 0 & -44.95\times10^6 \end{bmatrix}$$

$$\mathbf{K}_{22,2} = \begin{bmatrix} 1162\times10^3 & 0 & 0 \\ & 17.98\times10^3 & 0 \\ sym & & 17.98\times10^3 \end{bmatrix}$$

$$\mathbf{K}_{21,2} = \mathbf{K}_{12,2}^T$$

$$\mathbf{P}_2 = [-237.7\times10^6 \quad 0 \quad 62.57\times10^3 \quad 0 \quad -3092\times10^3 \quad 237.7\times10^6 \quad (P_7+3062\times10^3) \quad (P_8-11.29\times10^3) \quad (P_7-11.29\times10^3)]^T$$

Solving for the unknown displacements and reactions using Eqs 9.9 gives:

$$\mathbf{U}_{U,2} = \mathbf{K}_{11,2}^{-1}(\mathbf{P}_{K,2} - \mathbf{K}_{12,2}\mathbf{U}_{K,2}) = \begin{bmatrix} 4.71\times10^{-3} \\ -3.172 \\ 13.51 \\ 0 \\ -6.344 \\ -4.71\times10^{-3} \end{bmatrix} \quad \text{and}$$

$$\mathbf{P}_{U,2} = \begin{bmatrix} P_{7,2} \\ P_{8,2} \\ P_{9,2} \end{bmatrix} = \begin{bmatrix} 30\times10^6 \\ -20\times10^6 \\ -20\times10^6 \end{bmatrix}$$

Post-processing:

Element 1

The nodal displacements and the member actions associated with the end nodes of element 1 are:

$$\mathbf{u}_{(1),2} = [0 \quad 0 \quad 4.71\times10^{-3} \quad -3.172 \quad 13.51 \quad 0]^T$$

$$\mathbf{p}_{(1),2} = [30\times10^3 \quad -20\times10^3 \quad 0 \quad -30\times10^3 \quad 0 \quad -50\times10^6]^T$$

and with:

$$\beta_{1(1),2} = 45.46 \times 10^{-15} \text{ mm}$$

$$\beta_{2(1),2} = 6.247 \times 10^{-12} \text{ mm}^2$$

$$\beta_{3(1),2} = -6.542 \times 10^{-9} \text{ mm}^{-1}$$

$$\overline{C}_{1(1),2} = -334.1 \times 10^{-12} \text{ mm}^{-2}$$

$$\overline{C}_{2(1),2} = -385.1 \times 10^{-9} \text{ mm}^{-1}$$

$$\overline{C}_{3(1),2} = 4.71 \times 10^{-3}$$

$$\overline{C}_{4(1),2} = 0 \text{ mm}$$

$$\overline{C}_{5(1),2} = -563.0 \times 10^{-6}$$

$$\overline{C}_{6(1),2} = 0 \text{ mm}$$

The axial strain and curvature calculated at mid-span (i.e. at $z = 5000$ mm) at τ_2 are:

$$\varepsilon_{r(1),2} = u'_{(1),2} = -385.7 \times 10^{-6} \text{ and } \kappa_{(1),2} = -v''_{(1),2} = 0.841 \times 10^{-6} \text{ mm}^{-1}$$

Element 2

$$\mathbf{u}_{(2),2} = \begin{bmatrix} -3.172 & 13.51 & 0 & -6.344 & 0 & -4.71 \times 10^{-3} \end{bmatrix}^T$$

$$\mathbf{p}_{(2),2} = \begin{bmatrix} 30 \times 10^3 & 0 & 50 \times 10^6 & -30 \times 10^3 & -20 \times 10^3 & 0 \end{bmatrix}^T$$

9.6 Time analysis using the finite-element method

The finite-element method is a well-established method of analysis extensively used in all disciplines of engineering and its use is well reported in the literature (Ref. 2). Only the main aspects related to this method are outlined here, with particular attention placed on how time effects can be included in the analysis.

For illustrative purposes, a displacement-based line element similar to the one described for the stiffness method is considered in the following. Its detailed formulation is outlined in Appendix A. The proposed approach for the inclusion of time effects in finite-element analysis can be extended to any other type of finite element, i.e. displacement-based, force-based or mixed elements.

For ease of notation, subscripts defining particular time instants are included only when the instant considered is not clear from the context.

The weak form of the problem used to derive the proposed line finite element is obtained by means of the principle of virtual work as outlined in Section A.3. Restating Eq. A.18:

$$\int_L \mathbf{r}_i \cdot \mathbf{A}\,\hat{\mathbf{e}}\,dz = \int_L \mathbf{q} \cdot \hat{\mathbf{e}}\,dz \tag{9.19}$$

in which $\mathbf{q}$ contains the external distributed loads n and p (Fig. 9.3) and the vector $\mathbf{e}$ contains the independent variables defining the displacement field for the Euler–Bernoulli beam model consisting of the axial displacement measured at the level of the reference axis u and the deflection v. That is:

$$\mathbf{q} = \begin{bmatrix} n \\ p \end{bmatrix} \tag{9.20a}$$

$$\mathbf{e} = \begin{bmatrix} u \\ v \end{bmatrix} \tag{9.20b}$$

The variables defining the strain diagram, i.e. ε_r and κ, are obtained by differentiating $\mathbf{e}$ as follows (Eq. A.16):

$$\boldsymbol{\varepsilon} = \begin{bmatrix} \varepsilon_r \\ \kappa \end{bmatrix} = \begin{bmatrix} u' \\ -v'' \end{bmatrix} = \mathsf{A}\mathbf{e} \tag{9.21}$$

where the differential operator A is expressed (with $\partial \equiv \dfrac{d(\cdot)}{dz}$) by:

$$\mathsf{A} = \begin{bmatrix} \partial & 0 \\ 0 & -\partial^2 \end{bmatrix} \tag{9.22}$$

and the expression for the axial strain at a point in the member ε is:

$$\varepsilon = \varepsilon_r + y\kappa = u' - yv'' = \begin{bmatrix} 1 & y \end{bmatrix} \boldsymbol{\varepsilon} \tag{9.23}$$

The internal resultants collected in $\mathbf{r}_i$ (i.e. the internal axial force and moment) have been defined in Chapter 5 for different types of cross-sections and may be expressed as (Eqs A.11):

$$\mathbf{r}_i = \mathbf{D}\boldsymbol{\varepsilon} + \mathbf{f}_{cr} - \mathbf{f}_{sh} \tag{9.24}$$

where the terms related to creep and shrinkage vanish when performing an instantaneous analysis. The vectors and matrices in Eq. 9.24 are expanded in Eqs A.13 (and were used in Chapters 5 and 7). The vectors $\mathbf{f}_{cr}$ and $\mathbf{f}_{sh}$ depicting the creep and shrinkage response have been expressed in terms of their components f_{crN}, f_{crM}, f_{shN}, and f_{shM} to simplify the notation and, for the AEMM, are given by Eqs A.14 and, for the SSM, are given in Eqs A.15.

The proposed finite-element formulation relies on the approximation of the independent variables defining the kinematic response, i.e. u and v previously collected in $\mathbf{e}$ (Eq. 9.20b), as follows:

$$\mathbf{e} \approx \mathbf{N}_e \mathbf{d}_e \tag{9.25}$$

where $\mathbf{d}_e$ are the nodal displacements of the finite element under consideration and $\mathbf{N}_e$ contains the shape functions that describe the variations of these displacements along the element length.

Similarly to the stiffness relationship of Eq. 9.1, the governing system of equations describing the behaviour of a line element can be expressed in terms of nodal displacements $\mathbf{d}_e$ and actions $\mathbf{p}_e$:

$$\mathbf{p}_e = \mathbf{k}_e \mathbf{d}_e \tag{9.26}$$

where $\mathbf{k}_e$ is the finite element stiffness matrix (given in Eq. A.25a and reproduced here as Eq. 9.27) and $\mathbf{p}_e$ is the load vector accounting for the contributions of $\mathbf{p}_{e.m}$, $\mathbf{p}_{e.cr}$, and $\mathbf{p}_{e.sh}$ (given in Eqs A.25b and A.26):

$$\mathbf{k}_e = \int_L (\mathsf{AN_e})^T \mathbf{D}\, \boldsymbol{\varepsilon}\, dz = \int_L (\mathsf{AN_e})^T \mathbf{D}\, (\mathsf{AN_e})\, dz \tag{9.27}$$

The derivation for a particular finite element is carried out here for the 7 DOF element depicted in Fig. 9.6. In this case the axial and vertical displacements are approximated by parabolic and cubic polynomials, respectively. The nodal displacements and shape functions expressing this approximation are outlined in Eqs A.27–A.30 and the stiffness matrix and load vector related to member loads n and p are given in Eqs A.31 and A.32 (reproduced below):

$$\mathbf{k}_e = \begin{bmatrix} \frac{7R_A}{3L} & \frac{-4R_B}{L^2} & \frac{-3R_B}{L} & \frac{-8R_A}{3L} & \frac{R_A}{3L} & \frac{4R_B}{L^2} & \frac{-R_B}{L} \\ \frac{-4R_B}{L^2} & \frac{12R_I}{L^3} & \frac{6R_I}{L^2} & \frac{8R_B}{L^2} & \frac{-4R_B}{L^2} & \frac{-12R_I}{L^3} & \frac{6R_I}{L^2} \\ \frac{-3R_B}{L} & \frac{6R_I}{L^2} & \frac{4R_I}{L} & \frac{4R_B}{L} & \frac{-R_B}{L} & \frac{-6R_I}{L^2} & \frac{2R_I}{L} \\ \frac{-8R_A}{3L} & \frac{8R_B}{L^2} & \frac{4R_B}{L} & \frac{16R_A}{3L} & \frac{-8R_A}{3L} & \frac{-8R_B}{L^2} & \frac{4R_B}{L} \\ \frac{R_A}{3L} & \frac{-4R_B}{L^2} & \frac{-R_B}{L} & \frac{-8R_A}{3L} & \frac{7R_A}{3L} & \frac{4R_B}{L^2} & \frac{-3R_B}{L} \\ \frac{4R_B}{L^2} & \frac{-12R_I}{L^3} & \frac{-6R_I}{L^2} & \frac{-8R_B}{L^2} & \frac{4R_B}{L^2} & \frac{12R_I}{L^3} & \frac{-6R_I}{L^2} \\ \frac{-R_B}{L} & \frac{6R_I}{L^2} & \frac{2R_I}{L} & \frac{4R_B}{L} & \frac{-3R_B}{L} & \frac{-6R_I}{L^2} & \frac{4R_I}{L} \end{bmatrix} \tag{9.28}$$

$$\mathbf{p}_{e.m} = \left[\frac{L}{6}n \quad \frac{L}{2}p \quad \frac{L^2}{12}p \quad \frac{2L}{3}n \quad \frac{L}{6}n \quad \frac{L}{2}p \quad -\frac{L^2}{12}p \right]^T \tag{9.29}$$

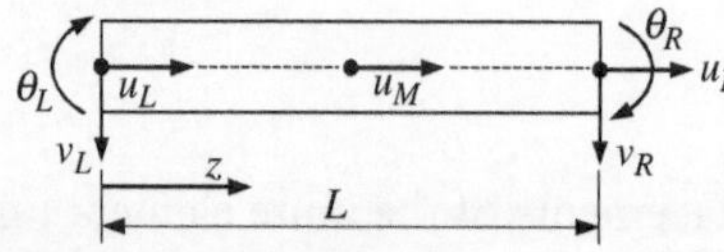

Figure 9.6 Nodal displacements of the 7 DOF finite element.

For illustrative purposes and to better outline all the steps involved in the solution process, the integrals in the expressions for $\mathbf{k}_e$ (Eq. 9.27) and $\mathbf{p}_{e.m}$ (Eq. A.26a) have been solved analytically throughout this chapter and in Appendix A. However, numerical integration could be easily implemented if preferred. Even if not carried out in the following, static condensation could be used to handle the internal freedom (Ref. 2).

The stiffness relationship of Eq. 9.26 is expressed in the local coordinates of the member (assuming the z-axis coincides with the member axis). This relationship must be transformed and expressed in global coordinates (i.e. coordinate system applicable to the whole structure) following the procedure previously outlined for the stiffness method in Eqs 9.3–9.6. The transformation matrix T for the 7 DOF finite element is:

$$\mathbf{T} = \begin{bmatrix} c & s & 0 & 0 & 0 & 0 & 0 \\ -s & c & 0 & 0 & 0 & 0 & 0 \\ 0 & 0 & 1 & 0 & 0 & 0 & 0 \\ 0 & 0 & 0 & 1 & 0 & 0 & 0 \\ 0 & 0 & 0 & 0 & c & s & 0 \\ 0 & 0 & 0 & 0 & -s & c & 0 \\ 0 & 0 & 0 & 0 & 0 & 0 & 1 \end{bmatrix} \tag{9.30}$$

The assembly procedure required to yield the stiffness relationship applicable to the whole structure is similar to the one utilised for the stiffness method. This produces relationships similar to those already presented in Eqs 9.7 and 9.8. The unknown displacements and reactions can be solved using Eqs 9.9.

When the analysis is completed, the solution is post-processed and, for each element, the different variables describing its structural response are determined. For example, the variables defining the strain diagram can be obtained by substituting the approximated displacements (Eq. 9.25) into Eq. 9.21 (Eqs A.34):

$$u' = -\frac{3}{L}u_L + \frac{4}{L}u_M - \frac{1}{L}u_R + \left(\frac{4}{L^2}u_L - \frac{8}{L^2}u_M + \frac{4}{L^2}u_R\right)z \tag{9.31a}$$

$$v'' = -\frac{6}{L^2}v_L - \frac{4}{L}\theta_L + \frac{6}{L^2}v_R - \frac{2}{L}\theta_R + \left(\frac{12}{L^3}v_L + \frac{6}{L^2}\theta_L - \frac{12}{L^3}v_R + \frac{6}{L^2}\theta_R\right)z \tag{9.31b}$$

where all nodal displacements are defined in Fig. 9.6.

Considering the post-processing of the instantaneous analysis, the stress resultants resisted by the concrete component can be expressed as (Eqs A.35a and d):

$$N_{c,0} = A_c E_{c,0} u_0' - B_c E_{c,0} v_0'' = a_{c0,0} + a_{c1,0} z \tag{9.32a}$$

$$M_{c,0} = B_c E_{c,0} u_0' - I_c E_{c,0} v_0'' = b_{c0,0} + b_{c1,0} z \tag{9.32b}$$

where the adopted subscript '0' has been introduced to highlight that these expressions are only applicable at time τ_0. The coefficients $a_{c0,0}$, $a_{c1,0}$, $b_{c0,0}$ and $b_{c1,0}$ are given in Eqs A.35.

As previously mentioned, for the calculation of the terms in $\mathbf{k}_e$ and $\mathbf{p}_{e.m}$ (Eqs 9.28 and 9.29) numerical integration could be used. For example, the axial force and moment resisted by the concrete at τ_0 could be calculated by numerical integration of the expressions presented in Eqs A.33. For clarity, an approach similar to the one already presented for the stiffness method has been adopted in which the variables are represented using polynomials that, in this case, are linear for both $N_{c,0}$ and $M_{c,0}$ (Eqs 9.32). Obviously, numerical integration could be easily implemented if preferred in the proposed solution process. The use of the linear polynomial has been determined by considering the degree of the polynomials adopted to approximate the displacements u and v in Eqs A.27–A.29. Based on these the variations of both u and v defining the strain diagram (Eq. 9.31) are linear functions of z and, as a consequence, so to are the expressions for $N_{c,0}$ and $M_{c,0}$ in Eqs 9.32.

9.6.1 Load vector to account for time effects using the AEMM

The calculation of the loading vector describing creep effects, i.e. $\mathbf{p}_{e.cr}$ introduced in Eq. A.26b, requires the knowledge of the axial force f_{crN} and moment f_{crM} included in $\mathbf{f}_{cr}$ which, when considering the AEMM, is defined in Eqs A.14a and c. Both f_{crN} and f_{crM} are found by multiplying the axial force and moment resisted by the concrete at time τ_0 by the material coefficient $\overline{F}_{e,0}$:

$$f_{crN} = \overline{F}_{e,0} N_{c,0} = \overline{F}_{e,0} \left(a_{c0,0} + a_{c1,0} z \right) \tag{9.33a}$$

$$f_{crM} = \overline{F}_{e,0} M_{c,0} = \overline{F}_{e,0} \left(b_{c0,0} + b_{c1,0} z \right) \tag{9.33b}$$

Equations 9.33 highlight that it is necessary to record the stress state of the concrete obtained from the instantaneous analysis before initiating the time analysis. When using numerical integration for the calculation of the stress resultants in the concrete, the values of the concrete axial force and moment at particular locations at time τ_0 along each member are stored. These locations are usually those required by the numerical integration of Eqs A.25 and A.27. Another option, used in the following (and in Appendix A), is to identify polynomials capable of describing the variations of the concrete axial force and moment along each element and to store the relevant coefficients. This has already been carried out in Eqs 9.32.

The loading vector describing shrinkage effects $\mathbf{p}_{e.sh}$ is expressed as Eq. A.26c. If it assumed that the shrinkage of a member does not vary along its length, the axial force f_{shN} and moment f_{shM} required to define $\mathbf{f}_{sh}$ in Eq. A.26c are given by:

$$f_{shN} = \int_{A_c} \overline{E}_{e,k} \varepsilon_{sh,k} dA = \overline{E}_{e,k} A_c \varepsilon_{sh,k} \tag{9.34a}$$

$$f_{shM} = \int_{A_c} y \overline{E}_{e,k} \varepsilon_{sh,k} dA = \overline{E}_{e,k} B_c \varepsilon_{sh,k} \tag{9.34b}$$

Based on the expressions for f_{crN} and f_{crM} (Eqs 9.33) and f_{shN} and f_{shM} (Eqs 9.34) and the definitions of $\mathbf{p}_{e.cr}$ and $\mathbf{p}_{e.sh}$ (Eqs A.26b and c), it is possible to determine the load vectors which account for creep and shrinkage. These are derived in Appendix A and given in Eqs A.36 and A.37.

9.6.2 *Load vector to account for time effects using the SSM*

Assuming the time-dependent behaviour of the concrete to be modelled by means of the SSM the expressions for f_{crN} and f_{crM} to be included in $\mathbf{f}_{cr}$ are (Eqs A.15):

$$f_{crN} = \sum_{i=0}^{k-1} \int_{A_c} F_{e,k,i}\sigma_{c,i}\, dA = \sum_{i=0}^{k-1} F_{e,k,i}N_{c,i} \tag{9.35a}$$

$$f_{crM} = \sum_{i=0}^{k-1} \int_{A_c} yF_{e,k,i}\sigma_{c,i}\, dA = \sum_{i=0}^{k-1} F_{e,k,i}M_{c,i} \tag{9.35b}$$

where the variation of the concrete stress resultants has also been described by a linear polynomial. For any time τ_i, the axial force and moment resisted by the concrete is calculated using the linear polynomials as given in Eqs A.41.

Substituting the polynomial describing the variations of $N_{c,i}$ and $M_{c,i}$ expressed in Eqs A.41 into Eqs 9.35 gives:

$$f_{crN,k} = \sum_{i=0}^{k-1} F_{e,k,i}N_{c,i} = \sum_{i=0}^{k-1} F_{e,k,i}\left(a_{c0,i} + a_{c1,i}z\right) \tag{9.36a}$$

$$f_{crM,k} = \sum_{i=0}^{k-1} F_{e,k,i}M_{c,i} = \sum_{i=0}^{k-1} F_{e,k,i}\left(b_{c0,i} + b_{c1,i}z\right) \tag{9.36b}$$

and the load vector accounting for creep is given as Eq. A.42.

Similar to the inclusion of shrinkage effects in the AEMM, shrinkage is here included using:

$$f_{shN} = \int_{A_c} E_{c,k}\varepsilon_{sh,k}\, dA = E_{c,k}A_c\varepsilon_{sh,k} \tag{9.37a}$$

$$f_{shM} = \int_{A_c} yE_{c,k}\varepsilon_{sh,k}\, dA = E_{c,k}B_c\varepsilon_{sh,k} \tag{9.37b}$$

9.6.3 *Remarks on the consistency requirements for finite elements*

The proposed 7 DOF finite element represents the simplest element that fulfils the consistency requirements approximating the displacements by means of polynomials and thereby avoids potential locking problems which may arise when the member local z-axis does not pass through the centroid of the member cross-section (Refs 5 and 7). The ability to select a reference system with the origin not necessarily coincident with the centroid of the section is fundamental when dealing with time effects as the location of the actual centroid of a reinforced cross-section usually varies with time due to creep of the concrete.

From a practical viewpoint, the consistency requirement is satisfied when the independent displacements (or their derivatives) present in the expressions of the strains of the model possess the same order (i.e. u' and v'' have the same order). Adopting a cubic function for the deflection v produces a linear contribution to the strain (i.e. due to v''). In order to produce the same (linear) contribution to the strain from u', it is necessary to have a parabolic function approximating the axial displacement u. For example, a linear function for u would not be able to achieve this as its first derivative (i.e. u') is constant. Other approaches could be used to address this problem, such as reduced integration.

To better illustrate this behaviour, the results obtained using the 7 DOF finite element (Fig. 9.6) are compared to those calculated using a 6 DOF element (Fig. 9.7). The 6 DOF element approximates u and v by means of linear and cubic polynomials, respectively. As previously discussed, this latter element does not satisfy the consistency requirements due to the orders of its polynomials.

The stiffness matrix of the 6 DOF finite element is:

$$\mathbf{k}_e = \begin{bmatrix} \frac{R_A}{L} & 0 & -\frac{R_B}{L} & -\frac{R_A}{L} & 0 & \frac{R_B}{L} \\ 0 & \frac{12R_I}{L^3} & \frac{6R_I}{L^2} & 0 & -\frac{12R_I}{L^3} & \frac{6R_I}{L^2} \\ -\frac{R_B}{L} & \frac{6R_I}{L^2} & \frac{4R_I}{L} & \frac{R_B}{L} & -\frac{6R_I}{L^2} & \frac{2R_I}{L} \\ -\frac{R_A}{L} & 0 & \frac{R_B}{L} & \frac{R_A}{L} & 0 & -\frac{R_B}{L} \\ 0 & -\frac{12R_I}{L^3} & -\frac{6R_I}{L^2} & 0 & \frac{12R_I}{L^3} & -\frac{6R_I}{L^2} \\ \frac{R_B}{L} & \frac{6R_I}{L^2} & \frac{2R_I}{L} & -\frac{R_B}{L} & -\frac{6R_I}{L^2} & \frac{4R_I}{L} \end{bmatrix} \quad (9.38)$$

For the case of an uncracked, simply supported, prismatic, concrete beam of rectangular section and subjected to a point load applied at mid-span, the instantaneous mid-span deflections calculated using the 6 and 7 DOF finite elements are shown in Fig. 9.8. These results have been obtained by considering two elements to clearly emphasise the implications of the different sets of polynomials. The instantaneous mid-span deflection has been plotted for different levels of the reference axis d_{ref} (measured from the top of the section) expressed as a function of the depth D. With this notation the reference axis is located at the level of the centroid when $d_{ref}/D = 0.5$, in which case both elements produce the same mid-span deflection (noted as v_{max} in Fig. 9.8).

Based on Fig. 9.8, it is apparent that when using the 6 DOF element with the origin of the reference system not coinciding with the cross-sectional centroid, a stiffer response than expected is obtained. A discussion on consistency requirements for different line elements is presented in Ref. 8.

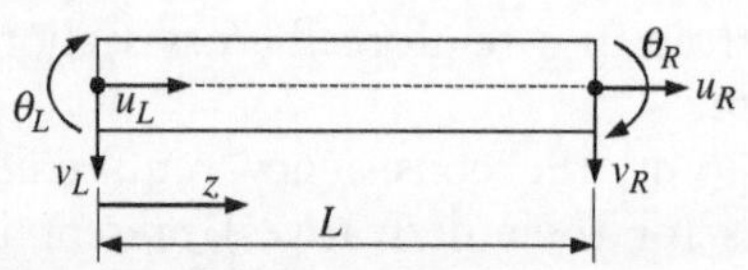

Figure 9.7 Nodal displacements of the 6 DOF finite element.

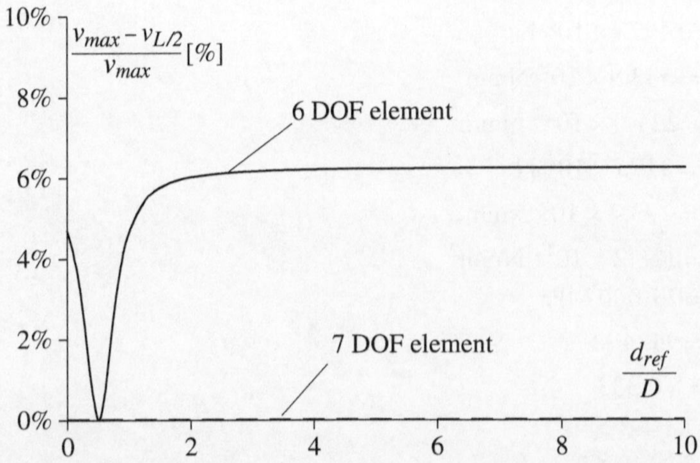

Figure 9.8 Comparisons between the results obtained using the 6 and 7 DOF finite elements.

Example 9.4

The short- and long-term values for the moment reaction at the fixed support and the vertical reactions at the roller support of the propped cantilever shown in Fig. 9.9 are to be determined. The member is subjected to a point load applied at mid-span as shown. The time-dependent behaviour of the concrete is to be modelled using the AEMM and the cross-sectional and material properties are the same as in Example 5.3. For ease of notation, the beam is modelled using two of the 7 DOF line elements depicted in Fig. 9.6.

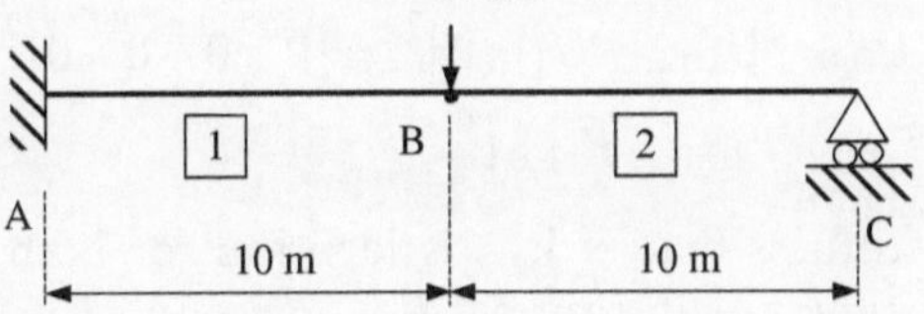

Figure 9.9 Loading, support conditions and discretisation for Example 9.4.

The numbering of the freedoms adopted in the proposed solution is outlined in Fig. 9.10. The numbering system has been selected to produce the partitioning introduced in Eqs 9.8 and also to avoid the need for transformations from local to global coordinates (as the local and global coordinate axes coincide for each element).

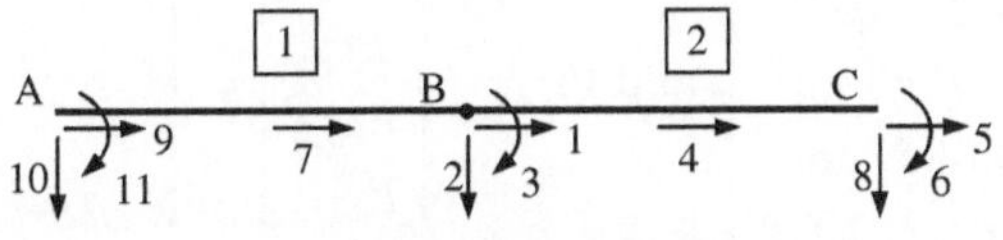

Figure 9.10 Freedom numbering for Example 9.4.

From Examples 5.1 and 5.3:

$$R_{A,0} = 4923 \times 10^6 \text{ N}$$
$$R_{B,0} = 543.9 \times 10^9 \text{ Nmm}$$
$$R_{I,0} = 221.0 \times 10^{12} \text{ Nmm}^2$$
$$R_{A,k} = 2175 \times 10^6 \text{ N}$$
$$R_{B,k} = 273.7 \times 10^9 \text{ Nmm}$$
$$R_{I,k} = 113.2 \times 10^{12} \text{ Nmm}^2$$
$$E_{c,0} = 25{,}000 \text{ MPa}$$
$$\overline{E}_{e,k} = 9524 \text{ MPa}$$
$$\overline{F}_{e,0} = -0.333$$
$$\varphi(\tau_k, \tau_0) = 2.5$$
$$\chi(\tau_k, \tau_0) = 0.65$$
$$\varepsilon_{sh}(\tau_k) = -600 \times 10^{-6}$$
$$A_c = 177.6 \times 10^3 \text{ mm}^2$$
$$B_c = 17.46 \times 10^6 \text{ mm}^3$$
$$I_c = 6966 \times 10^6 \text{ mm}^4$$

Instantaneous analysis (at time τ_0)

Based on Fig. 9.10, freedoms 1–7 are unrestrained, while no displacement is permitted for each of freedoms 8–11. Therefore:

$$\mathbf{U}_{U,0} = \begin{bmatrix} U_{1,0} & U_{2,0} & U_{3,0} & U_{4,0} & U_{5,0} & U_{6,0} & U_{7,0} \end{bmatrix}^T$$
$$\mathbf{U}_{K,0} = \begin{bmatrix} U_{8,0} & U_{9,0} & U_{10,0} & U_{11,0} \end{bmatrix}^T = \begin{bmatrix} 0 & 0 & 0 & 0 \end{bmatrix}^T$$
$$\mathbf{P}_{U,0} = \begin{bmatrix} P_{8,0} & P_{9,0} & P_{10,0} & P_{11,0} \end{bmatrix}^T$$

The finite-element stiffness matrix $\mathbf{k}_{e,0}$ is identical for both elements 1 and 2 and is calculated at time τ_0 substituting $R_{A,0}$, $R_{B,0}$ and $R_{I,0}$ into Eq. 9.28 and specifying an element length L equal to 10×10^3 mm.

The stiffness matrix of the whole structure is then assembled following similar steps to those carried out in previous examples. The unknown displacements and reactions are calculated using Eqs 9.9:

$$\mathbf{U}_{U,0} = \begin{bmatrix} 0.021 \text{ mm} \\ 4.531 \text{ mm} \\ 194.2 \times 10^{-6} \\ -0.059 \text{ mm} \\ -0.085 \text{ mm} \\ -776.7 \times 10^{-6} \\ 0.070 \text{ mm} \end{bmatrix} \quad \text{and } \mathbf{P}_{U,0} = \begin{bmatrix} P_{8,0} \\ P_{9,0} \\ P_{10,0} \\ P_{11,0} \end{bmatrix} = \begin{bmatrix} -3125 \text{ N} \\ 0 \text{ N} \\ -6875 \text{ N} \\ -37.5 \times 10^6 \text{ Nmm} \end{bmatrix}$$

The moment reaction at the fixed support (related to freedom 11) and the vertical reaction at the roller support (related to freedom 8) are equal to −37.5 kNm and −3125 N, respectively.

Post-processing

The solution is now post-processed to determine the coefficients describing the variation of the concrete stress resultants required subsequently to account for creep effects. The nodal displacements for elements 1 and 2 at time τ_0 are collected in vectors $\mathbf{u}_{(1),0}$ and $\mathbf{u}_{(2),0}$:

$$\mathbf{u}_{(1),0} = \begin{bmatrix} u_{L(1),0} \\ v_{L(1),0} \\ v'_{L(1),0} \\ u_{M(1),0} \\ u_{R(1),0} \\ v_{R(1),0} \\ v'_{R(1),0} \end{bmatrix} = \begin{bmatrix} 0 \text{ mm} \\ 0 \text{ mm} \\ 0 \\ 0.070 \text{ mm} \\ 0.021 \text{ mm} \\ 4.531 \text{ mm} \\ 194.2 \times 10^{-6} \end{bmatrix} \quad \text{and } \mathbf{u}_{(2),0} = \begin{bmatrix} u_{L(2),0} \\ v_{L(2),0} \\ v'_{L(2),0} \\ u_{M(2),0} \\ u_{R(2),0} \\ v_{R(2),0} \\ v'_{L(2),0} \end{bmatrix} = \begin{bmatrix} 0.021 \text{ mm} \\ 4.531 \text{ mm} \\ 194.2 \times 10^{-6} \\ -0.059 \text{ mm} \\ -0.085 \text{ mm} \\ 0 \text{ mm} \\ -776.7 \times 10^{-6} \end{bmatrix}$$

and from Eqs A.35:

$$a_{c0(1),0} = 12.56 \times 10^3 \text{ Nmm}^2$$

$$a_{c1(1),0} = -2.304 \text{ Nmm}^{-1}$$

$$b_{c0(1),0} = -29.34 \times 10^6 \text{ Nmm}$$

$$b_{c1(1),0} = 5.379 \times 10^3 \text{ N}$$

$$a_{c0(2),0} = -10.47 \times 10^3 \text{ Nmm}^2$$

$$a_{c1(2),0} = 1.047 \text{ Nmm}^{-1}$$

$$b_{c0(2),0} = 24.45 \times 10^6 \text{ Nmm}$$

$$b_{c1(2),0} = -2.45 \times 10^3 \text{ N}$$

Long-term analysis (at time τ_k)

The unknown and known displacements at time τ_k and the unknown reactions are:

$$\mathbf{U}_{U,k} = \begin{bmatrix} U_{1,k} & U_{2,k} & U_{3,k} & U_{4,k} & U_{5,k} & U_{6,k} & U_{7,k} \end{bmatrix}^T$$

$$\mathbf{U}_{K,k} = \begin{bmatrix} U_{8,k} & U_{9,k} & U_{10,k} & U_{11,k} \end{bmatrix}^T = \begin{bmatrix} 0 & 0 & 0 & 0 \end{bmatrix}^T$$

$$\mathbf{P}_{U,k} = \begin{bmatrix} P_{8,k} & P_{9,k} & P_{10,k} & P_{11,k} \end{bmatrix}^T$$

The finite-element stiffness matrix $\mathbf{k}_{e,k}$ for elements 1 and 2 is obtained from Eq. 9.28 using the cross-sectional properties $R_{A,k}$, $R_{B,k}$ and $R_{I,k}$.
The load vectors accounting for shrinkage effects for both elements 1 and 2 are calculated at time τ_k using Eq. A.37:

$$\mathbf{p}_{e.sh(1),k} = \mathbf{p}_{e.sh(2),k} = \begin{bmatrix} 1015 \times 10^3 \text{ N} \\ 0 \text{ N} \\ -99.79 \times 10^6 \text{ Nmm} \\ 0 \text{ N} \\ -1015 \times 10^3 \text{ N} \\ 0 \text{ N} \\ 99.79 \times 10^6 \text{ Nmm} \end{bmatrix}$$

The vectors describing the creep effects for elements 1 and 2 are calculated from Eq. A.36 using the coefficient $a_{cj(i),0}$ and $b_{cj(i),0}$ stored at the end of the instantaneous analysis:

$$\mathbf{p}_{e.cr(1),k} = \begin{bmatrix} 2909 \text{ N} \\ 1793 \text{ N} \\ 9779 \times 10^3 \text{ Nmm} \\ -5119 \text{ N} \\ 2211 \text{ N} \\ -1793 \text{ N} \\ 8150 \times 10^3 \text{ Nmm} \end{bmatrix} \text{ and } \mathbf{p}_{e.cr(2),k} = \begin{bmatrix} -2909 \text{ N} \\ -815.0 \text{ N} \\ -8150 \times 10^3 \text{ Nmm} \\ 2327 \text{ N} \\ 581.7 \text{ N} \\ 815.0 \text{ N} \\ 0 \text{ Nmm} \end{bmatrix}$$

The unknown displacements and reaction are then obtained using Eqs 9.9 following assembly of the global stiffness matrix and load vector (as was previously carried out at time τ_0):

$$\mathbf{U}_{U,k} = \begin{bmatrix} -4.544 \text{ mm} \\ 19.22 \text{ mm} \\ 948.3 \times 10^{-6} \\ -7.219 \text{ mm} \\ -9.814 \text{ mm} \\ -3.793 \times 10^{-3} \\ -2.051 \text{ mm} \end{bmatrix} \text{ and } \mathbf{P}_{U,k} = \begin{bmatrix} P_{8,k} \\ P_{9,k} \\ P_{10,k} \\ P_{11,k} \end{bmatrix} = \begin{bmatrix} -1033 \text{ N} \\ 0 \text{ N} \\ -8967 \text{ N} \\ -79.35 \times 10^6 \text{ Nmm} \end{bmatrix}$$

The moment reaction at the fixed support (related to freedom 11) and the vertical reaction at the roller support (related to freedom 8) are equal to −79.35 kNm and −1033 N, respectively.
For this lightly loaded member, the preponderance of reinforcement in the bottom fibres has caused the redistribution of internal actions illustrated in Fig. 8.5b(i).

9.7 Analysis of cracked members

A realistic analysis of concrete structures needs to account for the effects of cracking of the concrete. This is certainly important for calculations carried out at service conditions where significant errors in the evaluation of deformations and stresses will arise if the effects of cracking are not included.

In Chapter 7, cracking of the concrete is included in the cross-sectional analysis in one of two ways. In the first more approximate approach, the extent of cracking is determined from the instantaneous analysis assuming the concrete can carry no tension and, for the time analysis, the extent of cracking and the depth of the intact concrete on the cross-section are assumed to remain unchanged. The second approach to the time analysis of a cracked cross-section is a more refined method of analysis that traces the crack development with time by accounting for the change in the depth of the concrete compressive zone as time progresses. This latter approach is particularly suitable for the analysis and design of structural members where deflection or crack control at the serviceability limit states are the governing design considerations.

In the following, the application of these two approaches using the stiffness method and the finite-element approach is discussed. More advanced non-linear software for modelling cracking in concrete structures is available, but the two proposed methods provide a practical balance between the complexity adopted in the analysis and the variability of the concrete properties required for input into the analysis.

9.7.1 Approach 1

Cracking occurs when the stresses in the concrete produced by applied loads or deformations reach the tensile strength of the concrete. In the case of statically indeterminate members, the internal actions at a particular cross-section depend on the extent of cracking and this is usually unknown at the beginning of the analysis. A simple and practical method for including cracking in the analysis of a structure involves an iterative procedure that is illustrated here by considering the propped cantilever shown in Fig. 9.11.

The first iteration involves the analysis of the structure using either the stiffness or finite-element methods assuming the whole structure to be uncracked (Fig. 9.11b) following the procedures presented in the previous sections. Based on these results, the extent of cracking to be used in the second iteration is determined. In the propped cantilever of Fig. 9.11c, cracking occurs in that part of the beam adjacent to the fixed support where the applied moment is greater than the cracking moment M_{cr} (where M_{cr} was introduced in Chapter 3 and is the moment required to produce an extreme fibre tensile stress in the concrete equal to its tensile strength). The cross-sectional properties of the cracked region are next obtained following the procedure presented in Chapter 7 and the member is then re-analysed using the revised cross-sectional rigidities.

At the end of the second iteration, the lengths of the cracked and uncracked segments are adjusted based on the revised bending moment diagram, as shown in Fig. 9.11d, and the updated stiffnesses are adopted for the third iteration. The adjustments of the cracked and uncracked lengths continue until two subsequent iterations produce negligible differences.

The implementation of the convergence criteria can be carried out in different ways depending on the modelling technique adopted. For example, the beam shown in

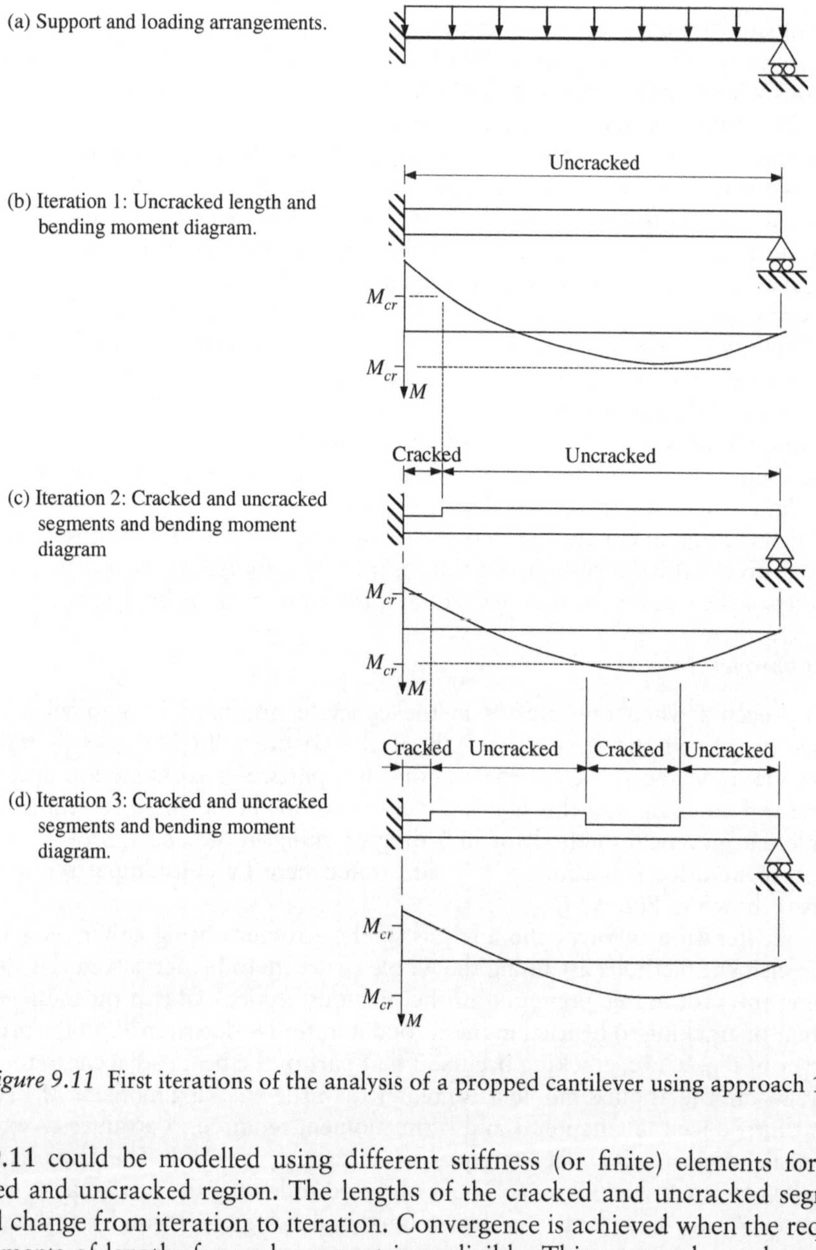

Figure 9.11 First iterations of the analysis of a propped cantilever using approach 1.

Fig. 9.11 could be modelled using different stiffness (or finite) elements for each cracked and uncracked region. The lengths of the cracked and uncracked segments would change from iteration to iteration. Convergence is achieved when the required adjustments of lengths for each segment is negligible. This approach can be applied only for simple members, such as the one shown in Fig. 9.11, as its use becomes prohibitive for large systems.

In the case of more complex structures, it is preferable to discretise the whole structure using a predefined element mesh and to assign to each element either cracked or uncracked properties based on the calculated internal actions. In this case, the first iteration is also carried out assuming all elements are uncracked. After the instantaneous solution at time τ_0 is completed, the time analysis is carried out assuming the extent of the cracked and uncracked regions does not change with time. For this purpose, the use of the AEMM to determine the time-dependent behaviour is recommended. In this way, only one additional analysis is required with no further iterations. Clearly, the method does not account for time-dependent cracking, as no attempt is made to discretise the time domain. This method is suitable for modelling structures when only linear-elastic analysis software is available.

9.7.2 *Approach 2*

The main advantage of this second approach relies on its ability to follow the occurrence and development of cracking throughout the structure. This is particularly useful for structural elements recognised to be critical at service conditions. In the more refined, second approach, the structure is discretised into a predetermined layout of elements, with the number and size of elements selected depending on the degree of accuracy required.

The extent of initial cracking in the structure is then determined by performing a non-linear analysis at time τ_0. This is usually based on well-known non-linear solution strategies such as the secant or Newthon–Raphson methods. Both can be implemented with the stiffness and finite element methods, even if the latter is naturally more suitable for these types of analysis. In fact, when using the finite element approach to analyse a concrete structure, the non-linear response of the concrete is accounted for in the numerical integration of the stiffness coefficients and of the load vectors. Considering the line element shown in Fig. 9.6, these approximations are applied to the integrals of Eqs A.25 and A.26 following procedures well established for the finite element method (Refs 2 and 4). In this process, the section is subdivided into smaller areas in order to better depict the non-linear response, such as the layered cross-sections already discussed in Section 7.4. With the stiffness method, the appropriate cross-sectional rigidities are calculated at one or more locations along the member length following the approach presented in Chapter 7.

When the instantaneous analysis is completed, the time analysis is carried out by discretising the time domain into a number of steps and modelling the concrete behaviour using the SSM. With this approach, the time-dependent behaviour of the concrete is carefully monitored at the integration points along each element.

The main non-linearity addressed in this analysis originates from the fact that the concrete is assumed not to be able to carry tensile stresses greater than its tensile strength. If at any time instant the concrete stress at any calculation point exceeds the tensile strength, the previously stored stress history at this point is set to zero, the contribution to element stiffness is adjusted and the structure is reanalysed. The process continues until no new cracking is identified at any point in the structure. Of course, if a cracked point is later subjected to compressive stresses during subsequent time steps, the point again becomes active and contributes to the element stiffness. Obviously, tension-stiffening could also be included in the analysis, even if its time-dependent behaviour is still the focus of current research (Ref. 6).

9.8 References

1. Hibbeler, R.C. (2009). *Structural Analysis*. Seventh edition, Pearson Prentice Hall.
2. Bathe, K.J. (2006). *Finite Element Procedures*. Prentice Hall, New Jersey.

3. Weaver, W. and Gere, J.M. (1990). *Matrix Analysis of Framed Structures*. Third edition, Van Nostrand Reinhold, New York.
4. CEB-FIB (2008). *Practitioners' Guide to Finite Element Modelling of Reinforced Concrete Structures*. International Federation for Structural Concrete, *fib* Bulletin 45, Lausanne, Switzerland.
5. Crisfield, M.A. (1991). The eccentricity issue in the design of plate and shell elements. *Communications in Applied Numerical Methods*, 7, 47–56.
6. Gilbert, R.I. and Wu, H.Q. (2009). Time-dependent stiffness of cracked reinforced concrete elements under sustained actions. *Australian Journal of Structural Engineering, Engineers Australia*, 9(2) 151–158.
7. Gupta, A.K. and Ma, P.S. (1977). Error in eccentric beam formulation. *International Journal for Numerical Methods in Engineering*, 11, 1473–1483.
8. Ranzi, G. and Zona, A. (2007). A composite beam model including the shear deformability of the steel component. *Engineering Structures*, 29, 3026–3041.

Appendix A Analytical formulations Euler–Bernoulli beam model

A.1 Introduction

The Euler–Bernoulli beam model is widely used for the analysis of structures. Designers use it on a daily basis in the form of either closed form solutions, such as those provided for the estimate of elastic deflections, or line (frame) elements for the prediction of deformations and internal actions using commercial structural analysis software. The analytical formulation at the basis of the Euler–Bernoulli beam model is described in the following sections for generic concrete members (such as those shown in Fig. A.1), where a perfect bond is assumed between the concrete and the reinforcement.

The weak form of the model (global balance condition) is presented to provide a basis for the finite element formulation briefly introduced in Chapter 9. The strong form (local balance condition) is then obtained by integrating this by parts. This has been used to derive the stiffness matrix and equivalent nodal actions required for the stiffness method described in Chapter 9.

The proposed formulations are derived for a typical instant in time and, because of this, could be applicable to both short- and long-term analyses. For ease of notation, additional subscripts are included in this appendix to identify the time step under

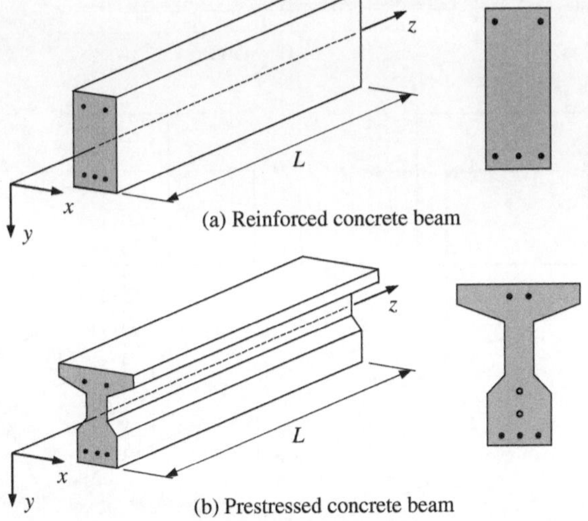

Figure A.1 Typical concrete members and cross-sections.

consideration only when it may not be clear from the context. A similar approach is used when defining the time dependency of variables. For example, the deflection along the member length that is a function of both member coordinate z and time t will be referred to as $v(z)$ (or simply v) instead of $v(t, z)$.

A.2 Kinematic model

In the undeformed state, the beam is assumed to be prismatic with the ortho-normal reference axis system $\{O; x, y, z\}$ as shown in Fig. A.1. The kinematic behaviour is illustrated in Fig. A.2b for a generic point P on the z-axis, highlighting both its axial displacement at the level of the reference axis $u(z)$ and its deflection $v(z)$. The formulation is derived for a beam segment of length L and the cross-section is assumed to be symmetric about the y-axis. Under these assumptions, no torsional and out-of-plane flexural effects are considered. The level of the reference x-axis is arbitrary.

Considering a point Q located away from the member axis, its final displacement can be expressed in terms of $u(z)$ and $v(z)$ as well as the rotation $\theta(z)$. In particular, the vertical and horizontal displacements, referred to as $d_y(y, z)$ and $d_z(y, z)$, can be expressed as:

$$d_y(y,z) = v(z) - y + y\cos\theta(z) \quad \text{(A.1a)}$$

$$d_z(y,z) = u(z) - y\sin\theta(z) \quad \text{(A.1b)}$$

and, for clarity, the kinematic response of the point Q is illustrated in Fig. A.2c.

The expressions of Eqs A.1 describe all possible displacements that the points of the beam can undergo. This set of displacements is usually referred to as the displacement field of the model. For structural engineering applications, it is usually sufficient and

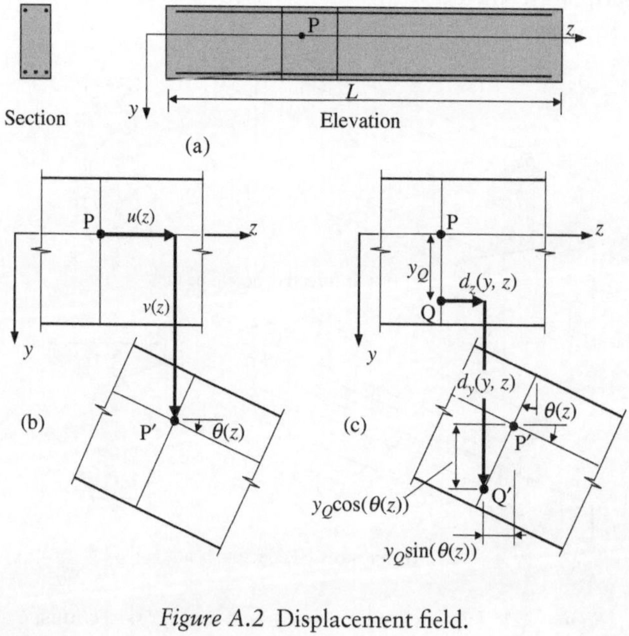

Figure A.2 Displacement field.

convenient to remain within the framework of small displacements. In this way, the cosine and sine of the angle $\theta(z)$ in Eqs A.1 can be approximated by: $\cos\theta(z) \approx 1$ and $\sin\theta(z) \approx \theta(z)$.

Consistent with the assumptions of the Euler-Bernoulli beam model, plane sections are assumed to remain plane and perpendicular to the beam axis before and after deformations. This condition implies that:

$$\theta(z) = v'(z) \tag{A.2}$$

where the prime represents differentiation with respect to z.

Based on these simplifications the displacement field of Eqs A.1 can be re-written as:

$$d_y(y,z) = v(z) \tag{A.3a}$$

$$d_z(y,z) = u(z) - y\,\theta(z) = u(z) - yv'(z) \tag{A.3b}$$

Equations A.3 show how the kinematic response of a point in the beam can be determined when the displacements $u(z)$ and $v(z)$ are known. For ease of notation, $u(z)$, $v(z)$ and $\theta(z)$ will be referred to as u, v and θ, respectively, in the following.

Based on the adopted displacement field, the only non zero strain is:

$$\varepsilon_z = u' - yv'' \tag{A.4}$$

where ε_z is the axial strain of the member. The expression for the curvature $\kappa(= -v'')$ can be obtained by differentiating the deflection v twice with respect to the coordinate z, and the negative sign in front of v'' is required in flexural members to produce a positive curvature in sagging moment regions (i.e. where compressive and tensile strains occur in the top and bottom fibres of the section in a horizontal beam, respectively, in accordance with the sign convention adopted in this book).

A.3 Weak formulation (global balance condition)

The weak form of the structural formulation for the frame element is obtained using the principle of virtual work considering a beam segment of length L with the free body diagram shown in Fig. A.3. In particular, $p(z)$ and $n(z)$ represent the vertical and horizontal distributed member loads which, for ease of notation, will be referred to as p and n in the following. The nodal actions at each end of the member represent external loads or reactions depending on the boundary conditions of the beam segment and have been referred to as S, N and M with the subscripts L and R specifying whether they relate to the left (at $z = 0$) or right (at $z = L$) ends of the member, respectively, as shown in Fig. A.3.

The derivation is carried out by equating the work of internal stresses to the work of external actions for each virtual admissible variation of the displacements and corresponding strains (which, by definition, represent all variations of

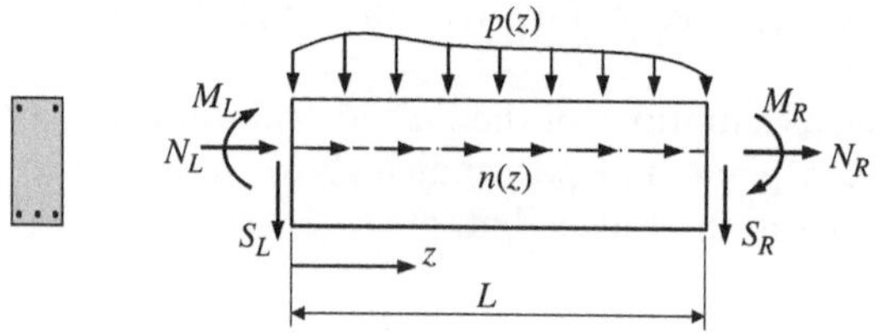

Figure A.3 Member loads and nodal actions.

possible displacements satisfying the kinematic boundary conditions of the problem) as follows:

$$\int_L \int_A \sigma_z \hat{\varepsilon}_z \, \mathrm{d}A \, \mathrm{d}z = \int_L (p\hat{v} + n\hat{u}) \, \mathrm{d}z + S_L\hat{v}_L + N_L\hat{u}_L + M_L\hat{\theta}_L + S_R\hat{v}_R + N_R\hat{u}_R + M_R\hat{\theta}_R \tag{A.5}$$

where the variables with the hat '^' represent virtual variations of displacements or strains. Substituting the expression for the axial strain ε_z of Eq. A.4 into Eq. A.5, the weak form of the derivation can be re-written as:

$$\int_L \int_A \sigma_z (\hat{u}' - y\hat{v}'') \, \mathrm{d}A \, \mathrm{d}z = \int_L (p\hat{v} + n\hat{u}) \, \mathrm{d}z + S_L\hat{v}_L + N_L\hat{u}_L + M_L\hat{\theta}_L + S_R\hat{v}_R + N_R\hat{u}_R + M_R\hat{\theta}_R \tag{A.6}$$

Recalling the definitions of internal axial force N and moment M (about the x-axis) introduced in Chapter 5 and reproduced here for ease of reference:

$$N_i = \int_A \sigma_z \, \mathrm{d}A \quad \text{and} \quad M_i = \int_A y\sigma_z \, \mathrm{d}A \tag{A.7a,b}$$

the integral at the cross-section (i.e. in $\mathrm{d}A$) present on the left-hand side of Eq. A.6 can be replaced by these internal actions as:

$$\int_L (N_i\hat{u}' - M_i\hat{v}'') \, \mathrm{d}z = \int_L (p\hat{v} + n\hat{u}) \, \mathrm{d}z + S_L\hat{v}_L + N_L\hat{u}_L + M_L\hat{\theta}_L + S_R\hat{v}_R + N_R\hat{u}_R + M_R\hat{\theta}_R \tag{A.8}$$

This relationship can be further rearranged to isolate the terms related to N_i and M_i as follows:

$$\int_L \begin{bmatrix} N_i \\ M_i \end{bmatrix} \cdot \begin{bmatrix} \hat{u}' \\ -\hat{v}'' \end{bmatrix} \mathrm{d}z = \int_L \begin{bmatrix} n \\ p \end{bmatrix} \cdot \begin{bmatrix} \hat{u} \\ \hat{v} \end{bmatrix} \mathrm{d}z + \begin{bmatrix} S_L \\ N_L \\ M_L \end{bmatrix} \cdot \begin{bmatrix} \hat{v}_L \\ \hat{u}_L \\ \hat{\theta}_L \end{bmatrix} + \begin{bmatrix} S_R \\ N_R \\ M_R \end{bmatrix} \cdot \begin{bmatrix} \hat{v}_R \\ \hat{u}_R \\ \hat{\theta}_R \end{bmatrix} \tag{A.9}$$

It is possible to ignore the nodal actions included on the right-hand side of Eq. A.9 as these can be easily included in the frame analysis during the assembly of the load vector. Based on this, Eq. A.9 can be simplified to:

$$\int_L \begin{bmatrix} N_i \\ M_i \end{bmatrix} \cdot \begin{bmatrix} \hat{u}' \\ -\hat{v}'' \end{bmatrix} dz = \int_L \begin{bmatrix} n \\ p \end{bmatrix} \cdot \begin{bmatrix} \hat{u} \\ \hat{v} \end{bmatrix} dz \tag{A.10}$$

In this form, the constitutive models for the materials are not explicitly specified as they are included in the definitions of the internal actions N_i and M_i. To remain consistent with the notation adopted in the rest of the book, the expressions already introduced in Chapter 5 (reproduced here for ease of reference) are adopted in the following to define the internal actions as a function of the material properties:

$$\mathbf{r}_i = \mathbf{D}\boldsymbol{\varepsilon} + \mathbf{f}_{cr} - \mathbf{f}_{sh} \tag{A.11}$$

which can be written highlighting the terms included in the matrix and vectors as:

$$\begin{bmatrix} N_i \\ M_i \end{bmatrix} = \begin{bmatrix} R_A & R_B \\ R_B & R_I \end{bmatrix} \begin{bmatrix} \varepsilon_r \\ \kappa \end{bmatrix} + \begin{bmatrix} f_{crN} \\ f_{crM} \end{bmatrix} - \begin{bmatrix} f_{shN} \\ f_{shM} \end{bmatrix} \tag{A.12}$$

and the cross-sectional rigidities R_A, R_B and R_I are defined in Chapter 5 for different types of cross-sections. In particular, $\mathbf{r}_i$ collects the internal axial force N_i and internal moment M_i, $\mathbf{D}$ specifies the geometric and material properties of the cross-section, $\boldsymbol{\varepsilon}$ includes the strain calculated at the level of the reference axis ε_r and the curvature κ which can be expressed in terms of the horizontal and vertical displacements (i.e. u and v) as shown below, while $\mathbf{f}_{cr}$ and $\mathbf{f}_{sh}$ account for creep and shrinkage effects, respectively:

$$\mathbf{r}_i = \begin{bmatrix} N_i \\ M_i \end{bmatrix} = \begin{bmatrix} \int_A \sigma_z \, dA \\ \int_A y\sigma_z \, dA \end{bmatrix}; \quad \mathbf{D} = \begin{bmatrix} R_A & R_B \\ R_B & R_I \end{bmatrix}; \quad \boldsymbol{\varepsilon} = \begin{bmatrix} \varepsilon_r \\ \kappa \end{bmatrix} = \begin{bmatrix} u' \\ -v'' \end{bmatrix} \tag{A.13a,b,c}$$

$$\mathbf{f}_{cr} = \begin{bmatrix} f_{crN} \\ f_{crM} \end{bmatrix} \quad \text{and} \quad \mathbf{f}_{sh} = \begin{bmatrix} f_{shN} \\ f_{shM} \end{bmatrix} \tag{A.13d,e}$$

This representation of $\mathbf{f}_{cr}$ and $\mathbf{f}_{sh}$ is particularly useful as it enables the proposed derivation to be applicable when using either the age-adjusted effective modulus method (AEMM) or the step-by-step method (SSM). When using the AEMM the components of $\mathbf{f}_{cr}$ and $\mathbf{f}_{sh}$ are determined as:

$$f_{crN} = \int_{A_c} \overline{F}_{e,0}\, \sigma_{c,0}\, dA = \overline{F}_{e,0}\, N_{c,0} \quad \text{and} \quad f_{shN} = \overline{E}_{e,k}\, A_c\, \varepsilon_{sh,k} \tag{A.14a,b}$$

$$f_{crM} = \int_{A_c} \overline{F}_{e,0}\, y\sigma_{c,0}\, dA = \overline{F}_{e,0}\, M_{c,0} \quad \text{and} \quad f_{shM} = \overline{E}_{e,k}\, B_c\, \varepsilon_{sh,k} \tag{A.14c,d}$$

where $\overline{E}_{e,k}$ and $\overline{F}_{e,0}$ are defined in Eqs 4.35 and 4.46. When using the SSM, these components at time τ_k become:

$$f_{crN} = \sum_{i=0}^{k-1} \int_{A_c} F_{e,k,i}\, \sigma_{c,i}\, dA = \sum_{i=0}^{k-1} F_{e,k,i}\, N_{c,i} \;\text{ and }\; f_{shN} = E_{c,k}\, A_c\, \varepsilon_{sh,k} \qquad \text{(A.15a,b)}$$

$$f_{crM} = \sum_{i=0}^{k-1} \int_{A_c} F_{e,k,i}\, y\sigma_{c,i}\, dA = \sum_{i=0}^{k-1} F_{e,k,i}\, M_{c,i} \;\text{ and }\; f_{shM} = E_{c,k}\, B_c\, \varepsilon_{sh,k} \qquad \text{(A.15c,d)}$$

where $\sigma_{c,i}$ is the stress in the concrete at time τ_i, while $N_{c,i}$ and $M_{c,i}$ represent the axial force and moment, respectively, resisted by the concrete component at time τ_i.

The vector $\boldsymbol{\varepsilon}$ can also be re-written in terms of the independent displacements u and v as:

$$\boldsymbol{\varepsilon} = \begin{bmatrix} \varepsilon_r \\ \kappa \end{bmatrix} = \begin{bmatrix} u' \\ -v'' \end{bmatrix} = \begin{bmatrix} \partial & 0 \\ 0 & -\partial^2 \end{bmatrix} \begin{bmatrix} u \\ v \end{bmatrix} = \mathsf{A}\,\mathbf{e} \qquad \text{(A.16)}$$

where A is a differential operator and the symbol ∂ defines the derivative with respect to the member coordinate z and the displacements u and v are collected in vector $\mathbf{e}$ as:

$$\mathbf{e} = \begin{bmatrix} u \\ v \end{bmatrix} \qquad \text{(A.17)}$$

At this point it is useful to re-write Eq. A.10 in terms of vectors $\mathbf{r}_i$ and $\boldsymbol{\varepsilon}$ $(= \mathsf{A}\,\mathbf{e})$:

$$\int_L \mathbf{r}_i \cdot \mathsf{A}\hat{\mathbf{e}}\, dz = \int_L \mathbf{q} \cdot \hat{\mathbf{e}}\, dz \qquad \text{(A.18)}$$

where the member loads n and p have been collected in the vector $\mathbf{q}$ as:

$$\mathbf{q} = \begin{bmatrix} n \\ p \end{bmatrix} \qquad \text{(A.19a,b)}$$

Substituting the constitutive properties defined in Eq. A.11 into Eq. A.18 produces the general expression for the weak form accounting for time effects:

$$\int_L (\mathbf{D}\boldsymbol{\varepsilon} + \mathbf{f}_{cr} - \mathbf{f}_{sh}) \cdot \mathsf{A}\hat{\mathbf{e}}\, dz = \int_L \mathbf{q} \cdot \hat{\mathbf{e}}\, dz \qquad \text{(A.20)}$$

A.4 Finite element formulation

The formulation proposed in the following is applicable to displacement-based finite elements. If necessary, the approach can be extended to other elements, i.e. force-based or mixed elements. The basis of the finite element formulation relies on the approximation of the independent displacements u and v, previously collected in the vector $\mathbf{e}$. This approximation can be expressed as:

$$\mathbf{e} \cong \mathbf{N_e d_e} \tag{A.21}$$

in which $\mathbf{N_e}$ is the matrix that specifies the adopted shape functions and $\mathbf{d_e}$ is the vector of the nodal displacements of the finite element. Substituting the approximation of Eq. A.21 into Eq. A.20 produces the weak form of the problem expressed in terms of nodal displacements:

$$\int_L (\mathbf{D\varepsilon} + \mathbf{f}_{cr} - \mathbf{f}_{sh}) \cdot \mathbf{A N_e \hat{d}_e}\, dz = \int_L \mathbf{q} \cdot \mathbf{N_e \hat{d}_e}\, dz \tag{A.22}$$

This relationship can be re-arranged to isolate the virtual nodal displacements $\mathbf{\hat{d}_e}$ on one side of the dot product:

$$\int_L (\mathbf{A N_e})^T (\mathbf{D\varepsilon} + \mathbf{f}_{cr} - \mathbf{f}_{sh})\, dz \cdot \mathbf{\hat{d}_e} = \int_L \mathbf{N}_e^T \mathbf{q}\, dz \cdot \mathbf{\hat{d}_e} \tag{A.23}$$

from which the stiffness relationship of the finite element can be obtained:

$$\mathbf{k}_e \mathbf{d_e} = \mathbf{p}_e \tag{A.24}$$

where $\mathbf{k}_e$ is the finite element stiffness matrix and $\mathbf{p}_e$ represents the loading vector related to both member loads and time effects. In particular, these are defined as:

$$\mathbf{k}_e = \int_L (\mathbf{A N_e})^T \mathbf{D}\, \boldsymbol{\varepsilon}\, dz = \int_L (\mathbf{A N_e})^T \mathbf{D}\, (\mathbf{A N_e})\, dz \tag{A.25a}$$

$$\mathbf{p}_e = \mathbf{p}_{e.m} - \mathbf{p}_{e.cr} + \mathbf{p}_{e.sh} \tag{A.25b}$$

where the loading vector $\mathbf{p}_e$ has been separated into $\mathbf{p}_{e.m}$, $\mathbf{p}_{e.cr}$, and $\mathbf{p}_{e.sh}$ relating to member loads (n and p), creep effects and shrinkage effects, respectively, and given by:

$$\mathbf{p}_{e.m} = \int_L \mathbf{N}_e^T \mathbf{q}\, dz; \quad \mathbf{p}_{e.cr} = \int_L (\mathbf{A N_e})^T \mathbf{f}_{cr}\, dz; \quad \text{and} \quad \mathbf{p}_{e.sh} = \int_L (\mathbf{A N_e})^T \mathbf{f}_{sh}\, dz \tag{A.26a–c}$$

Conventional finite element procedures are then utilized to assemble the vectors and matrices for the whole structure and to perform the structural analysis (Refs 1 and 2). The derivations are outlined in the following by means of worked examples.

A.4.1 Age-adjusted effective modulus method

Expressions for the stiffness matrix and loading vectors related to member loads, creep effects and shrinkage effects for the 7 DOF finite element shown in Fig. 9.6 are derived here. For this element, the axial (u) and transverse displacements (v) are approximated using parabolic and cubic functions, respectively. The time-dependent behaviour of the concrete is modelled using the AEMM specified in Eq. 4.45 (Approach 2 in Section 4.4.4).

The parabolic and cubic approximation for the axial and vertical displacements of the 7 dof element are:

$$u = N_{u1}u_L + N_{u2}u_M + N_{u3}u_R \quad \text{and} \quad v = N_{v1}v_L + N_{v2}\theta_L + N_{v3}v_R + N_{v4}\theta_R \tag{A.27a,b}$$

where:

$$N_{u1} = 1 - \frac{3z}{L} + \frac{2z^2}{L^2}; \quad N_{u2} = \frac{4z}{L} - \frac{4z^2}{L^2}; \quad N_{u3} = -\frac{z}{L} + \frac{2z^2}{L^2} \tag{A.28a–c}$$

$$N_{v1} = 1 - \frac{3z^2}{L^2} + \frac{2z^3}{L^3}; \quad N_{v2} = z - \frac{2z^2}{L} + \frac{z^3}{L^2}; \tag{A.29a,b}$$

$$N_{v3} = \frac{3z^2}{L^2} - \frac{2z^3}{L^3}; \quad N_{v4} = -\frac{z^2}{L} + \frac{z^3}{L^2} \tag{A.29c,d}$$

It is usually convenient to represent these approximations by means of the matrix of shape functions $\mathbf{N}_e$ and a vector $\mathbf{d}_e$ collecting the nodal displacements:

$$\begin{bmatrix} u \\ v \end{bmatrix} = \begin{bmatrix} N_{u1} & 0 & 0 & N_{u2} & N_{u3} & 0 & 0 \\ 0 & N_{v1} & N_{v2} & 0 & 0 & N_{v3} & N_{v4} \end{bmatrix} \begin{bmatrix} u_L \\ v_L \\ \theta_L \\ u_M \\ u_R \\ v_R \\ \theta_R \end{bmatrix} = \mathbf{N}_e\mathbf{d}_e \tag{A.30}$$

The stiffness matrix of the 7 DOF finite element can be calculated based on Eq. A.25a. This requires the calculation of $\mathsf{A}\,\mathbf{N}_e$ which, recalling the definition of the differential operator A specified in Eq. A.16, is obtained as:

$$\mathsf{A}\mathbf{N}_e = \begin{bmatrix} N'_{u1} & 0 & 0 & N'_{u2} & N'_{u3} & 0 & 0 \\ 0 & -N''_{v1} & -N''_{v2} & 0 & 0 & -N''_{v3} & -N''_{v4} \end{bmatrix}$$

$$= \begin{bmatrix} -\frac{3}{L} + \frac{4z}{L^2} & 0 & 0 & \frac{4}{L} - \frac{8z}{L^2} & -\frac{1}{L} + \frac{4z}{L^2} & 0 & 0 \\ 0 & \frac{6}{L^2} - \frac{12z}{L^3} & \frac{4}{L} - \frac{6z}{L^2} & 0 & 0 & -\frac{6}{L^2} + \frac{12z}{L^3} & \frac{2}{L} - \frac{6z}{L^2} \end{bmatrix}$$

Substituting the expression obtained for $\mathbf{A}\,\mathbf{N}_e$ into Eq. A.25a and carrying out the integration along the member length produces the element stiffness matrix:

$$\mathbf{k}_e = \begin{bmatrix} \frac{7R_A}{3L} & \frac{-4R_B}{L^2} & \frac{-3R_B}{L} & \frac{-8R_A}{3L} & \frac{R_A}{3L} & \frac{4R_B}{L^2} & \frac{-R_B}{L} \\ \frac{-4R_B}{L^2} & \frac{12R_I}{L^3} & \frac{6R_I}{L^2} & \frac{8R_B}{L^2} & \frac{-4R_B}{L^2} & \frac{-12R_I}{L^3} & \frac{6R_I}{L^2} \\ \frac{-3R_B}{L} & \frac{6R_I}{L^2} & \frac{4R_I}{L} & \frac{4R_B}{L} & \frac{-R_B}{L} & \frac{-6R_I}{L^2} & \frac{2R_I}{L} \\ \frac{-8R_A}{3L} & \frac{8R_B}{L^2} & \frac{4R_B}{L} & \frac{16R_A}{3L} & \frac{-8R_A}{3L} & \frac{-8R_B}{L^2} & \frac{4R_B}{L} \\ \frac{R_A}{3L} & \frac{-4R_B}{L^2} & \frac{-R_B}{L} & \frac{-8R_A}{3L} & \frac{7R_A}{3L} & \frac{4R_B}{L^2} & \frac{-3R_B}{L} \\ \frac{4R_B}{L^2} & \frac{-12R_I}{L^3} & \frac{-6R_I}{L^2} & \frac{-8R_B}{L^2} & \frac{4R_B}{L^2} & \frac{12R_I}{L^3} & \frac{-6R_I}{L^2} \\ \frac{-R_B}{L} & \frac{6R_I}{L^2} & \frac{2R_I}{L} & \frac{4R_B}{L} & \frac{-3R_B}{L} & \frac{-6R_I}{L^2} & \frac{4R_I}{L} \end{bmatrix} \qquad (A.31)$$

Recalling the expression previously obtained for $\mathbf{A}\,\mathbf{N}_e$, the loading vectors are then calculated based on Eqs A.26. In particular, the one related to the member loads n and p can be derived as:

$$\mathbf{p}_{e.m} = \int_L \mathbf{N}_e^T \mathbf{q}\, dz = \int_L \begin{bmatrix} N_{u1} & 0 \\ 0 & N_{v1} \\ 0 & N_{v2} \\ N_{u2} & 0 \\ N_{u3} & 0 \\ 0 & N_{v3} \\ 0 & N_{v4} \end{bmatrix} \begin{bmatrix} n \\ p \end{bmatrix} dz = \begin{bmatrix} \frac{L}{6}n \\ \frac{L}{2}p \\ \frac{L^2}{12}p \\ \frac{2L}{3}n \\ \frac{L}{6}n \\ \frac{L}{2}p \\ -\frac{L^2}{12}p \end{bmatrix} \qquad (A.32)$$

As highlighted in Eq. A.26b, the calculation of the loading vector describing creep effects $\mathbf{p}_{e.cr}$ requires the knowledge of the axial force and moment for the calculation of f_{crN} and f_{crM} included in $\mathbf{f}_{cr}$. These depend to the loading history of the concrete as both f_{crN} and f_{crM} are determined by multiplying the axial force and moment resisted by the concrete at one or more instants in time by particular material coefficients. The definition of these coefficients varies depending on the constitutive relationship for concrete and, for the AEMM, it is given in Eqs A.14. The analysis at a particular instant in time can be carried out only if the instantaneous analysis has been completed and the concrete stress resultants recorded. The actual recording of these actions can be carried out by storing the values of the concrete axial force and moment at particular locations along each member. These locations are usually those required by the numerical integration of Eq. A.26b. Another option is to identify polynomials capable of describing the variation of the concrete axial force and moment along each element and to store the relevant coefficients. This option is preferred here as it enables closed form solutions to be obtained.

The orders of the polynomial describing f_{crN} and f_{crM} (Eqs A.14) depend on the orders of the expressions for the axial force and moment resisted by the concrete at time τ_0, $N_{c.0}$ and $M_{c.0}$. Recalling that:

$$N_{c,0} = \int_{A_c} \sigma_{c,0}\, dA = \int_{A_c} E_{c,0}\left(u_0' - yv_0''\right) dA = A_c E_{c,0} u_0' - B_c E_{c,0} v_0'' \tag{A.33a}$$

$$M_{c,0} = \int_{A_c} y\sigma_{c,0}\, dA = \int_{A_c} yE_{c,0}\left(u_0' - yv_0''\right) dA = B_c E_{c,0} u_0' - I_c E_{c,0} v_0'' \tag{A.33b}$$

and considering that, in the 7 DOF finite element (Fig. 9.6), u and v are approximated by a parabolic and a cubic polynomial, respectively, the actual functions describing $N_{c,0}$ and $M_{c,0}$ determined using Eqs A.33 become linear and can be expressed as:

$$N_{c.0} = a_{c0,0} + a_{c1,0} z \quad \text{and} \quad M_{c.0} = b_{c0,0} + b_{c1,0} z$$

where the coefficients $a_{c0,0}$, $a_{c1,0}$, $b_{c0,0}$, and $b_{c1,0}$ are determined from Eqs A.33. Considering Eqs A.27–A.30, the expressions for u_0' and v_0'' can be written in terms of the nodal displacements of the finite element as:

$$\begin{aligned}
u_0' &= N_{u1}' u_{L,0} + N_{u2}' u_{M,0} + N_{u3}' u_{R,0}\\
&= \left(-\frac{3}{L} + \frac{4z}{L^2}\right) u_{L,0} + \left(\frac{4}{L} - \frac{8z}{L^2}\right) u_{M,0} + \left(-\frac{1}{L} + \frac{4z}{L^2}\right) u_{R,0}\\
&= -\frac{3}{L} u_{L,0} + \frac{4}{L} u_{M,0} - \frac{1}{L} u_{R,0} + \left(\frac{4}{L^2} u_{L,0} - \frac{8}{L^2} u_{M,0} + \frac{4}{L^2} u_{R,0}\right) z
\end{aligned} \tag{A.34a}$$

$$\begin{aligned}
v_0'' &= N_{v1}'' v_{L,0} + N_{v2}'' \theta_{L,0} + N_{v3}'' v_{R,0} + N_{v4}'' \theta_{R,0}\\
&= \left(-\frac{6}{L^2} + \frac{12z}{L^3}\right) v_{L,0} + \left(-\frac{4}{L} + \frac{6z}{L^2}\right) \theta_{L,0} + \left(\frac{6}{L^2} - \frac{12z}{L^3}\right) v_{R,0} + \left(-\frac{2}{L} + \frac{6z}{L^2}\right) \theta_{R,0}\\
&= -\frac{6}{L^2} v_{L,0} - \frac{4}{L} \theta_{L,0} + \frac{6}{L^2} v_{R,0} - \frac{2}{L} \theta_{R,0} + \left(\frac{12}{L^3} v_{L,0} + \frac{6}{L^2} \theta_{L,0} - \frac{12}{L^3} v_{R,0} + \frac{6}{L^2} \theta_{R,0}\right) z
\end{aligned} \tag{A.34b}$$

The '0' subscript in the nodal freedoms in Eqs A34 is to specify that these are calculated at time τ_0.

The expression for the concrete axial force $N_{c,0}$ can then be obtained by substituting Eqs A.34 into Eq. A.33a:

$$\begin{aligned}
N_{c,0} &= A_c E_{c,0} u_0' - B_c E_{c,0} v_0'' \\
&= A_c E_{c,0}\left[-\frac{3}{L}u_{L,0} + \frac{4}{L}u_{M,0} - \frac{1}{L}u_{R,0} + \left(\frac{4}{L^2}u_{L,0} - \frac{8}{L^2}u_{M,0} + \frac{4}{L^2}u_{R,0}\right)z\right] \\
&- B_c E_{c,0}\left[-\frac{6}{L^2}v_{L,0} - \frac{4}{L}\theta_{L,0} + \frac{6}{L^2}v_{R,0} - \frac{2}{L}\theta_{R,0}\right. \\
&\left. + \left(\frac{12}{L^3}v_{L,0} + \frac{6}{L^2}\theta_{L,0} - \frac{12}{L^3}v_{R,0} + \frac{6}{L^2}\theta_{R,0}\right)z\right] \\
&= a_{c0,0} + a_{c1,0} z
\end{aligned} \tag{A.35a}$$

where:

$$\begin{aligned}
a_{c0,0} &= A_c E_{c,0}\left(-\frac{3}{L}u_{L,0} + \frac{4}{L}u_{M,0} - \frac{1}{L}u_{R,0}\right) \\
&- B_c E_{c,0}\left(-\frac{6}{L^2}v_{L,0} - \frac{4}{L}\theta_{L,0} + \frac{6}{L^2}v_{R,0} - \frac{2}{L}\theta_{R,0}\right)
\end{aligned} \tag{A.35b}$$

$$\begin{aligned}
a_{c1,0} &= A_c E_{c,0}\left(\frac{4}{L^2}u_{L,0} - \frac{8}{L^2}u_{M,0} + \frac{4}{L^2}u_{R,0}\right) \\
&- B_c E_{c,0}\left(\frac{12}{L^3}v_{L,0} + \frac{6}{L^2}\theta_{L,0} - \frac{12}{L^3}v_{R,0} + \frac{6}{L^2}\theta_{R,0}\right)
\end{aligned} \tag{A.35c}$$

Similarly, substituting Eqs A.34 into Eq. A.33b leads to the expression for $M_{c,0}$:

$$\begin{aligned}
M_{c,0} &= B_c E_{c,0} u_0' - I_c E_{c,0} v_0'' \\
&= B_c E_{c,0}\left[-\frac{3}{L}u_{L,0} + \frac{4}{L}u_{M,0} - \frac{1}{L}u_{R,0} + \left(\frac{4}{L^2}u_{L,0} - \frac{8}{L^2}u_{M,0} + \frac{4}{L^2}u_{R,0}\right)z\right] \\
&- I_c E_{c,0}\left[-\frac{6}{L^2}v_{L,0} - \frac{4}{L}\theta_{L,0} + \frac{6}{L^2}v_{R,0} - \frac{2}{L}\theta_{R,0}\right. \\
&\left. + \left(\frac{12}{L^3}v_{L,0} + \frac{6}{L^2}\theta_{L,0} - \frac{12}{L^3}v_{R,0} + \frac{6}{L^2}\theta_{R,0}\right)z\right] \\
&= b_{c0,0} + b_{c1,0} z
\end{aligned} \tag{A.35d}$$

where:

$$b_{c0,0} = B_c E_{c,0}\left(-\frac{3}{L}u_{L,0} + \frac{4}{L}u_{M,0} - \frac{1}{L}u_{R,0}\right)$$
$$- I_c E_{c,0}\left(-\frac{6}{L^2}v_{L,0} - \frac{4}{L}\theta_{L,0} + \frac{6}{L^2}v_{R,0} - \frac{2}{L}\theta_{R,0}\right) \quad \text{(A.35e)}$$

$$b_{c1,0} = B_c E_{c,0}\left(\frac{4}{L^2}u_{L,0} - \frac{8}{L^2}u_{M,0} + \frac{4}{L^2}u_{R,0}\right)$$
$$- I_c E_{c,0}\left(\frac{12}{L^3}v_{L,0} + \frac{6}{L^2}\theta_{L,0} - \frac{12}{L^3}v_{R,0} + \frac{6}{L^2}\theta_{R,0}\right) \quad \text{(A.35f)}$$

At this point, it is possible to derive the expressions for f_{crN} and f_{crM} substituting Eqs A.35 into Eqs A.14a and c:

$$f_{crN} = \overline{F}_{e,0}N_{c,0} = \overline{F}_{e,0}\left(a_{c0,0} + a_{c1,0}z\right)$$
$$f_{crM} = \overline{F}_{e,0}M_{c,0} = \overline{F}_{e,0}\left(b_{c0,0} + b_{c1,0}z\right)$$

and $\mathbf{p}_{e.cr}$ can then be obtained as follows:

$$\mathbf{p}_{e.cr} = \int_L (\mathbf{AN}_e)^T \mathbf{f}_{cr}\, dz$$

$$= \int_L \begin{bmatrix} -\frac{3}{L} + \frac{4z}{L^2} & 0 \\ 0 & \frac{6}{L^2} - \frac{12z}{L^3} \\ 0 & \frac{4}{L} - \frac{6z}{L^2} \\ \frac{4}{L} - \frac{8z}{L^2} & 0 \\ -\frac{1}{L} + \frac{4z}{L^2} & 0 \\ 0 & -\frac{6}{L^2} + \frac{12z}{L^3} \\ 0 & \frac{2}{L} - \frac{6z}{L^2} \end{bmatrix} \begin{bmatrix} \overline{F}_{e,0}\left(a_{c0,0} + a_{c1,0}z\right) \\ \overline{F}_{e,0}\left(b_{c0,0} + b_{c1,0}z\right) \end{bmatrix} dz = \overline{F}_{e,0} \begin{bmatrix} -a_{c0,0} - \frac{L}{6}a_{c1,0} \\ -b_{c1,0} \\ b_{c0,0} \\ -\frac{2}{3}La_{c1,0} \\ \frac{5}{6}La_{c1,0} + a_{c0,0} \\ b_{c1,0} \\ -b_{c1,0}L - b_{c0,0} \end{bmatrix} \quad \text{(A.36)}$$

The loading vector describing shrinkage effects $\mathbf{p}_{e.sh}$ can be determined based on Eq. A.26c. If shrinkage does not vary along a member, f_{shN} and f_{shM} required to define $\mathbf{f}_{sh}$ in Eq. A.26c are constant and for the AEMM are given by:

$$f_{shN} = \int_{A_c} \overline{E}_{e,k}\, \varepsilon_{sh,k}\, dA = \overline{E}_{e,k} A_c \varepsilon_{sh,k} \quad \text{and} \quad f_{shM} = \int_{A_c} y\overline{E}_{e,k}\, \varepsilon_{sh,k}\, dA = \overline{E}_{e,k}\, B_c\, \varepsilon_{sh,k}$$

The assumption of uniform shrinkage is usually acceptable for most practical applications.

The shrinkage loading vector $\mathbf{p}_{e.sh}$ can then be calculated as:

$$\mathbf{p}_{e.sh} = \int_L (\mathbf{A}\mathbf{N}_e)^T \mathbf{f}_{sh}\, dz$$

$$\times \int_L \begin{bmatrix} -\frac{3}{L}+\frac{4z}{L^2} & 0 \\ 0 & \frac{6}{L^2}-\frac{12z}{L^3} \\ 0 & \frac{4}{L}-\frac{6z}{L^2} \\ \frac{4}{L}-\frac{8z}{L^2} & 0 \\ -\frac{1}{L}+\frac{4z}{L^2} & 0 \\ 0 & -\frac{6}{L^2}+\frac{12z}{L^3} \\ 0 & \frac{2}{L}-\frac{6z}{L^2} \end{bmatrix} \begin{bmatrix} \overline{E}_{e,k} A_c \varepsilon_{sh,k} \\ \overline{E}_{e,k} B_c \varepsilon_{sh,k} \end{bmatrix} dz = \begin{bmatrix} -\overline{E}_{e,k} A_c \varepsilon_{sh,k} \\ 0 \\ \frac{\overline{E}_{e,k} B_c \varepsilon_{sh,k}}{L} \\ 0 \\ \overline{E}_{e,k} A_c \varepsilon_{sh,k} \\ 0 \\ -\frac{\overline{E}_{e,k} B_c \varepsilon_{sh,k}}{L} \end{bmatrix} \qquad (A.37)$$

A.4.2 Step-by-step method

The 7 DOF finite element of the previous section is reconsidered here with the time-dependent behaviour of the concrete described using the SSM.

The change in the representation of the concrete time-dependent behaviour only affects the calculation of the loading vectors accounting for creep and shrinkage effects. In fact, the stiffness matrix and the vectors depicting member loads and shrinkage are always defined using the cross-sectional properties relevant to the time step under consideration. In particular, with the SSM the concrete contributions are determined using $E_{c,k}$ introduced in Eqs 4.25 and 4.26.

Similarly to the AEMM, the loading vector describing creep effects $\mathbf{p}_{e.cr}$ can be determined once the values of f_{crN} and f_{crM} included in $\mathbf{f}_{cr}$ are determined. The expressions of these terms calculated using the SSM are defined in Eqs A.15a and c.

Also in this case, polynomials are utilized to represent the variation of the concrete axial force and moment along the element length measured at different time steps. For this purpose, the coefficients included in these polynomials need to be recorded at each instant in time. Another possible approach would rely on the use of a numerical integration for the calculation of $f_{crN,k}$ and $f_{crM,k}$. In this case, the concrete axial force and moment obtained at particular locations along each element (and required for the numerical integration) need to be stored.

As the orders of the functions representing the concrete axial force and moment do not vary with time, the polynomials already identified for the AEMM in the previous example to describe the concrete instantaneous response are now adopted for each time instant. The concrete axial force and moment at time τ_i may be expressed as:

$$N_{c.i} = a_{c0,i} + a_{c1,i} z \quad \text{and} \quad M_{c.i} = b_{c0,i} + b_{c1,i} z \qquad (A.38a,b)$$

and the coefficients $a_{c0,i}$, $a_{c1,i}$, $b_{c0,i}$, and $b_{c1,i}$ are determined as outlined in the following.

Considering the constitutive relationship for concrete introduced in Eq. 4.25 (reproduced here for ease of reference):

$$\sigma_{c,i} = E_{c,i}\left(\varepsilon_i - \varepsilon_{sh}\right) + \sum_{n=0}^{k-1} F_{e,k,n}\sigma_{c,n}$$

the definitions of both axial force and moment resisted by the concrete component at any time τ_i can be expressed as:

$$N_{c,i} = \int_{A_c} \sigma_{c,i}\, \mathrm{d}A = \int_{A_c} \left[E_{c,i}\left(u_i' - yv_i''\right) - E_{c,i}\varepsilon_{sh,i} + \sum_{n=0}^{i-1} F_{e,i,n}\sigma_{c,n} \right] \mathrm{d}A$$

$$= A_c E_{c,i} u_i' - B_c E_{c,i} v_i'' - A_c E_{c,i}\varepsilon_{sh,i} + \sum_{n=0}^{i-1} F_{e,i,n} N_{c,n} \tag{A.39a}$$

$$M_{c,i} = \int_{A_c} y\sigma_{c,i}\, \mathrm{d}A = \int_{A_c} y \left[E_{c,i}\left(u_i' - yv_i''\right) - E_{c,i}\varepsilon_{sh,i} + \sum_{n=0}^{i-1} F_{e,i,n}\sigma_{c,n} \right] \mathrm{d}A$$

$$= B_c E_{c,i} u_i' - I_c E_{c,i} v_i'' - B_c E_{c,i}\varepsilon_{sh,i} + \sum_{n=0}^{i-1} F_{e,i,n} M_{c,n} \tag{A.39b}$$

Considering Eqs A.27–A.30, the expressions for u_i' and v_i'' required in Eqs A.39 are determined in terms of nodal displacements as:

$$u_i' = -\frac{3}{L}u_{L,i} + \frac{4}{L}u_{M,i} - \frac{1}{L}u_{R,i} + \left(\frac{4}{L^2}u_{L,i} - \frac{8}{L^2}u_{M,i} + \frac{4}{L^2}u_{R,i}\right)z \tag{A.40a}$$

$$v_i'' = -\frac{6}{L^2}v_{L,i} - \frac{4}{L}\theta_{L,i} + \frac{6}{L^2}v_{R,i} - \frac{2}{L}\theta_{R,i} + \left(\frac{12}{L^3}v_{L,i} + \frac{6}{L^2}\theta_{L,i} - \frac{12}{L^3}v_{R,i} + \frac{6}{L^2}\theta_{R,i}\right)z \tag{A.40b}$$

Substituting Eqs A.40 into Eq. A.39a produces the expression for $N_{c,i}$ in terms of nodal displacements, shrinkage and previous loading history:

$$\begin{aligned} N_{c,i} &= A_c E_{c,i} u_i' - B_c E_{c,i} v_i'' - A_c E_{c,i}\varepsilon_{sh,i} + \sum_{n=0}^{i-1} F_{e,i,n} N_{c,n} \\ &= A_c E_{c,i}\left[-\frac{3}{L}u_{L,i} + \frac{4}{L}u_{M,i} - \frac{1}{L}u_{R,i} + \left(\frac{4}{L^2}u_{L,i} - \frac{8}{L^2}u_{M,i} + \frac{4}{L^2}u_{R,i}\right)z\right] \\ &\quad - B_c E_{c,i}\left[-\frac{6}{L^2}v_{L,i} - \frac{4}{L}\theta_{L,i} + \frac{6}{L^2}v_{R,i} - \frac{2}{L}\theta_{R,i}\right. \\ &\quad \left. + \left(\frac{12}{L^3}v_{L,i} + \frac{6}{L^2}\theta_{L,i} - \frac{12}{L^3}v_{R,i} + \frac{6}{L^2}\theta_{R,i}\right)z\right] \\ &\quad - A_c E_{c,i}\varepsilon_{sh,i} + \sum_{n=0}^{i-1} F_{e,i,n}\left(a_{c0,n} + a_{c1,n}\, z\right) \\ &= a_{c0,i} + a_{c1,i}\, z \end{aligned} \tag{A.41a}$$

where:

$$a_{c0,i} = A_c E_{c,i}\left(-\frac{3}{L}u_{L,i} + \frac{4}{L}u_{M,i} - \frac{1}{L}u_{R,i}\right)$$
$$- B_c E_{c,i}\left(-\frac{6}{L^2}v_{L,i} - \frac{4}{L}\theta_{L,i} + \frac{6}{L^2}v_{R,i} - \frac{2}{L}\theta_{R,i}\right)$$
$$- A_c E_{c,i}\,\varepsilon_{sh,i} + \sum_{n=0}^{i-1} F_{e,i,n}\,a_{c0,n}$$

$$a_{c1,i} = A_c E_{c,i}\left(\frac{4}{L^2}u_{L,i} - \frac{8}{L^2}u_{M,i} + \frac{4}{L^2}u_{R,i}\right)$$
$$- B_c E_{c,i}\left(\frac{12}{L^3}v_{L,i} + \frac{6}{L^2}\theta_{L,i} - \frac{12}{L^3}v_{R,i} + \frac{6}{L^2}\theta_{R,i}\right) + \sum_{n=0}^{i-1} F_{e,i,n}\,a_{c1,n}$$

Similarly, for $M_{c,i}$:

$$M_{c,i} = B_c E_{c,i}\,u_i' - I_c E_{c,i}\,v_i'' - B_c E_{c,i}\,\varepsilon_{sh,i} + \sum_{n=0}^{i-1} F_{e,i,n}\,M_{c,n} = b_{c0,i} + b_{c1,i}\,z \quad \text{(A.41b)}$$

where:

$$b_{c0,i} = B_c E_{c,i}\left(-\frac{3}{L}u_{L,i} + \frac{4}{L}u_{M,i} - \frac{1}{L}u_{R,i}\right)$$
$$- I_c E_{c,i}\left(-\frac{6}{L^2}v_{L,i} - \frac{4}{L}\theta_{L,i} + \frac{6}{L^2}v_{R,i} - \frac{2}{L}\theta_{R,i}\right) - B_c E_{c,i}\varepsilon_{sh,i} + \sum_{n=0}^{i-1} F_{e,i,n}\,b_{c0,n}$$

$$b_{c1,i} = B_c E_{c,i}\left(\frac{4}{L^2}u_{L,i} - \frac{8}{L^2}u_{M,i} + \frac{4}{L^2}u_{R,i}\right)$$
$$- I_c E_{c,i}\left(\frac{12}{L^3}v_{L,i} + \frac{6}{L^2}\theta_{L,i} - \frac{12}{L^3}v_{R,i} + \frac{6}{L^2}\theta_{R,i}\right) + \sum_{n=0}^{i-1} F_{e,i,n}\,b_{c1,n}$$

The expressions for $f_{crN,k}$ and $f_{crM,k}$ can then be re-written as:

$$f_{crN,k} = \sum_{i=0}^{k-1} F_{e,k,i} N_{c,i} = \sum_{i=0}^{k-1} F_{e,k,i}\left(a_{c0,i} + a_{c1,i} z\right)$$

$$f_{crM,k} = \sum_{i=0}^{k-1} F_{e,k,i} M_{c,i} = \sum_{i=0}^{k-1} F_{e,k,i}\left(b_{c0,i} + b_{c1,i} z\right)$$

and the loading vector related to creep effects $\mathbf{p}_{e.cr}$ is obtained from:

$$\mathbf{p}_{e.cr} = \int_L (\mathbf{A}\mathbf{N}_e)^T \mathbf{f}_{cr}\, dz$$

$$= \int_L \begin{bmatrix} -\frac{3}{L}+\frac{4z}{L^2} & 0 \\ 0 & \frac{6}{L^2}-\frac{12z}{L^3} \\ 0 & \frac{4}{L}-\frac{6z}{L^2} \\ \frac{4}{L}-\frac{8z}{L^2} & 0 \\ -\frac{1}{L}+\frac{4z}{L^2} & 0 \\ 0 & -\frac{6}{L^2}+\frac{12z}{L^3} \\ 0 & \frac{2}{L}-\frac{6z}{L^2} \end{bmatrix} \begin{bmatrix} \sum_{i=0}^{k-1} F_{e,k,i}(a_{c0,i}+a_{c1,i}z) \\ \sum_{i=0}^{k-1} F_{e,k,i}(b_{c0,i}+b_{c1,i}z) \end{bmatrix} dz = \sum_{i=0}^{k-1} F_{e,k,i} \begin{bmatrix} -a_{c0,i}-\frac{L}{6}a_{c1,i} \\ -b_{c1,i} \\ b_{c0,i} \\ -\frac{2}{3}La_{c1,i} \\ \frac{5}{6}La_{c1,i}+a_{c0,i} \\ b_{c1,i} \\ -b_{c1,i}L-b_{c0,i} \end{bmatrix} \quad \text{(A.42)}$$

The loading vector related to shrinkage effects $\mathbf{p}_{e.sh}$ is obtained from Eq. A.37, except that $E_{c,k}$ replaces $\overline{E}_{e,k}$.

A.5 Strong formulation (local balance condition)

The strong form of the problem, consisting of the governing system of differential equations and corresponding boundary conditions, is obtained by integrating by parts the weak form presented in Eq. A.8, as follows:

$$[\mathrm{N}_i\hat{u}]_0^L - \int_L \mathrm{N}_i'\hat{u}\, dz - [\mathrm{M}_i\hat{v}']_0^L + [\mathrm{M}_i'\hat{v}]_0^L - \int_L \mathrm{M}_i''\hat{v}\, dz$$

$$= \int_L (p\hat{v}+n\hat{u})\, dz + S_L\hat{v}_L + N_L\hat{u}_L + M_L\hat{\theta}_L + S_R\hat{v}_R + N_R\hat{u}_R + M_R\hat{\theta}_R \quad \text{(A.43)}$$

Enforcing horizontal and vertical equilibrium, the system of differential equations obtained from Eq. A.43 can then be expressed as:

$$\mathrm{N}_i' + n = 0 \quad \text{and} \quad \mathrm{M}_i'' + p = 0 \quad \text{(A.44a,b)}$$

From Eq. A.43, the corresponding boundary conditions to be applied at the left and right ends of the beam segment (i.e. at $z=0$ and L) are:

$$(\mathrm{N}_{iL}+N_L)\,\hat{u}_L = 0; \quad (\mathrm{N}_{iR}-N_R)\,\hat{u}_R = 0 \quad \text{(A.45a,b)}$$

$$(\mathrm{M}_{iL}'+S_L)\,\hat{v}_L = 0; \quad (\mathrm{M}_{iR}'-S_R)\,\hat{v}_R = 0 \quad \text{(A.45c,d)}$$

$$(\mathrm{M}_{iL}-M_L)\,\hat{\theta}_L = 0; \quad (\mathrm{M}_{iR}+M_R)\,\hat{\theta}_R = 0 \quad \text{(A.45e,f)}$$

The differential equations and their boundary conditions can also be defined in terms of the independent displacements u and v by recalling the expressions for N_i and M_i of Eq. A.12:

$$N_i = R_A u' - R_B v'' + f_{crN} - f_{shN} \quad \text{(A.46a)}$$

$$M_i = R_B u' - R_I v'' + f_{crM} - f_{shM} \quad \text{(A.46b)}$$

where the rigidities R_A, R_B and R_I are defined in Chapter 5 for different types of cross-sections.

Substituting Eqs A.46 into Eqs A.44 and A.45 the differential equations can be re-written as:

$$R_A u'' - R_B v''' + f'_{crN} - f'_{shN} + n = 0 \quad \text{(A.47a)}$$

$$R_B u''' - R_I v'''' + f''_{crM} - f''_{shM} + p = 0 \quad \text{(A.47b)}$$

and the boundary conditions at the left (L) and right (R) ends of the member become:

$$\left(R_A u'_L - R_B v''_L + f_{crNL} - f_{shNL} + N_L\right) \hat{u}_L = 0 \quad \text{(A.48a)}$$

$$\left(R_A u'_R - R_B v''_R + f_{crNR} - f_{shNR} - N_R\right) \hat{u}_R = 0 \quad \text{(A.48b)}$$

$$\left(R_B u''_L - R_I v'''_L + f'_{crML} - f'_{shML} + S_L\right) \hat{v}_L = 0 \quad \text{(A.48c)}$$

$$\left(R_B u''_R - R_I v'''_R + f'_{crMR} - f'_{shMR} - S_R\right) \hat{v}_R = 0 \quad \text{(A.48d)}$$

$$\left(R_B u'_L - R_I v''_L + f_{crML} - f_{shML} - M_L\right) \hat{\theta}_L = 0 \quad \text{(A.48e)}$$

$$\left(R_B u'_R - R_I v''_R + f_{crMR} - f_{shMR} + M_R\right) \hat{\theta}_R = 0 \quad \text{(A.48f)}$$

The system of differential equations (Eqs A.47) is solved in the following, initially for the instantaneous material properties (i.e. no time effects) and then for the time analysis using the AEMM and the SSM.

A.5.1 Instantaneous analysis

Ignoring creep and shrinkage effects (i.e. $f_{crN} = 0, f_{crM} = 0, f_{shN} = 0$ and $f_{shM} = 0$), the general expressions for u and v which satisfy Eqs A.47 can be written as:

$$v = \alpha_1 p \frac{z^4}{24} + C_1 \frac{z^3}{6} + C_2 \frac{z^2}{2} + C_3 z + C_4 \quad \text{(A.49a)}$$

$$u = \alpha_2 p \frac{z^3}{6} + (\alpha_3 C_1 + \alpha_4 n) \frac{z^2}{2} + C_5 z + C_6 \quad \text{(A.49b)}$$

while the following coefficients are been introduced to simplify the notation:

$$\alpha_0 = R_1 R_A - R_B^2; \quad \alpha_1 = \frac{R_A}{\alpha_0}; \quad \alpha_2 = \frac{R_B}{\alpha_0}; \quad \alpha_3 = \frac{R_B}{R_A}; \quad \text{and} \quad \alpha_4 = \frac{-1}{R_A} \quad \text{(A.50a–e)}$$

The constants of integration C_j (with $j = 1, \ldots, 6$) can be determined based on the boundary conditions (Eqs A.48). In the following these analytical solutions have been used to derive the stiffness matrix and loading vector to be used in the stiffness method outlined in Chapter 9.

A.5.1.1 Stiffness matrix

Consider the unloaded fixed ended beam shown in Fig. A.4a. The beam is modelled using a single 6 DOF stiffness element similar to that shown in Fig. 9.1 (i.e. three freedoms at each end node). The freedom numbering for the beam is shown in Fig. A.4b. The direct stiffness method is used here to derive the coefficients of the stiffness matrix.

The direct stiffness method is a procedure which enables the determination of the stiffness coefficients by enforcing a unit displacement along one freedom while all other freedoms are restrained. The actual stiffness coefficients are then equal to the reactions of the fixed-ended beam required to keep the displaced configuration. For example, enforcing a unit displacement along freedom 3 (i.e. rotation at the left support in Fig. A.4b), the reactions of the fixed-ended beam correspond to the coefficients included in the third column of the stiffness matrix. This approach is useful only when there is an analytical or numerical solution available to calculate the sought end reactions shown in Fig. A.4c. In the following, this will be carried out using the analytical solutions presented in Eqs A.49 combined with the boundary conditions of Eqs A.48.

In order to calculate the coefficients of the first column of the stiffness matrix, a unit displacement is enforced along freedom 1 (Fig. A.4b), which corresponds to the axial displacement of the left node. For clarity, this deformation is shown in Fig. A.5.

Considering that both member ends are fully restrained by the fixed support conditions, the constants of integration included in the general solutions for u and v presented in Eqs A.49 can be determined by enforcing the following kinematic conditions:

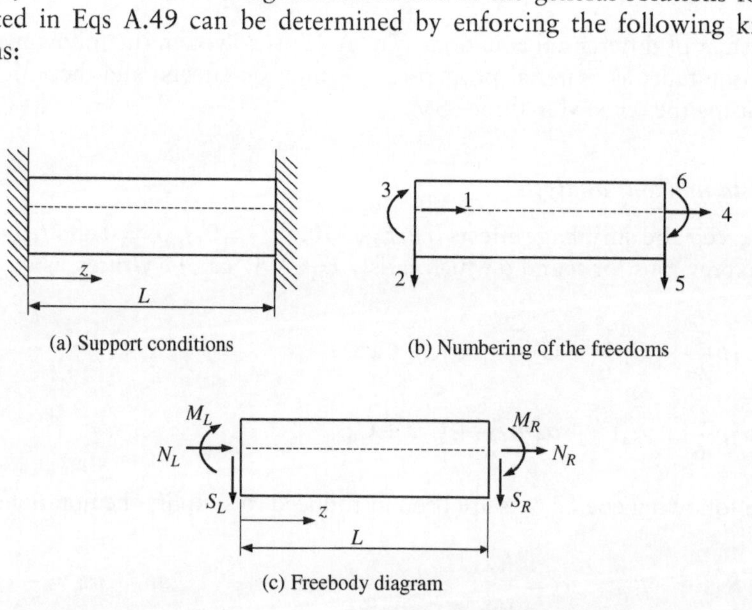

Figure A.4 Unloaded, fixed-ended beam.

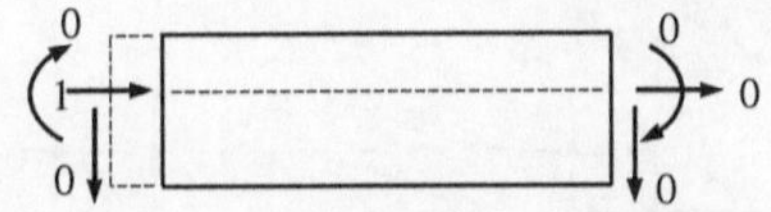

Figure A.5 A unit displacement enforced along freedom 1.

$u(z=0)=1;\quad u(z=L)=0;\quad v(z=0)=0;\quad v(z=L)=0;\quad \theta(z=0)=0;$
and $\theta(z=L)=0.$

The calculated constants of integration can then be written as:

$$C_1=0;\quad C_2=0;\quad C_3=0;\quad C_4=0;\quad C_5=-\frac{1}{L};\text{ and } C_6=1.$$

Adopting the static boundary conditions of Eqs A.48, the reactions of the fixed-ended beam can be determined by solving the following equations for the unknown nodal actions (which represent the desired reactions):

$$R_A u'_L - R_B v''_L + N_L = 0;\quad R_A u'_R - R_B v''_R - N_R = 0$$
$$R_B u''_L - R_I v'''_L + S_L = 0;\quad R_B u''_R - R_I v'''_R - S_R = 0$$
$$R_B u'_L - R_I v''_L - M_L = 0;\quad R_B u'_R - R_I v''_R + M_R = 0$$

where the expressions related to creep and shrinkage have been omitted as not relevant to the calculations of the stiffness coefficients.

The reactions required to maintain this deformed configuration (Fig. A.5) can be collected in the following vector:

$$\mathbf{p}_1 = \begin{bmatrix} N_L \\ S_L \\ M_L \\ N_R \\ S_R \\ M_R \end{bmatrix} = \begin{bmatrix} R_A/L \\ 0 \\ R_B/L \\ -R_A/L \\ 0 \\ -R_B/L \end{bmatrix}$$

which defines the first column of the stiffness matrix (identified by the subscript '1').

Considering a unit displacement along freedom 2 (Fig. A.6), the boundary conditions necessary to determine the constants of integration are:

$u(z=0)=0;\quad u(z=L)=0;\quad v(z=0)=1;\quad v(z=L)=0;\quad \theta(z=0)=0;$

and $\theta(z=L)=0.$

and these produce the following expressions:

$$C_1=\frac{12}{L^3};\quad C_2=-\frac{6}{L^2};\quad C_3=0;\quad C_4=1;\quad C_5=-\frac{6}{L^2}\frac{R_B}{R_A};\text{ and } C_6=0.$$

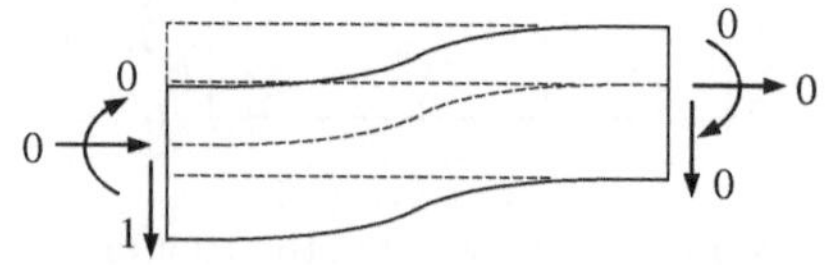

Figure A.6 A unit displacement enforced along freedom 2.

Based on these constants of integration, the analytical solution for this displaced configuration is described by a cubic function for the deflection v and a parabolic one for the axial displacement u, therefore providing consistent contributions to the strain field from both v and u.

The reactions, which form the second column of the stiffness matrix, are determined using the static boundary conditions of Eqs A.48 as:

$$\mathbf{p}_2 = \begin{bmatrix} N_L \\ S_L \\ M_L \\ N_R \\ S_R \\ M_R \end{bmatrix} = \begin{bmatrix} 0 \\ \dfrac{12}{L^3}\dfrac{R_I R_A - R_B^2}{R_A} \\ -\dfrac{6}{L^2}\dfrac{R_I R_A - R_B^2}{R_A} \\ 0 \\ -\dfrac{12}{L^3}\dfrac{R_I R_A - R_B^2}{R_A} \\ -\dfrac{6}{L^2}\dfrac{R_I R_A - R_B^2}{R_A} \end{bmatrix} = \begin{bmatrix} 0 \\ 12/\alpha_1 L^3 \\ -6/\alpha_1 L^2 \\ 0 \\ -12/\alpha_1 L^3 \\ -6/\alpha_1 L^2 \end{bmatrix}$$

Considering the remaining freedoms one at the time, the coefficients of the whole stiffness matrix are determined:

$$\mathbf{k}_e = \begin{bmatrix} \dfrac{R_A}{L} & 0 & -\dfrac{R_B}{L} & -\dfrac{R_A}{L} & 0 & \dfrac{R_B}{L} \\ 0 & \dfrac{12}{\alpha_1 L^3} & \dfrac{6}{\alpha_1 L^2} & 0 & -\dfrac{12}{\alpha_1 L^3} & \dfrac{6}{\alpha_1 L^2} \\ -\dfrac{R_B}{L} & \dfrac{6}{\alpha_1 L^2} & \dfrac{4}{\alpha_1 L} + \dfrac{R_B^2}{R_A L} & \dfrac{R_B}{L} & -\dfrac{6}{\alpha_1 L^2} & \dfrac{2}{\alpha_1 L} - \dfrac{R_B^2}{R_A L} \\ -\dfrac{R_A}{L} & 0 & \dfrac{R_B}{L} & \dfrac{R_A}{L} & 0 & -\dfrac{R_B}{L} \\ 0 & -\dfrac{12}{\alpha_1 L^3} & -\dfrac{6}{\alpha_1 L^2} & 0 & \dfrac{12}{\alpha_1 L^3} & -\dfrac{6}{\alpha_1 L^2} \\ \dfrac{R_B}{L} & \dfrac{6}{\alpha_1 L^2} & \dfrac{2}{\alpha_1 L} - \dfrac{R_B^2}{R_A L} & -\dfrac{R_B}{L} & -\dfrac{6}{\alpha_1 L^2} & \dfrac{4}{\alpha_1 L} + \dfrac{R_B^2}{R_A L} \end{bmatrix} \tag{A.51}$$

A.5.1.2 Equivalent nodal actions for member loads

Consider the fixed-ended beam of the previous section subjected to constant member loads n and p (Fig. A.7). Eqs A.49 are used here to determine the equivalent nodal loads required to be used in the stiffness method of analysis. Note that this set of actions corresponds to the opposite of the reactions of the fixed ended beam illustrated in Fig. A.7.

As both member ends are fully restrained, the kinematic boundary conditions are:

$$u(z=0)=0;\quad u(z=L)=0;\quad v(z=0)=0;\quad v(z=L)=0;\quad \theta(z=0)=0;$$
$$\text{and}\quad \theta(z=L)=0$$

and the expressions for the constants of integration are:

$$C_1=-\frac{L}{2}\alpha_1 p;\quad C_2=\frac{L^2}{12}\alpha_1 p;\quad C_3=0;\quad C_4=0;$$
$$C_5=\frac{3\alpha_1\alpha_3-2\alpha_2}{12}L^2 p-\frac{L}{2}\alpha_4 n;\ \text{and}\ C_6=0.$$

The reactions are obtained by solving the static boundary conditions of Eqs A.48 (ignoring time effects):

$$R_A u'_L - R_B v''_L + N_L = 0;\quad R_B u''_L - R_I v'''_L + S_L = 0;\quad R_B u'_L - R_I v''_L - M_L = 0$$
$$R_A u'_R - R_B v''_R - N_R = 0;\quad R_B u''_R - R_I v'''_R - S_R = 0;\quad R_B u'_R - R_I v''_R + M_R = 0$$

and are given by:

$$\mathbf{p}_{F.m}=\begin{bmatrix} N_{L.m}\\ S_{L.m}\\ M_{L.m}\\ N_{R.m}\\ S_{R.m}\\ M_{R.m}\end{bmatrix}=\begin{bmatrix} 0\\ -L/2\\ -L^2/12\\ 0\\ -L/2\\ L^2/12\end{bmatrix}p+\begin{bmatrix} -L/2\\ \alpha_3\\ L\alpha_3/2\\ -L/2\\ \alpha_3\\ L\alpha_3/2\end{bmatrix}n \tag{A.52}$$

where $\mathbf{p}_{F.m}$ represents the reactions of the fixed-ended beam subjected to constant longitudinal and transverse distributed loads, i.e. n and p. The equivalent nodal loads are equal and opposite to the reactions of $\mathbf{p}_{F.m}$.

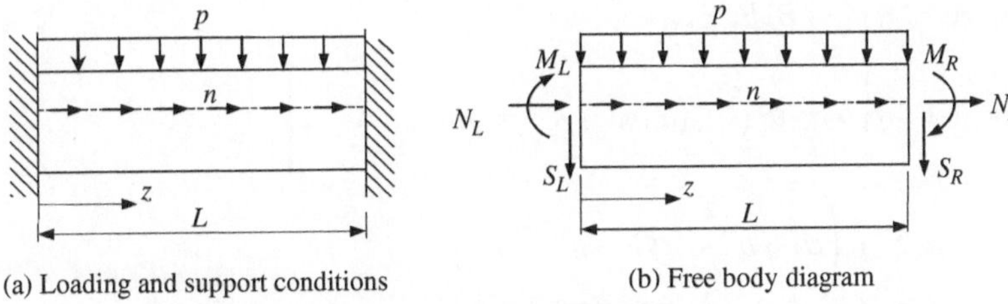

Figure A.7 Fixed-ended beam subjected to constant member loads.

A.5.1.3 Post-processing of the instantaneous solution

From the instantaneous analysis, the axial force and moment resisted by the concrete at time τ_0 are obtained from Eq. A.33. The deflection v_0 and axial deformation u_0 at any point z along a member are:

$$v_0 = \alpha_{1,0}\, p\frac{z^4}{24} + \overline{C}_{1,0}\frac{z^3}{6} + \overline{C}_{2,0}\frac{z^2}{2} + \overline{C}_{3,0}z + \overline{C}_{4,0} \tag{A.53a}$$

$$u_0 = \alpha_{2,0}\, p\frac{z^3}{6} + \left(\alpha_{3,0}\overline{C}_{1,0} + \alpha_{4,0}n\right)\frac{z^2}{2} + \overline{C}_{5,0}\, z + \overline{C}_{6,0} \tag{A.53b}$$

where the constants $\alpha_{i,0}$ ($i = 0$ to 4) are given in Eqs A.50 using the cross-sectional rigidities at time τ_0. Using the nodal displacements at each end of the member, the constants of integration are given by:

$$\overline{C}_{1,0} = -\frac{\alpha_{1,0}L}{2}p + \frac{6}{L^2}\left(\theta_{R,0} + \theta_{L,0}\right) - \frac{12}{L^3}\left(v_{R,0} - v_{L,0}\right) \tag{A.54a}$$

$$\overline{C}_{2,0} = \frac{\alpha_{1,0}L^2}{12}p - \frac{2}{L}\left(\theta_{R,0} + 2\theta_{L,0}\right) + \frac{6}{L^2}\left(v_{R,0} - v_{L,0}\right) \tag{A.54b}$$

$$\overline{C}_{3,0} = \theta_{L,0}; \quad \overline{C}_{4,0} = v_{L,0} \tag{A.54c,d}$$

$$\overline{C}_{5,0} = \frac{3\alpha_{1,0}\alpha_{3,0} - 2\alpha_{2,0}}{12}L^2 p - \frac{\alpha_{4,0}L}{2}n - \frac{3\alpha_{3,0}}{L}\left(\theta_{R,0} + \theta_{L,0}\right)$$
$$- \frac{6\alpha_{3,0}}{L^2}\left(v_{L,0} - v_{R,0}\right) + \frac{u_{R,0} - u_{L,0}}{L} \tag{A.54e}$$

$$\overline{C}_{6,0} = u_{L,0} \tag{A.54f}$$

The expressions for u_0' and v_0'' can then be obtained by differentiating Eqs A.53:

$$u_0' = \alpha_{2,0}\, p\frac{z^2}{2} + \left(\alpha_{3,0}\overline{C}_{1,0} + \alpha_{4,0}n\right)z + \overline{C}_{5,0} \tag{A.55a}$$

$$v_0'' = \alpha_{1,0}\, p\frac{z^2}{2} + \overline{C}_{1,0}z + \overline{C}_{2,0} \tag{A.55b}$$

The internal actions resisted by the concrete component can then be re-written as:

$$\begin{aligned} N_{c,0} &= A_c E_{c,0}\, u_0' - B_c E_{c,0}\, v_0'' \\ &= A_c E_{c,0}\left[\alpha_{2,0}\, p\frac{z^2}{2} + \left(\alpha_{3,0}\overline{C}_{1,0} + \alpha_{4,0}n\right)z + \overline{C}_{5,0}\right] \\ &\quad - B_c E_{c,0}\left(\alpha_{1,0}\, p\frac{z^2}{2} + \overline{C}_{1,0}\, z + \overline{C}_{2,0}\right) \\ &= a_{c0,0} + a_{c1,0}\, z + a_{c2,0}\, z^2 \end{aligned} \tag{A.56}$$

where:

$$a_{c0,0}=A_cE_{c,0}\overline{C}_{5,0}-B_cE_{c,0}\overline{C}_{2,0},\quad a_{c1,0}=A_cE_{c,0}\alpha_{3,0}\overline{C}_{1,0}+A_cE_{c,0}\alpha_{4,0}n-B_cE_{c,0}\overline{C}_{1,0} \tag{A.57a,b}$$

$$a_{c2,0}=\frac{A_cE_{c,0}\,\alpha_{2,0}\,p-B_cE_{c,0}\,\alpha_{1,0}\,p}{2} \tag{A.57c}$$

and

$$\begin{aligned} M_{c,0} &= B_cE_{c,0}u_0' - I_cE_{c,0}v_0'' \\ &= B_cE_{c,0}\left[\alpha_{2,0}p\frac{z^2}{2}+(\alpha_{3,0}\overline{C}_{1,0}+\alpha_{4,0}n)z+\overline{C}_{5,0}\right] \\ &\quad - I_cE_{c,0}\left(\alpha_{1,0}p\frac{z^2}{2}+\overline{C}_{1,0}z+\overline{C}_{2,0}\right) \\ &= b_{c0,0}+b_{c1,0}z+b_{c2,0}z^2 \end{aligned} \tag{A.58}$$

where:

$$b_{c0,0}=B_cE_{c,0}\overline{C}_{5,0}-I_cE_{c,0}\overline{C}_{2,0};\quad b_{c1,0}=B_cE_c\alpha_{3,0}\overline{C}_{1,0}+B_cE_{c,0}\alpha_{4,0}n-I_cE_{c,0}\overline{C}_{1,0} \tag{A.59a,b}$$

$$b_{c2,0}=\frac{B_cE_{c,0}\alpha_{2,0}p-I_cE_{c,0}\alpha_{1,0}p}{2} \tag{A.59c}$$

A.5.2 Age-adjusted effective modulus method (AEMM)

The differential equations and boundary conditions to be used for the time analysis require the knowledge of the functions f_{crN} and f_{crM} in Eqs A.47 and A.48. Expressions for these terms to be used with the AEMM can be obtained substituting Eqs A.56 and A.58 into Eqs A.14:

$$f_{crN}=\overline{F}_{e,0}N_{c,0}=\overline{F}_{e,0}\left(a_{c0,0}+a_{c1,0}\,z+a_{c2,0}\,z^2\right) \tag{A.60a}$$

$$f_{crM}=\overline{F}_{e,0}M_{c,0}=\overline{F}_{e,0}\left(b_{c0,0}+b_{c1,0}\,z+b_{c2,0}\,z^2\right) \tag{A.60b}$$

and their first and second derivatives, included in the system of differential equations, become:

$$f'_{crN}=\overline{F}_{e,0}\left(a_{c1,0}+2a_{c2,0}z\right);\quad f'_{crM}=\overline{F}_{e,0}\left(b_{c1,0}+2b_{c2,0}z\right) \tag{A.61a,b}$$

$$f''_{crN}=2a_{c2,0}\overline{F}_{e,0};\quad f''_{crM}=2b_{c2,0}\overline{F}_{e,0} \tag{A.61c,d}$$

Shrinkage effects are included using the same approach adopted for the finite element formulation in Section A4.1 where both f_{shN} and f_{shM} (related to shrinkage) are assumed to remain constant along the member length and when used with the AEMM are:

$$f_{shN} = \int_{A_c} \overline{E}_{e,k}\varepsilon_{sh,k}\,dA = \overline{E}_{e,k}A_c\varepsilon_{sh,k} \quad \text{and} \quad f_{shM} = \int_{A_c} y\overline{E}_{e,k}\varepsilon_{sh,k}dA = \overline{E}_{e,k}B_c\varepsilon_{sh,k} \qquad \text{(A.62a,b)}$$

It is now possible to solve the differential equations (Eqs A.47) as a function of the axial displacement and deflection at time τ_k by substituting Eqs A.60 and A.62 into Eqs A.47:

$$u_k = (\alpha_{2,k}p + \beta_{2,k})\frac{z^3}{6} + (\alpha_{3,k}C_{1,k} + \alpha_{4,k}n + \beta_{3,k})\frac{z^2}{2} + C_{5,k}z + C_{6,k} \qquad \text{(A.63a)}$$

$$v_k = (\alpha_{1,k}p + \beta_{1,k})\frac{z^4}{24} + C_{1,k}\frac{z^3}{6} + C_{2,k}\frac{z^2}{2} + C_{3,k}z + C_{4,k} \qquad \text{(A.63b)}$$

where the constants $\alpha_{i,k}$ $(i = 0$ to $4)$ are given in Eqs A.50 using the cross-sectional rigidities at time τ_k and the constants $\beta_{i,k}$ $(i = 1$ to $3)$ are given by:

$$\beta_{1,k} = \frac{2R_A b_{c2,0}\overline{F}_{e,0} - 2R_B a_{c2,0}\overline{F}_{e,0}}{\alpha_{0,k}} \qquad \text{(A.64a)}$$

$$\beta_{2,k} = \frac{R_B}{R_A}\beta_{1,k} - \frac{2a_{c2,0}\overline{F}_{e,0}}{R_A} \quad \text{and} \quad \beta_{3,k} = -\frac{a_{c1,0}\overline{F}_{e,0}}{R_A} \qquad \text{(A.64b,c)}$$

The equivalent nodal loads required to account for member loads, creep and shrinkage are determined in the following worked example.

A.5.2.1 *Equivalent nodal actions at time τ_k*

Consider the fixed-ended beam of the previous sections (shown in Fig. A.7) subjected to constant member loads n and p, as well as creep and shrinkage effects. The equivalent nodal loads required in the stiffness method to simulate member loads, creep and shrinkage effects are here determined separately using the general solution in Eqs A.63.

From the boundary conditions for the fixed ended beam (i.e. $u_{L,k} = u_{R,k} = v_{L,k} = v_{R,k} = \theta_{L,k} = \theta_{R,k} = 0$), the following expressions for the constants of integration are determined:

$$C_{1,k} = -\frac{\alpha_{1,k}p + \beta_{1,k}}{2}L; \quad C_{2,k} = \frac{\alpha_{1,k}p + \beta_{1,k}}{12}L^2; \quad C_{3,k} = 0; \quad C_{4,k} = 0 \qquad \text{(A.65a–d)}$$

$$C_{5,k} = \frac{3\alpha_{1,k}\alpha_{3,k} - 2\alpha_{2,k}}{12}L^2 p - \frac{\alpha_{4,k}}{2}n + \frac{\alpha_{3,k}L^2}{4}\beta_{1,k} - \frac{L^2}{6}\beta_{2,k} - \frac{L}{2}\beta_{3,k}; \quad C_{6,k} = 0. \qquad \text{(A.65e,f)}$$

and the support reactions are then obtained solving (Eqs A.48):

$$R_A u'_{L,k} - R_B v''_{L,k} + f_{crNL} - f_{shNL} + N_L = 0; \quad R_A u'_{R,k} - R_B v''_{R,k} + f_{crNR} - f_{shNR} - N_R = 0;$$

$$R_B u''_{L,k} - R_I v'''_{L,k} + f'_{crML} - f'_{shML} + S_L = 0; \quad R_B u''_{R,k} - R_I v'''_{R,k} + f'_{crMR} - f'_{shMR} - S_R = 0;$$

$$R_B u'_{L,k} - R_I v''_{L,k} + f_{crML} - f_{shML} - M_L = 0; \quad R_B u'_{R,k} - R_I v''_{R,k} + f_{crMR} - f_{shMR} + M_R = 0.$$

The results have been summarized below for each loading condition considered separately. The three sets of reactions $\mathbf{p}_{F.m}$, $\mathbf{p}_{F.sh}$ and $\mathbf{p}_{F.cr}$ are the opposite of the equivalent nodal loads required in the stiffness method to handle these loading conditions:

$$\mathbf{p}_{F.m} = \begin{bmatrix} N_{L.m} \\ S_{L.m} \\ M_{L.m} \\ N_{R.m} \\ S_{R.m} \\ M_{R.m} \end{bmatrix} = \begin{bmatrix} 0 \\ -L/2 \\ -L^2/12 \\ 0 \\ -L/2 \\ L^2/12 \end{bmatrix} p + \begin{bmatrix} -L/2 \\ \alpha_{3,k} \\ L\alpha_{3,k}/2 \\ -L/2 \\ -\alpha_{3,k} \\ L\alpha_{3,k}/2 \end{bmatrix} n; \quad \mathbf{p}_{F.sh} = \begin{bmatrix} \bar{E}_{e,k} A_c \varepsilon_{sh,k} \\ 0 \\ -\bar{E}_{e,k} B_c \varepsilon_{sh,k} \\ -\bar{E}_{e,k} A_c \varepsilon_{sh,k} \\ 0 \\ \bar{E}_{e,k} B_c \varepsilon_{sh,k} \end{bmatrix} \tag{A.66a,b}$$

$$\mathbf{p}_{F.cr} = \begin{bmatrix} -\bar{F}_{e,0}\left(\dfrac{a_{c2,0}L^2}{3} + \dfrac{a_{c1,0}L}{2} + a_{c0,0}\right) \\ \bar{F}_{e,0}\left(\alpha_{3,k}a_{c1,0} + L\alpha_{3,k}a_{c2,0} - b_{c1,0} - b_{c2,0}L\right) \\ -\bar{F}_{e,0}\left(-\dfrac{a_{c1,0} + a_{c2,0}L}{2}\alpha_{3,k}L - b_{c0,0} + b_{c2,0}\dfrac{L^2}{6}\right) \\ \bar{F}_{e,0}\left(\dfrac{a_{c2,0}L^2}{3} + \dfrac{a_{c1,0}L}{2} + a_{c0,0}\right) \\ -\bar{F}_{e,0}\left(\alpha_{3,k}a_{c1,0} + L\alpha_{3,k}a_{c2,0} - b_{c1,0} - b_{c2,0}L\right) \\ -\bar{F}_{e,0}\left(-\dfrac{a_{c1,0} + a_{c2,0}L}{2}\alpha_{3,k}L + b_{c0,0} + b_{c1,0}L + b_{c2,0}\dfrac{5L^2}{6}\right) \end{bmatrix} \tag{A.66c}$$

A.5.2.2 Post-processing of the solution at time τ_k

The expressions for u_k and v_k to be used in the post-processing are given in Eqs A.63. Considering that both creep and shrinkage have been included in the analysis using equivalent nodal actions it is necessary to solve the governing system of differential equations at time τ_k (Eqs A.47) using the nodal displacements at that time as boundary conditions. The relevant constants of integration (for Eqs A.63) can be obtained using Eqs A.48 as:

$$\bar{C}_{1,k} = -\frac{\alpha_{1,k}p + \beta_{1,k}}{2}L + \frac{6}{L^2}(\theta_{R,k} + \theta_{L,k}) - \frac{12}{L^3}(v_{R,k} - v_{L,k}); \quad \bar{C}_{3,k} = \theta_{R,k} \tag{A.67a,b}$$

$$\overline{C}_{2,k}=\frac{\alpha_{1,k}p+\beta_{1,k}}{12}L^2-\frac{2}{L}(\theta_{R,k}+2\theta_{L,k})+\frac{6}{L^2}(v_{R,k}-v_{L,k});\quad \overline{C}_{4,k}=v_{L,k};\quad \overline{C}_{6,k}=u_{L,k}. \tag{A.67c–e}$$

$$\overline{C}_{5,k}=\frac{3\alpha_{1,k}\alpha_{3,k}-2\alpha_{2,k}}{12}L^2p-\frac{\alpha_{4,k}}{2}n+\frac{\alpha_{3,k}L^2}{4}\beta_{1,k}-\frac{L^2}{6}\beta_{2,k}-\frac{L}{2}\beta_{3,k}$$
$$-\frac{3\alpha_{3,k}}{L}(\theta_{R,k}+\theta_{L,k})-\frac{6\alpha_{3,k}}{L^2}(v_{L,k}-v_{R,k})+\frac{u_{R,k}-u_{L,k}}{L} \tag{A.67f}$$

Based on these, the expressions for u'_k and v''_k which define the strain diagram at τ_k can be written as:

$$u'_k=(\alpha_{2,k}p+\beta_{2,k})\frac{z^2}{2}+(\alpha_{3,k}\overline{C}_{1,k}+\alpha_{4,k}n+\beta_{3,k})z+\overline{C}_{5,k} \tag{A.68a}$$

$$v''_k=(\alpha_{1,k}p+\beta_{1,k})\frac{z^2}{2}+\overline{C}_{1,k}z+\overline{C}_{2,k} \tag{A.68b}$$

For completeness, the expressions for the internal actions resisted by the concrete at τ_k can be determined as follows:

$$N_{c.k}=A_c\overline{E}_{e,k}u'_k-B_c\overline{E}_{e,k}v''_k-A_c\overline{E}_{e,k}\varepsilon_{sh,k}+\overline{F}_{e,0}N_{c,0}$$
$$=A_c\overline{E}_{e,k}\left[(\alpha_{2,k}p+\beta_{2,k})\frac{z^2}{2}+(\alpha_{3,k}\overline{C}_{1,k}+\alpha_{4,k}n+\beta_{3,k})z+\overline{C}_{5,k}\right]$$
$$-B_c\overline{E}_{e,k}\left[(\alpha_{1,k}p+\beta_{1,k})\frac{z^2}{2}+\overline{C}_{1,k}z+\overline{C}_{2,k}\right]$$
$$-A_c\overline{E}_{e,k}\varepsilon_{sh,k}+\overline{F}_{e,0}\left(a_{c0,0}+a_{c1,0}z+a_{c2,0}z^2\right)$$
$$=a_{c0,k}+a_{c1,k}z+a_{c2,k}z^2 \tag{A.69a}$$

$$M_{c.k}=B_c\overline{E}_{e,k}u'_k-I_c\overline{E}_{e,k}v''_k-B_c\overline{E}_{e,k}\varepsilon_{sh,k}+\overline{F}_{e,0}M_{c,0}$$
$$=B_c\overline{E}_{e,k}\left[(\alpha_{2,k}p+\beta_{2,k})\frac{z^2}{2}+(\alpha_{3,k}\overline{C}_{1,k}+\alpha_{4,k}n+\beta_{3,k})z+\overline{C}_{5,k}\right]$$
$$-I_c\overline{E}_{e,k}\left[(\alpha_{1,k}p+\beta_{1,k})\frac{z^2}{2}+\overline{C}_{1,k}z+\overline{C}_{2,k}\right]$$
$$-B_c\overline{E}_{e,k}\varepsilon_{sh,k}+\overline{F}_{e,0}\left(b_{c0,0}+b_{c1,0}z+b_{c2,0}z^2\right)$$
$$=b_{c0,k}+b_{c1,k}z+b_{c2,k}z^2 \tag{A.69b}$$

where:

$$a_{c0,k}=A_c\overline{E}_{e,k}\overline{C}_{5,k}-B_c\overline{E}_{e,k}\overline{C}_{2,k}-A_c\overline{E}_{e,k}\varepsilon_{sh,k}+\overline{F}_{e,0}a_{c0,0} \tag{A.70a}$$

$$a_{c1,k}=A_c\overline{E}_{e,k}(\alpha_{3,k}\overline{C}_{1,k}+\alpha_{4,k}n+\beta_{3,k})-B_c\overline{E}_{e,k}\overline{C}_{1,k}+\overline{F}_{e,0}a_{c1,0} \tag{A.70b}$$

$$a_{c2,k} = A_c\overline{E}_{e,k}\frac{\alpha_{2,k}p + \beta_{2,k}}{2} - B_c\overline{E}_{e,k}\frac{\alpha_{1,k}p + \beta_{1,k}}{2} + \overline{F}_{e,0}a_{c2,0} \tag{A.70c}$$

$$b_{c0,k} = B_c\overline{E}_{e,k}\overline{C}_{5,k} - I_c\overline{E}_{e,k}\overline{C}_{2,k} - B_c\overline{E}_{e,k}\varepsilon_{sh,k} + \overline{F}_{e,0}b_{c0,0} \tag{A.70d}$$

$$b_{c1,k} = B_c\overline{E}_{e,k}\left(\alpha_{3,k}\overline{C}_{1,k} + \alpha_{4,k}n + \beta_{3,k}\right) - I_c\overline{E}_{e,k}\overline{C}_{1,k} + \overline{F}_{e,0}b_{c1,0} \tag{A.70e}$$

$$b_{c2,k} = B_c\overline{E}_{e,k}\frac{\alpha_{2,k}\,p + \beta_{2,k}}{2} - I_c\overline{E}_{e,k}\frac{\alpha_{1,k}\,p + \beta_{1,k}}{2} + \overline{F}_{e,0}b_{c2,0} \tag{A.70f}$$

A.5.3 Step-by-step method (SSM)

The time-dependent response of a beam segment calculated using the SSM is obtained by solving the system of differential (Eqs A.47) and corresponding boundary conditions (Eqs A.48) at time τ_k using the following expressions for $f_{crN,k}$ and $f_{crM,k}$ (Eqs A.15):

$$f_{crN,k} = \sum_{i=0}^{k-1} \int_{A_c} F_{e,k,i}\,\sigma_{c,i}\,dA = \sum_{i=0}^{k-1} F_{e,k,i}\,N_{c,i} \tag{A.71a}$$

$$f_{crM,k} = \sum_{i=0}^{k-1} \int_{A_c} yF_{e,k,i}\,\sigma_{c,i}\,dA = \sum_{i=0}^{k-1} F_{e,k,i}\,M_{c,i} \tag{A.71b}$$

where the coefficient $F_{e,k,i}$ are defined in Eq. 4.25, while the internal actions resisted by the concrete at times τ_i ($N_{c,i}$ and $M_{c,i}$) are given in Eqs A.39.

It is assumed that the internal actions (i.e. $N_{c,n}$ and $M_{c,n}$) resisted by the concrete in previous time steps τ_n (with $n = 0, \ldots, i-1$) are known and available. Based on the fact that the format of the polynomials or functions describing the concrete internal actions do not vary with time, it is possible to derive the basic expressions for $N_{c,n}$ and $M_{c,n}$ from the results of the instantaneous analysis. This was already carried out when considering the AEMM leading to the use of parabolic expressions for both $N_{c,n}$ and $M_{c,n}$ (Eqs A.56 and A.58) assuming both member loads n and p to remain constant. These can be written for the generic time step τ_n required in Eqs A.39 as:

$$N_{c,n} = a_{c0,n} + a_{c1,n}\,z + a_{c2,n}\,z^2 \tag{A.72a}$$

$$M_{c,n} = b_{c0,n} + b_{c1,n}\,z + b_{c2,n}\,z^2 \tag{A.72b}$$

where the coefficients $a_{c0,n}$ and $b_{c0,n}$ are assumed to be known for $n < i$. The coefficients describing the internal actions at time τ_i are obtained in the following by substituting u'_i and v''_i determined at time τ_i into Eqs A.39. For this purpose, the expressions for u'_i and v''_i are obtained by solving the system of differential equations of Eqs A.47 enforcing the kinematic boundary conditions consistent with the nodal displacements calculated from the stiffness method at time τ_i.

The functions $f_{crN,i}$ and $f_{crM,i}$ are obtained substituting $N_{c,n}$ and $M_{c,n}$ into Eqs A.71:

$$f_{crN,i} = \sum_{n=0}^{i-1} \int_{A_c} F_{e,i,n}\sigma_{c,n}\, dA = \sum_{n=0}^{i-1} F_{e,i,n}N_{c,n} = \sum_{n=0}^{i-1} F_{e,i,n}\left(a_{c0,n} + a_{c1,n}z + a_{c2,n}z^2\right) \tag{A.73a}$$

$$f_{crM,i} = \sum_{n=0}^{i-1} \int_{A_c} yF_{e,i,n}\sigma_{c,n}\, dA = \sum_{n=0}^{i-1} F_{e,i,n}M_{c,n} = \sum_{n=0}^{i-1} F_{e,i,n}\left(b_{c0,n} + b_{c1,n}z + b_{c2,n}z^2\right) \tag{A.73b}$$

and their first and second derivatives become:

$$f'_{crN,i} = \sum_{n=0}^{i-1} F_{e,i,n}\left(a_{c1,n} + 2a_{c2,n}\, z\right); \quad f'_{crM,i} = \sum_{n=0}^{i-1} F_{e,i,n}\left(b_{c1,n} + 2b_{c2,n}\, z\right) \tag{A.74a}$$

$$f''_{crN,i} = \sum_{n=0}^{i-1} 2a_{c2,n}\, F_{e,i,n}; \quad f''_{crM,i} = \sum_{n=0}^{i-1} 2b_{c2,n}\, F_{e,i,n} \tag{A.74b}$$

Under the assumptions of constant shrinkage exhibited along the member length, the expressions for $f_{shN,i}$ and $f_{shM,i}$ are:

$$f_{shN,i} = \int_{A_c} E_{c,i}\varepsilon_{sh,i}\, dA = E_{c,i}\, A_c\, \varepsilon_{sh,i} \quad \text{and} \quad f_{shM,i} = \int_{A_c} yE_{c,i}\, \varepsilon_{sh,i}\, dA = E_{c,i}B_c\varepsilon_{sh,i} \tag{A.75a,b}$$

Substituting Eqs A.73 to A.75 into Eqs A.47 the system of differential equations can be written at time τ_i as:

$$R_A u_i'' - R_B v_i''' + \sum_{n=0}^{i-1} F_{e,i,n}\left(a_{c1,n} + 2a_{c2,n}\, z\right) + n = 0 \tag{A.76a}$$

$$R_B u_i''' - R_I v_i'''' + \sum_{n=0}^{i-1} F_{e,i,n}\left(b_{c1,n} + 2b_{c2,n}\, z\right) + p = 0 \tag{A.76b}$$

This can be solved for u_i and v_i giving:

$$u_i = \left(\alpha_{2,i}p + \beta_{2,i}\right)\frac{z^3}{6} + \left(\alpha_{3,i}\overline{C}_{1,i} + \alpha_{4,i}n + \beta_{3,i}\right)\frac{z^2}{2} + \overline{C}_{5,i}z + \overline{C}_{6,i} \tag{A.77a}$$

$$v_i = \left(\alpha_{1,i}p + \beta_{1,i}\right)\frac{z^4}{24} + \overline{C}_{1,i}\frac{z^3}{6} + \overline{C}_{2,i}\frac{z^2}{2} + \overline{C}_{3,i}z + \overline{C}_{4,i} \tag{A.77b}$$

where the constants $\alpha_{j,i}$ ($j = 0$ to 4) are given in Eqs A.50 using the cross-sectional rigidities at time τ_i and the constants $\beta_{j,i}$ ($j = 1$ to 3) are given by:

$$\beta_{1,i} = \frac{2R_A \sum_{n=0}^{i-1} b_{c2,n}F_{e,i,n} - 2R_B \sum_{n=0}^{i-1} a_{c2,n}F_{e,i,n}}{\alpha_{0,i}} \tag{A.78a}$$

$$\beta_{2,i} = \frac{R_B}{R_A}\beta_{1,i} - \frac{2\sum_{n=0}^{i-1} a_{c2,n}F_{e,i,n}}{R_A} \quad \text{and} \quad \beta_{3,i} = -\frac{\sum_{n=0}^{i-1} a_{c1,n}F_{e,i,n}}{R_A} \tag{A.78b,c}$$

The constants of integration required to define the displacements along the member lengths based on the solution at time τ_i (i.e. u_i and v_i) are obtained by enforcing the boundary conditions at that time and are given by:

$$\overline{C}_{1,i} = -\frac{\alpha_{1,i}p+\beta_{1,i}}{2}L + \frac{6}{L^2}(\theta_{R,i}+\theta_{L,i}) - \frac{12}{L^3}(v_{R,i}-v_{L,i}); \quad \overline{C}_{3,i} = \theta_{L,i} \tag{A.79a,b}$$

$$\overline{C}_{2,i} = \frac{\alpha_{1,i}p+\beta_{1,i}}{12}L^2 - \frac{2}{L}(\theta_{R,i}+2\theta_{L,i}) + \frac{6}{L^2}(v_{R,i}-v_{L,i}); \quad \overline{C}_{4,i} = v_{L,i}; \quad \overline{C}_{6,i} = u_{L,i} \tag{A.79c–e}$$

$$\begin{aligned}\overline{C}_{5,i} = {} & \frac{3\alpha_{1,i}\alpha_{3,i} - 2\alpha_{2,i}}{12}L^2 p - \frac{\alpha_{4,i}}{2}n + \frac{\alpha_{3,i}L^2}{4}\beta_{1,i} - \frac{L^2}{6}\beta_{2,i} - \frac{L}{2}\beta_{3,i} \\ & - \frac{3\alpha_{3,i}}{L}(v'_{R,i}+v'_{L,i}) - \frac{6\alpha_{3,i}}{L^2}(v_{L,i}-v_{R,i}) + \frac{u_{R,i}-u_{L,i}}{L}\end{aligned} \tag{A.79f}$$

For the time instant τ_i the strain diagram can be defined by substituting the constants of integration of Eqs A.79 into Eqs A.77 and differentiating:

$$u'_i = (\alpha_{2,i}p + \beta_{2,i})\frac{z^2}{2} + (\alpha_{3,i}\overline{C}_{1,i} + \alpha_{4,i}n + \beta_{3,i})z + \overline{C}_{5,i} \tag{A.80a}$$

$$v''_i = (\alpha_{1,i}p + \beta_{1,i})\frac{z^2}{2} + \overline{C}_{1,i}z + \overline{C}_{2,i} \tag{A.80b}$$

Substituting Eqs A.80 with the calculated constants of integrations (Eqs A.79) into Eqs A.39 gives:

$$\begin{aligned}N_{c.i} &= A_cE_{c,i}u'_i - B_cE_{c,i}v''_i - A_cE_{c,i}\varepsilon_{sh,i} + \sum_{n=0}^{i-1} F_{e,i,n}N_{c,n} \\ &= A_cE_{c,i}\left[(\alpha_{2,i}p + \beta_{2,i})\frac{z^2}{2} + (\alpha_{3,i}\overline{C}_{1,i} + \alpha_{4,i}n + \beta_{3,i})z + \overline{C}_{5,i}\right] \\ &\quad - B_cE_{c,i}\left[(\alpha_{1,i}p + \beta_{1,i})\frac{z^2}{2} + \overline{C}_{1,i}z + \overline{C}_{2,i}\right]\end{aligned}$$

$$-A_cE_{c,i}\varepsilon_{sh,i}+\sum_{n=0}^{i-1}F_{e,i,n}\left(a_{c0,n}+a_{c1,n}z+a_{c2,n}z^2\right)$$

$$=a_{c0,i}+a_{c1,i}z+a_{c2,i}z^2 \tag{A.81}$$

where:

$$a_{c0,i}=A_cE_{c,i}\overline{C}_{5,i}-B_cE_{c,i}\overline{C}_{2,i}-A_cE_{c,i}\varepsilon_{sh,i}+\sum_{n=0}^{i-1}F_{e,i,n}a_{c0,n} \tag{A.82a}$$

$$a_{c1,i}=A_cE_{c,i}\left(\alpha_{3,i}\overline{C}_{1,i}+\alpha_{4,i}n+\beta_{3,i}\right)-B_cE_{c,i}\overline{C}_{1,i}+\sum_{n=0}^{i-1}F_{e,i,n}a_{c1,n} \tag{A.82b}$$

$$a_{c2,i}=A_cE_{c,i}\frac{\alpha_{2,i}p+\beta_{2,i}}{2}-B_cE_{c,i}\frac{\alpha_{1,i}p+\beta_{1,i}}{2}+\sum_{n=0}^{i-1}F_{e,i,n}a_{c2,n} \tag{A.82c}$$

Similarly, for $M_{c,i}$:

$$M_{c,i}=B_cE_{c,i}\,u_i'-I_cE_{c,i}\,v_i''-B_cE_{c,i}\,\varepsilon_{sh,i}+\sum_{n=0}^{i-1}F_{e,i,n}M_{c,n}$$

$$=B_cE_{c,i}\left[\left(\alpha_{2,i}\,p+\beta_{2,i}\right)\frac{z^2}{2}+\left(\alpha_{3,i}\overline{C}_{1,i}+\alpha_{4,i}n+\beta_{3,i}\right)z+\overline{C}_{5,i}\right]$$

$$-I_cE_{c,i}\left[\left(\alpha_{1,i}p+\beta_{1,i}\right)\frac{z^2}{2}+\overline{C}_{1,i}z+\overline{C}_{2,i}\right]$$

$$-B_cE_{c,i}\varepsilon_{sh,i}+\sum_{n=0}^{i-1}F_{e,i,n}\left(b_{c0,n}+b_{c1,n}\,z+b_{c2,n}\,z^2\right)$$

$$=b_{c0,i}+b_{c1,i}\,z+b_{c2,i}\,z^2 \tag{A.83}$$

where:

$$b_{c0,i}=B_cE_{c,i}\overline{C}_{5,i}-I_cE_{c,i}\overline{C}_{2,i}-B_cE_{c,i}\varepsilon_{sh,i}+\sum_{n=0}^{i-1}F_{e,i,n}b_{c0,n} \tag{A.84a}$$

$$b_{c1,i}=B_cE_{c,i}\left(\alpha_{3,i}\overline{C}_{1,i}+\alpha_{4,i}n+\beta_{3,i}\right)-I_cE_{c,i}\overline{C}_{1,i}+\sum_{n=0}^{i-1}F_{e,i,n}b_{c1,n} \tag{A.84b}$$

$$b_{c2,i}=B_cE_{c,i}\frac{\alpha_{2,i}\,p+\beta_{2,i}}{2}-I_cE_{c,i}\frac{\alpha_{1,i}\,p+\beta_{1,i}}{2}+\sum_{n=0}^{i-1}F_{e,i,n}b_{c2,n} \tag{A.84c}$$

It is now possible to address the solution of the problem at time τ_k which will be used in the following example to determine the equivalent nodal actions to be used with the SSM. This is now possible as the loading history of the concrete can be calculated and recorded for each previous time step τ_i $(i < k)$ based on Eqs A.81 and A.83. Also, these can be used to define the expressions for $f_{crN.k}$ and $f_{crM.k}$ (Eqs A.14) required in the solution based on the creep effects due to stress increments at all previous time instants:

$$f_{crN,k} = \sum_{i=0}^{k-1} F_{e,k,i} N_{c,i} = \sum_{i=0}^{k-1} F_{e,k,i}\left(a_{c0,i} + a_{c1,i}\, z + a_{c2,i}\, z^2\right) \tag{A.85a}$$

$$f_{crM,k} = \sum_{i=0}^{k-1} F_{e,k,i} M_{c,i} = \sum_{i=0}^{k-1} F_{e,k,i}\left(b_{c0,i} + b_{c1,i}\, z + b_{c2,i}\, z^2\right) \tag{A.85b}$$

with corresponding derivatives:

$$f'_{crN,k} = \sum_{i=0}^{k-1} F_{e,k,i}\left(a_{c1,i} + 2a_{c2,i} z\right) \quad \text{and} \quad f'_{crM} = \sum_{i=0}^{k-1} F_{e,k,i}\left(b_{c1,i} + 2b_{c2,i} z\right) \tag{A.86a,b}$$

$$f''_{crN} = \sum_{i=0}^{k-1} 2a_{c2,i} F_{e,k,i} \quad \text{and} \quad f''_{crM} = \sum_{i=0}^{k-1} 2b_{c2,i}\, F_{e,k,i} \tag{A.86c,d}$$

Using the notation adopted for the SSM, the shrinkage effects are:

$$f_{shN,k} = \int_{A_c} E_{c,k}\varepsilon_{sh,k} \mathrm{d}A = E_{c,k} A_c \varepsilon_{sh,k} \quad \text{and} \quad f_{shM,k} = \int_{A_c} y E_{c,k}\varepsilon_{sh,k} \mathrm{d}A = E_{c,k} B_c \varepsilon_{sh,k} \tag{A.87a,b}$$

Substituting Eqs A.85 and A.87 into Eqs A.47 produces the governing system of differential equations:

$$R_A u''_k - R_B v'''_k + \sum_{i=0}^{k-1} F_{e,k,i}\left(a_{c1,i} + 2a_{c2,i} z\right) + n = 0 \tag{A.88a}$$

$$R_B u'''_k - R_I v''''_k + \sum_{i=0}^{k-1} F_{e,k,i}\left(b_{c1,i} + 2b_{c2,i} z\right) + p = 0 \tag{A.88b}$$

whose general solution can be expressed as:

$$u_k = \left(\alpha_{2,k} p + \beta_{2,k}\right)\frac{z^3}{6} + \left(\alpha_{3,k} C_{1,k} + \alpha_{4,k} n + \beta_{3,k}\right)\frac{z^2}{2} + C_{5,k} z + C_{6,k} \tag{A.89a}$$

$$v_k = \left(\alpha_{1,k} p + \beta_{1,k}\right)\frac{z^4}{24} + C_{1,k}\frac{z^3}{6} + C_{2,k}\frac{z^2}{2} + C_{3,k} z + C_{4,k} \tag{A.89b}$$

where the constants $\alpha_{j,k}$ ($j = 0$ to 4) are given in Eqs A.50 using the cross-sectional rigidities at time τ_k and the constants $\beta_{j,i}$ ($j = 1$ to 3) are given by:

$$\beta_{1,k} = \frac{2R_A \sum_{i=0}^{k-1} b_{c2,i}F_{e,k,i} - 2R_B \sum_{i=0}^{k-1} a_{c2,i}F_{e,k,i}}{\alpha_{0,k}}; \quad \beta_{2,k} = \frac{R_B}{R_A}\beta_{1,k} - \frac{2\sum_{i=0}^{k-1} a_{c2,i}F_{e,k,i}}{R_A} \tag{A.90a,b}$$

$$\beta_{3,k} = -\frac{\sum_{i=0}^{k-1} a_{c1,i}F_{e,k,i}}{R_A} \tag{A.90c}$$

A.5.3.1 *Equivalent nodal actions at time τ_k*

Based on the fixed-ended beam of the previous section (Fig. A.7), the expressions for the equivalent nodal loads required to account for member loads n and p, creep effects and shrinkage effects in the stiffness method are here determined using the SSM. The constants of integration to define the expressions for the axial displacement u_k and deflection v_k at time τ_k are identical to those obtained for the AEMM (and given in Eqs A.65) as the material properties are included in the coefficients $\alpha_{j,k}$ and $\beta_{j,k}$.

The end reactions at the fixed supports are then obtained applying the static boundary conditions of Eqs A.48:

$$R_A u'_{L,k} - R_B v''_{L,k} + f_{crNL,k} - f_{shNL,k} + N_L = 0$$

$$R_A u'_{R,k} - R_B v''_{R,k} + f_{crNR,k} - f_{shNR,k} - N_R = 0$$

$$R_B u''_{L,k} - R_I v'''_{L,k} + f'_{crML,k} - f'_{shML,k} + S_L = 0$$

$$R_B u''_{R,k} - R_I v'''_{R,k} + f'_{crMR,k} - f'_{shMR,k} - S_R = 0$$

$$R_B u'_{L,k} - R_I v''_{L,k} + f_{crML,k} - f_{shML,k} - M_L = 0$$

$$R_B u'_{R,k} - R_I v''_{R,k} + f_{crMR,k} - f_{shMR,k} + M_R = 0$$

The vectors $\mathbf{p}_{F.m}$, $\mathbf{p}_{F.sh}$ and $\mathbf{p}_{F.cr}$ can be defined as:

$$\mathbf{p}_{F.m} = \begin{bmatrix} N_{L.m} \\ S_{L.m} \\ M_{L.m} \\ N_{R.m} \\ S_{R.m} \\ M_{R.m} \end{bmatrix} = \begin{bmatrix} 0 \\ -L/2 \\ -L^2/12 \\ 0 \\ -L/2 \\ L^2/12 \end{bmatrix} p + \begin{bmatrix} -L/2 \\ \alpha_{3,k} \\ L\alpha_{3,k}/2 \\ -L/2 \\ -\alpha_{3,k} \\ L\alpha_{3,k}/2 \end{bmatrix} n; \quad \mathbf{p}_{F.sh} = \begin{bmatrix} E_{c,k}A_c\varepsilon_{sh,k} \\ 0 \\ -E_{c,k}B_c\varepsilon_{sh,k} \\ -E_{c,k}A_c\varepsilon_{sh,k} \\ 0 \\ E_{c,k}B_c\varepsilon_{sh,k} \end{bmatrix} \tag{A.91a,b}$$

$$\mathbf{p}_{F.cr} = \begin{bmatrix} -\sum_{i=0}^{k-1} F_{e,k,i}\left(\frac{a_{c2,i}L^2}{3} + \frac{a_{c1,i}L}{2} + a_{c0,i}\right) \\ \sum_{i=0}^{k-1} F_{e,k,i}\left(\alpha_{3,k}a_{c1,i} + L\alpha_{3,k}a_{c2,i} - b_{c1,i} - b_{c2,i}L\right) \\ -\sum_{i=0}^{k-1} F_{e,k,i}\left(-\frac{a_{c1,i}+a_{c2,i}L}{2}\alpha_{3,k}L - b_{c0,i} + b_{c2,i}\frac{L^2}{6}\right) \\ \sum_{i=0}^{k-1} F_{e,k,i}\left(\frac{a_{c2,i}L^2}{3} + \frac{a_{c1,i}L}{2} + a_{c0,i}\right) \\ -\sum_{i=0}^{k-1} F_{e,k,i}\left(\alpha_{3,k}a_{c1,i} + L\alpha_{3,k}a_{c2,i} - b_{c1,i} - b_{c2,i}L\right) \\ -\sum_{i=0}^{k-1} F_{e,k,i}\left(-\frac{a_{c1,i}+a_{c2,i}L}{2}\alpha_{3,k}L + b_{c0,i} + b_{c1,i}L + b_{c2,i}\frac{5L^2}{6}\right) \end{bmatrix} \quad \text{(A.91c)}$$

As expected these loading vectors are very similar to those already calculated using the AEMM. For example, Eqs A.91 are identical to Eqs A.66 with the only difference relying on the way the instantaneous concrete modulus is determined with the AEMM and SSM. The equivalent nodal actions for use in the stiffness method are equal and opposite to $\mathbf{p}_{F.m}$, $\mathbf{p}_{F.cr}$ and $\mathbf{p}_{F.sh}$.

A.5.3.2 Post-processing of the solution at time τ_i

In the case of the SSM the expressions for the axial displacement u_i and deflection v_i required for the post-processing of the solution at any time instant have already been derived in Eqs A.77. For a time τ_i, the constants of integration defining u_i and v_i are obtained using Eqs A.79. The expressions for u_i' and v_i'' defining the strain diagram at τ_i are given in Eqs A.80. For completeness, the internal actions resisted by the concrete component can be determined using Eqs A.81 and A.83.

A.6 References

1. Bathe, K.-J. (2006). *Finite Element Procedures.* Prentice Hall, Pearson Education, New Jersey.
2. Reddy, J.N. (2004). *An Introduction to the Finite Element Method.* Third edition, McGraw-Hill, New York.
3. Gupta, A.K. and Ma, P.S. (1977). Error in eccentric beam formulation. *International Journal for Numerical Methods in Engineering*, 11, 1473–1483.
4. Crisfield, M.A. (1991). The eccentricity issue in the design if plate and shell elements. *Communications in applied Numerical Methods*, 7, 47–56.
5. Ranzi, G. and Zona, A. (2007). A composite beam model including the shear deformability of the steel component. *Engineering Structures*, 29, 3026–3041.

Index